# Nanotechnology in the Life Sciences

**Series Editor**

Ram Prasad
School of Environmental Science and Engineering,
Sun Yat-Sen University, Guangzhou, China

Amity Institute of Microbial Technology,
Amity University, Noida, UP, India

Nano and biotechnology are two of the 21st century's most promising technologies. Nanotechnology is demarcated as the design, development, and application of materials and devices whose least functional make up is on a nanometer scale (1 to 100 nm). Meanwhile, biotechnology deals with metabolic and other physiological developments of biological subjects including microorganisms. These microbial processes have opened up new opportunities to explore novel applications, for example, the biosynthesis of metal nanomaterials, with the implication that these two technologies (i.e., thus nanobiotechnology) can play a vital role in developing and executing many valuable tools in the study of life. Nanotechnology is very diverse, ranging from extensions of conventional device physics to completely new approaches based upon molecular self-assembly, from developing new materials with dimensions on the nanoscale, to investigating whether we can directly control matters on/in the atomic scale level. This idea entails its application to diverse fields of science such as plant biology, organic chemistry, agriculture, the food industry, and more.

Nanobiotechnology offers a wide range of uses in medicine, agriculture, and the environment. Many diseases that do not have cures today may be cured by nanotechnology in the future. Use of nanotechnology in medical therapeutics needs adequate evaluation of its risk and safety factors. Scientists who are against the use of nanotechnology also agree that advancement in nanotechnology should continue because this field promises great benefits, but testing should be carried out to ensure its safety in people. It is possible that nanomedicine in the future will play a crucial role in the treatment of human and plant diseases, and also in the enhancement of normal human physiology and plant systems, respectively. If everything proceeds as expected, nanobiotechnology will, one day, become an inevitable part of our everyday life and will help save many lives.

More information about this series at http://www.springer.com/series/15921

Ram Prasad • Elisabet Aranda
Editors

# Approaches in Bioremediation

## The New Era of Environmental Microbiology and Nanobiotechnology

Springer

*Editors*
Ram Prasad
School of Environmental Science
and Engineering
Sun Yat-sen University
Guangzhou, China

Amity Institute of Microbial Technology
Amity University
Noida, UP, India

Elisabet Aranda
Department of Microbiology
Faculty of Pharmacy, and Institute
of Water Research
University of Granada
Granada, Spain

ISSN 2523-8027 ISSN 2523-8035 (electronic)
Nanotechnology in the Life Sciences
ISBN 978-3-030-02368-3 ISBN 978-3-030-02369-0 (eBook)
https://doi.org/10.1007/978-3-030-02369-0

Library of Congress Control Number: 2018962875



This Springer imprint is published by the registered company Springer Nature Switzerland AG
The registered company address is: Gewerbestrasse 11, 6330 Cham, Switzerland

# Foreword

The environmental and health impacts of contaminants mostly of anthropogenic origin are of an increasing interest due to their persistency and toxicity. Many of these compounds and materials were deliberately or inadvertently released into the environment. Others organic contaminants are personal care compounds, pharmaceuticals, or undesired chemical by-products that resist against microbial or chemical processes, either during wastewater treatment or in soil. As a consequence, there is a great demand for new technologies that can be employed for decontamination of polluted areas and environmental matrices. In this context, bioremediation stands for environmental-friendly decontamination technologies based on the application of organisms (or their parts) in order to decontaminate various compartments of the environment. This environmental biotechnology approach employs microorganisms and/or plants that represent essential components of the global carbon cycle. Apart from this, it was documented that most xenobiotic industrial chemicals can be decomposed by microorganisms and plants, either via cometabolism phenomenon yielding partial degradation, or by serving as growth substrate, which is accompanied by mineralization of at least part of the molecule. Contrary to expensive physicochemical approaches, bioremediation stands for methods typically less economically demanding. Nevertheless, the main advantage of these technologies is conservation of the treated matrix, e.g., soil structure including microbial biota. Noteworthy, the phenomenon of bioremediation has been studied since the 1980s; however, not all of the methods have been introduced into practice. The problem lies in the complexity of environmental matrices and interactions of organisms among each other. This book very greatly reflects these difficulties and brings an overview of advanced modern approaches within this research field. Attention is paid to the whole overview of application of microbes in soil and water with respect to new available research methods at the beginning of this book. Together with three following chapters, it shows the importance of modern omics approaches that could greatly enhance the understanding of transformation and decontamination processes and could enable new insights into the involved mechanisms. The whole book also reflects various types of pollution, and, besides classical contaminants, several chapters also describe problems related to the new types including fluorinated hydrocarbons, industrial dyes, and microplastics. Other chapters focus on roles and applications of fungi in environmental biotechnology processes, which is a subject often overlooked in other books and literature. Attention is also paid to possible applications of plants, namely, constructed wetlands and algae. A substantial part of this book also elaborates on topics related to a novel research field of nanotechnology and its possible applications together with bioremediation methods.

This book generally brings a contemporary outlook on the modern aspects of environmental decontamination technologies and will be greatly helpful to all the readers interested in the subject of bioremediation of the environment and related subjects.

**Tomáš Cajthaml** is an environmental chemist, the director of the Institute for Environmental Studies, Faculty of Science, Charles University, Prague, and the head of Laboratory of Environmental Biotechnology, Institute of Microbiology. He obtained his PhD in 2002 from the Charles University in the field of Applied and Landscape Ecology. Among his scientific interests are environmental biotechnology, microbial bioremediation and biodegradation mechanisms of organopollutants, toxicology, nanotoxicology and ecotoxicology, and bioavailability and degradation of relevant pollutants. In these topics, he has published more than 170 research papers, with an h-index of 34, and he has 5 patents, 6 verified technologies, and 1 utility model related to this field. He is a member of the scientific council of Faculty of Science, Charles University, Prague. He is a member of the editorial board of several journals – Folia Microbiologica, *European Journal of Environmental Sciences*, and *Frontiers in Microbiology*. He was awarded the Prize of Otto Wichterle for young researchers and the Prize of Carolinum Association for the research on the environment.

Tomáš Cajthaml
Institute for Environmental Studies, Faculty of Science,
Charles University, Prague 2, Czech Republic
Institute of Microbiology, Czech Academy of Sciences,
Prague 4, Czech Republic

# Preface

The Earth is alarmingly polluted, and the current economic model and population growth do not give a progressive expectation. Thousands of new chemical compounds are released every day into the environment, producing changes in the environment, particularly in microbial populations and global effects which still remain unknown.

The knowledge of the microbial processes in the environment as well as microbial communities and their interactions with other living organisms and the environment are the basis of bioremediation. Microorganisms have long been the subject of bioremediation studies. These started in the 1950s using selective pressures to select microorganisms with the capacity to use different xenobiotic compounds as a source of carbon and energy, particularly. Nevertheless, fungi have been traditionally poorly studied, due to their slow growth, complex and often wrong classification, and scarcity of genome database in comparison with bacteria. In this book, we address different topics related to biodegradation processes involving bacteria and fungi, marine-derived fungi (Chap. 12), the binomial fungi-plant (Chap. 5), and Microalgae-bacteria consortia (Chap. 8). Among them, the degradation of plastics and polythene (Chap. 6), phenolic compounds (Chap. 8), distillery effluents (Chap. 9), residues from the petroleum refining industry (Chap. 10), and fluorinated compounds (Chap. 11) are discussed.

A new step was taken when the research community realized the importance of the interaction between communities for efficient bioremediation processes. Sequencing technologies of metagenomes have dealt with an ever-increasing rate of thousands of bacterial and hardly any fungal genomes, which are often studied as isolated communities. During this era, advances in new approaches have made significant progress to provide solutions to different environmental problems. The study of key gene expression during bioremediation processes or the regulation of proteins is helping to map these complex processes that occur under changes in the environment. Thus, the full understanding of the complexity of these interrelationships is still in its infancy. In this book, some critical reviews about advances on bacterial omics (Chap. 1), fungal transcriptomics (Chaps. 2 and 3), and fungal proteomics (Chap. 3) are given, including a revision of microbial dynamics during bioremediation of petroleum hydrocarbon stimulation (Chap. 7).

In order to understand the role that different genes play in environmental pathways, researchers need to be able to modify levels of gene expression. New technologies are currently being developed by using CRISPR-CAS9 technology, which is being tentatively applied to environmental problems. In this book, a critical overview on CRISPR-Cas in fungi is provided (Chap 4).

Finally, another growing field is nanobiotechnology, which integrates biotechnology at nanoscience scale, is addressed in the last chapters of this book. This discipline is being rapidly developed as a resource to support bioremediation processes and soil health (Chaps. 13 and 14) and energy and environmental remediation (Chap. 17). Here we address nanobioremediation and an innovative approach for fluoride contamination (Chap. 15) and dye effluents (Chap. 16).

We would like to express our gratitude to all the authors for their valuable contributions. We also thank the Springer Nature team, especially Eric Stannard, Anthony Dunlap, and Rahul Sharma for their continuous support during the preparation of this book. And finally, this book is dedicated to all of our family and friends, because sometimes it is difficult to understand this strange but fascinating profession.

Guangzhou, China
Noida, UP, India
Ram Prasad

Granada, Spain
Elisabet Aranda

# Contents

# Contributors

**José Manuel Adame-Rodríguez** Unidad de Manipulación Genética, Laboratorio de Micología y Fitopatología, Departamento de Microbiología, Facultad de Ciencias Biológicas, Universidad Autónoma de Nuevo León, Cd. Universitaria, San Nicolás de los Garza, NL, Mexico

**Shagufta Afreen** Dayanand Girls Post Graduate College, Kanpur, India

**Catarina L. Amorim** Universidade Católica Portuguesa, CBQF-Centro de Biotecnologia e Química Fina, Laboratório Associado, Escola Superior de Biotecnologia, Porto, Portugal

Biology Department and CESAM, University of Aveiro, Campus Universitário de Santiago, Aveiro, Portugal

**Elisabet Aranda** Department of Microbiology, Faculty of Pharmacy, and Institute of Water Research, University of Granada, Granada, Spain

**Elva T. Aréchiga-Carvajal** Unidad de Manipulación Genética, Laboratorio de Micología y Fitopatología, Departamento de Microbiología, Facultad de Ciencias Biológicas, Universidad Autónoma de Nuevo León, Cd. Universitaria, San Nicolás de los Garza, NL, Mexico

**Mohammad Ashfaq** School of Life Science, BS Abdur Rahaman Institute of Science and Technology, Chennai, India

**Ramón Alberto Batista-García** Centro de Investigación en Dinámica Celular-IICBA, Universidad Autónoma del Estado de Morelos, Cuernavaca, Morelos, Mexico

**Geeta Bhandari** Sardar Bhagwan Singh Post Graduate Institute of Biomedical Sciences and Research, Dehradun, Uttarakhand, India

**Adriana Otero-Blanca** Centro de Investigación en Dinámica Celular-IICBA, Universidad Autónoma del Estado de Morelos, Cuernavaca, Morelos, Mexico

**Paula M. L. Castro** Universidade Católica Portuguesa, CBQF-Centro de Biotecnologia e Química Fina, Laboratório Associado, Escola Superior de Biotecnologia, Porto, Portugal

**Divya Chauhan** Department of Chemistry, Punjab University, Chandigarh, India

**Li-Na Chen** College of Plant Sciences, Jilin University, Changchun, People's Republic of China

**Verónica L. Colin** Planta Piloto de Procesos Industriales Microbiológicos (PROIMI-CONICET), Tucumán, Argentina

**Bharti P. Dave** Department of Life Sciences, Maharaja Krishnakumarsinhji Bhavnagar University, Bhavnagar, India

School of Sciences, Indrasheel University, Rajpur, Chhattisgarh, India

**Jorge Luis Folch-Mallol** Centro de Investigación en Dinámica Celular-IICBA, Universidad Autónoma del Estado de Morelos, Cuernavaca, Morelos, Mexico

**Mercedes García-Sánchez** UMR Eco&Sols, Ecologie fonctionelle and Biogéochimie des sols and Agro-écosystémes, Montpellier, France

**Luciana Melisa Del Gobbo** Facultad de Bioquímica, Química y Farmacia, Universidad Nacional de Tucumán, Tucumán, Argentina

**Jesús González-López** Department of Microbiology, Faculty of Pharmacy, and Institute of Water Research, University of Granada, Granada, Spain

**Alejandro González-Martínez** Department of Microbiology, Faculty of Pharmacy, and Institute of Water Research, University of Granada, Granada, Spain

**Verónica Irazusta** Instituto de Investigaciones para la Industria Química (INIQUI) – Consejo Nacional de Investigaciones Científicas y Técnicas (CONICET), Universidad Nacional de Salta (UNSa), Salta, Argentina

Facultad de Ciencias Naturales, UNSa, Salta, Argentina

**Belén Juárez-Jiménez** Department of Microbiology, Faculty of Pharmacy, and Institute of Water Research, University of Granada, Granada, Spain

**Suphiya Khan** Department of Bioscience and Biotechnology, Banasthali Vidyapith Tonk, Rajasthan, India

**Alejandro Ledezma-Villanueva** Unidad de Manipulación Genética, Laboratorio de Micología y Fitopatología, Departamento de Microbiología, Facultad de Ciencias Biológicas, Universidad Autónoma de Nuevo León, Cd. Universitaria, San Nicolás de los Garza, NL, Mexico

**Piet N. L. Lens** UNESCO-IHE, Institute for Water Education, Delft, The Netherlands

**Verónica Lira-Ruan** Centro de Investigación en Dinámica Celular-IICBA, Universidad Autónoma del Estado de Morelos, Cuernavaca, Morelos, Mexico

**Liliana Martínez-Ávila** Centro de Investigación en Dinámica Celular-IICBA, Universidad Autónoma del Estado de Morelos, Cuernavaca, Morelos, Mexico

**Fabiana Lilian Martínez** Instituto de Investigaciones para la Industria Química (INIQUI) – Consejo Nacional de Investigaciones Científicas y Técnicas (CONICET), Universidad Nacional de Salta (UNSa), Salta, Argentina

**Paula Maza-Márquez** Department of Microbiology, Faculty of Pharmacy, and Institute of Water Research, University of Granada, Granada, Spain

**Norma Beatriz Moraga** Instituto de Investigaciones para la Industria Química (INIQUI) – Consejo Nacional de Investigaciones Científicas y Técnicas (CONICET), Universidad Nacional de Salta (UNSa), Salta, Argentina

Facultad de Ingeniería, UNSa, Salta, Argentina

**Irina S. Moreira** Universidade Católica Portuguesa, CBQF-Centro de Biotecnologia e Química Fina, Laboratório Associado, Escola Superior de Biotecnologia, Porto, Portugal

**Cormac D. Murphy** UCD School of Biomolecular and Biomedical Science, University College Dublin, Dublin, Ireland

**Hassana Ibrahim Mustapha** UNESCO-IHE, Institute for Water Education, Delft, The Netherlands

Department of Agricultural and Bioresources Engineering, Federal University of Technology Minna, Minna, Nigeria

**Rafael Navajas-Pérez** Departamento de Genética, Facultad de Ciencias, Universidad de Granada, Granada, Spain

**Rishabh Anand Omar** Centre for Environmental Science and Engineering, Indian Institute of Technology Kanpur, Kanpur, India

**Yordanis Pérez-Llano** Centro de Investigación en Dinámica Celular-IICBA, Universidad Autónoma del Estado de Morelos, Cuernavaca, Morelos, Mexico

**Verónica Beatriz Rajal** Instituto de Investigaciones para la Industria Química (INIQUI) – Consejo Nacional de Investigaciones Científicas y Técnicas (CONICET), Universidad Nacional de Salta (UNSa), Salta, Argentina

Facultad de Ingeniería, UNSa, Salta, Argentina

Singapore Centre for Environmental Life Sciences Engineering (SCELSE), School of Biological Sciences, Nanyang Technological University, Singapore, Singapore

**Maria del Rayo Sánchez Carbente** Centro de Investigación en Biotecnología, Universidad Autónoma del Estado de Morelos, Cuernavaca, Morelos, Mexico

**Belén Rodelas** Department of Microbiology, Faculty of Pharmacy, and Institute of Water Research, University of Granada, Granada, Spain

**Neli Romano-Armada** Instituto de Investigaciones para la Industria Química (INIQUI) – Consejo Nacional de Investigaciones Científicas y Técnicas (CONICET), Universidad Nacional de Salta (UNSa), Salta, Argentina

Facultad de Ingeniería, UNSa, Salta, Argentina

**Bhumi Sachaniya** Department of Life Sciences, Maharaja Krishnakumarsinhji Bhavnagar University, Bhavnagar, India

**Shubha Rani Sharma** Department of Bio-engineering, Birla Institute of Technology, Mesra, Ranchi, India

**José A. Siles** Department of Plant and Microbial Biology, University of California, Berkeley, Berkeley, CA, USA

**Neha Singh** Department of Bioscience and Biotechnology, Banasthali Vidyapith Tonk, Rajasthan, India

**Haley P. Stein** Institute of Water Research, University of Granada, Granada, Spain

**Neetu Talreja** Department of Bio-nanotechnology, Gachon University, Seongnam, South Korea

**Anjana K. Vala** Department of Life Sciences, Maharaja Krishnakumarsinhji Bhavnagar University, Bhavnagar, India

**Yi Wei** College of Plant Sciences, Jilin University, Changchun, People's Republic of China

**Monika Yadav** Department of Bioscience and Biotechnology, Banasthali Vidyapith Tonk, Rajasthan, India

**María Florencia Yañez-Yazlle** Instituto de Investigaciones para la Industria Química (INIQUI) – Consejo Nacional de Investigaciones Científicas y Técnicas (CONICET), Universidad Nacional de Salta (UNSa), Salta, Argentina

**Shi-Hong Zhang** College of Plant Sciences, Jilin University, Changchun, People's Republic of China

**Zi-Yu Zhang** College of Plant Sciences, Jilin University, Changchun, People's Republic of China

**Chi Zhu** College of Plant Sciences, Jilin University, Changchun, People's Republic of China

# About the Editors

**Ram Prasad, PhD** is associated with the Amity Institute of Microbial Technology, Amity University, Uttar Pradesh, India, since 2005. His research interests includes plant-microbe interactions, sustainable agriculture, and microbial nanobiotechnology. Dr. Prasad has more than hundred publications to his credit, including research papers, review articles, and book chapters, and five patents issued or pending and edited or authored several books. Dr. Prasad has 12 years of teaching experience, and has been awarded the Young Scientist Award (2007) and Prof. J. S. Datta Munshi Gold Medal (2009) by the International Society for Ecological Communications; FSAB Fellowship (2010) by the Society for Applied Biotechnology; the American Cancer Society UICC International Fellowship for Beginning Investigators, USA (2014); Outstanding Scientist Award (2015) in the field of microbiology by Venus International Foundation; BRICPL Science Investigator Award (ICAABT-2017); and Research Excellence Award (2018). He serves as editorial board member in *Frontiers in Microbiology*, *Frontiers in Nutrition*, and *Academia Journal of Biotechnology* including series editor of *Nanotechnology in the Life Sciences*, Springer Nature, USA. Previously, Dr. Prasad served as visiting assistant professor in Whiting School of Engineering, Department of Mechanical Engineering, Johns Hopkins University, USA, and is presently working as research associate professor at School of Environmental Science and Engineering, Sun Yat-sen University, Guangzhou, China.

**Elisabet Aranda, PhD** is a Ramon y Cajal researcher at the Microbiology Department of the University of Granada (UGR) and member of the Institute of Water Research (UGR). She has more than 16 years of research experience. She has several pre- and postdoctoral research stays (Estación Experimental del Zaidín, Spanish National Research Council (CSIC), Spain; the University of Naples Federico II, Italy; Institute of Mass Spectrometry, Proteomic and Molecular Biology, Italy; IHI Zittau, Technical University of Dresden, Germany; and Lawrence Berkeley National Laboratory, California, USA, among others). Her main research expertise is in the field of fungal bioremediation; fungal degradation mechanisms of priority micropollutants and emerging contaminants, in water and soil systems; and application of molecular tools such as NGS (Illumina) and proteomic approaches in this field. She has published more than 50 publications in this topic (including research papers, review articles, and book chapters). She is the inventor of two patents. She is associate editor in *Frontiers in Microbiology* and member of board of directors of the specialized group "Biodeterioration, Biodegradation and Bioremediation" of the Spanish Society of Microbiology (SEM). Among the prizes with which it has been awarded stands out the "Innova Sustainable" Award from the Aquae Foundation.

# Chapter 1
# Stepwise Strategies for the Bioremediation of Contaminated Soils: From the Microbial Isolation to the Final Application

**Fabiana Lilian Martínez, Norma Beatriz Moraga, Neli Romano-Armada, María Florencia Yañez-Yazlle, Verónica Beatriz Rajal, and Verónica Irazusta**

## Contents



---

F. L. Martínez · M. F. Yañez-Yazlle
Instituto de Investigaciones para la Industria Química (INIQUI) –Consejo Nacional de Investigaciones Científicas y Técnicas (CONICET), Universidad Nacional de Salta (UNSa), Salta, Argentina

N. B. Moraga · N. Romano-Armada
Instituto de Investigaciones para la Industria Química (INIQUI) –Consejo Nacional de Investigaciones Científicas y Técnicas (CONICET), Universidad Nacional de Salta (UNSa), Salta, Argentina

Facultad de Ingeniería, UNSa, Salta, Argentina

V. B. Rajal
Instituto de Investigaciones para la Industria Química (INIQUI) –Consejo Nacional de Investigaciones Científicas y Técnicas (CONICET), Universidad Nacional de Salta (UNSa), Salta, Argentina

Facultad de Ingeniería, UNSa, Salta, Argentina

Singapore Centre for Environmental Life Sciences Engineering (SCELSE), School of Biological Sciences, Nanyang Technological University, Singapore, Singapore

V. Irazusta (✉)
Instituto de Investigaciones para la Industria Química (INIQUI) –Consejo Nacional de Investigaciones Científicas y Técnicas (CONICET), Universidad Nacional de Salta (UNSa), Salta, Argentina

Facultad de Ciencias Naturales, UNSa, Salta, Argentina



R. Prasad, E. Aranda (eds.), *Approaches in Bioremediation*, Nanotechnology in the Life Sciences, https://doi.org/10.1007/978-3-030-02369-0_1

## 1.1 Introduction

As a matter of fact, increasingly widespread soil pollution has caused vast areas of land to become unproductive for agriculture and hazardous for both wildlife and human populations (Garbisu and Alkorta 2003). This pollution can be caused by a wide range of contaminants as heavy metals, pesticides, xenobiotics, hydrocarbons, wastes from mining industry, agrochemicals, poor disposal of hazardous waste, and poor management of urban solid wastes, among others.

Soil is a nonrenewable resource because the rates of formation and self-restoration exceed the human life span. Despite the inability to completely recover a soil to its previous state to a contamination event, a partial recovery can be carried out by microorganisms or microorganisms' products, such as polymers and enzymes.

Bioremediation processes consist in using microorganisms and/or manipulating the metabolic activities or compounds produced by them, to eliminate, decrease, or, at least, turn pollutants into less aggressive chemical species, minimizing environmental commitment by facilitating the biodegradation processes (EPA 2004).

## 1.2 Niche: Selection of the Right Place for Microbial Isolation

### *1.2.1 Bioremediation Strategies and Microorganisms*

When microorganisms are subjected to an environment that contains toxic concentrations of a recalcitrant element or compound, special mechanisms to allow them tolerate those conditions are developed. In that context, the concept of resistance can be defined as the microorganism's ability to survive stress, using response mechanisms that are activated by the presence of the stressor (Gadd 2010).

As not all microorganisms are able to grow, survive, or tolerate every polluted environment, to ameliorate contaminated soils, each kind of pollutant requires specific microorganisms (Fuentes et al. 2010). So, to ensure this, the most common strategy is to use autochthonous microorganisms, which means to isolate microorganisms from the same polluted environment to be recovered (Fuentes et al. 2011). These species will have already been able to adapt to such stressful conditions in that environment by developing different intra- and/or extracellular mechanisms (Prasad 2017, 2018) (Tables 1.1 and 1.2).

Examples of the strategies developed by microorganisms are excreting the metals through transport systems and sequestering compounds through cytosol agents which may bind to the metal and detoxicate the interior of cells (Majzlik et al. 2011). Bioaccumulation, performed by bacteria and fungi (Suarez 2002; Elangovan et al. 2006; Chojnacka 2010), is a mixture of intra- and extracellular mechanisms, because it requires metabolic activity of living cells involving intracellular sequestration, extracellular precipitation, metal accumulation, and complex formation. It comprises two stages: metabolism-independent passive biosorption (e.g., physical and chemical adsorption, metal ion exchange, chelation, coordination, surface complexation, and microprecipitation) and metabolism-dependent active bioaccumulation (e.g., transport of metal ions into microbial cells including complex permeation, carrier-mediated ion pumps, and endocytosis) (Ma et al. 2015, 2016).

Other adaptive mechanisms proposed are the organic acid excretion to make metals soluble (Gadd 1999) and developing cytoplasmatic protection mechanisms through inclusion bodies which retain a great number of metal cations (González and Jensen 1998; Gentili et al. 2006), biomineralization (Lowenstam 1981; Pérez-González et al. 2011), and EPS formation (Ryder et al. 2007).

**Table 1.1** Intracellular mechanisms involved in detoxification aim to reduce metal burden in the cytosol

| Mechanism | Description | Organism | Reference |
|---|---|---|---|
| Organellar location/chelation | Metal sequestration in intracellular compartments (mainly cell vacuole) or binding to nonprotein thiols and transport into intracellular compartments | Ectomycorrhizal Fungi and bacteria | Bellion et al. (2006) and Ma et al. (2016) |
| Efflux or enzymatic detoxification | Exclusion of metal chelates in to the extracellular space | Bacteria and fungi | Ma et al. (2016) |
| Biologically controlled mineralization (BCM) | Formation of minerals on or inside organic matrices or vesicles within the cell | Bacteria | Frankel and Bazylinski (2003) and Moraga et al. (2017) |
| Bioleaching | Extraction of heavy metals from sludge, sediments, and soils, therefore alleviating metal phytotoxicity directly or indirectly through various metabolic activities such as oxidation, reduction, and complexation | Bacteria and fungi | Mulligan and Galvez-Cloutier (2003), Kletzin (2006), Navarro et al. (2013), and Ma et al. (2016) |
| Bioexclusion | Active transport or efflux of toxic metals from the cytoplasm | Bacteria | Ma et al. (2016) |

**Table 1.2** Extracellular mechanisms involved in detoxification mainly in avoidance of metal entry

| Mechanism | Description | Organism | Reference |
|---|---|---|---|
| Complexation | Secretion of extracellular polymeric substances (EPSs) protects against harmful effect of metal | Bacteria | Guibaud et al. (2005), Zhang et al. (2006), Vodnik et al. (2008), Slaveykova et al. (2010), and Hou et al. (2013) |
| | Production of insoluble metal-sorbing glycoprotein (glomalin) reduces metal mobility or sequesters metals | | |
| Precipitation | Formation of insoluble precipitates by the rapid reaction of inorganic acid with certain dissolved metals (such as Cu, Fe, Zn, and Pb) | Bacteria | Park et al. (2011) and Zhou et al. (2013) |
| Chelation | Binding of organic molecules (particularly di- and tricarboxylic acids), metallothionein, and siderophores within fungi walls and outer layers | Ectomycorrhizal fungi and bacteria | Bellion et al. (2006) and Dimkpa et al. (2008) |
| Redox transformations | Enzymatic redox reaction to convert a metal ion into a nontoxic or less toxic state | Ectomycorrhizal fungi and bacteria | Ma et al. (2016), Chatterjee et al. (2009), Olegario et al. (2010), Majumder et al. (2013), and Oves et al. (2013) |
| Biologically induced mineralization (BIM) | Mineral formation as a consequence of changes in the supersaturated system due to the intake or excretion of different metabolites and to the contribution of substances that can act as nuclei of crystallization, such as cell surfaces (cell wall, membranes, excreted organic compounds, cell debris, etc.) | Bacteria | Lowenstam (1981) and Moraga et al. 2017 |
| Release of metabolites | Mobilization, which helps the removal from contaminated soils, of metals facilitated by metabolites such as metallophores, biosurfactants (BSs), sophorolipids, and rhamnolipids | Bacteria | Mulligan et al. (2001), Juwarkar et al. (2007), Venkatesh and Vedaraman (2012), and Deicke et al. (2013) |

As different species may have different adaptive mechanisms, the use of autochthonous consortiums or mixed cultures is a widely applied strategy to enhance remediation processes (Fuentes et al. 2011, 2013a, b). In particular, extreme environments are a major source of microbial diversity for potential uses in biotechnological processes and products; thus, to study in different bioremediation processes their microbial diversity would enhance the challenges in understanding their involvement in (Nesme et al. 2016). Direct access to the genomic DNA of coexisting microbial species can give a better understanding of evolution, lifestyle, and diversity of the microorganisms and expose more of the hidden world of microbes (Krause et al. 2008). In order to find out "who is there" and with which frequency, metagenomics rises nowadays as a very powerful tool.

Metagenomic analysis of nucleic acids provides direct access to the genomes of the "uncultivated majority." Using amplicon surveys or metagenomic approaches for comparing soil microbial communities and correlating indicator species with specific environmental perturbations or specific land usage tend to produce statistically valid trends whether the selection of the different methods minimizes the bias of subsequent results or not (Nesme et al. 2016). Thus, as suggested by Lynch and Neufeld (2015), the objectives should include minimally biased methods (or combinations of methods) for soil characterization, differentiating between active soil microorganisms and dormant cells (and extracellular DNA). They should also cover seasonal variability, quantification of the full extent, and scale of functional diversity and microbial soil taxonomy, together with the diversity of "biosphere rare" microorganisms that typically dominate assessments of soil microbial diversity.

Step-by-step, from sampling a contaminated soil to identifying potential bioremediation, strategies are listed below (Fig. 1.1):

1. Sampling
2. Isolation of microorganisms and metagenomic analysis
3. Physicochemical soil analysis
4. Identification of toxic tolerance levels by isolated bacteria
5. Analysis of tolerance mechanisms
6. Identification of compounds of interest involved in tolerance
7. Identification of genes or proteins differentially expressed in the presence of toxic

In the next sections, each one will be developed.

## 1.3 Microbial Isolation

The standard sampling design for the characterization of a soil differs from that carried out for microbial isolation. In the first, it is of utmost relevance to collect sufficient samples to have a representative overview of the situation of the soil, i.e., as close to reality as possible. However, in the case of soil sampling for microbial isolation, with the final aim of bioremediation or biotechnological purposes, the priority is to target the site and sampling point, minding the final application. Such is the case, for example, of studies that focus on the search of enzymatic activities in extremophiles to optimize existing industrial processes. One of the best-known examples is the application in biotechnology of the thermostable DNA polymerases Taq, Pfu, and Vent, all of bacterial origin (from the thermophile bacteria *Thermus aquaticus*, *Pyrococcus furiosus*, and *Thermococcus litoralis*, respectively).

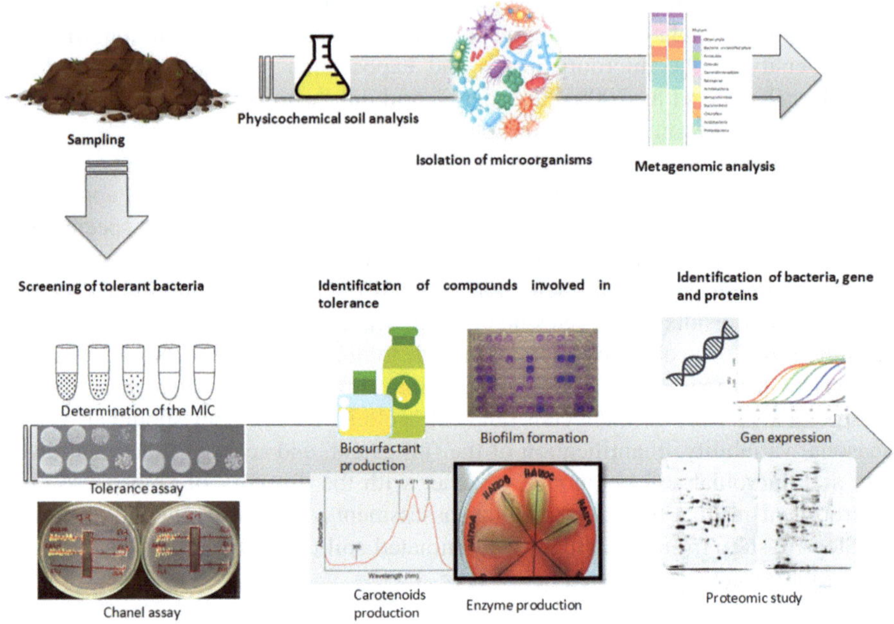


**Fig. 1.1** Scheme of the steps to follow from sampling a contaminated soil to identifying potential bioremediation strategies

## 1.3.1 Soil Sampling Design

The soil matrix is a complex heterogeneous living system, influenced by many biotic and abiotic factors. For example, on a hillside, there will be differences between the soil in the lower, middle, and upper part of the slope. However, on the surface of different points on flatland, it is possible to find differences as well. The soil is heterogeneous in space and time, mostly as the product of variations in topography, original material, environmental forming processes, and human intervention. Thus, this heterogeneity must be considered in the planning of the sampling.

From a horizontal point of view (i.e., in the extension of the surface), the soil has a patchy nature. Slight variations of terrain elevation, soil texture, or surface vegetation can produce essential differences in the environment for the soil microbial community. Also, from a vertical point of view (i.e., in depth), the different horizons that comprise the soil will present specific properties regarding air and water movement and overall transport of specific compounds through the soil profile.

Therefore, always minding the final goal of the isolates, a specific point in the selected location can be targeted for sampling. For such selection, the premise to consider is that the bacterial community in a specific point is the fittest for the prevailing conditions there. This strategy was followed in several bioremediation studies that aimed to degrade hydrocarbon compounds (Zheng et al. 2018) and pesticides (Campos et al. 2015) or sequester toxic compounds such as those with boron or heavy metals (Moraga et al. 2014b; Gillan et al. 2017). Also, when the aim is to positively enhance plant-microorganism interactions, this approach is followed by isolating bacteria from the plant's surfaces like the rhizosphere and phyllosphere and also from the interior of the plant in the search for endophytic bacteria (Karnwal 2017).

For rigorous systematization of the sampling procedure, normalization guidelines for soil quality assessment can be obtained from the International Standardization Organization (ISO). They provide guidance on the collection, handling, and storage of soil samples on its norms (both part of the former ISO 10381-6:2009) ISO 18400-105:2017 and ISO/FDIS 18400-206 (under development).

#### 1.3.1.1 Sample Collection and Transportation

Once the specific point and depth of sampling are determined, the collection and transport of samples must be planned considering two essential requisites: (1) avoid cross-contamination and (2) ensure the integrity of the bacterial community until its arrival to the laboratory for processing. The minimal set of elements, for the collection of soil aiming to the isolation of aerobic mesophilic microorganisms, include black plastic bags, transparent plastic bags, tags, a cord, a plastic knife, a shovel, gloves, a nice box, and a cooling agent (ice or ice packs). These elements can be complemented with a GPS (to mark the exact location of the sample collection point) and portable devices to measure environmental or soil variables (temperature, humidity, and pH in a soil suspension in water), depending on the objective of the study and the availability of time and resources. Also, it is useful to keep a log of the information about the site: details on topography; weather conditions at the moment of sampling, e.g., if it rained the day before; and any other data relevant to the study, e.g., surface cover (crops, vegetation, roadside, brine, frost, burning signs, oil) and land use history (application of pesticides, oil spill, mining tailing), must be taken account, among others.

Briefly, before collecting the sample, clean the ground surface by removing those elements that are not part of the soil (stones, salt crust, not decomposed organic matter such as branches, leaves, and any other strange material), and dig a hole with the shovel. The hole should be at least 10 cm deeper than the previously set sampling depth and of a diameter such that allows comfortable hand and arm movement inside. Then, inside the hole, scrape a section of the wall with a plastic knife to remove the soil surface that was in contact with the shovel. Scraping is required to eliminate any metal contamination or toxicity from the shovel, and it also provides visibility for qualitative identification of the horizons in the sampled soil profile, as well as the texture, structure, humidity, the presence of macrofauna, and others. To collect the sample, first clean the hole as thoroughly as possible removing the loose soil, and place a black plastic bag inside. Make 5-cm-deep cuts in the sides and bottom (at the previously determined sampling depth) of the scraped section in the hole's wall, and place one hand against it, cutting the back carefully with the knife to obtain a slice of soil. Place the soil slice in the bag with a tag identifying the sample, and close it with a cord to ensure gas diffusion. Place the black bag inside a transparent bag with a second tag, and close again with a piece of cord. Finally, place the bagged sample inside the cold box with a refrigerating agent to transport it readily to the laboratory for processing.

The operation described can be more or less complicated depending on the texture of the soil (heavy soils are too hard, and light texture soils are too lose), the water content, the structure, and the density. Therefore, other sampling tools like a manual or electrical augers and soil core samplers can be more appropriate to drill the soil and collect samples. There is a large variety of augers in the market depending on the specific characteristics of sampling site, for example, augers designed for sand, clay, and mud and of one piece, open-faced, screw type, or gauged; cutters for hard soil; and others. These tools, commercialized in different sizes, many times can be customized according to the requirements of the study.

The amount of sample to collect will be given by the accessibility to the sampling site, the transportation and storage capability, fundamentally the minimum required amount to carry out the previously planned analysis, and the availability of the sample. The two latter must be well identified, since sometimes a sample can only be taken at a specific moment or opportunity, after which the surrounding conditions vary, changing as well the intrinsic properties of the sample itself.

Although during the microbial isolation the culture conditions will mimic those of the sampling site, it is useful to maintain them during the sampling and transportation. Thus, the surrounding conditions should resemble those in the site of origin regarding temperature and oxygen availability. Thence, specific elements and conditions for manipulation and transportation are to be considered for particular cases. For the isolation of anaerobic microorganisms, airtight containers and gas (argon or a mix of $CO_2$ and $N_2$) are required to remove the air in them (Wiegel et al. 1979; Song et al. 2000; Kim et al. 2013). Instead for the isolation of psychrophiles, a portable cooler with temperature control is necessary (Vishnivetskaya et al. 2000; Panikov and Sizova 2007).

#### 1.3.1.2 Samples Processing and Storage

The samples can be either processed in situ or taken to the laboratory for storage. For immediate processing after the collection, the soil is placed in containers with specific culture media that will allow the survival of the microbial community targeted (Kim et al. 2013). Thus, the samples can be stored in vials with culture media for further microbial extraction and isolation from the soil. The most common procedures to process the samples in the laboratory, after collecting and transporting them, involve freezing or drying for storage of the soil in cold temperatures ranging from −70 to 4 °C.

The processing of the samples for frozen storage only involves sieving (<4 mm) and fractioning of the soil. Then the homogenized soil is placed in thin layers inside zipper bags that can be readily stored by freezing at −20 °C or ultra-freezing at −70 °C (Wallenius et al. 2010). For the storage of dried soil, the sample is let to dry at room temperature for 24 h. To avoid contamination, a large volume of soil can be spread as a layer inside a laminar flow hood or similar or by fractioning the soil and placing it inside Petri dishes loosely covered. Then, after removing any remaining organic debris and stones, the dry soil is sieved through a 2 mm mesh to be stored in black plastic bags at 4 °C until use (Romano-Armada et al. 2017). In the case of aggregates larger than 2 mm, the soil is ground before sieving. The bags, filled with 1 kg of dry homogenized soil (maximum), are closed with a cord and must not be stacked one over another. These conditions allow gas diffusion inside the bag, which will preserve the aerobic microbial community in the soil sample.

There are some advantages on freezing the soil over drying it regarding the preservation of the soil's enzymatic and microbial community activity (Wallenius et al. 2010). However, in the case of microbial isolation, it is advisable to process the samples and extract the microorganisms of interest as soon as possible, regardless the storage method. If the target of isolation does not comprise the aerobic mesophilic microbial community, specific care, concerning temperature and oxygen availability, must be considered for the manipulation of the sample in the different steps (Vishnivetskaya et al. 2000; Panikov and Sizova 2007).

### *1.3.2 Isolation of Microorganisms and Metagenomic Analysis*

#### 1.3.2.1 Metagenomic Analysis

A total soil metagenomic DNA extraction must be performed and then an amplification of the hypervariable regions V1-3 and V4-7 of the 16S rRNA belonging to the bacterial microbial community using PRBA63f and UN518r primers. Then a hypervariable region search in databases of other report community members must be done (Tekere et al. 2011).

## 1.4 Bacterial Population Extraction from the Soil Matrix

Overall, the extraction of microorganisms from the soil consists of the suspension of the latter in a liquid. However, all the procedures aim to accomplish the dispersion of the soil and their microbial aggregates with minimal reduction in the microbial viability. According to the literature, there is little consensus in the ratios of soil-to-liquid for the suspensions (1 g in 10 ml; 5 g in 15 ml, 20 ml, 50 ml, and 250 ml; 10 g in 90 ml and 100 ml; 30 g in 90 ml) and also in the used liquids (0.9% sodium chloride (NaCl), 1% sodium hexametaphosphate ($(NaPO_3)_6$), 0.1% peptone water, specific culture media) (Wiegel et al. 1979; Vishnivetskaya et al. 2000; Panikov and Sizova 2007; Kim et al. 2013; Moraga et al. 2014a; Gillan et al. 2017; Romano-Armada et al. 2017; Zheng et al. 2018). The many combinations for the extraction can encompass steps of suspension, agitation, centrifugation, sonication, and the addition of chemical dispersing agents.

The most straightforward procedure is the extraction for direct plating and isolation from the first extract. Briefly, to break down the soil aggregates without the addition of a chemical dispersing agent, the soil suspension can be agitated at 250 rpm for 30 min with a subsequent 10 min sonication to disperse the microbial aggregates (Romano-Armada et al. 2017). Then the soil particles can be let to settle for 30 min (Moraga et al. 2014b), or a centrifugation step can be carried out to eliminate the soil and collect the microbial cells from the supernatant (Gillan et al. 2017; Zheng et al. 2018).

Also, in the processing of the sample previous to the isolation step, culture media enrichment can be carried out to apply selective pressure to favor the growth of the microorganisms with the targeted metabolic activities. The soil samples and its extract are suspended in culture media added with the compounds to be degraded as sole carbon or nitrogen sources (Song et al. 2000; Zheng et al. 2018) or with heavy metals (Moraga et al. 2014b; Gillan et al. 2017).

Some procedures or reagents can be avoided by taking into account the targeted bacterial community to isolate. While the $(NaPO_3)_6$ method is time and cost effective for dispersing soil aggregates for fast texture analysis, it may be too aggressive for the fraction of microorganisms highly dependent on the soil organic carbon. Also, the saline solution (0.9% NaCl) can interfere with the desired microbial isolation. Both solutions can be replaced by 0.1% peptone water or by sterile distilled water (as long as the salts in the soil sample allow maintaining the osmotic balance in the solution).

### *1.4.1 Bioremediation-Oriented Microbial Isolation*

Following the microbial extraction from the soil samples, the supernatants are plated in culture media for isolation. Aliquots (0.1–1 ml) of the serial dilutions of the supernatants are spread plated in specific media, commonly enriched with the compound to be remediated, culturing under the environmental conditions of interest. Most soil isolates are cultured in the dark at close to neutral pH and between 25 and 30 °C (Song et al. 2000; Moraga et al. 2014b; Gillan et al. 2017; Zheng et al. 2018). Once the microbial colonies were identified, they must be isolated by streak plating following the aseptic technique. The single colonies must be re-streaked until the purity of the culture can be confirmed by microscopy or observation of the colonies morphology, after which they can be stored for further studies.

Note that the isolation procedures are conducted in solid media, which allows differentiation of single colonies. Therefore, particular attention must be paid to the used gelling agents in the case of extremophiles isolation regarding temperature and pH. While for the isolation of thermophiles that grow above 60 °C (Wiegel et al. 1979) agar-agar can be used since its fusion temperature is 85 °C, for the isolation of acidophiles, the use of agarose is necessary (Nancucheo et al. 2016). However, in the case of cryophiles that grow under the freezing point in supercooled media, the addition of up to 20% glycerol is required as an antifreezing agent (Panikov and Sizova 2007).

Also, in the case of anaerobic microorganisms, the oxygen must be removed from the sealed vials for proper culture (Wiegel et al. 1979; Song et al. 2000; Panikov and Sizova 2007; Kim et al. 2013). Moreover, the culture manipulations should be held inside the anaerobic chambers (Kim et al. 2013).

## *1.4.2 Selection and Characterization of Microorganisms for Soil Bioremediation*

The selection and subsequent characterization of adequate microorganisms are the main steps that will determine the success of the entire bioremediation procedure and the recovery of the affected soil.

Once the selection is done, a good characterization and evaluation in different conditions are particularly important. This is due to the fact that selected microorganisms have, as final destination field application, a much larger scale than laboratory, and therefore all their characteristics and the expected behavior in the conditions they will face in field must be perfectly known. In this sense, while some authors consider that the first part of the selection should be to identify strains which can degrade the target contaminant, others suggest that it is essential to base the selection on a priori knowledge of the population dynamics and their distribution in sampled habitats in the first place, to ensure the success of the bioremediation technique. After that, the second phase of the selection procedure would be to identify strains which can degrade the target contaminant, having better persistence and colonization rates (Thompson et al. 2005).

Beyond the taken position, once the pure cultures are obtained, it is necessary to select the microorganisms of interest according to the contaminant to be remediated. After this, it should proceed to the characterization of the selected microorganisms, in search mainly of those that present greater tolerance and degradative capacity or well looking for specific molecules that promote and favor bioremediation, such as enzymes, biosurfactants, etc., depending on the contaminant of interest. Regarding the characterization, it can be mentioned in the first place basic techniques such as morphological characterization and, later, metabolic characterization, among others.

### 1.4.2.1 Selection of Microorganisms for Bioremediation

**Selective Enrichment** This involves the selective culture of some strains over others from polluted samples, with a metabolic advantage, for example, using the target contaminant as the sole carbon or nitrogen source. The technique results in the selection of strains that express the required degradation or metabolic capacity

(Thompson et al. 2005). Its importance lies in the fact that it is a very selective technique where nutritional conditions are manipulated, using the contaminant of interest as energy source as example.

**Selection by Channels in Plates** This technique involves cutting small channels in the solid medium of the plate. Then, in the channel, a solution with the contaminant of interest in the desired concentration is discharged, and isolated strains are sown around it. The inoculation pattern could vary making the channel in the middle of the plate or at both or one side. The solution diffuses through the channels, and the strains grow more or less near the channels according to their tolerance to the pollutant (Moraga et al. 2014b; Martinez et al. 2018).

**Selection by Agar Diffusion Test** This technique allows the identification of tolerant strains to the target contaminant. It consists in using paper filter discs embedded in the contaminant of interest in different concentrations. In the first place, the plate has to be filled with the 50% of the final volume and is left until the medium partially solidified. The remaining 50% is subsequently inoculated with the microorganisms of interest and poured, forming a second layer, over the first semi-gelled layer on the plate. After that sterile filter paper discs, impregnated in a solution of the contaminant of interest, are placed on the agar. The appearance of an inhibition growth halo around the filter paper disc is expected in the case of susceptible bacteria (Rosas Hernández 2009).

#### 1.4.2.2 Characterization of Selected Microorganisms

Morphological Characterization

Bacterial colonies that grew on agar plates can be initially characterized according to their morphology in each particular media. Macroscopically, size, shape, pigmentation, elevation, surface, appearance, edges, etc. can be determined in solid medium, and if it is an evaluation in liquid medium, the formation of pellets, exocellular polymeric substances (EPS), and precipitates can also be visually evaluated. In this step, microbial movement can also be evaluated in semisolid medium.

Regarding microscopic evaluation, it allows differentiating structural characteristics of bacteria, for which different stains are commonly used:

*Gram staining*: This technique is used to classify bacteria into two large groups, Gram-positive and Gram-negative, on the basis of differential staining with a crystal violet-iodine complex and a safranin counterstain. This is due to differences in the cell wall structure: while Gram-positives retain the complex after the treatment with alcohol and appear purple, Gram-negative organisms decolorize and after the treatment with safranin become pink. This staining also facilitates the determination of cell morphology (Madigan et al. 2009). As known, the age of the reagents for the stain as well as that of the microorganisms is crucial for the proper classification.

*Endospore staining*: Some bacterial genres, like *Clostridium*, produce resistance structures called endospores. The position of endospores in the cell (in the center or in the pole) may be indicative of some species. Malachite green dye is used for this staining (Rosas Hernández 2009).

*Staining with methylene blue*: This is useful for the identification of morphological characteristics as bacterial motility; for such staining the smear is stained with blue methylene for 1 or 2 min and then washed with tap water and observed at the microscope (Rosas Hernández 2009). On the other hand, it can be used as a quick method to know the viability of yeasts.

### Metabolic Characterization

After the general morphological description is done, it is important to proceed to the metabolic characterization in the search of the specific molecules, and if possible the mechanisms involved, that allow microorganisms to degrade or modify the target contaminant or help them in the decontamination process.

### Enzymatic Characterization

The process of bioremediation depends on microorganisms that enzymatically attack the pollutants and convert them to innocuous products (Karigar and Rao 2011). However, some enzymatic products can be toxic, in some cases indeed more toxic than the initial pollutant. In this sense, the search can be oriented toward enzymes that play a key role in the degradation of specific pollutants.

**Hydrolytic Enzymes** These enzymes disrupt major chemical bonds in the toxic molecules resulting in the reduction of their toxicity (Karigar and Rao 2011). Due to their intrinsic low substrate specificity, hydrolases play a pivotal role in the bioremediation of several pollutants including insoluble wastes, oil spill and organophosphate and carbamate insecticides, cellulose materials, chitin, keratin, kraft pulp, sewage sludge, starch materials, polyacrylate, and polyurethane, among others (Gianfreda and Rao 2004). This group of enzymes includes proteases, carbohydratases (ascellulases, amylases, xylanases, etc.), esterases, phosphatases and phytases, lipases, and proteases, among others (Karigar and Rao 2011).

**Dehydrogenase and Oxidoreductase** These enzymes take part in the metabolism (catabolic and also anabolic reactions) in microbial cells. During catabolism they oxidize organic compounds by the transfer of electron pairs from a substrate to nicotinamide adenine dinucleotide (NAD+) or nicotinamide adenine dinucleotide phosphate (NADP), and then they also act during anabolism to regenerate the electron transporters. Different investigations revealed that oxidoreductase enzymes were involved in removing toxic heavy metals like chromium or cooper. There are microorganisms resistant to chromium and showed Cr (VI) reduction to be undetectable (Irazusta et al. 2018). Experiments with cell-free extracts, mitochondrial and extracellular yeast extract of *Cyberlindnera jadinii* M9 and *Wickerhamomyces anomalus* M10, indicated that a soluble intracellular type of enzymes was responsible for Cr(VI) reduction. These enzymes usually show a NADH/flavin oxidoreductase activity and have the ability to reduce chromate as a secondary function (Irazusta et al. 2018).

**Laccases** They are a type of ligninolytic enzymes, polyphenol oxidases, that catalyze the oxidation of various phenolic and non-phenolic compounds, particularly those with electron-donating groups such as phenols (−OH) and anilines (−NH2), by using molecular oxygen as an electron acceptor (Karigar and Rao 2011). They participate in the cross-linking of monomers, involved in the degradation of a wide range of industrial pollutants. Their low substrate specificity makes these enzymes interesting for bioremediation not only for industrial pollutants and effluents but also for wastes of paper, pulp, textile, and distillery industries. Also, and particularly in soils, these enzymes, in conjunction with peroxidases, enhance the natural degradation of xenobiotic and organochlorides, conversion and mineralization of polycyclic aromatic hydrocarbons and pentachlorophenols, oxidation of aromatic hydrocarbon and pentachlorophenol, and degradation of herbicides as bentazon and diuron, among others (Duran and Esposito 2000). They are encoded by different genes and expressed in different organelles and can be readily detected by gel electrophoresis (Chandra and Chowdhary 2015).

**Peroxidases** These enzymes are oxidoreductases which catalyze reactions in the presence of hydrogen peroxide and act in the oxidation of a variety of organic and inorganic compounds (Duran and Esposito 2000). They are useful in decomposition of pollullants: textil dye degradation, ligning degradation, sewage treatment, and also as biosensor (Bansal and Kanwar 2013). Considering their bioremediation importance, three main enzymes have been studied the most, lignin peroxidase (LiP), manganese-dependent peroxidase (MnP), and versatile peroxidase (VP), most of them due to their high potential to degrade toxic substances in nature (Karigar and Rao 2011).

## Biosurfactant Production

Biosurfactants are amphiphilic compounds produced by microorganisms with pronounced surface and emulsifying activities (Singh et al. 2007). They consist of a hydrophilic and a hydrophobic group, and due to this structure, they are able to increase the surface area of hydrophobic water-insoluble substances (Pacwa-Płociniczak et al. 2011). They act diminishing the surface tension ($\sigma$) between two liquids with different polarity, making the emulsification of one liquid in the other easier (Iustman et al. 2013).

Their importance on bioremediation lays on the fact that biodegradation rate of a contaminant depends on its bioavailability to the metabolizing organisms, and they increase the water bioavailability of such substances and change the properties of the bacterial cell surface (Pacwa-Płociniczak et al. 2011). In this sense, biosurfactants have two main functions: in one hand, they increase substrate availability, and in the other hand, they enhance biodegradation by mobilization, solubilization, and emulsification (Ahmad Khan et al. 2015).

These compounds are specially used for decomposition of heavy metals and hydrocarbons including polynuclear aromatic hydrocarbons (PAHs) (Rangarajan and Narayanan 2018). They are also used in the oil industry to clean up oil spills and to enhance oil recovery (Jones 1997).

Siderophore Synthesis

These molecules are specially used for detoxification of heavy metals as a consequence of chelation, since they form stable metal-ligand complexes and influence the metal mobility in the environment (Naik and Dubey 2013).

## 1.5 Molecular Omics Technologies in Microorganisms' Selection and Characterization

Traditionally, studies on bioremediation of target pollutants were made using conventional isolation strategies based on culturing microorganisms. However, these techniques exclude all microorganisms that cannot be cultured, which represent a huge proportion. Culture-independent methods as molecular techniques in general and omics technologies in particular are extremely important to discover a wide range of unidentified pollutant-degrading microorganisms that could have crucial roles in bioremediation (Watanabe 2001). In this sense, "omics"-based approaches can help investigate the genome, transcriptome, proteome, and metabolome of single organisms and even mixed communities, opening new opportunities to decipher molecular mechanisms in pollutant biodegradation (El Amrani et al. 2015).

### *1.5.1 Metagenomics*

It involves the study of entirely genetic information contained within an environmental sample, making possible to identify the functional potential and the taxonomic identity of all organisms in the sample, but without information about the actual active members of the community (El Amrani et al. 2015). Metagenomic approaches often take two forms: targeted metagenomics or shotgun metagenomics. In targeted ones, environmental DNA is extracted, and the gene of interest is PCR amplified using primers designed to amplify the greatest diversity of sequences for that gene. This approach is regularly used to investigate the diversity of small subunit rRNA sequences (16S/18S/ITSrRNA) in a sample. However, it is employed not only to investigate both the phylogenetic diversity and relative abundance of a particular gene in a sample but also as a tool to investigate the impact of environmental contaminants in altering microbial community structure (Techtmann and Hazen 2016).

In shotgun metagenomics, the total genomic complement of an environmental community is probed through genomic sequencing (Techtmann and Hazen 2016). A typical shotgun study comprises five steps: (i) the collection, processing, and sequencing of the samples; (ii) preprocessing of the sequencing reads; (iii) sequence analysis to profile taxonomic, functional, and genomic features of the microbiome; (iv) statistical and biological postprocessing analysis; and (v) validation (Quince et al. 2017). Its main limitation is the depth of sequencing because gaining a complete inventory of the genes in an environmental sample often requires extremely deep sequencing. Secondly, oftentimes shotgun metagenomics samples the dominant microbes in a community and only sparsely covers the genomic content of the low-abundance members of that community (Techtmann and Hazen 2016).

### *1.5.2 Metatranscriptomics*

It consists in a community RNA analysis that provides an inventory of the actively expressed genes in a sample. RNA is extracted from an environmental sample, converted into cDNA, and sequenced in a similar way than metagenomics (Techtmann and Hazen 2016). This approach facilitates insight into the potential expression of genes at the time of sampling. Its importance relies in the fact that while posttranscriptional and posttranslational gene expression can regulate protein synthesis, transcriptional-level control of gene expression enables organisms such as bacteria to rapidly adapt to changing environmental conditions (Moran 2009). Therefore, immediate regulatory responses to environmental changes may be better reflected by the metatranscriptome (Carvalhais et al. 2012).

### *1.5.3 Metaproteomics*

Metaproteomics consists in the community protein analysis. This approach provides insights into the complement of proteins found in an environmental sample including posttranslational modifications in proteins that may impact their activity (Techtmann and Hazen 2016). It is useful mainly in determining changes in the composition and abundance of proteins, as well as in the identification of key proteins involved in the physiological response of microorganisms when exposed to pollutants (Desai et al. 2010). The first step in any proteomics-based approach is protein extraction, for which, there are numerous mechanisms and the available instruments (Singh 2006). For protein separation and identification, two strategies have been established: one is the gel-based method. After protein extraction, highly efficient methods of separation based on two-dimensional polyacrylamide gel electrophoresis (2-DE) and modern tools of bioinformatics in conjunction with mass spectrometry (MS) are used. Matrix-associated laser desorption/ionization time-of-flight MS (MALDI-TOF-MS) is the most commonly used MS approach to identify proteins of interest excised from 2-DE gels, by generation of peptide-mass fingerprinting (Singh and Nagaraj 2006). The other strategy is the liquid chromatography (LC)-based method, where the whole proteome is digested into a more complex peptide mixture using proteases without prior protein separation in gel. Then the resulting peptides are separated using strong cation exchange chromatography or microcapillary reverse-phase. In general, the separated peptides are analyzed using liquid chromatography coupled with MS/MS (LC-MS/MS). The produced MS data are interpreted for protein identification and then bioinformatic analysis. The second approach circumvents the limitations of the gel-based approach and greatly increases the proteome coverage compared with the gel-based method, allowing high-throughput identification of thousands of proteins within a short time and especially making detection of insoluble membrane proteins possible. Thus, the LC-based approach has become the main stream of microbial community proteomic studies, although it still suffers from problems of reproducibility, dynamic range, and database availability (Wang et al. 2016).

### *1.5.4 Metabolomics*

Metabolomic addresses the whole complement of metabolites in an environmental sample (El Amrani et al. 2015). Metabolome analysis covers the identification and quantification of all intracellular and extracellular metabolites with molecular mass lower than 1000 Da, using different analytical techniques (Villas-Bôas et al. 2005). This analysis encompasses sampling, sample preparation, metabolite separation and detection, data analysis, and interpretation. Some particular advantages of metabolomics is that it allows monitoring changes in an organism as it is exposed to environmental pollutants, to follow the degradative pathways and to track their intermediates and responses during mineralization (Singh 2006). However, one of its main limitations is the complexity. The convoluted nature of cell metabolism, in which the same metabolite can participate in many different pathways, complicates the interpretation of metabolite data (Villas-Bôas et al. 2005), so it is difficult if not impossible to establish a direct link between genes and metabolites. Likewise, the metabolome consists of extremely diverse chemical compounds with a large variance in chemical structures and properties (Villas-Bôas and Bruheim 2007).

## 1.6 Microbial Features Involved in Resistance Against Pollutants

Microorganisms able to survive in extreme conditions harbor a wide variety of properties. The processes developed by these organisms often include several characteristics that are coded in genes, which are activated or silenced when needed. Some of the features that characterize extreme microorganisms and that allow them to survive in such hostile environments also provide them other features that are very promising from a biotechnological perspective. As an example, some bacteria able to grow in highly salted environments have been proved to use polycyclic hydrocarbons as a source of carbon and energy (Isaac et al. 2015). They are able to grow in highly salted environments by producing EPS that also functions solubilizing hydrophobic compounds. The solubilization of these compounds increases the interaction of the bacteria with the surrounding organic components, thus favoring its degradation (Isaac et al. 2015). This example shows that a single bacterium owns many interesting features with applicability in biotechnological processes. This is the main goal for the study of extreme organisms, other than their characterization, the search for novel molecules that could be used as an alternative to conventional technologies.

This section aims to highlight some of the mechanisms used by microorganisms when they interact with different contaminants. It is widely known that some organisms have the ability to interact with different contaminants affecting the concentration of the contaminant in the surrounding environment. When microbial interaction produces a decrease in the concentration of the pollutant under study, the biological system can be used to remediate contaminated sites. Some of the examples that have been mentioned involve different metals. In the case of uranium, many researchers have studied the way microorganisms adapt toward its presence. They have encountered biosorption, biomineralization, accumulation inside the cell, and direct reaction with the ions, having them changed their oxidation state, as the main mechanisms (Sánchez-Castro et al. 2017).

In general, microorganisms possess a huge amount of information condensed and coded in their genome. Depending on the environmental conditions, only some of this information (the essential which will allow survival) is expressed. All of the mechanisms involved in DNA replication, expression, and translation are regulated. When environmental and nutritional conditions change, microorganisms adapt to this variation by modifying gene expression and, as a consequence, protein synthesis. The newly synthesized proteins may be involved directly in the interaction with the contaminant or indirectly with participation in synthesis of compounds that interact with the contaminant. Other typical mechanism is the over synthesis of proteins involved in the stress cell protection (Irazusta et al. 2013, 2016, 2018). One example of the latter is the production of EPS. Some extreme bacteria which inhabit environments exposed to UV radiation produce carotenoids, to avoid mutations as a consequence of the highly environmental radiation.

Depending on the genetic properties of microorganisms inhabiting certain niches, different actions can lead to interaction with a contaminant. The contaminant can be tolerated through intra- and extracellular mechanisms, extracted from the cell through active transport systems, and the contaminant can also interact with the membrane preventing the influx.

Differential protein synthesis may lead to the activation of pathways involved in the interaction of the microorganism with the contaminant. These pathways can lead to the direct interaction of the microorganism with the contaminant or can be used to produce some substances as a response against the presence of the contaminant in the surroundings of the cell.

The mechanisms that microorganisms use may lead to:

### *1.6.1 Removal of the Pollutant by Biomineralization*

Biomineralization is a general term for the processes by which living organisms form minerals resulting in the remotion of the pollutant and providing a means of detoxification as well as biorecovery depending on the type of pollutant (Gadd and Pan 2016). There are two main mechanisms: biologically controlled mineralization (BCM) and biologically induced mineralization (BIM) (Pérez-González et al. 2011). In controlled ones, the mechanisms are closely regulated, and organisms precipitate minerals that serve physiological and structural roles. This process can include the development of intracellular or epicellular organic matrices into which specific ions are actively introduced and their concentrations regulated such that appropriate mineral saturation states are achieved. Accordingly, minerals can be formed within the organism even when conditions in the bulk solution are thermodynamically unfavorable (Konhauser and Riding 2012). In induced ones, minerals are formed without any apparent regulatory control, as a consequence of changes in the supersaturated system due to the intake or excretion of different metabolites (active mechanism) (Lowenstam 1981) and to the contribution of substances that can act as crystallization nuclei, such as cell surfaces which favor precipitation (passive mechanisms).

The interaction of microorganisms with different pollutants can occur in the surface of the cell. Depending on the nature of the pollutant, the interaction can lead to the modification of the oxidative state of the element. If it is an ion, the variation of the oxidation number may result in its mineralization, and the process is known as biomineralization, as it is a consequence of living organisms. This phenomenon includes the chemical reaction of the ions with membrane transporters, thus inducing the variation of their state to a mineral. The final effect in the surrounding environment is a decrease in the concentration of the soluble ions. This process then can be used as a mechanism to decontaminate polluted sites. This mechanism, though studied in several organisms, is still very interesting in respect of the benefits it provides to the microorganisms involved. Microorganisms able to perform biomineralization may stay trapped between its own biomineral, thus producing its death.

Numerous bacterial genera produce biominerals that have been used in the remediation of soil contaminants, especially in bioremediation of heavy metals such as chromium (Crean et al. 2012), nickel (Haferburg et al. 2008), cobalt (Handley-Sidhu et al. 2016), and arsenic (Achal et al. 2012), among others.

It is noticeable that mineralization mediated by microorganisms can be a consequence of a fully regulated process or microorganisms can also act as inductor of the biomineralization. For each case, minerals with different characteristics are formed. If microorganisms control the biomineralization, its shape is regular, and it turns into a heterogeneous structure when the cells induce the biomineralization. It is also important to highlight that when this process is highly regulated, the biomineral can be formed extra-, inter-, or intracellularly in respect of its location.

The most common biominerals precipitated by microbes include oxides, phosphates, sulfides, and oxalates, and these can have special chemical properties such as high metal sorption capacities and redox catalysis (Gadd and Pan 2016).

Some marine bacteria are capable of bioaccumulating pollutants from the surrounding environment, like in the case of the bacterial consortium tested for mercury removal by Canstein et al. (2002). They have also been reported to degrade hydrocarbons, chelate heavy metals, and remove pollutants from the environment by production of EPS. Also, *Streptomyces* have been studied for its ability to accumulate heavy metals (Majzlik et al. 2011).

### 1.6.2 Synthesis of Carotenoids

Microorganisms able to produce carotenoids are widely distributed in nature. It is known that this feature confers some abilities to survive in environments with high levels of radiation (Sandmann 2001). Some researchers have found for microorganisms able to withstand high concentrations of contaminants that the ability to synthetize carotenoids is strongly related to the concentration of the pollutant in the surrounding environment (Ortega-Cabello et al. 2017). An example occurs with the effect of ferrous ions toward the response of *Rhodococcus* and *Gordonia* strains (Ortega-Cabello et al. 2017).

The green algae *Dunaliella salina* and *Dunaliella bardawil* are known for their properties in synthetizing β-carotene in adverse conditions. Some of the carotenoids produced by these algae have great importance for its wide applications and because of the characteristics of the compounds produced, such as liposolubility and antioxidant capacity (Gómez et al. 2003). In the research developed by Gómez et al. (2003), *Dunaliella salina* showed a clear tendency in accumulating high amounts of carotenoids per volume when salinity increased.

Some microorganisms have also been described as able to produce carotenoids when grown under other stresses such as high radiation and high temperature or in the presence of certain compounds like heavy metals (Irazusta et al. 2013). Irazusta et al. (2013) have studied the behavior of the pigmented yeast *Rhodotorula mucilaginosa* RCL-11 in the presence of copper salts and determined that the synthesis of carotenoids increases in the presence of the heavy metal. However, they found an inverse relation between the accumulation of cooper and the carotenoid biosynthesis, indicating that these processes may be related.

### 1.6.3 Production of Exocellular Polymeric Substances

The components of the extracellular matrixes vary among microorganisms, and this composition is closely related to the function they develop. In general, EPSs are composed of carbohydrates (responsible of the texture of the EPS), proteins (related to the function of the EPS, since they can be enzymes which participate in extracellular degradation processes), extracellular DNA (of main importance in biofilm processes, since the community forming the biofilm may exchange genetic information through this path), and lipids and surfactants (responsible of the EPS hydrophobicity) (More et al. 2014). They play a major role in the architecture of the biofilm matrix being responsible for adhesion to surfaces and for cohesion in the biofilm. They immobilize biofilm cells and keep them in close proximity promoting interactions, including cell-cell communication and the formation of synergistic microconsortia, among other functions (Flemming and Wingender 2010). Considering its bioremediation concern, cells in a biofilm have a better chance of adaptation and survival (especially during periods of stress) as they are protected within the matrix (Singh et al. 2006), so they are especially useful in the immobilization and degradation of pollutants at relatively high rates because of the multiple interactions and high contact surface in the matrix with the target contaminant. EPS produced by algae, bacteria, fungi, and yeasts can be used for bioremoval of toxic compounds from the environment, and they have the characteristics of being a low-cost, nonhazardous, and effective method (Amoozegar et al. 2012).

Many halophilic organisms have been studied in order to describe the composition of the matrices they synthetize in the outside of the cell, since they may be involved in several biotechnological applications. They are also used in decontamination of chlorophenols (such as 2-chlorophenol,2,4-dichlorophenol, pyrene, phenanthrene, o-cresol, naphthalene, phenol, 1,2,3-trimethylbenzene), toluene, azo dyes, and herbicides, among others (Singh et al. 2006).

Amoozegar et al. (2012) have described that a moderately halophilic bacterial strain isolated from saline environments in Iran is capable of removing lead and cadmium from the surrounding environment. They have studied the ability of this microorganism and of the EPS it produces to interact with these heavy metals. The strain, characterized phylogenetically as belonging to the genus *Halomonas*, was able to produce a decrease of 90% and 50% of lead and cadmium, respectively. Since this strain is also known for its EPS formation capacity, the uptake of the heavy metals as a consequence of the interaction with this biological product was also evaluated.

Other than the function they are produced for in the natural environment, EPSs have many other applications. They are being studied to replace chemical polymers and have many advantages in comparison to them. The main advantages include biodegradability, high efficiency, and reduced or absence of toxicity. These features have made them attractive for many industries, including food, pharmaceutical, and medical.

## 1.7 Taking Advantage of Special Features Observed in Microorganisms That Tolerate Pollutants and Extreme Conditions

All of the microbial processes are highly regulated in microorganisms' cells. These regulations generally occur at gene expression level, modifying protein synthesis. In general, microorganisms with special features may be used for biotechnological applications. In some cases, it is even better to clone a specific gene of interest in a different biological system to optimize the production of the component of interest. For example, in the case of halophilic microorganisms, *Archaea* are known to produce proteins with high functionality at high levels of salt. Enzymes such as hydrolases (proteases, lipases, esterases), biopolymers, and surfactants are some of the components that can be obtained from halophilic *Archaea* (Litchfield 2011). If the element of interest for research or production is a protein, it is possible to clone the corresponding gene in other bacteria easier to grow at the laboratory and obtain the protein of interest in augmented amounts and with improved properties. For example, acute lymphoblastic leukemia (ALL) is a disease that has been treated with L-asparaginase for more than 30 years. Several studies have been developed to obtain an enzyme with better properties. Ghasemi et al. (2017) have found that a recombinant L-asparaginase from the halophilic microorganism *Halomonas elongata* is very promising regarding cancer therapy, because it has more desirable properties than those currently has been used. Proteins from microorganisms isolated from extreme environments are the target of study for any research groups around the world. Enzymes like hydrolases (esterases, glucosidases, and azoreductases, among others) are being the target for research as a consequence of increased enzymatic activities and for being a better alternative compared to the traditional enzymes currently in use (Castilla et al. 2017; Eslami et al. 2016). These enzymes can be used, as previously mentioned, in medical industry, for remediation of azo dyes and in processes for obtaining degraded compounds from complex mixtures (like obtaining sugar residues from cellulose).

## 1.8 Final Comments

Microorganisms have the ability to colonize a wide variety of environments. Those which are able to survive in extreme conditions often possess several properties which allow not only the colonization of extreme environments but also turn them into targets for biotechnological applications.

The features developed by microorganisms to tolerate adverse conditions can be exploited at an industrial level and turn them into possible candidates for bioremediation.

# References


Achal V, Pan X, Fu Q, Zhang D (2012) Biomineralization based remediation of As (III) contaminated soil by Sporosarcina ginsengisoli. J Hazard Mater 201:178–184

Ahmad Khan MS, Singh B, Singh Cameotra S (2015) Biological applications of biosurfactants and strategies to potentiate commercial production. In: CRC press (ed) Biosurfactants: production and utilization: processes, technologies and economics, vol 159. Taylor & Francis Group, Boca Raton

Amoozegar MA, Ghazanfari N, Didari M (2012) Lead and cadmium bioremoval by *Halomonas* sp, an exopolysaccharide-producing halophilic bacterium. Prog Biol Sci 2:1–11

Bansal N, Kanwar SS (2013) Peroxidase(s) in environment protection. Sci World J 2013:714639

Bellion M, Courbot M, Jacob C, Blaudez D, Chalot M (2006) Extracellular and cellular mechanisms sustaining metal tolerance in ectomycorrhizal fungi. FEMS Microbiol Lett 254(2):173–181

Campos M, Perruchon C, Vasilieiadis S, Menkissoglu-Spiroudi U, Karpouzas DG, Diez MC (2015) Isolation and characterization of bacteria from acidic pristine soil environment able to transform iprodione and 3,5-dichloraniline. Int Biodeterior Biodegrad 104:201–211

Canstein H, Kelly S, Li Y, Wagner-Dobler I (2002) Species diversity improves the efficiency of mercury-reducing biofilms under changing environmental conditions. Appl Environ Microbiol 68:2829–2837

Carvalhais LC, Dennis PG, Tyson GW, Schenk PM (2012) Application of metatranscriptomics to soil environments. J Microbiol Methods 91(2):246–251

Castilla A, Panizza P, Rodríguez D, Bonino L, Díaz P, Irazoqui G, Rodríguez Giordano S (2017) A novel thermophilic and halophilic esterase from Janibacter sp. R02, the first member of a new lipase family (family XVII). Enzym Microb Technol 98:86–95. https://doi.org/10.1016/j.enzmictec.2016.12.010

Chandra R, Chowdhary P (2015) Properties of bacterial laccases and their application in bioremediation of industrial wastes. Environ Sci Process Impact 17(2):326–342

Chatterjee S, Sau GB, Mukherjee SK (2009) Plant growth promotion by a hexavalent chromium reducing bacterial strain, *Cellulosimicrobium cellulans* KUCr3. World J Microbiol Biotechnol 25(10):1829–1836

Chojnacka K (2010) Biosorption and bioaccumulation–the prospects for practical applications. Environ Int 36:299–307

Crean DE, Coker VC, Van der Laan G, Lloyd JR (2012) Engineering biogenic magnetite for sustained Cr(VI) remediation in flow-through systems. Environ Sci Technol 46:3352–3359

Deicke M, Bellenger JP, Wichard T (2013) Direct quantification of bacterial molybdenum and iron metallophores with ultra-high-performance liquid chromatography coupled to time-of-flight mass spectrometry. J Chromatogr A 1298:50–60

Desai C, Pathak H, Madamwar D (2010) Advances in molecular and "-omics" technologies to gauge microbial communities and bioremediation at xenobiotic/anthropogen contaminated sites. Bioresour Technol 101(6):1558–1569

Dimkpa CO, Svatoš A, Dabrowska P, Schmidt A, Boland W, Kothe E (2008) Involvement of siderophores in the reduction of metal-induced inhibition of auxin synthesis in *Streptomyces* spp. Chemosphere 74:19–25

Duran N, Esposito E (2000) Potential applications of oxidative enzymes and phenoloxidase-like compounds in wastewater and soil treatment: a review. Appl Catal B 28(2):83–99

El Amrani A, Dumas AS, Wick LY, Yergeau E, Berthomé R (2015) "Omics" insights into PAH degradation toward improved green remediation biotechnologies. Environ Sci Technol 49(19):11281–11291. https://doi.org/10.1021/acs.est.5b01740

Elangovan R, Abhipsa S, Rohit B, Ligy P, Chandraraj K (2006) Reduction of Cr(VI)by a *Bacillus* sp. Biotechnol Lett 28:247–252

EPA (Environmental Protection Agency) (2004) Introduction to phytoremediation EPA-report. http://clu-in.org/techfocus/default.focus/sec/Bioremediation/cat/Overview/. Accessed Jan 2014

Eslami M, Amoozegar MA, Asad S (2016) Isolation, cloning and characterization of an azoreductase from the halophilic bacterium *Halomonas elongata*. Int J Biol Macromol 85:111–116
Flemming HC, Wingender J (2010) The biofilm matrix. Nat Rev Microbiol 8(9):623–633
Frankel RB, Bazylinski DA (2003) Biologically induced mineralization by bacteria. Rev Mineral Geochem 54(1):95–114
Fuentes MS, Benimeli CS, Cuozzo SA, Amoroso MJ (2010) Isolation of pesticide-degrading actinomycetes from a contaminated site: bacterial growth, removal and dechlorination of organochlorine pesticides. Int Biodeterior Biodegrad 64:434–441
Fuentes MS, Sáez JM, Benimeli CS, Amoroso MJ (2011) Lindane biodegradation by defined consortia of indigenous *Streptomyces* strains. Water Air Soil Pollut 222:217–231
Fuentes MS, Alvarez A, Sáez JM, Benimeli CS, Amoroso MJ (2013a) Methoxychlor bioremediation by defined consortium of environmental *Streptomyces* strains. Int J Environ Sci Technol 11(4):1147–1156
Fuentes MS, Briceño GE, Sáez JM, Benimeli CS, Diez MC, Amoroso MJ (2013b) Enhanced removal of a pesticides mixture by single cultures and consortia of free and immobilized Streptomyces strains. BioMed Res Int. https://doi.org/10.1155/2013/392573
Gadd GM (1999) Fungal production of citric and oxalic acid: importance in metal speciation, physiology and biogeochemical processes. Adv Microb Physiol 41:47–92
Gadd GM (2010) Metals, minerals and microbes: geomicrobiology and bioremediation. MBio 156(3):609–643
Gadd GM, Pan X (2016) Biomineralization, bioremediation and biorecovery of toxic metals and radionuclides. Geomicrobiol J 33(3–4):175–178
Garbisu C, Alkorta I (2003) Basic concepts on heavy metal soil bioremediation. EJMP & EP (Eur J Mineral Process Environ Protect) 3(1):58–66
Gentili AR, Cubitto MA, Ferrero M, Rodríguez MS (2006) Bioremediation of crude oil polluted seawater by a hydrocarbon-degrading bacterial strain immobilized on chitin and chitosan flakes. Int Biodeterior Biodegrad 57:222–228
Ghasemi A, Asad S, Kabiri M, Dabirmanesh B (2017) Cloning and characterization of *Halomonas elongata* L-asparaginase, a promising chemotherapeutic agent. Appl Microbiol Biotechnol 101:7227–7238
Gianfreda L, Rao MA (2004) Potential of extra cellular enzymes in remediation of polluted soils: a review. Enzym Microb Technol 35(4):339–354
Gillan DC, Van Camp C, Mergeay M, Provoost A, Thomas N, Vermard L, Billon G, Wattiez R (2017) Paleomicrobiology to investigate copper resistance in bacteria: isolation and description of *Cupriavidus necator* B9 in the soil of a medieval foundry. Environ Microbiol 19:770–787
Gómez PI, Barriga A, Cifuentes AS, González MA (2003) Effect of salinity on the quantity and quality of carotenoids accumulated by *Dunaliella salina* (strain CONC-007) and *Dunaliella bardawil* (strain ATCC 30861) chlorophyta. Biol Res 36:185–192
González H, Jensen T (1998) Nickel sequestering by polyphosphate bodies in *Staphylococcus aureus*. Microbios 93:179–185
Guibaud G, Comte S, Bordas F, Dupuy S, Baudu M (2005) Comparison of the complexation potential of extracellular polymeric substances (EPS), extracted from activated sludges and produced by pure bacteria strains, for cadmium, lead and nickel. Chemosphere 59(5):629–638
Haferburg G, Kloess G, Schmitz W, Kothe E (2008) "Ni-struvite"–a new biomineral formed by a nickel resistant *Streptomyces acidiscabies*. Chemosphere 72(3):517–523
Handley-Sidhu S, Mullan TK, Grail Q, Albadarneh M, Ohnuki T, Macaskie LE (2016) Influence of pH, competing ions, and salinity on the sorption of strontium and cobalt onto biogenic hydroxyapatite. Sci Rep 6:23361
Hou W, Ma Z, Sun L, Han M, Lu J, Li Z, Mohamad OA, Wei G (2013) Extracellular polymeric substances from copper-tolerance *Sinorhizobium meliloti* immobilize $Cu^{2+}$. J Hazard Mater 261:614–620
Irazusta V, Nieto-Peñalver CG, Cabral ME, Amoroso MJ, De Figueroa LIC (2013) Relationship among carotenoid production, copper bioremediation and oxidative stress in *Rhodotorula mucilaginosa* RCL-11. Process Biochem 48:803–809

Irazusta V, Michel L, De Figueroa L (2016) Biomineralización de cobre en *Candida fukuyamaensis* RCL-3. Rev Argent Microbiol 48(2):166–170

Irazusta V, Bernal AR, Estévez MC, De Figueroa L (2018) Proteomic and enzymatic response under Cr (VI) overload in yeast isolated from textile-dye industry effluent. Ecotoxicol Environ Saf 148:490–500

Isaac P, Martínez FL, Bourguignon N, Sánchez LA, Ferrero MA (2015) Improved PAHs removal performance by a defined bacterial consortium of indigenous *Pseudomonas* and actinobacteria from Patagonia, Argentina. Int Biodeterior Biodegrad 101:23–31

Iustman LR, López NI, Ruzal SM, Vullo DL (2013) Bioremediation approaches in a laboratory activity for the industrial biotechnology and applied microbiology (IBAM) course. J Microbiol Biol Educ 14(1):131

Jones RK (1997) A simplified pseudo-component oil evaporation model. In: Proceedings of the 20th Arctic and Marine Oil Spill Program (AMOP) Technical Seminar, Environment Canada, Ottawa, pp 43–61

Juwarkar AA, Nair A, Dubey KV, Singh SK, Devotta S (2007) Biosurfactant technology for remediation of cadmium and lead contaminated soils. Chemosphere 10:1996–2002

Karigar CS, Rao SS (2011) Role of microbial enzymes in the bioremediation of pollutants: a review. Enzym Res 2011. https://doi.org/10.4061/2011/805187

Karnwal A (2017) Isolation and identification of plant growth promoting rhizobacteria from maize (*Zea mays* L) rhizosphere and their plant growth promoting effect on rice (*Oryza sativa* L). J Plant Prot Res 57:144–151

Kim HM, Chae N, Jung JY, Lee YK (2013) Isolation of facultatively anaerobic soil bacteria from Ny-Alesund, Svalbard. Polar Biol 36(6):787–796

Kletzin A (2006) Metabolism of inorganic sulfur compounds in archaea. In: Garret RA, Klenk HP (eds) Archaea: evolution, physiology, and molecular biology. Blackwell Publishing, Oxford, pp 262–274

Konhauser K, Riding R (2012) Bacterial biomineralization. In: Knoll H, Canfield DE, Konhauser KO (eds) Fundamentals of geobiology. Wiley, Somerset, pp 105–130

Krause L, Diaz NN, Goesmann Z, Kelley S, Nattkemper TW, Rohwer F, Edwards RA, Stoye J (2008) Phylogenetic classification of short environmental DNA fragments. Nucleic Acids Res 36(7):2230–2239

Litchfield CD (2011) Potential for industrial products from halophilic Archaea. J Int Microbiol Biotechnol 38(10):1635–1647

Lowenstam HA (1981) Minerals formed by organisms. Science 211(4487):1126–1131

Lynch MD, Neufeld JD (2015) Ecology and exploration of the rare biosphere. Nat Rev Microbiol 13(4):217

Ma Y, Oliveira RS, Nai F, Rajkumar M, Luo Y, Rocha I, Freitas H (2015) The hyperaccumulator Sedum plumbizincicola harbors metal-resistant endophytic bacteria that improve its phytoextraction capacity in multi-metal contaminated soil. J Environ Manag 156:62–69

Ma Y, Oliveira RS, Freitas H, Zhang C (2016) Biochemical and molecular mechanisms of plant-microbe-metal interactions: relevance for phytoremediation. Front Plant Sci 7:918

Madigan MT, Martinko JM, Parker J (2009) Brock biology of microorganisms, 12th edn. Pearson Benjamin Cummings, New York

Majumder A, Bhattacharyya K, Bhattacharyya S, Kole SC (2013) Arsenic-tolerant, arsenite-oxidising bacterial strains in the contaminated soils of West Bengal, India. Sci Total Environ 46:1006–1014

Majzlik P, Strasky A, Adam V, Nemec M, Trnkova L, Zehnalek J, Hubalek J, Provaznik I, Kizek R (2011) Influence of zinc (II) and copper(II) ions on Streptomyces bacteria revealed by electrochemistry. Int J Electrochem Sci 6:2171–2191

Martinez F, Orce IG, Rajal VB, Irazusta V (2018) Salar del Hombre Muerto, source of lithium-tolerant bacteria. Environ Geochem Health 1–15. https://doi.org/10.1007/s10653-018-0148-2

Moraga NB, Amoroso MJ, Rajal VB (2014a) Strategies to ameliorate soils contaminated with boron compounds. In: Bioremediation in Latin America. Springer, Cham, pp 41–51

Moraga NB, Poma HR, Amoroso MJ, Rajal VB (2014b) Isolation and characterization of indigenous *Streptomyces* and *Lentzea* strains from soils containing boron compounds in Argentina. J Basic Microbiol 54(6):568–577
Moraga NB, Irazusta V, Amoroso MJ, Rajal VB (2017) Bio-precipitates produced by two autochthonous boron tolerant *Streptomyces* strains. J Environ Chem Eng 5(4):3373–3383
Moran MA (2009) Metatranscriptomics: eavesdropping on complex microbial communities-large-scale sequencing of mRNAs retrieved from natural communities provides insights into microbial activities and how they are regulated. Microbe 4(7):329
More TT, Yadav JSS, Yan S, Tyagi RD, Surampalli RY (2014) Extracellular polymeric substances of bacteria and their potential environmental applications. J Environ Manag 144:1–25
Mulligan CN, Galvez-Cloutier R (2003) Bioremediation of metal contamination. Environ Monit Assess 84:45–60
Mulligan CN, Yong RN, Gibbs BF (2001) Surfactant-enhanced remediation of contaminated soil: a review. Eng Geol 60:371–380
Naik MM, Dubey SK (2013) Lead resistant bacteria: lead resistance mechanisms, their applications in lead bioremediation and biomonitoring. Ecotoxicol Environ Saf 98:1–7
Nancucheo I, Rowe OF, Hedrich S, Johnson DB (2016) Solid and liquid media for isolating and cultivating acidophilic and acid-tolerant sulfate-reducing. FEMS Microbiol Lett 363:fnwo83. https://doi.org/10.1093/femsle/fnwo83
Navarro CA, Von Bernath D, Jerez CA (2013) Heavy metal resistance strategies of acidophilic bacteria and their acquisition: importance for biomining and bioremediation. Biol Res 46(4):363–371
Nesme J, Achouak W, Agathos SN, Bailey M, Baldrian P, Brunel D, Frostegård Å, Heulin T, Jansson JK, Jurkevitch E et al (2016) Back to the future of soil metagenomics. Front Microbiol 7:73
Olegario J, Yee N, Miller M et al (2010) Reduction of Se(VI) to Se (-II) by zero valent iron nanoparticle suspensions. J Nanopart Res 12(6):2057–2068
Ortega-Cabello L, Pérez-Méndez HI, Manjarrez-Alvarez N, Solís-Oba A, López-Luna A (2017) Effect of iron salts on *Rhodococcus* sp. and *Gordonia* sp. on carotenoid production. Rev Mex Ing Quim 16(1):1–10
Oves M, Saghir Khan M, Zaidi A (2013) Chromium reducing and plant growth promoting novel strain *Pseudomonas aeruginosa* OSG41 enhance chickpea growth in chromium amended soils. Eur J Soil Biol 56:72–83
Pacwa-Płociniczak M, Płaza GA, Piotrowska-Seget Z, Cameotra SS (2011) Environmental applications of biosurfactants: recent advances. Int J Mol Sci 12(1):633–654
Panikov NS, Sizova MV (2007) Growth kinetics of microorganisms isolated from Alaskan soil and permafrost in solid media frozen down to −35°C. FEMS Microbiol Ecol 59(2):500–512
Park JH, Bolan N, Megharaj M, Naidu R (2011) Isolation of phosphate solubilizing bacteria and their potential for lead immobilization in soil. J Hazard Mater 185:829–836
Pérez-González T, Valverde-Tercedor C, Jiménez-López C (2011) Biomineralización bacterianad e magnetita y aplicaciones. http://www.ehues/sem/seminario_pdf/SEMINARIO_SEM_7_058pdf. Accessed Nov 2013
Prasad R (2017) Mycoremediation and environmental sustainability, vol 1. Springer International Publishing, Cham, Switzerland https://doi.org/10.1007/978-3-319-68957-9
Prasad R (2018) Mycoremediation and environmental sustainability, vol 2. Springer International Publishing, Cham, Switzerland https://www.springer.com/us/book/9783319773858
Quince C, Walker AW, Simpson JT, Loman NJ, Segata N (2017) Shotgun metagenomics, from sampling to analysis. Nat Biotechnol 35(9):833–844
Rangarajan V, Narayanan M (2018) Biosurfactants in soil bioremediation. In: Advances in soil microbiology: recent trends and future prospects. Springer, Singapore, pp 193–204
Romano-Armada N, Amoroso MJ, Rajal VB (2017) Effect of glyphosate application on soil quality and health under natural and zero tillage field conditions. Soil Environ 36(2):141–154

Rosas Hernández I (2009) Identification and characterization of microorganisms with resistance to mercurial compounds. Master's thesis, Interdisciplinary Center for Research and Studies on the Environment and Development (CIIEMAD)

Ryder C, Byrd M, Wozniak DJ (2007) Role of polysaccharides in *Pseudomonas aeruginosa* biofilm development. Curr Opin Microbiol 10(6):644–648

Sánchez-Castro I, Amador-García A, Moreno-Romero C et al (2017) Screening of bacterial strains isolated from uranium mill tailings porewaters for bioremediation purposes. J Environ Radioact 166:130–141

Sandmann G (2001) Carotenoid biosynthesis and biotechnological application. Arch Biochem Biophys 385:4–12

Singh OV (2006) Proteomics and metabolomics: the molecular make-up of toxic aromatic pollutant bioremediation. Proteomics 6(20):5481–5492

Singh OV, Nagaraj NS (2006) Transcriptomics, proteomics and interactomics: unique approaches to track the insights of bioremediation. Brief Funct Genomics 4(4):355–362

Singh R, Paul D, Jain RK (2006) Biofilms: implications in bioremediation. Trends Microbiol 14(9):389–397

Singh A, Van Hamme JD, Ward OP (2007) Surfactants in microbiology and biotechnology: part 2. Application aspects. Biotechnol Adv 25(1):99–121

Slaveykova VI, Parthasarathy N, Dedieu K, Toescher D (2010) Role of extracellular compounds in Cd-sequestration relative to Cd uptake by bacterium *Sinorhizobium meliloti*. Environ Pollut 158:2561–2565. https://doi.org/10.1016/j.envpol.2010.05.016

Song B, Palleroni NJ, Häggblom MM (2000) Isolation and characterization of diverse halobenzoate-degrading denitrifying bacteria from soils and sediments. Appl Environ Microbiol 66:3446–3453

Suarez P (2002) Heavy metal incorporation in bacteria and its environmental significance. Interciencia 27:160–172

Techtmann SM, Hazen TC (2016) Metagenomic applications in environmental monitoring and bioremediation. J Ind Microbiol Biotechnol 43(10):1345–1354

Tekere M, Lötter A, Olivier J, Jonker N, Venter S (2011) Metagenomic analysis of bacterial diversity of Siloam hot water spring, Limpopo, South Africa. Afr J Biotechnol 10(78):18005–18012

Thompson IP, Van Der Gast CJ, Ciric L, Singer AC (2005) Bioaugmentation for bioremediation: the challenge of strain selection. Environ Microbiol 7(7):909–915

Venkatesh NM, Vedaraman N (2012) Remediation of soil contaminated with copper using rhamnolipids produced from *Pseudomonas aeruginosa* MTCC2297 using waste frying rice bran oil. Ann Microbiol 62:85–91

Villas-Bôas SG, Bruheim P (2007) The potential of metabolomics tools in bioremediation studies. OMICS 11(3):305–313

Villas-Bôas SG, Mas S, Åkesson M, Smedsgaard J, Nielsen J (2005) Mass spectrometry in metabolome analysis. Mass Spectrom Rev 24(5):613–646

Vishnivetskaya T, Kathariou S, McGrath J, Gilichinsky D, Tiedje JM (2000) Low-temperature recovery strategies for the isolation of bacteria from ancient permafrost sediments. Extremophiles 4:165–173

Vodnik D, Grčman H, Maček I, van Elteren JT, Kovacevic M (2008) The contribution of glomalin-related soil protein to Pb and Zn sequestration in polluted soil. Sci Total Environ 392(1):130–136

Wallenius K, Rita H, Simpanen S, Mikkonen A, Niemi RM (2010) Sample storage for soil enzyme activity and bacterial community profiles. J Microbiol Methods 81(1):48–55

Wang D-Z, Kong L-F, Li Y-Y, Xie Z-X (2016) Environmental microbial community proteomics: status, challenges and perspectives. Int J Mol Sci 17(8):1275. https://doi.org/10.3390/ijms17081275

Watanabe K (2001) Microorganisms relevant to bioremediation. CurrOpinBiotechnol 12(3): 237–241

Wiegel J, Ljungdahl LG, Rawson JH (1979) Isolation from soil and properties of the extreme thermophile *Clostridium thermohydrosulfuricum*. J Bacteriol 139:800–810

Zhang DY, Wang JL, Pan XL (2006) Cadmium sorption by EPSs produced by anaerobic sludge under sulfate-reducing conditions. J Hazard Mater 138:589–593

Zheng J, Feng J-Q, Zhou L, Mbadinga SM, Gu JD, Mu BZ (2018) Characterization of bacterial composition and diversity in a long-term petroleum contaminated soil and isolation of high-efficiency alkane-degrading strains using an improved medium. World J Microbiol Biotechnol 34:1–11

Zhou Q, Chen Y, Yang M, Li W, Deng L (2013) Enhanced bioremediation of heavy metal from effluent by sulfate-reducing bacteria with copper-iron bimetallic particles support. Bioresour Technol 136:413–417

# Chapter 2
# Transcriptomics as a First Choice Gate for Fungal Biodegradation Processes Description

**Alejandro Ledezma-Villanueva, José Manuel Adame-Rodríguez, and Elva T. Aréchiga-Carvajal**

## Contents



## 2.1 Introduction: Fungal Bioremediation (Mycoremediation)

Increasing industrial activities around the world is essential to sustain all countries' economy and provide career options. Nowadays, there is a fourth industrial revolution taking place, and little is known about the details of how this new industrial approach will stablish to manage, contain, and dispose chemical and energy wastes (McMaster 2018). By these days, different kinds of pollutants have forever altered natural environments; it is in these affected areas that bioremediation has proven to be a valuable and adaptable set of techniques that can degrade, reduce, oxidize, or encapsulate hazardous materials such as hydrocarbons, heavy metals, pesticides, and radionuclide elements (Das 2014). Such techniques are focused on the stimulation of either macroorganisms (plants) (Bell et al. 2014) or microorganisms that possess specific metabolic pathways to create enzymes that would interact with the xenobiotic compounds to either cleave C:H, C:N, C:C, and C:S bonds for organic compounds (Ladino-Orjuela et al. 2016) or to reduce, precipitate, accumulate, sorb, or even leach metallic atoms and inorganic molecules by reducing its electrons (Nancharaiah et al. 2016). They do this around their surrounding environments.

A. Ledezma-Villanueva · J. M. Adame-Rodríguez · E. T. Aréchiga-Carvajal (✉)
Unidad de Manipulación Genética, Laboratorio de Micología y Fitopatología, Departamento de Microbiología, Facultad de Ciencias Biológicas, Universidad Autónoma de Nuevo León, Cd. Universitaria, San Nicolás de los Garza, NL, Mexico
e-mail: elva.arechigacr@uanl.edu.mx

 
R. Prasad, E. Aranda (eds.), *Approaches in Bioremediation*, Nanotechnology in the Life Sciences, https://doi.org/10.1007/978-3-030-02369-0_2

Microorganisms are vast; we call them to englobe the Protista, Fungi, and Protozoa kingdoms where mostly bacteria and fungi excel at being extensively used because of their good abundancy in soil and water and their efficient remediation rates related to their metabolic plasticity (Whitby 2010). Bioremediation studies had always focused on enhancing these rates by providing specific growth factors applying oxygen, nitrogen, and/or mineral sources and controlling pH, $H_2O$ levels, and temperature depending on the pollutant and the details of the environment. Bacteria such as *Pseudomonas* and *Bacillus* have proven to be very efficient with degradation of aromatic compounds, BTEX (benzene, toluene, ethylbenzene, and xylene), and short to medium hydrocarbon chains compounds. But they drop degradation rates at pollutants with higher molecular complexity such as crude oil, creosote, and tar. It is in these hard to break pollutants that fungal strains are used to provide better results for their eukaryotic genetics that produce higher enzymatic compounds (20–60 kDa) that can cleave these kinds of molecules (Deshmukh et al. 2016).

Within the Fungi kingdom, there is an estimation of 1.5 million species (Hawksworth 2001) classified in four or five phyla, being those belonging to the *Ascomycota* and to the *Basidiomycota* the most frequent in soil and employed for bioremediation (Tortella et al. 2005; Sankaran et al. 2010). Biological features of these fungi are the secretion of cellulolytic enzymes such as cellobiohydrolases (CBH), evolved beta-galactosidases (EBG), and β-glucosidases. Some of these cellulase enzyme complexes such as the lytic polysaccharide monooxygenases (LMPO) are widely described among the white-, brown-, and soft-rot fungi as the extracellular enzymes that decompose complex nutrients (pollutants) into simpler substances that can be easily assimilated through the cell wall (Obeng et al. 2017). By doing so, the fungus might produce antibiotics or other suppressive metabolites. This is difficult to demonstrate at the scale of individual hyphae, but Burton and Coley-Smith (1993) reported that antibacterial compounds were released by hyphae of *Rhizoctonia* species that are members of the *Basidiomycota* known to degrade cellulose. Another special aspect of fungi is that they do not fix nitrogen from the atmosphere; they rather use amino acids or, in extreme polluted soils with ammonia or ammonium ($NH_4$) environments, they use them as a nitrogen source. After uptake, ammonia/ammonium is combined with organic acids, usually to produce either glutamic acid (from α-ketoglutaric acid) or aspartic acid; then the other amino acids needed can be formed by transamination reactions. Heavy metal bioremediation is an extended field as their concentration in mg/kg of many elements like Zn, Ni, Hg, Cr, Pb, Cu, Cd, As, Co, Sn, Au, Pd, Pt, Ag, Ru, Th, U, Am, and Ra could mean a serious environmental pollution, risking ecosystem and human health. Fortunately, organisms such as *Saccharomyces cerevisiae* and *Rhizopus arrhizus* have been used as a promising cleaning technology because of its high removal capacity and eco-friendly and cost-effective properties. Such cleaning effect relies on their exopolysaccharide (EPS) production that ends up excluding heavy metals (Sağ et al. 1995; Ozer and Ozer 2003; Mohite et al. 2017). Other helpful enzymes like laccases, lignin peroxidases (LiP), and manganese peroxidases (MnP) had been reported to interact with other complex aromatic pollutants such as dyes (Vats and Mishra 2018). Nevertheless, all these approaches have not yet explained which metabolic routes are active at every specific moment of the bioremediation process, in order to guide (once well-known) the biodegradation process with elevated efficiency. Before the achievement of this goal, two main objectives must be contemplated: first, an efficient implementation of nucleic acid extraction technique, making it possible for -omics studies to develop this area in the presence of many other recalcitrant substances, and second a precise description of the fungi or fungal communities related with the process (this can be achieved by a metagenomic approach).

Nucleic acid extraction (DNA-RNA) techniques focus on creating a lysis effect on the membranes or cell walls from a pure culture to environmental samples. Fungal cell walls contain chitin, β-(1,3)-β-(1,4)- β-(1,6) glucans, and other proteins like melanin and mannan (Gow et al. 2017); such osmotic pressure effect can be applied on samples to release nucleic acids from such well-protected fungi cells by the use of chemical (SDS or phenol), physical (glass or zirconia beads, microwaves, sonication, or frozen/thaw cycles), and enzymatic methods (lysozyme, proteinase K). They can be performed by preparing all reactives or using commercial kits, like TRI Reagent kit (Sigma) that pre-ensembles this desired solutions in an easy-to-follow procedure (Moré et al. 1994; Boon et al. 2000; Nelson et al. 2007; Jiang et al. 2011). Sampling methodology and sample treatment are essential of a good nucleic acid yielding. Depending on the origin of the sample, we could find many inhibitors such as humic acids, metals, and xenobiotics that interfere with the extraction process (Wintzingerode et al. 1997; He et al. 2009). Sampling procedures for toxic environments may be contemplated (WHO 2004) to then proceed with the nucleic acid extraction protocols mentioned before. The best quality (less impurities) and quantity (μg of nucleic acid yield) of DNA and RNA come from commercial kits; nevertheless fungal nucleic acid may be better obtained from enzymatic/phenol-chloroform extraction methods as it is shown in the study by Guillén-Navarro et al. (2015) as they compare different extraction procedures. Preferred RNA extraction protocol could be decided by the sample circumstance, ribonucleic acid from a fresh mycelium on an enriched medium may be well obtained from a classic liquid nitrogen-buffer-phenol-chloroform extraction, but the ribonucleic acid from living fungi in a polluted environmental sample may get a better yield by any commercial RNA extraction kits.

## 2.2 Molecular Approaches in Fungal Bioremediation

It is clear that microbial metabolism can achieve specific pollutant bioremediation, thus enhancing polluted site diagnoses is vital to stablish the most adequate treatment. For accidental spillage of pollutants or catastrophic incidents that end up with impacted waters, atmosphere, and soils of natural or urban environments, most countries have specific contingency and normativity to be followed (National Research Council 2014). The use of microorganisms that can degrade and restore wellness is allowed following all protocols that clearly stipulate the use of native microbial communities or, if a biological product may be applied, is necessary to demonstrate the avoidance of GMOs; even these have yield excellent results in lab conditions (Ruta et al. 2017; Zhang et al. 2015). Recently mentioned genomic DNA extraction protocols can be applied at recent or longtime polluted sites to get a lecture of gene targets most commonly used for sequence-based fungal identification; these are the internal transcribed spacer regions ITS-1 and ITS-2, which are variable regions (the ITS region) and can be located between conserved genes encoding the 18S, 5.8S, and 28S ribosomal subunits. Another variable region is called the D1/D2 region, located toward the 5′ end of the large nuclear ribosomal subunit (28S rDNA). These regions can be amplified using ITS-1, ITS-4, NL-1, and NL-4 primers (Romanelli et al. 2014). Total microbial communities are seriously deplenished when toxic pollutants get at pristine water and soils (Rodrigues et al. 2015; Morais et al. 2016). If there is no good mixture of helpful microorganisms, then biological products may be of use; there are plenty of fungal products in the market (http://www.epa.gov/emergencies/content/ncp/index.htm). To mention some examples, there are the US patent numbers, 6,143,549; 6,383,800; 6,387,689; 6,387,691; 6,495,134; 6,664,102; 6,727,087; and 6,972,169, that belong to fungal product inventions to target pollutants like chromated copper arsenate (CCA), pentachlorophenol, ammoniacal copper quat (ACQ), and creosote. The modus operandi is that the fungal inoculum is applied to the polluted site and maintained in an aerated and hydrated environment having temperature conditions sufficient to allow the inoculum to grow and metabolize the xenobiotic compounds (Lamar et al. 2000; Illman et al. 2002a, b, c, d, 2003, 2004, 2005). Also, it is worthy to mention that fungal community, if it is the case, can change among the bioremediation process, so it is a good idea to follow up these changes associated with the expression of genes related to biodegradation process. If a single fungal strain is used, this description can be avoided.

For general bioremediation process then, having efficient molecular approaches for both cases, time and cost, can give us a deep understanding of how microbial communities shift depending on the molecules that are on treatment. So we could be in the correct status to start an expression analysis. Expanding the understanding of how we can fix biological treatments to our favor, several molecular tools can be used with a total nucleic acid from a polluted site, like using specific microorganism probes, and if there is an RNA extraction instead, we can scale it up to a complete new scale of genomic possibilities, like transcriptomic analysis (Fig. 2.1).

RNA extraction protocols mentioned before set an advent of next-generation sequencing methods. This has made possible to sequence whole microbial genomes of polluted sites at a much lower cost providing opportunities to examine gene expression patterns under adverse environmental conditions. Genomic and transcriptomic analyses have recently been used to characterize bacteria and fungi that have the ability to degrade multiple xenobiotic compounds (Maghsoudi et al. 2016). Strategies in which this is applied can be described in two fundamentally different ways: (a) sequence-based (involving RNA extraction to then synthesize sequenciable cDNA) and (b) function-based. The basis for both approaches has been the construction of metagenomic clone libraries (Warnecke and Hess 2009). Several new technologies in sequencing equipment had to progress and lower its cost so scientist

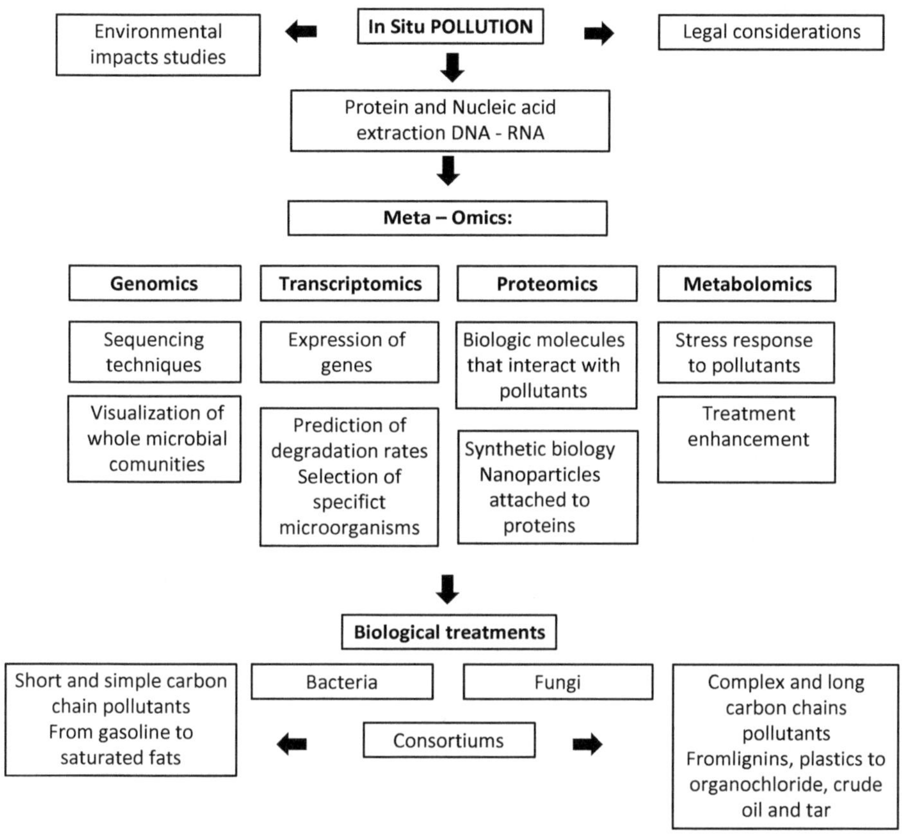


**Fig. 2.1** Workflow of applied meta-omics technologies

could have access to equipment like Roche's 454, Illumina's Solexa, and ABI's SOLiD to mention some examples and begin to build genomic libraries that we could relay to compare and identify metabolic profiles of natural environments. Some of these investigations are addressed in Table 2.1.

The most well-known yeast *Saccharomyces cerevisiae* has already been proven to have positive cleaning effects in water and soil treatments against heavy metals (Soares and Soares 2012). Transcriptome studies on these universal biological models have revealed that its production of sulfur compounds comes from the expression of glutathione (GSH) pathways under As (III) exposure. This was proven by RNA isolation, cDNA synthesis, microarray hybridization, and analysis of the transcriptional response to arsenite (Thorsen et al. 2007). *Saccharomyces* genome expression still needs to be proven when confronting other recalcitrant molecules. In particular, one example of the use of brewing yeast strains to remediate heavy metal pollution can be taken into account; due to the autoaggregation properties that this fungus possesses, they can be quickly and easily separated from the treated effluent. This intrinsic property avoids the use of cell immobilization techniques or

**Table 2.1** Summary of published transcriptome studies of fungal bioremediation process

| Fungi | Bioremediation model | RNA extraction method | Main genomic/ metabolic findings | Analytical software | Database ID | Reference |
|---|---|---|---|---|---|---|
| *Saccharomyces cerevisiae* | Arsenite | Cold 5% trichloroacetic acid | Sulfur/GSH pathways | LIMMA package (http://www.bioconductor.org) in the statistical language R (http://www.R-project.org) | GSE6129 | Thorsen et al. (2007) |
| *Westerdykella aurantiaca* | Arsenic | Cold 5% trichloroacetic acid | Arsenic methyltransferase | BLASTp, MUSCLE software, MEGA 6.0.1 | KP165533.1 | Verma et al. (2016) |
| *Phanerochaete chrysosporium* | Cellulose | Liquid nitrogen, cold homogenization buffers | Endo-1,4-β-glucanase, exo-1,4-β-glucanase, and b-glucosidases | BLASTN, bioconductor package graph (http://bioconductor.org/packages/2.3/bioc/html/graph.html), LWP Perl module, InterProScan 4.3.1 software, Sixpack software, ImageJ | PF00734, PF00331, PF00840 | Sato et al. (2009) |
| *Trichoderma atroviride* | Dichlorvos | Frozen powdered mycelia using Trizol | ABC transporters | SOAP2, Blast2GO, Cytoscape software with the ClueGO plug-in, T-MeV<br>4.9 and Cluster 3.0 | IMI 206040 | Zhang et al. (2015) |
| *Postia placenta* and *Phanerochaete chrysosporium* | Cellulose and hemicelluloses | Grounded frozen fungal pellets with extraction<br>Buffer and phenol-chloroform | Glycosyl hydrolases | DNASTAR ArrayStar v2.1 software (Madison, WI), BLOSUM62 matrix | GSE14736 and GSE12540 | Wymelenberg et al. (2010) |
| *Cyathus bulleri* | Lignocellulosics | Fungal mycelium wash with diethyl pyrocarbonate. Snap-frozen in liquid nitrogen and then extracted using TRI Reagent (Sigma) | Laccase isoforms | Trinity v2.1.1, BWA v0.7.12., BlastX, Clustal Omega, InterPro, Modeller 9.1, PrediSi and SignalP 4.1, ClustalW, MEGA 7.0 software, BLASTp | SRR5208586 | Vats and Mishra (2018) |

| | | | | | | |
|---|---|---|---|---|---|---|
| *Phanerochaete chrysosporium* | Anthracene and anthrone | Phenol-guanidinium thiocyanate-chloroform extraction | Cytochrome P450 monooxygenases | Array Vision software | CYP5138A1 | Chigu et al. (2010) |
| *Neosartorya fischeri* | Petroleum asphaltenes | TRI Reagent kit (Sigma) | Monooxygenase enzyme | GenePix, Array-Pro Analyzer, genArise, Blast2GO tool | GSE68146 | Hernández-López et al. (2015) |
| *Paxillus involutus* | Soil organic matter | RNeasy Plant Mini Kit (Qiagen) | CAZymes | Blast2GO tool, BLASTN tool, Mira assembler version 3.0.3, gsAssembler software, CAP3 assembler, ssaha2 algorithm, NimbleScan software v. 2.5 | SRA046093 | Rineau et al. (2012) |
| *Phanerochaete chrysosporium* | Cellulose and hemicellulose | GEO database (no RNA extraction) | CAZymes | GEO2R, Cluster 3.0 software, Java Treeview, Venny 2.1, Jvenn | GSE14734, GSE14735, GSE54542, GSE2794, GSE52922, GSE69008, GSE69461 | Kameshwar and Qin (2017) |
| *Aspergillus niger*, *Trichoderma harzianum*, *Talaromyces purpurogenum*, and *Aspergillus flavus* | Crude oil | CTAB procedure with 3 M Lithium chloride to precipitate | Catalase, peroxidase and laccase | BLAST | KY473958, KY488466, KY488463, and KY488467 | Asemoloye et al. (2018) |

solid-liquid separation processes. After treating the effluent, the convenient management of the contaminated biomass and the selective recovery of metals to ensure the minimization of waste production and low operating costs become important (Soares and Soares 2012). Cellulose and hemicellulose compounds are very persistent in natural habitats, and human's excessive use of recipient products had resulted in much garbage generated and disposed at landfills (Mathews et al. 2015). *Cyathus bulleri*, *Postia placenta*, and *Phanerochaete chrysosporium* are white rot fungi that show promising results in providing diverse glucanases, glycosyl hydrolases, carbohydrate-active enzyme (CAZymes), and laccase isoforms that work in depolymerization (Rineau et al. 2012). These complex molecules make significant reduction in the number of side chains of the pollutant molecule. This was observed during transcriptome analysis of these substrates (like wood) submerged into liquid medium composed of other trace elements (Kameshwar and Qin 2017). Other hazardous pollutants that have been addressed are petrochemical products such as anthracene, anthrone, petroleum asphaltenes, and even crude oil; for these substances fungal species like *Neosartorya fischeri*, *Phanerochaete chrysosporium*, *Aspergillus niger*, *Trichoderma harzianum*, *Talaromyces purpurogenum*, and *Aspergillus flavus* have good bioremediation activity. RNA extraction procedures at different times of treatments with these fungal species which were carried out by using cetyltrimethylammonium bromide (CTAB) or TRI Reagent kit (Sigma) have led to the possibility to evaluate transcriptomic responses of catalase, peroxidase, laccase, and cytochrome P450 monooxygenases encoding genes that could present beneficial activities in accidental oil spillage by stimulating the inhabiting rhizospheric fungal strains (Hernández-López et al. 2015; Asemoloye et al. 2018). A major xenobiotic compound like dichlorvos (2,2-dichlorovinyl dimethyl phosphate) is an organophosphate widely used as an insecticide, to control and protect houses and stored product from arthropods. It has been commercially available since 1961, and its use has become controversial because of its prevalence in urban waterways and the fact that its toxicity extends well beyond insects (Gao et al. 2009; Kazemi et al. 2012). Exploration of possible resolutions to this impact relies on the use of *Trichoderma atroviride*; this filamentous cosmopolitan fungus commonly found in soil has shown the expression of 5382 differential genes in response to exposure of dichlorvos stress for 2, 6, and 24 h in lab conditions. At these times, RNA extraction with cold 5% trichloroacetic acid was performed to show a transcriptional profile in which metabolic pathways were found to be regulated by the super families hex1 encoding cytochrome P450, glutathione-S-transferase, flavoprotein, Hsp70, and Hsp90. After reads mapping and gene clustering analysis, it was found that ABC transporters were affected by the disruption of hex1 gene. This deletion was made in order to prove that ABC transporter genes might play a vital role in the tolerance process. Expression patterns of seven selected ABC transporter genes were confirmed by qRT-PCR of *Trichoderma atroviride* and a *T. atroviride* hex1-deleted mutant (Zhang et al. 2015).

## 2.3 Fungal Transcriptomic Perspectives

Further research in other environmental liability sites will be a powerful resolutive tool that could position fungal bioremediation treatments as most effective and safe suggested options of restoration procedures as the fourth industrial age settles. Recent and longtime polluted environments can be first evaluated by a DNA and RNA extraction procedure to screen native microbial communities (DNA sequencing) and their metabolic activity (transcriptomic). In this way there could be evidence that a polluted site already possesses genomic potential for self-cleaning with an appropriate bioremediation technique. If a microbial community does not outright this, there can always be other potential biological options like many fungal species that can be inoculated. Their use can be forever preferred around commercial bioremediation products mentioned before. Transcriptomics and metatranscriptomics are influential techniques that let us know what are the active genes essential for the continuance of microbial populations in adverse conditions, highlighting the genomic potential of a particular organism or a microbial community using this discipline of applied genomics (Warnecke et al. 2007). Other various expression profile studies in other species have been performed over the last decades to focus on central interrogates in fields like medicine and microbiology, providing valuable insight into the understanding of how certain phenotypes such as radiation resistance, pathogenesis, or heat-shock resistance are correlated with gene expression (Qiu et al. 2008; Liu et al. 2003; Audia et al. 2008). These transcriptomic studies were usually performed using DNA microarrays. Initially these experiments were restricted to a single organism grown in pure culture but were soon expanded to target several organisms at once (You et al. 2008; Parro et al. 2007; Bulow et al. 2008). A disadvantage of the DNA microarray-based expression profiling is that gene or even complete genome sequences and the corresponding annotations are required before a functional microarray chip can be manufactured (Darby and Hall 2008). Hence, DNA microarrays are rather ineffective for the discovery of novel biocatalysts from the environment. Other challenges in using DNA-based microarrays include their limited detection sensitivity and quantification reliability (Darby and Hall 2008). Bioremediation techniques stimulate microorganisms with genes that are transcribed under specific environmental conditions. For this, direct extraction techniques of bacterial, archaeal, and eukaryotic mRNA from environmental samples were reported recently (Bailly et al. 2007; Poretsky et al. 2005). In both studies, cDNA libraries from environmental RNA were constructed and 400 and 119 clones sequenced, respectively. Most of the obtained sequences had no significant hit to any protein sequence that was previously deposited in public databases. Thus, these studies demonstrate the potential of discovering novel proteins. Furthermore, less sequencing capacity is required for transcripts as compared to the analysis of genomic or metagenomic DNA. This is especially true for the low coding density of eukaryotic genomes. Substantial progress for the efficient analysis of more complex expression profiles has become available with the development of next-generation sequencing technologies mentioned before. These new technologies allow not only the direct sequencing of DNA or likewise cDNA without any cloning step (Medini et al. 2008) but also increased the throughput in terms of numbers of base pairs sequenced per run and decreased cost per sequenced base. Although all three of these new technologies produced extremely short reads when they initially entered the market, the performance of the third generation of the 454 pyrosequencer increased substantially. An average read length of ~400 bp has been reported for the recently released 454FLX Titanium platform (Hugenholtz and Tyson 2008), and it is not unlikely that future releases will produce reads comparable to those produced with traditional dye terminator sequencing technology (i.e., also referred to as Sanger methodology). The first report of using 454 pyrosequencing for studying the metatranscriptome of a complex microbial community was published in 2006. Using this highly parallelized sequencing technology, Leininger et al. (2006) could show that archaeal transcripts of the key enzyme (amoA) for ammonia oxidation were several magnitudes more abundant in soils than the bacterial version of it, suggesting archaea as the numerical dominant ammonia oxidizers in soil (Leininger et al. 2006). The obtained 25 Mbp of sequence data of this transcriptome study were further analyzed by Urich et al. (2008). They reported that 8% of the initial >250 k reads could be identified as mRNA tags. In 2008, Frias-Lopez et al. produced >50 Mbp by 454 pyrosequencing, still using the first generation of this technology (Frias-Lopez et al. 2008). Gilbert et al. in 2008 followed shortly afterward with >300 Mbp of sequence data now using the second generation called GS-FLX (Gilbert et al. 2008). All three studies demonstrated how high-throughput sequencing technologies can be applied with ease to access information stored in known and unknown transcripts that have been isolated directly from complex environments such as marine and soil microbial communities.

## 2.4 Concluding Remarks

A continual progress of the genomic fields ensuring metabolic profiles of native or inducted microbial communities during bioremediation process is essential for coming pollution adversities and definition of environmental authorities' procedures. One of the main problems to overcome in transcriptomic studies is the correct ribonucleic acid extraction protocol, as interference molecules like humic acids, metals, and xenobiotic compounds may compromise the quality and quantity obtained; thus a different extraction protocol may be chosen depending on the integrity of the fungal bioremediation technique. RNA from fungi exposed to pollutant-enriched media in lab conditions may go well by a traditional liquid nitrogen-buffer-phenol-chloroform extraction, but in situ bioremediation treatments of soil and water may contemplate the use of a commercial kit as these products had proven to be efficient against interference molecules. Once there is a good RNA recovery from fungal biological treatments, transcriptomic studies will help to continue the expansion of database transcripts. This is an urgent matter as fungal genomes are not quite well annotated, and there is a lack of understanding on how fungal genes become functional even though there are predicted gene function suggestions for *Saccharomyces* and *Aspergillus.* Thus, transcriptomics applied in mycoremediation treatments succeed when known genes demonstrate new functions against xenobiotic compounds and also when structural or functional unknown genes are discovered. In this case, transcriptomic studies may greatly benefit science by describing new metabolic pathways to later build punctual bioremediation strategies. Deep understanding of fungal genomic structure and expressed metabolism pathways is a key concept to be approached in time. Scientific community had made significant efforts to have bigger and well-annotated databases as there are ongoing projects like 1000 fungal genomes (http://1000.fungalgenomes.org/home). This initiative promotes and facilitates user community participation to gather contributions from taxonomists, mycologists, and those focusing on fungal genomics contributing to a better understanding on how fungal genomes behave under recalcitrant chemical exposure like heavy metals, hydrocarbons, and pesticides. It is motivating that all around the globe, many research groups accept the challenge of immersing themselves in the analysis of transcriptomic metadata in order to advance in the description of degradation processes of polluting compounds, which brings us closer to the installation of industrial processes of generalized bioremediation.

## References


Asemoloye MD, Ahmad R, Jonathan SG (2018) Transcriptomic responses of catalase, peroxidase and laccase encoding genes and enzymatic activities of oil spill inhabiting rhizospheric fungal strains. Environ Pollut 235:55–64

Audia JP, Patton MC, Winkler HH (2008) DNA microarray analysis of the heat shock transcriptome of the obligate intracytoplasmic pathogen *Rickettsia prowazekii.* Appl Environ Microbiol 74:7809–7812

Bailly J, Fraissinet-Tachet L, Verner MC, Debaud JC, Lemaire M, Wesolowski-Louvel M, Marmeisse R (2007) Soil eukaryotic functional diversity, a metatranscriptomic approach. ISME J 1:632

Bell TH, Joly S, Pitre FE, Yergeau E (2014) Increasing phytoremediation efficiency and reliability using novel omics approaches. Trends Biotechnol 32(5):271–280

Boon N, Marlé C, Top EM, Verstraete W (2000) Comparison of the spatial homogeneity of physico-chemical parameters and bacterial 16S rRNA genes in sediment samples from a dumping site for dredging sludge. Appl Microbiol Biotechnol 53:742–747

Bulow SE, Francis CA, Jackson GA, Ward BB (2008) Sediment denitrifier community composition and nirS gene expression investigated with functional gene microarrays. Environ Microbiol 10:3057–3069

Burton RJ, Coley-Smith JR (1993) Production and leakage of antibiotics by *Rhizoctonia cerealis, R. oryzae-sativae* and *R. tuliparum.* Mycol Res 97:86–90

Chigu NL, Hirosue S, Nakamura C, Teramoto H, Ichinose H, Wariishi H (2010) Cytochrome P450 monooxygenases involved in anthracene metabolism by the white-rot basidiomycete *Phanerochaete chrysosporium.* Appl Microbiol Biotechnol 87(5):1907–1916

Darby AC, Hall N (2008) Fast forward genetics. Nat Biotechnol 26:1248–1249

Das S (ed) (2014) Microbial biodegradation and bioremediation. Elsevier, Rourkela Odisha, pp 167–201

Deshmukh R, Khardenavis AA, Purohit HJ (2016) Diverse metabolic capacities of fungi for bioremediation. Indian J Microbiol 56(3):247–264

Frias-Lopez J, Shi Y, Tyson GW, Coleman ML, Schuster SC, Chisholm SW, Delong EF (2008) From the cover: microbial community gene expression in ocean surface waters. Proc Natl Acad Sci U S A 105:3805–3810

Gao J, Liu L, Liu X, Zhou H, Lu J, Huang S, Wang Z (2009) The occurrence and spatial distribution of organo phosphorous pesticides in Chinese surface water. Bull Environ Contam Toxicol 82:223–229

Gilbert JA, Field D, Huang Y, Edwards R, Li W, Gilna P, Joint FI (2008) Detection of large numbers of novel sequences in the metatranscriptomes of complex marine microbial communities. PLoS One 3:3042

Gow NA, Latge JP, Munro CA (2017) The fungal cell wall: structure, biosynthesis, and function. Microbiol Spectr 5:1–25

Guillén-Navarro K, Herrera-López D, López-Chávez MY, Cancino-Gómez M, Reyes-Reyes AL (2015) Assessment of methods to recover DNA from bacteria, fungi and archaea in complex environmental samples. Folia Microbiol 60(6):551–558

Hawksworth DL (2001) The magnitude of fungal diversity: the 1.5 million species estimate revisited. Mycol Res 105(12):1422–1432

He Y, Zhao Y, Zhou G, Huang M (2009) Evaluation of extraction and purification methods for obtaining PCR-amplifiable DNA from aged refuse for microbial community analysis. Word J Microbiol Biotechnol 25(11):2043–2051

Hernández-López EL, Ramírez-Puebla ST, Vazquez-Duhalt R (2015) Microarray analysis of *Neosartorya fischeri* using different carbon sources, petroleum asphaltenes and glucose-peptone. Genom Data 5:235–237

Hugenholtz P, Tyson GW (2008) Microbiology: metagenomics. Nature 455:481

Illman BL, Yang VW, Ferge LA (2002a) US Patent No. 6,383,800. US Patent and Trademark Office, Washington, DC

Illman BL, Yang VW, Ferge LA (2002b) US Patent No. 6,387,689. US Patent and Trademark Office, Washington, DC

Illman BL, Yang VW, Ferge LA (2002c) US Patent No. 6,387,691. US Patent and Trademark Office, Washington, DC

Illman BL, Yang VW, Ferge LA (2002d) US Patent No. 6,495,134. US Patent and Trademark Office, Washington, DC

Illman BL, Yang VW, Ferge LA (2003) US Patent No. 6,664,102. US Patent and Trademark Office, Washington, DC

Illman BL, Yang VW, Ferge LA (2004) US Patent No. 6,727,087. US Patent and Trademark Office, Washington, DC

Illman BL, Yang VW, Ferge LA (2005) US Patent No. 6,972,169. US Patent and Trademark Office, Washington, DC

Jiang YX, Wu JG, Yu KQ, Ai CX, Zou F, Zhou HW (2011) Integrated lysis procedures reduces extraction biases of microbial DNA from mangrove sediments. J Biosci Bioeng 111(2):153–157

Kameshwar AKS, Qin W (2017) Metadata analysis of *Phanerochaete chrysosporium* gene expression data identified common CAZymes encoding gene expression profiles involved in cellulose and hemicellulose degradation. Int J Biol Sci 13(1):85

Kazemi M, Tahmasbi A, Valizadeh R, Naserian A, Soni A (2012) Organophosphate pesticides: a general review. Agric Sci Res J 2:512–522

Ladino-Orjuela G, Gomes E, da Silva R, Salt C, Parsons JR (2016) Metabolic pathways for degradation of aromatic hydrocarbons by bacteria. Rev Environ Contam Toxicol 237:105–121

Lamar RT, Lestan D, Smith CE, Dietrich DM (2000) US Patent No. 6,143,549. US Patent and Trademark Office, Washington, DC

Leininger S, Urich T, Schloter M, Schwark L, Qi J, Nicol GW, Prosser JI, Schuster SC, Schleper C (2006) Archaea predominate among ammonia-oxidizing prokaryotes in soils. Nature 442:806

Liu Y, Zhou J, Omelchenko MV, Beliaev AS, Venkateswaran A, Stair J, Wu L, Thompson DK, Xu D, Rogozin IB, Gaidamakova EK, Zhai M, Makarova KS, Koonin EV, Daly MJ (2003) Transcriptome dynamics of *Deinococcus radiodurans* recovering from ionizing radiation. Proc Natl Acad Sci U S A 100:4191–4196

Maghsoudi E, Fortin N, Greer C, Maynard C, Pagé A, Duy SV, Dorner S (2016) Cyanotoxin degradation activity and mlr gene expression profiles of a *Sphingopyxis sp.* isolated from Lake Champlain, Canada. Environ Sci Process Impact 18(11):1417–1426

Mathews SL, Pawlak J, Grunden AM (2015) Bacterial biodegradation and bioconversion of industrial lignocellulosic streams. Appl Microbiol Biotechnol 99(7):2939–2954

McMaster R (2018) Is the fourth industrial revolution relevant to you? Nurs Health Sci 20(2):139–141

Medini D, Serruto D, Parkhill J, Relman DA, Donati C, Moxon R, Falkow S, Rappuoli R (2008) Microbiology in the post-genomic era. Nat Rev Microbiol 6:419

Mohite BV, Koli SH, Narkhede CP, Patil SN, Patil SV (2017) Prospective of microbial exopolysaccharide for heavy metal exclusion. Appl Biochem Biotechnol 183(2):582–600

Morais D, Pylro V, Clark IM, Hirsch PR, Tótola MR (2016) Responses of microbial community from tropical pristine coastal soil to crude oil contamination. Peer J 4:1733

Moré MI, Herrick JB, Silva MC, Ghiorse WC, Madsen EL (1994) Quantitative cell lysis of indigenous microorganisms and rapid extraction of microbial DNA from sediment. Appl Environ Microbiol 60(5):1572–1580

Nancharaiah YV, Mohan SV, Lens PNL (2016) Biological and bioelectrochemical recovery of critical and scarce metals. Trends Biotechnol 34(2):137–155

National Research Council (2014) Review of EPA's integrated risk information system (IRIS) process. National Academies Press, Washington, DC

Nelson DM, Ohene-Adjei S, Hu FS, Cann IKO, Mackie RI (2007) Bacterial diversity and distribution in the holocene sediments of a northern temperate lake. Microbial Ecol 54(2):252–263

Obeng EM, Adam SNN, Budiman C, Ongkudon CM, Maas R, Jose J (2017) Lignocellulases: a review of emerging and developing enzymes, systems, and practices. Bioresour Bioprocess 4(1):16

Ozer A, Ozer D (2003) Comparative study of the biosorption of Pb(II), Ni(II) and Cr(VI) ions onto *S. cerevisiae*: determination of biosorption heats. J Hazard Mater 100:219–229

Parro V, Moreno-Paz M, Gonzalez-Toril E (2007) Analysis of environmental transcriptomes by DNA microarrays. Environ Microbiol 9:453–464

Poretsky RS, Bano N, Buchan A, Lecleir G, Kleikemper J, Pickering M, Pate WM, Moran MA, Hollibaugh JT (2005) Analysis of microbial gene transcripts in environmental samples. Appl Environ Microbiol 71:4121–4126

Qiu J, Guo Z, Liu H, Zhou D, Han Y, Yang R (2008) DNA microarray-based global transcriptional profiling of *Yersinia pestis* in multicellularity. J Microbiol 46:557–563

Rineau F, Roth D, Shah F, Smits M, Johansson T, Canbäck B, Grigoriev IV (2012) The ectomycorrhizal fungus *Paxillus involutus* converts organic matter in plant litter using a trimmed brown-rot mechanism involving Fenton chemistry. Environ Microbiol 14(6):1477–1487

Rodrigues EM, Kalks KH, Fernandes PL, Tótola MR (2015) Bioremediation strategies of hydrocarbons and microbial diversity in the Trindade Island shoreline. Mar Pollut Bull 101(2):517–525

Romanelli AM, Fu J, Herrera ML, Wickes BL (2014) A universal DNA extraction and PCR amplification method for fungal rDNA sequence-based identification. Mycoses 57(10):612–622

Ruta LL, Kissen R, Nicolau I, Neagoe AD, Petrescu AJ, Bones AM, Farcasanu IC (2017) Heavy metal accumulation by *Saccharomyces cerevisiae* cells armed with metal binding hexapeptides targeted to the inner face of the plasma membrane. Appl Microbiol Biotechnol 101(14):5749–5763

Sağ Y, Özer D, Kutsal T (1995) A comparative study of the biosorption of lead(II) ions to *Z. ramigera* and *R. arrhizus*. Process Biochem 30:169–174

Sankaran S, Khanal SK, Jasti N, Jin B, Pometto AL, Van Leeuwen JH (2010) Use of filamentous fungi for wastewater treatment and production of high value fungal byproducts: a review. Crit Rev Environ Sci Technol 40(5):400–449

Sato S, Feltus FA, Iyer P, Tien M (2009) The first genome-level transcriptome of the wood-degrading fungus *Phanerochaete chrysosporium* grown on red oak. Curr Genet 55(3):273–286

Soares EV, Soares HM (2012) Bioremediation of industrial effluents containing heavy metals using brewing cells of *Saccharomyces cerevisiae* as a green technology: a review. Environ Sci Pollut R 19(4):1066–1083

Thorsen M, Lagniel G, Kristiansson E, Junot C, Nerman O, Labarre J, Tamás MJ (2007) Quantitative transcriptome, proteome, and sulfur metabolite profiling of the Saccharomyces cerevisiae response to arsenite. Physiol Genomics 30(1):35–43

Tortella GR, Diez MC, Durán N (2005) Fungal diversity and use in decomposition of environmental pollutants. Crit Rev Microbiol 31(4):197–212

Urich T, Lanzen A, Qi J, Huson DH, Schleper C, Schuster FSC (2008) Simultaneous assessment of soil microbial community structure and function through analysis of the meta-transcriptome. PLoS One 3:2527

Vats A, Mishra S (2018) Identification and evaluation of bioremediation potential of laccase isoforms produced by *Cyathus bulleri* on wheat bran. J Hazard Mater 344:466–479

Verma S, Verma PK, Meher AK, Dwivedi S, Bansiwal AK, Pande V, Chakrabarty D (2016) A novel arsenic methyltransferase gene of Westerdykella *aurantiaca* isolated from arsenic contaminated soil: phylogenetic, physiological, and biochemical studies and its role in arsenic bioremediation. Metallomics 8(3):344–353

Warnecke F, Hess M (2009) A perspective: metatranscriptomics as a tool for the discovery of novel biocatalysts. J Biotechnol 142(1):91–95

Warnecke F, Luginbühl P, Ivanova N, Ghassemian M, Richardson TH, Stege JT, Cayouette M, Mchardy AC, Djordjevic G, Aboushadi N, Sorek R, Tringe SG, Podar M, Garcia-Martin H, Kunin V, Dalevi D, Madejska J, Kirton E, Platt D, Szeto E, Salamov A, Barry K, Mikhailova N, Kyrpides NC, Matson EG, Ottesen EA, Zhang X, Hernández M, Murillo C, Acosta LG, Rigoutsos I, Tamayo G, Green BD, Chang C, Rubin EM, Mathur EJ, Robertson DE, Hugenholtz P, Leadbetter JR (2007) Metagenomic and functional analysis of hindgut microbiota of a wood-feeding higher termite. Nature 450:560–565

Whitby C (2010) Microbial naphthenic acid degradation. Adv Appl Microbiol 70:93–125

Wintzingerode FV, Göbel UB, Stackebrandt E (1997) Determination of microbial diversity in environmental samples: pitfalls of PCR-based rRNA analysis. FEMS Microbiol Rev 21(3):213–229

World Health Organization (2004) Guidelines for drinking-water quality: recommendations, vol 1. World Health Organization, Geneva

Wymelenberg AV, Gaskell J, Mozuch M, Sabat G, Ralph J, Skyba O, Kersten PJ (2010) Comparative transcriptome and secretome analysis of wood decay fungi *Postia placenta* and *Phanerochaete chrysosporium*. Appl Environ Microbiol 76(11):3599–3610

You Y, Fu C, Zeng X, Fang D, Yan X, Sun B, Xiao D, Zhang J (2008) A novel DNA microarray for rapid diagnosis of enteropathogenic bacteria in stool specimens of patients with diarrhea. J Microbiol Methods 75:566–571

Zhang T, Tang J, Sun J, Yu C, Liu Z, Chen J (2015) Hex1-related transcriptome of *Trichoderma atroviride* reveals expression patterns of ABC transporters associated with tolerance to dichlorvos. Biotechnol Lett 37(7):1421–1429

# Chapter 3
# Omics Approaches: Impact on Bioremediation Techniques

**Yordanis Pérez-Llano, Liliana Martínez-Ávila, and Ramón Alberto Batista-García**

## Contents



## 3.1 The Uprising of the "Omics"

One major feature of the "omics" techniques is that they are top-down holistic methodologies. This means that the research is focused on the properties of the systems rather than on those of the individual parts. Thus, instead of investigating a single molecular entity, the "omics" approaches aim to characterize a collection of molecules from the same category. Depending on the target molecule under investigation, different technologies have arisen. The genomic and transcriptomic approaches measure nucleic acids (DNA and RNA, respectively) and have advanced rapidly with the advent of next-generation sequencing technologies. Genomics deal with the gene composition of an organism, which shed light on the potentiality for adaptation to different environments and allows the construction of its evolutionary history. Transcriptomics, on the other hand, can unravel the genome function under particular conditions, allowing to determine which genes are expressed and enable the organism to thrive in an environment or circumstance. However, the expression


Y. Pérez-Llano · L. Martínez-Ávila · R. A. Batista-García (✉)
Centro de Investigación en Dinámica Celular-IICBA, Universidad Autónoma del Estado de Morelos, Cuernavaca, Morelos, Mexico
e-mail: rabg@uaem.mx





R. Prasad, E. Aranda (eds.), *Approaches in Bioremediation*, Nanotechnology in the Life Sciences, https://doi.org/10.1007/978-3-030-02369-0_3

level of a gene might not correlate with the protein levels or activity, nor it indicates protein localization within the cells or metabolic status. Therefore, recent advances in proteomics have allowed not only the genome-wide identification of proteins but also the characterization of their primary structure (in terms of posttranslational modifications and maturation), their turnover or degradation, their regulation, and their interactions with other proteins (interatomics). Similarly, assessing the metabolic status of cells is also possible by the well-established methodologies that measure low molecular weight organic compounds (metabolomics) and their temporal dynamics (fluxomics).

In environmental research, though, the techniques that analyze the microbial populations and their interactions in a complex sample (e.g., soil, water, organic waste, sludge, etc.) have become a milestone (Fig. 3.1). These techniques avoid the bottleneck of culturing specific microorganisms, thus allowing to recover information of the uncultured microbiota in an environmental sample. Although different terms have been used for these methodologies, the most widely used are the ones that use the preposition "meta" modifying an already established omics technique, i.e., metagenomics, metatranscriptomics, and metaproteomics. Handelsmanl et al. (1998) were the first in the use of the term "metagenomics" to refer to the totality of the genomes found in a certain environment. The main drawback of these methodologies is that recovering the target molecules (nucleic acids or proteins) from a complex source can pose a technical challenge. Another important problem is that the "omics" techniques generate a relatively large amount

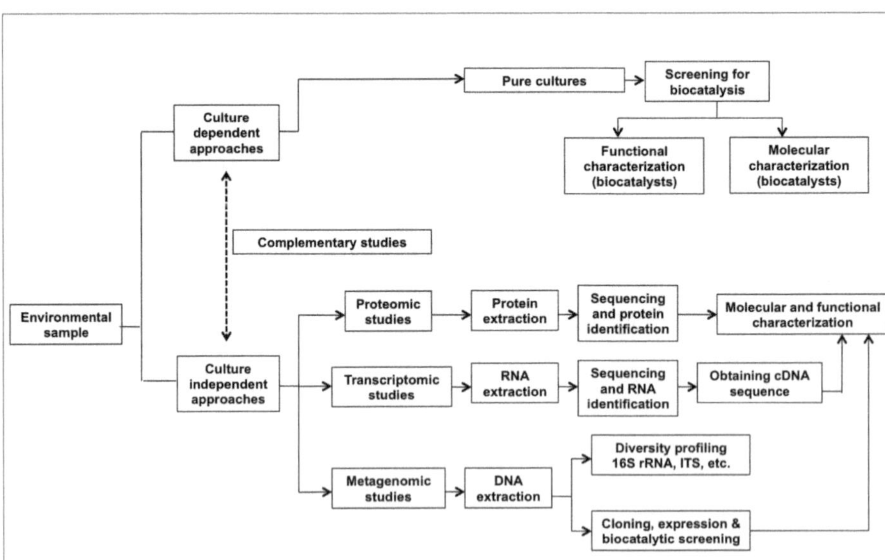


**Fig. 3.1** Culture-dependent and culture-independent approaches for environmental sample screening. (Figure was taken from Batista-García et al. 2016)

of information and processing this information requires specific tools and frameworks. Recently, methods and software designed to deal with metagenomic data have become more accessible (or friendly) to biology scientist, who generally lacks professional programming skills. Also, databases gathering "omics" information have been continuously growing and are publicly available, which facilitate the exchange and integration of information.

## 3.2 The Promises of Metagenomics

The first task of an environmental management project is assessing the contamination level in a site and the extent of the detrimental effect on ecosystem dynamics. Currently, most environments are polluted to some degree with natural or synthetic compounds that are hazardous to animals, plants, and the ecosystems. The traditional approaches during site investigation in environmental risk assessment are determining the physical-chemical properties of the matrix (soil, water, or air) and the presence of suspected contaminants. However, the presence of a toxic compound in a site does not directly implicate that the health of the system is compromised nor that an intervention is required to eliminate the pollutant. In many occasions, the naturally occurring microorganisms are able to degrade the exotic compounds and restore the balance to the ecological unit. The goal in environmental management is to preserve or restore the natural functions of an ecosystem with the minimal intervention and economical cost. Therefore, the bottom line is that a holistic and comprehensive analysis is required to define the most convenient form of intervention in a site-dependent manner.

The discharge of a contaminant can shape the composition of the microbial communities in an ecosystem. Other effects might be the net increase in biomass, the genetic diversity, or the activity (Pulleman et al. 2012). The species able to degrade the pollutant or its intermediary degradation compounds are generally enriched, while the species sensitive to the pollutant are lessened. These species are regarded as bioindicators and have been used widely in the diagnosis of contaminated places (Pulleman et al. 2012). In this way, metagenomics can help to measure the deterioration of the natural function of ecosystems as a result of contamination. By targeting phylogenetic markers, such as 16S rDNA for bacteria and 18S rDNA or ITS for fungi and other eukaryotes, metagenomics can portray the phylogenetic composition of a microbial community. Samples from uncontaminated sites or pre-contamination samples from the same site can serve as controls to determine if particular phyla have varied in the community.

Through a meta-analysis pipeline, Oliveira et al. (2017) analyzed metagenome samples from biomes with crude oil contamination across different geographies and obtained by different research projects. Firstly, they found that distributions of genes involved in the degradation of petroleum hydrocarbons and genes involved in the biosynthesis of biosurfactants are strongly correlated, which might seem intuitive but is not generally analyzed jointly in metagenomic studies. Also, their results show that terrestrial ecosystems present more degradation-related genes and fewer biosurfactant genes when compared to water biomes. Taking this into consideration, it would imply that treatment strategies to bioremediate oil-contaminated waters would differ from those used for terrestrial spills in terms of the activities that need fostering to achieve a particular rate of hydrocarbon degradation. Lastly, but not less important, these authors found that latitude has a significant influence on the diversity of genes involved in biodegradation and biosurfactant production. According to this, microbiomes near the equator have higher diversity in genes involved in hydrocarbon degradation when compared to other geographical zones. This information can be used in the design of bioremediation techniques, as microbial consortia from these areas can be particularly suited for application to bioaugmentation, enzyme discovery, and degradation pathway analysis.

In addition, the knowledge of the microbial interactions and capabilities of the native microorganisms in a contaminated site has allowed the application of various biostimulation and bioaugmentation strategies. The former pursues the goal of stimulating the capacity of resident microbial communities to degrade pollutants by modifying biochemical parameters such as nutrient availability, carbon sources, pH, temperature, aeration and oxygenation, water content, etc. Here, the systems biology approach is essential to determine the limiting reactions that can affect the biodegradation rates of toxic compounds. Bioaugmentation, on the other hand, is a more intrusive strategy by which nonnative microorganisms are added to the contaminated site, generally to supply pathways that are not present or optimal in the indigenous microbiome. For such purpose, the consortia and synergistic interactions between different microbial species should have been characterized a *priori*. This strategy has some drawback: for example, the foreign microorganisms may not be competitive in the ecological environment that is treated, or the microorganisms degrade the pollutant but remain in the site long after the contaminant has disappeared, modifying the dynamics of the natural ecosystem. In this case, metagenomics comes back into play as a monitoring tool that allows assessing the effect of the bioremediation treatment on the native microbial communities.

Finally, many of the microorganisms that inhabit the heavily contaminated sites represent a source of potential biotechnological products to restore environmental health. Therefore, metagenomics has served to investigate microbial communities that are actively degrading a compound, under the notion that the genes with more variability or representation in the metagenomic libraries might be involved in the biodegradation of such contaminant. This so-called gene-centric approach has gained much attention because it can serve for the prospection of new enzyme activities or new genes that are accessory to the degradative pathways, such as the aforementioned biosurfactants in the case of hydrocarbon degradation. Also, this gene-centric approach has served as a monitoring tool, because some genes known to be responsive to contamination levels can be screened in the investigated sites to assess the effect of a pollutant in the microbiome without the need to sequence all the genetic content of the samples.

Taking all into consideration, metagenomics has properly served to various purposes on the bioremediation field and is currently a powerful tool during decision-making for treatment strategies and management of contaminated environments. The continuous reduction of the sequencing prices makes these methodologies affordable for industrial applications. Nonetheless, significant efforts remain to be done to change the perception of these techniques in the industrial community and to expand their use to a daily basis.

## 3.3 Transcriptomics

Transcriptomic approaches are successfully used to obtain insights into microbial functions in contaminated sites. These techniques preferentially study the total environmental mRNA. On the other hand, metatranscriptomic studies the gene expression profile of the entire microbial communities, being the RNA extraction one of the major challenges (Desai et al. 2010). These methods allow analyzing the gene expression dynamic at the transcriptional level of one organism or microbial community inhabiting a specific habitat in a specific time. Transcriptomics also reveal the plasticity of the entire cell repertoire when microorganisms are exposed to pollutants.

Since transcriptomics studies the mRNA profile, it can offer a general idea about the enzymatic mechanisms involved in pollutant degradation. Thus, these omics tools generate information related with subsequent protein profile of the cell. However, the posttranscriptional modification could drastically modulate this information.

Transcriptomic studies are not abundant to understand the enzymatic strategies used for microorganisms. For example, transcriptional analyses associated with the polycyclic aromatic hydrocarbons degradations are scarce. However, the degradative mechanisms involved in asphaltene biotransformation by *Neosartorya fischeri* were studied at the transcriptional level. It has been found that genes encoding for monooxygenase enzymes were upregulated (Hernández-López et al. 2015). Another transcriptional analysis in *Exophiala pisciphila* obtained that glutathione S-transferase genes were upregulated during its growth in soil contaminated with heavy metals (Morel et al. 2013).

## 3.4 Proteomic in a Degradation Concept

In this chapter, we highlight proteomic approaches since the presence of proteins determines the final phenotype of the cell. In the context of this chapter, it is obligatory that enzymes can finally be detected to fully understand biodegradation pathways.

Proteome refers to the entire protein profiles encoded by the genome, and proteomic, on another hand, is the tool that allows the study of the proteome, their expression, function, and structure, including the posttranslational modifications. The analysis of the proteome plays an important role to fully understand any biological process, including biodegradative processes. In this context, secretome refers to the entire set of extracellular proteins encoded by the genome, and it is very relevant to hypothesize extracellular pathways related with the catabolism of any pollutant. Thus, the analysis of sub-proteomes and secretomes allows the identification of proteins with relevant functions in the degradation or removal of organic contaminants and reveals those differentially expressed proteins involved in the biodegradation. The proteome is very dynamic since the cell quickly regulates its expression according to specific living conditions.

In recent years, proteomics has become important as a discipline in environmental biotechnology. "The application of proteomics to the field of biodegradation is in its infancy" (Chauhan and Jain 2010). However, proteomics is now completely established as a robust and powerful methodology to characterize functional proteins relevant for environmental biotechnology. In general terms, even up today, proteomics is an expensive technique, and it is required to develop a new experimental methodology with reasonable costs for environmental applications. Gel-free approaches are not yet at the hand for many groups around the world, and, in consequence, there are no notable advances in the environmental proteomics. For this reason, gel-based approaches are still useful. Figure 3.2 shows both approaches, 2-DE-based proteomics and gel-free quantitative proteomics (Fig. 3.2 was taken from Loh and Cao 2008).

The information generated by proteomics could be supplemented with information obtained from other global studies such as transcriptomics, metagenomics, and genomics, as well as with those results recovered from other emergent omics such as phenomics, ionomics, lipidomics, culturomics, etc. The synergy of all these molecular and morphological tools could drastically transform the point of view to study the biotransformation of xenobiotics. The new omics era has big challenges to generate new knowledge in the border of the science and, in consequence, propose novel eco-friendly ideas to remove xenobiotics from the environment.

Polycyclic aromatic hydrocarbons (PAHs) are model compounds to study metabolic pathways involved in the xenobiotic biodegradation. Proteomics has been successfully used to elucidate catabolic and regulatory mechanisms involved in PAH degradation (Kim et al. 2011; Vandera et al. 2015; Liu et al. 2017). These mechanisms have been largely studied in bacteria, while the fungal mechanisms are

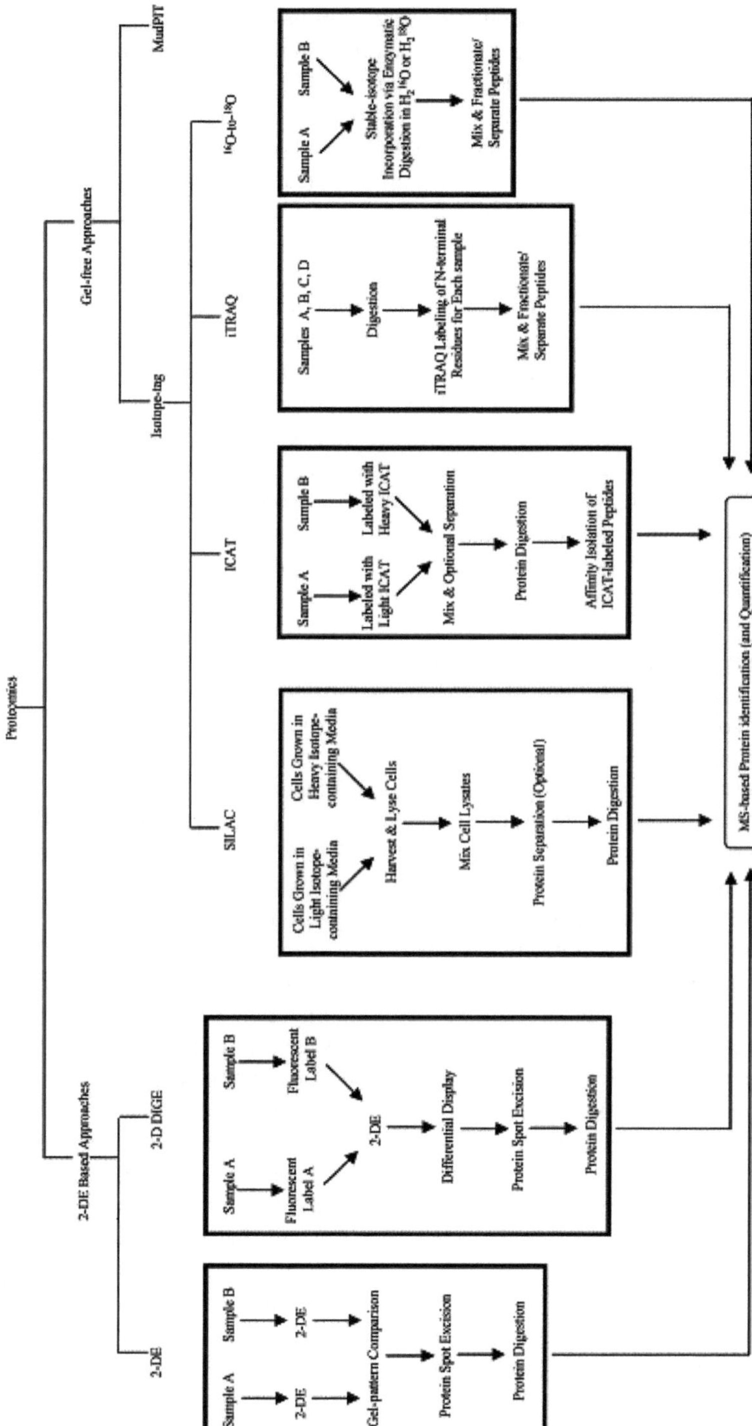


**Fig. 3.2** Schematic illustration of representative proteomic approaches. 2-DE-based proteomics and gel-free quantitative proteomic are shown. (Figure and legend were taken from Loh and Cao 2008)

even unclear. Figure 3.3 shows a schematic model published by Liu and co-workers (Liu et al. 2017) that was obtained from proteomic studies in *Sphingomonas* sp. strain GY2B during phenanthrene degradation (Fig. 3.3 was taken from Liu et al. 2017). A proteomic analysis conducted in *Achromobacter xylosoxidans* during pyrene biodegradation informed that proteins involved in cell homeostasis, genetic information synthesis and storage, and chemical stress were upregulated. 4-Hydroxyphenylpyruvate dioxygenase and homogentisate 1,2-dioxygenase were particularly important during pyrene transformation (Nzila et al. 2018).

Regarding fungi, the progress in environmental proteomics is discreet with a slight takeoff in the last years. A recent study in *Penicillium oxalicum* identified via shotgun proteomics 158 upregulated and 174 downregulated unique protein species differentially expressed during anthracene biotransformation. Cytochrome P450s, epoxide hydrolases, and transferase enzymes were the main upregulated proteins involved in anthracene biotransformation (Lucero Camacho-Morales et al. 2018). In this study, 77.1% of the proteins were annotated as hypothetical uncharacterized proteins. The protein annotation is crucial during the proteomic analysis. It currently represents an important bottleneck, and it is one of the major challenges of fungal studies. Verdin and co-workers (2005) used a 2-D gel-based approach to study the benzo[a]pyrene degradation by *Fusarium solani*, and an overexpression of P450 was found.

Another proteomic analysis was addressed to understand the metabolic pathway preferentially used by the fungus *Metarhizium robertsii* in presence of 4-*n*-nonylphenol, an endocrine-disrupting compound, as a carbon source (Szewczyk et al. 2014). Oxidation-reduction system related to nitroreductase-like proteins, reactive oxygen species defense systems (peroxiredoxin and superoxide dismutase), the tricarboxylic acid cycle, and energy-related systems were the main proteins and metabolic pathways upregulated in the proteome of *M. robertsii*. In this study, a huge group of unidentified proteins was observed.

Proteomics has also been applied to investigate the fungal degradation of herbicide. The upregulated protein profile of *Paecilomyces marquandii* revealed that in presence of alachlor, an herbicide used to protect plant crops against broadleaf weeds and annual grasses, the proteins related with energy and sugar metabolism, as well as reactive oxygen species, are key enzymes involved in the alachlor biotransformation pathway (Szewczyk et al. 2015). Authors identified cyanide hydratase as the main enzyme implicated in the biotransformation of alachlor.

The previous studies demonstrate that proteomics is a useful and valuable tool to analyze those open reading frames induced by the presence of xenobiotics. But, other aspect related with proteins can also be studied by proteomics since it can answer other questions. For example, what happens with the isoenzymes of key proteins in biotransformation processes? Are the posttranslational modifications important during the xenobiotic biotransformation? How is the dynamic of the protein degradation? These questions remain absolutely understood. Further researches and efforts are needed to clarify these points.

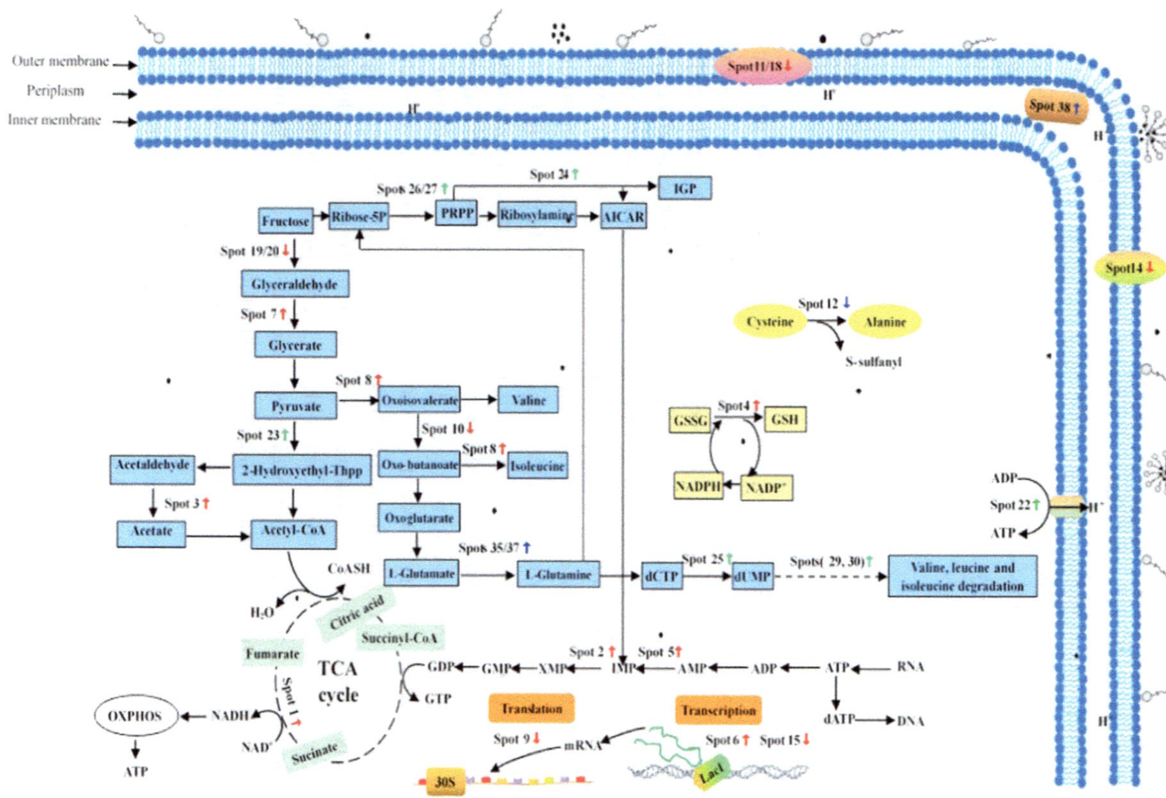


**Fig. 3.3** Schematic model of the protein expression of *Sphingomonas* sp. GY2B in response to Tween 80 during phenanthrene (100 mg/L) biodegradation. The upregulated and downregulated proteins are marked by ↑ and ↓, respectively. Red arrows represent the protein with altered expression in all three concentrations of Tween 80 (0.2, 1, and 8 CMC); the green and blue arrows represent the differentially expressed proteins detected in two concentrations (1 and 8 CMC) and only in 8 CMC treated sample, respectively. (Figure and legend were taken from Liu et al. 2017)

Xenobiotics also elicit molecular mechanisms related to stress responses. In this context, proteomics has brought relevant information about the proteins involved in the stress physiology. When microorganisms are exposed at recalcitrant pollutants, changes occur in their protein profiles including membrane proteins and those enzymes related with DNA repair mechanisms, among others. Proteomic approaches have shown in *Pseudomonas putida* KT2440 exposed to phenol that proteins participating in oxidative stress response, general stress response, energetic metabolism, fatty acid biosynthesis, inhibition of cell division, cell envelope biosynthesis, transcriptional regulation, and transport of small molecules are totally upregulated. These proteomic studies have also revealed that the protein profile involved in cell morphology and cell surface hydrophobicity also drastically changes (Santos et al. 2004).

### 3.4.1 Metaproteomic

Finally, metaproteomic is also a powerful methodology to study the relationship between microorganisms in polluted environments and their possible role in xenobiotic degradation. Metaproteomic is an emerging tool in the post-genomic era and cannot distinguish between proteins from different microorganisms since it analyzes at high-throughput scale, the pull of proteins produced by microorganisms inhabiting a specific environment and time (Keller and Hettich 2009; Wilmes et al. 2008). This tool is based on the extraction, purification, identification, and annotation of all proteins present in an environment, being the first step one of the major challenges. The meta-analysis at protein levels brings relevant information related to metabolic pathways in nature and allows the hypothesis postulation about the ecological role of microbial communities in the xenobiotic degradation. The metaproteomic studies are very scarce in contaminated environments. For example, metaproteomic application to understand the biodegradation processes in wastewater biotreatment is still in its infancy. However, new research efforts are needed due to that metaproteomics could strongly contribute to identifying microbial proteins with key functions in wastewater biotreatments as well as to elucidate the microbial metabolic pathways involved in wastewater biotransformation.

Metaproteomics could also bring knowledge related with the availability to access pollutants to microorganisms, how is the growth control and microbial homeostasis in polluted sites, how xenobiotics could affect the nutrient acquisition in microbes, and how is the dynamic of the stress response at the microbial ecosystem level, among others. As it is supposed that microbial networks are environmentally necessary for xenobiotic degradation (Aydin et al. 2017), these points are crucial for the understanding of the bioprocesses in nature. Thus, the omics era and its molecular biological tools can be used to understand the dynamics of the microbial networks in the biotransformation of pollutants. Additionally, metaproteomics can also generate information about the regulatory mechanisms in these microbial networks (Aydin et al. 2017) and, interestingly, how is the communication, the chemical talks, between different species and populations in polluted ecosystems. Metagenomics have studied the functioning of microbial networks in soils (Benndorf et al. 2007; Wang et al. 2011), activate sludges (Kuhn et al. 2011; Osman et al. 2007; Wilmes et al. 2008), groundwater (Benndorf et al. 2007; Kan et al. 2005), wool fabrics (Solazzo et al. 2013), wastewater bioreactor (Lacerda et al. 2007), psychrophilic anaerobic wastewater treatment bioreactor (Abram et al. 2011), and thermophilic anaerobic treatment of agriculture biomass (Hanreich et al. 2012). Figure 3.4 represents a schematic model of the active metabolic pathways involved in a swine manure-based thermophilic anaerobic digester obtained from metaproteomic results (Fig. 3.4 was taken from Lin et al. 2016).

Another metaproteomic study also proposed the carbohydrate metabolic pathways of microbial communities during a large-scale aerobic composting plant (Fig. 3.5) (Fig. 3.5 was taken from Liu et al. 2015). Thus, metaproteomics opens new perspectives at the protein level to understand molecular mechanisms of xenobiotic degradation at the community level (El Amrani et al. 2015; Shahi et al. 2016a, b).

In conclusion, (meta)proteomics represent an unique opportunity to study protein profiles produced during pollutant degradations. This approach comprises structural and quantitive analysis. As 90% of the microorganisms in nature cannot be cultivated, a broad range of obtained peptides is probably from those uncultivable microbes. Thus, (meta)proteomics is a source of novel proteins with relevant impact in environmental science.

Finally, the "omics" application to generate novel knowledge related with polluted environments is an area of big opportunities to strongly impact the biotechnological applications. The integrative omics approaches represent a methodological challenge from sampling to analysis. Different disciplines play a key role in omics application. Figure 3.6 shows an overview of molecular and omics technologies applied to contaminated sites (Fig. 3.6 was taken from Desai et al. 2010).

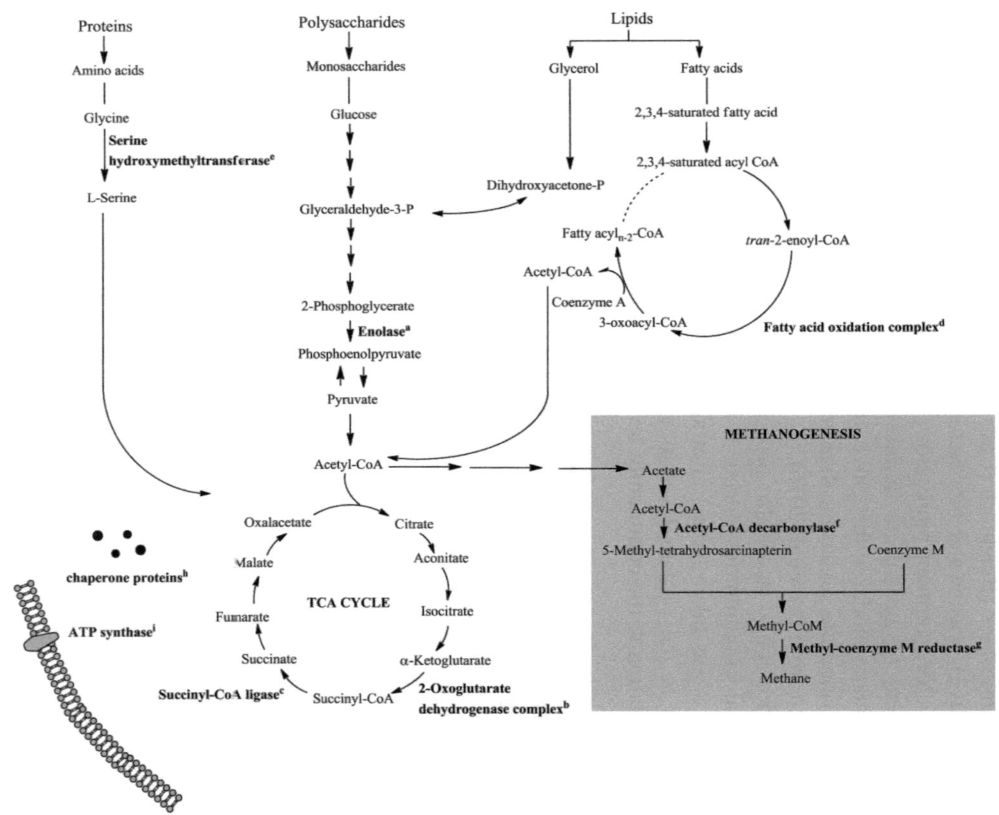


**Fig. 3.4** The proposed metabolic pathways active in a swine manure-based thermophilic anaerobic digester as inferred from the metaproteomic results. Enzymes identified in this study are shown in bold. Not all intermediates are shown in this figure. *indicates source organisms that are not confirmed because

**Fig. 3.4** (continued) the identified peptide sequences are conserved across many species. The identified proteins belong to eubacteria and archaea. The source organisms are assigned by the identified peptides: [a]*Pseudomonas fluorescens* SBW25; [b]*Bdellovibrio bacteriovorus*, *Buchnera aphidicola* str. Bp, *Clostridium novyi*-NT, *Idiomarina loihiensis*, *Neocallimastix frontalis*, *Pseudomonas aeruginosa*, *Pseudomonas fluorescens*, *Pseudomonas putida*, *Rubrobacter xylanophilus* DSM 9941, or *Thiomicrospira crunogena* XCL-2; [c]*Pseudomonas putida*; [d]**Dictyostelium discoideum*, *Bartonella* sp., or *Pseudomonas aeruginosa*; [e]*Azotobacter vinelandii*, *Buchnera aphidicola*, *Mycobacterium bovis*, *Pseudomonas aeruginosa*, or *Staphylococcus haemolyticus* JCSC1435; [f]*Actinobacillus succinogenes* 130Z, *Acinetobacter* sp. ADP1, *Escherichia coli* BW2952, *Marinobacter aquaeolei* VT8, *Pseudomonas aeruginosa*, *Pseudomonas fluorescens*, *Pseudomonas mendocina* ymp, *Pseudomonas putida*, or *Vibrio cholerae*; [g]**Escherichia coli* or *Salmonella typhimurium*; [h]*Colwellia psychrerythraea* 34H; [i]*Pseudomonas aeruginosa*; [j]**Mycobacterium bovis*, *M. marinum*, or *M. tuberculosis*; [k]fatty acid oxidation complex including dodecenoyl-CoA d-isomerase, enoyl-CoA hydratase, 3-hydroxybutyryl-CoA epimerase, and 3-hydroxyacyl-CoA dehydrogenase; [l]**Pseudomonas fluorescens* Pf-5; *Pseudomonas fluorescens* Pf0-1 or *Pseudomonas syringae* sp.; [m]*Pseudomonas aeruginosa*; [n]*Streptococcus pneumonia*; [o]*Pseudomonas fluorescens* or *Pseudomonas putida*; [p]*Methanocaldococcus jannaschii* DSM 2661; [q]*Methanosarcina* sp.; [r]*Methanothermobacter thermautotrophicus* str. Delta H, *Methanosarcina barkeri* str. Fusaro, or *Methanothermus fervidus*; [s]*Acidithiobacillus ferrooxidans*, *Pseudomonas fluorescens*, *Pseudomonas syringae*, *Edwardsiella ictaluri* 93e146, or *Chromobacterium violaceum*; [t]*Pseudomonas putida*; [u]*Candidatus solibacter usitatus* Ellin6076, *Leifsonia xyli* subsp. xyli, *Dechloromonas aromatica* RCB, *Geobacter lovleyi* SZ, *Candidatus blochmannia floridanus*, *Colwellia psychrerythraea* 34H, *Idiomarina loihiensis*, *Pseudomonas fluorescens* Pf0-1, *Pseudomonas fluorescens* Pf-5, *Pseudomonas putida* W619, *Pseudomonas entomophila* L48, *Petrotoga mobilis* SJ95, *Methylobacillus flagellatus* KT, *Methylococcus capsulatus*, *Pelobacter carbinolicus* DSM 2380, *Polaromonas naphthalenivorans* CJ2, *Psychromonas ingrahamii* 37, or *Thioalkalivibrio sulfidophilus* HL-EbGr7; [v]*Bacillus licheniformis* ATCC 14580; [w]*Pseudomonas fluorescens*. (Figure and legend were taken from Lin et al. 2016)

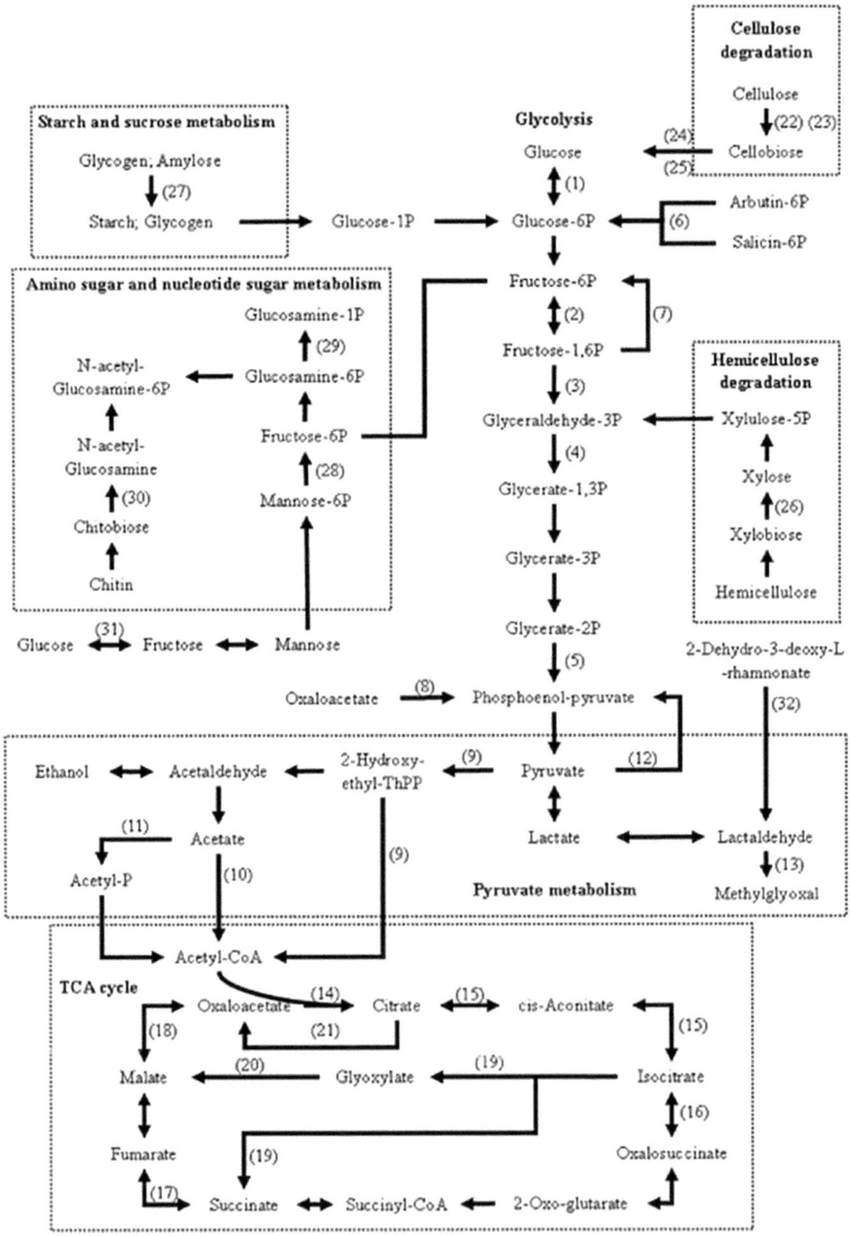


**Fig. 3.5** Depiction of the carbohydrate metabolic characteristics of microbial communities inferred from the metaproteome. Proteins shown are (1) hexokinase; (2) 6-phosphofructokinase; (3) triose-phosphate isomerase; (4) glyceraldehyde-3-phosphate dehydrogenase; (5) enolase; (6) aryl-phospho-beta-D-glucosidase; (7) fructose-1,6-bisphosphatase; (8) phosphoenolpyruvate carboxykinase; (9) pyruvate dehydrogenase; (10) acetyl-coenzyme A synthetase; (11) acetate kinase; (12) pyruvate, phosphate dikinase; (13) glyoxylate/hydroxypyruvate reductase; (14) citrate synthase; (15) aconitate hydratase; (16) isocitrate dehydrogenase; (17) succinate dehydrogenase; (18) malate dehydrogenase; (19) isocitrate lyase; (20) malate synthase; (21) citrate lyase; (22) exoglucanase; (23) endoglucanase; (24) cellobiose dehydrogenase; (25) beta-glucosidase; (26) endo-1,4-beta-xylanase B; (27) 1,4-alpha-glucan-branching enzyme GlgB; (28) mannose-6-phosphate isomerase; (29) phosphoglucosamine mutase; (30) beta-N-acetylhexosaminidase; (31) 2-keto-3-deoxy-L-rhamnonate aldolase; (32) xylose isomerase. (Figure and legend were taken from Liu et al. 2015)

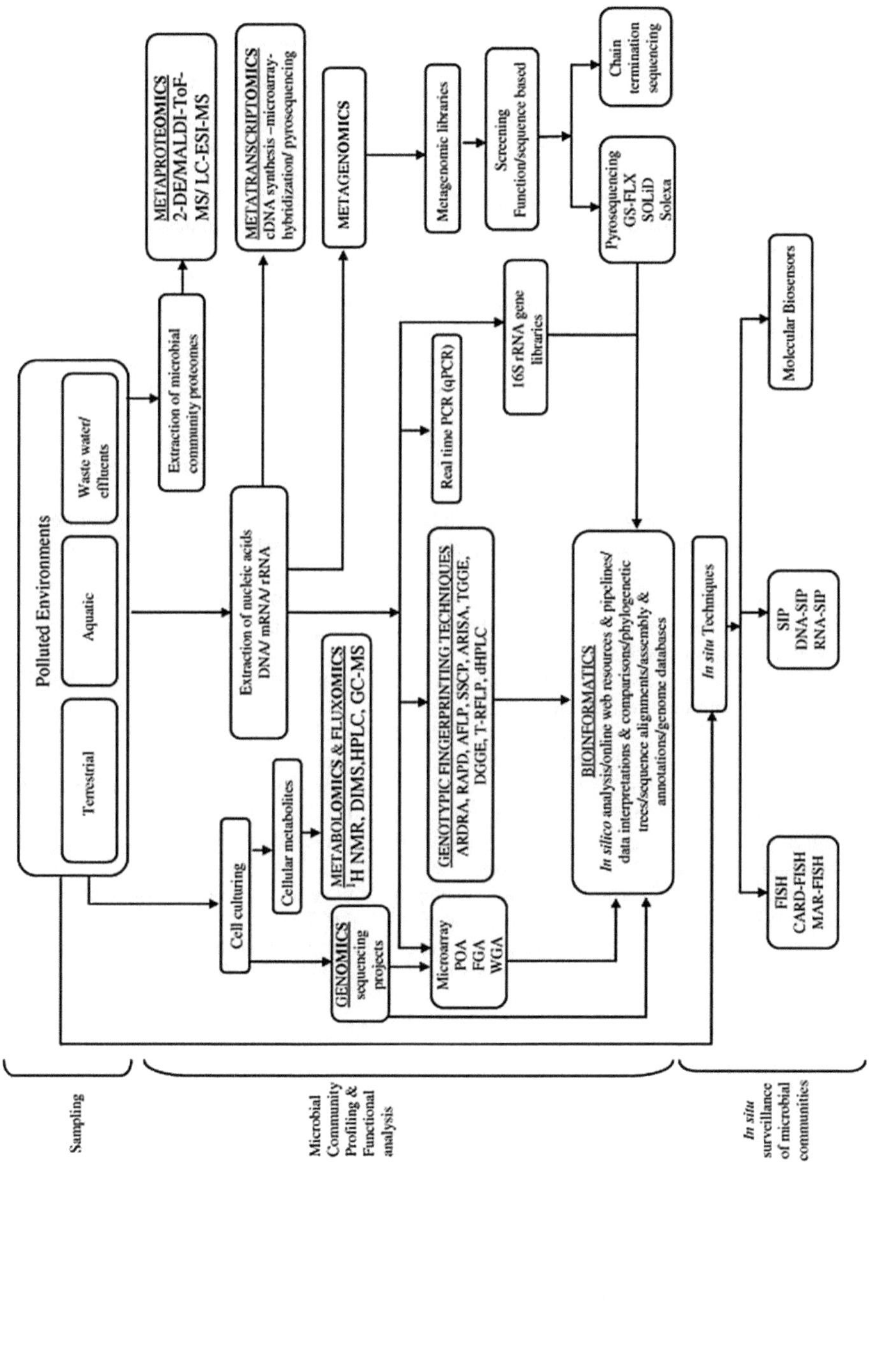


**Fig. 3.6** An overview of molecular and omics technologies employed to survey intrinsic microbial communities underlying bioremediation at contaminated sites. (Figure and legend were taken from Desai et al. 2010)

## References

Abram F, Enright AM, O'Reilly J, Botting CH, Colling G, O'Flaherty V (2011) A metaproteomic approach gives functional insights into anaerobic digestion. J Appl Microbiol 110:1550–1560

Aydin S, Karaçay HA, Shahi A, Gökçe S, Ince B, Ince O (2017) Aerobic and anaerobic fungal metabolism and Omics insights for increasing polycyclic aromatic hydrocarbons biodegradation. Fungal Biol Rev 31:61–72

Batista-García RA, Sánchez-Carbente MR, Talia P, Jackson SA, O'Leary ND, Dobson ADW, Folch-Mallol JL (2016) From lignocellulosic metagenomes to lignocellulolytic genes: trends, challenges and future prospects. Biofuels, Bioproducts and Biorefining 10 (6):864-882

Benndorf D, Balcke GU, Harms H, von Bergen M (2007) Functional metaproteome analysis of protein extracts from contaminated soil and groundwater. ISME J 1:224–234

Chauhan A, Jain RK (2010) Biodegradation: gaining insight through proteomics. Biodegradation 21:861–879

Desai C, Pathak H, Madamwar D (2010) Advances in molecular and "-omics" technologies to gauge microbial communities and bioremediation at xenobiotic/anthropogen contaminated sites. Bioresour Technol 101:1558–1569

El Amrani A, Dumas AS, Wick LY, Yergeau E, Berthomé R (2015) "Omics" insights into PAH degradation toward improved green remediation biotechnologies. Environ Sci Technol 49:11281–11291

Handelsmanl J, Rondon MR, Goodman RM, Brady SF, Clardy J (1998) Molecular biological access to the chemistry of unknown soil microbes: a new frontier for natural products. Chem Biol 5:245–R249

Hanreich A, Heyer R, Benndorf D, Rapp E, Pioch M, Reichl U, Klocke M (2012) Metaproteome analysis to determine the metabolically active part of a thermophilic microbial community producing biogas from agricultural biomass. Can J Microbiol 58:917–922

Hernández-López EL, Ramírez-Puebla ST, Vazquez-Duhalt R (2015) Microarray analysis of *Neosartorya fischeri* using different carbon sources, petroleum asphaltenes and glucose-peptone. Genomics Data 5:235–237

Kan J, Hanson TE, Ginter JM, Wang K, Chen F (2005) Metaproteomic analysis of Chesapeake Bay microbial communities. Saline Syst 1:7–21

Keller M, Hettich R (2009) Environmental proteomics: a paradigm shift in characterizing microbial activities at the molecular level. Microbiol Mol Biol Rev 73:62–70

Kim SJ, Kweon O, Cerniglia CE (2011) Proteomic applications to elucidate bacterial aromatic hydrocarbon metabolic pathways. Compr Biotechnol 6:105–114

Kuhn R, Benndorf D, Rapp E, Reichl U, Palese LL, Pollice A (2011) Metaproteome analysis of sewage sludge from membrane bioreactors. Proteomics 11:2738–2744

Lacerda CMR, Choe LH, Reardon KF (2007) Metaproteomic analysis of a bacterial community response to cadmium exposure. J Proteome Res 6:1145–1152

Lin YW, Tuan NN, Huang SL (2016) Metaproteomic analysis of the microbial community present in a thermophilic swine manure digester to allow functional characterization: a case study. Int Biodeterior Biodegrad 115:64–73

Liu D, Li M, Xi B, Zhao Y, Wei Z, Song C, Zhu C (2015) Metaproteomics reveals major microbial players and their biodegradation functions in a large-scale aerobic composting plant. Microb Biotechnol 8:950–960

Liu S, Guo C, Dang Z, Liang X (2017) Comparative proteomics reveal the mechanism of Tween80 enhanced phenanthrene biodegradation by *Sphingomonas* sp. GY2B. Ecotoxicol Environ Saf 137:256–264

Loh KC, Cao B (2008) Paradigm in biodegradation using *Pseudomonas putida*-a review of proteomics studies. Enzym Microb Technol 43:1–12

Lucero Camacho-Morales R, García-Fontana C, Fernández-Irigoyen J, Santamaría E, González-López J, Manzanera M, Aranda E (2018) Anthracene drives sub-cellular proteome-wide alterations in the degradative system of *Penicillium oxalicum*. Ecotoxicol Environ Saf 159:127–135

Morel M, Meux E, Mathieu Y, Thuillier A, Chibani K, Harvengt L, Jacquot JP, Gelhaye E (2013) Xenomic networks variability and adaptation traits in wood decaying fungi. Microb Biotechnol 6:248–263

Nzila A, Ramirez CO, Musa MM, Sankara S, Basheer C, Li QX (2018) Pyrene biodegradation and proteomic analysis in *Achromobacter xylosoxidans,* PY4 strain. Int Biodeterior Biodegrad 130:40–47

Oliveira JS, Araújo WJ, Figueiredo RM, Silva-Portela RB, Guerra ADB, Carla S, Minnicelli C, Carlos AC, Tereza A, Vasconcelos RD, Freitas AT, Agnez-Lima L (2017) Biogeographical distribution analysis of hydrocarbon degrading and biosurfactant producing genes suggests that near- equatorial biomes have higher abundance of genes with potential for bioremediation. BMC Microbiol 17:1–10

Osman C, Wilmes C, Tatsuta T, Langer T (2007) Prohibitins interact genetically with Atp23, a novel processing peptidase and chaperone for the F1Fo-ATP synthase. Mol Biol Cell 18:627–635

Pulleman M, Creamer R, Hamer U, Helder J, Pelosi C, Peres G, Rutgers M (2012) Soil biodiversity, biological indicators and soil ecosystem services – an overview of European approaches. Curr Opin Environ Sustain 4:529–538

Santos PM, Benndorf D, Sá-Correia I (2004) Insights into *Pseudomonas putida* KT2440 response to phenol-induced stress by quantitative proteomics. Proteomics 4:2640–2652

Shahi A, Aydin S, Ince B, Ince O (2016a) Evaluation of microbial population and functional genes during the bioremediation of petroleum-contaminated soil as an effective monitoring approach. Ecotoxicol Environ Saf 125:153–160

Shahi A, Aydin S, Ince B, Ince O (2016b) Reconstruction of bacterial community structure and variation for enhanced petroleum hydrocarbons degradation through biostimulation of oil contaminated soil. Chem Eng J 306:60–66

Solazzo C, Dyer JM, Clerens S, Plowman J, Peacock EE, Collins MJ (2013) Proteomic evaluation of the biodegradation of wool fabrics in experimental burials. Int Biodeterior Biodegrad 80:48–59

Szewczyk R, Soboń A, Sylwia R, Dzitko K, Waidelich D, Długoński J (2014) Intracellular proteome expression during 4-*n*-nonylphenol biodegradation by the filamentous fungus Metarhizium robertsii. Int Biodeterior Biodegrad 93:44–53

Szewczyk R, Soboń A, Słaba M, Długoński J (2015) Mechanism study of alachlor biodegradation by Paecilomyces marquandii with proteomic and metabolomic methods. J Hazard Mater 291:52–64

Vandera E, Samiotaki M, Parapouli M, Panayotou G, Koukkou AI (2015) Comparative proteomic analysis of Arthrobacter phenanthrenivorans Sphe3 on phenanthrene, phthalate and glucose. J Proteomics 113:73–89

Verdin A, Lounès-Hadj Sahraoui A, Newsam R, Robinson G, Durand R (2005) Polycyclic aromatic hydrocarbons storage by *Fusarium solani* in intracellular lipid vesicles. Environ Pollut 133:283–291

Wang HB, Zhang ZX, Li H, He HB, Fang CX, Zhang AJ, Li QS, Chen RS, Guo XK, Lin HF, Wu LK, Lin S, Chen T, Lin RY, Peng XX, Lin WX (2011) Characterization of metaproteomics in crop rhizospheric soil. J Proteome Res 10:932–940

Wilmes P, Wexler M, Bond PL (2008) Metaproteomics provides functional insight into activated sludge wastewater treatment. PLoS One 3:e1778

# Chapter 4
# Potential for CRISPR Genetic Engineering to Increase Xenobiotic Degradation Capacities in Model Fungi

**Haley P. Stein, Rafael Navajas-Pérez, and Elisabet Aranda**

## Contents



## 4.1 Xenobiotic Compounds

The Industrial Revolution marked a major turning point in human history, producing changes in economy, politics, society, and especially the environment. Since the 1980s, there has been a shift in attention to recognize the increasing presence of stable and often toxic, anthropogenic compounds. These synthetically derived compounds are called "xenobiotics" and are chemically distinct from molecules found in nature derived from biological and abiotic process.

The classification of xenobiotic compounds can be made according to (i) the chemical structure (amine, aromatic, halogenated hydrocarbon; metals; etc.), (ii) the industrial origin (agricultural chemistry, textile processing industry, leather


H. P. Stein
Institute of Water Research, University of Granada, Granada, Spain

R. Navajas-Pérez
Departamento de Genética, Facultad de Ciencias, Universidad de Granada, Granada, Spain

E. Aranda (✉)
Department of Microbiology, Faculty of Pharmacy, and Institute of Water Research, University of Granada, Granada, Spain
e-mail: earanda@ugr.es


 
R. Prasad, E. Aranda (eds.), *Approaches in Bioremediation*, Nanotechnology in the Life Sciences, https://doi.org/10.1007/978-3-030-02369-0_4

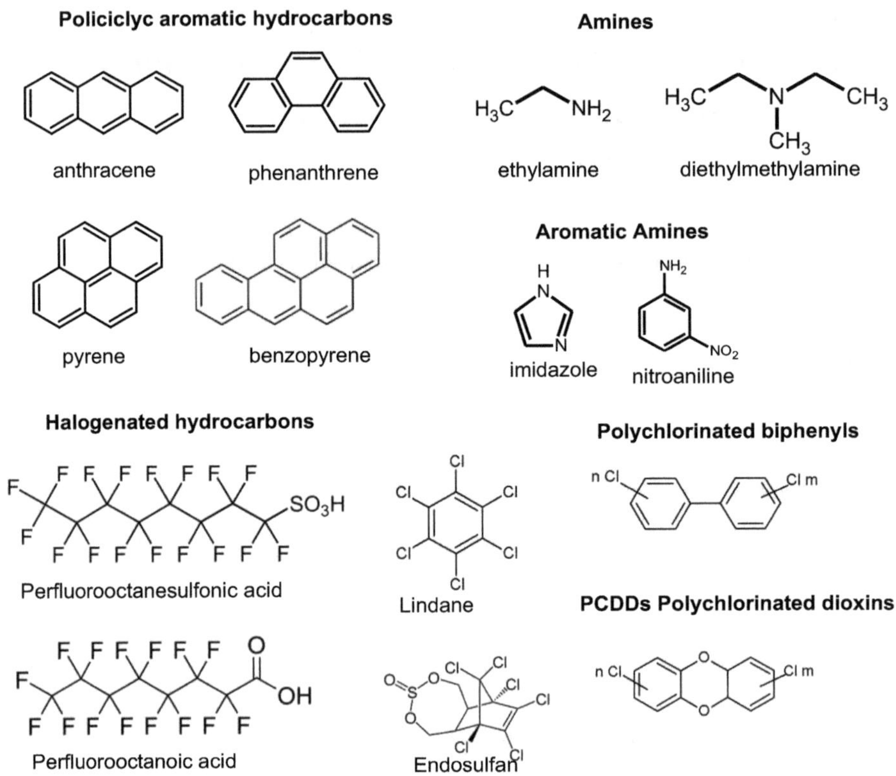


**Fig. 4.1** Some of the most representative groups of xenobiotic compounds found in nature, according to their chemical structure. Polycyclic aromatic hydrocarbons, amines, aromatic amines, halogenated hydrocarbon polychlorinated biphenyl, polychlorinated dioxin

industry, personal domestic care, pulp and paper industry, etc.), (iii) the function/use (absorbents, adhesives, coloring agents, cosmetics, flame retardants, fuel additives, pharmaceutical active compounds, etc.), or (iv) the regulation (priority or emerging pollutants) (Iovdijová and Bencko 2010) (Fig. 4.1).

Xenobiotic compounds are often found in sources of chemical waste, for example, excessive use of chemical fertilizers in agriculture contributes to refuse pollution. In addition, surface runoff, transport emission, heating, urban wastes, and natural disaster increase the presence of these substances in the environment. Industry is the main source responsible for large-scale contamination. The fate of these compounds is determined by the metabolism of the microorganism in addition to abiotic processes, such as photooxidation.

Though microorganisms tend to funnel xenobiotic compounds into the natural metabolic pathways based on certain structures (aromatic) and substituents (halogen, nitro groups, etc.), xenobiotics are less susceptible to degradation, and consequently they tend to persist in the environment (Knackmuss 1996). The accumulation of xenobiotic compounds in the environment is problematic, specifically as it relates to bioaccumulation in living organisms.

Bioaccumulation in living organisms can upset every level of the biosphere, from changes in the diversity of microbial communities to endocrine disruption in fish. In addition, bioaccumulation is not only problematic for the natural environment. Through consumption of food products and freshwater from contaminated sites, long-term problems in human health result (Baun et al. 2004).

Public concern about the widespread presence and the effect of these substances in the environment was first incited by the 1962 publication of Rachel Carson's *Silent Spring*.

The first experiments of toxic mineralization began in the 1950s, and until today, researchers continue searching for solutions to reduce health hazards and pollution within the discipline of environmental biotechnology (Wittich and González 2016; Connell 2018).

## 4.2 Environmental Biotechnology Using Fungi

Bioremediation exploits organisms in contaminated settings to transform, neutralize, or remove toxic compounds. Species of bacteria, archaea, fungi, nematodes, plants, and even insects have demonstrated the ability to remediate contamination. Processes such as immobilization or adsorption are considered useful tools for bioremediation yet may still result in accumulation of the compounds in question. The most desirable effect in bioremediation events is the avoidance of the accumulation of xenobiotic compounds in the environment. Biomineralization takes place through catabolic reactions and the use of this energy for anabolisms to break down the toxic xenobiotics. In some cases, biomineralization can occur using only the contaminant. In other cases, the contaminant is not available for use as the sole source of carbon and energy. In these cases, the pollutant can still be transformed by the induction of specific enzymes in response to the presence of other utilizable compounds in a process known as cometabolism. Both bioremediation processes occur in nature and in engineered systems alike, though it remains unclear whether pathways are predominantly metabolic or cometabolic (Tran et al. 2013).

For bacteria, which can more quickly adapt to the presence of xenobiotics in the environment, xenobiotics can be more readily applied as source of carbon and energy in catabolic pathways. This adaptation is generally attributed to spontaneously arising mutants that allow the use of these compounds as a carbon and/or energy source. Genetic sequences of benefit can potentially circulate through bacterial communities through donation and further horizontal gene transference mechanisms in the form of genomic island, islets, and profages (Perna et al. 2001). Horizontal gene transfer has, for example, allowed for the insertion of the TN4371 biphenyl transposon from *R. oxalatica*, which circulated genes for biphenyl degradation (Toussaint et al. 2003).

In fungi, these kinds of adaptive events are more complex, due to the eukaryotic cell type characterized by the presence of a true membrane-bound nucleus and the compartmentalization functions. DNA is blocked in chromosomes. Beneficial mutations can be transferred to the progeny when mutations are produced within the cell. In addition, if a mutation partially affects an intron region, this can be removed during RNA splicing during maturation of the RNA and may result in the loss of the mutation in the next generation.

Regardless, fungi have special features that make them suitable microorganisms for bioremediation processes (reviewed by Harms et al. 2011). They can spatially grow through several hectares by forming hyphae, do not require continuous water phases for dispersion, and allow the transport and growth of other microorganism such as bacteria (reviewed by Harms et al. 2011). Fungi can attenuate pollutant concentrations by physically adsorbing different contaminants through the presence of a thick cell wall composed of polymers such as cellulose and chitin. On the other hand, fungi are heterotrophic microorganisms that produce a set of enzymes involved in the decomposition of the organic matter for further mineralization to $CO_2$ and $H_2O$, thereby contributing to global carbon cycling. Enzymes that degrade, modify, or create glycosidic bonds are classified according to the carbohydrate-active enzyme (CAZYme; http://www.cazy.org/) database. Among these enzymes, auxiliary enzymes, including the lignin-modifying enzymes (LME), have been well studied in the transformation of xenobiotic compounds (Jiang et al. 2014). LMEs can transform aromatic substrates in cometabolic processes, due to the characteristically lenient substrate specificity. From that point, transformation of the target substances can result in less toxic or nontoxic compounds, which can then be further metabolized by the same fungi or by the microbial community. Lignin-modifying enzymes include laccases, tyrosinases, lignin peroxidases, manganese peroxidases, versatile peroxidases, *Coprinopsis cinerea* peroxidases, and others (Table 4.1). These enzymes are oxidoreductases; thus the capability to transform a specific compound depends on the redox potential of the enzymes as well as the ionization potential of the target compound.

Several studies concerning the fungal metabolism of xenobiotic compounds have shown the importance of the participation of the intracellular enzymatic system during aerobic xenobiotic transformation (Cerniglia 1997; Marco-Urrea et al. 2015; Aranda 2016; Olicón-Hernández et al. 2017). This intracellular system is comprised of a multifamily of enzymes widely distributed in the fungal kingdom and is involved in the detoxification systems. The xenobiotic degradation process involves the internalization of the compounds and the preliminary attack by cytochrome P450 (CYP) enzymes, followed by further transformations by epoxide hydrolases. This transformation produces hydroxylated metabolites that can lead to the ring cleave and mineralization through the β-ketoadipate pathway (Fuchs et al. 2011). In phase II of the fungal xenobiotic metabolism, miscellaneous transferase enzymes can include functional groups to the xenobiotic compounds, leading to the formation of more soluble compounds (Cerniglia and Sutherland 2010).

This intracellular enzymatic system is not well-known and may be more diverse than initially thought, as recent studies reveal the high diversity in the CYP family (Morel et al. 2013). In addition, the possibility of an alternative anaerobic pathway in fungi remains understudied. Thus, genetic engineering experiments with these fungal microorganisms can present an important tool for the future of the bioremediation.

**Table 4.1** Main fungal enzymes involved in the metabolism of xenobiotic

| Enzyme | Known gene examples | Regulation of fungal production | Bibliography |
|---|---|---|---|
| *Caldariomyces fumago* haem chloroperoxidase (CPO) | *Htp1– htp8 Caldariomyces fumago* | Production in complex liquid media | Kellner et al. (2016) |
| Cytochrome P450 monooxygenases | *CYP52X2*, *CYP53A1*, *CYP53A11*, *CYP53A24* | Differential transcriptional regulation in response to xenobiotic compounds and carbon sources | Huarte-Bonnet et al. (2018) |
| Dye decolorizing peroxidases (DyP) | *AauDyP (Auricularia auricula-judae)*, *MepDyP (Exidia glandulosa)* | Fungal production in phenolic complex liquid media | Liers et al. (2013) |
| Laccases | *laccl (pox4)*; *lacc2 (Lpoxa3) Pleurotus ostreatus* | Differential transcriptional regulation in response to Cu, Mn, Cd, Ag, various organic compounds, N, and C | Pezzella et al. (2013) |
| Lignin peroxidases (LiP) | *Pclip1 Phanerochaete chrysosporium* | | Holzbaur and Tien (1988) |
| Mn peroxidases (MnP) | Mnp1, mnp, mnp6, mnp7, mnp8 (*Pleurotus ostreatus*) | Differential transcriptional regulation in response to Mn | Knop et al. (2014) |
| Quinone reductases | qr1 (*Gloeophyllum trabeum*); qr (*Phanerochaete chrysosporium*) | Differential transcriptional regulation in response to heat shock and chemicals | Jensen et al. (2001) |
| Unspecific peroxygenase (UPO) | *UPO1 Agrocybe aegerita* | Fungal production in soya media | Pecyna et al. (2009)) |
| Versatile peroxidases (VP) | vpl2 (*Pleurotus eryngii*); vp (*Bjerkandera adusta*); mnp2, mnp4, mnp5 (*Pleurotus ostreatus*) | | Ruiz-Dueñas et al. (2008), Martínez (2002), and Knop et al. (2014) |

Adapted from Harms et al. (2011)

## 4.3 The Relevance of Gene Databases in Fungal Engineering

Recent advancements in biotechnology have made new opportunities for research in genetic engineering with respect to biodegradation. The idea of engineering a microbial metabolism to break down toxic compounds in soil and water is an appealing possibility for remediating heavily polluted areas. The application of fungi as tools in bioremediation can present many advantages over manual removal of toxic chemicals in terms of cost and time efficiency. Strains of fungi isolated from heavily contaminated soils have already proved to play a role in the degradation of xenobiotics such as polycyclic aromatic hydrocarbons (PAHs) and polychlorinated hydrocarbons and polychlorinated biphenyls (PCBs) (Potin et al. 2004; Mineki et al. 2015).

To exploit this element of fungal metabolism that results in PAH and PCB degradation requires a greater depth of understanding the genetics at play in the system. Identifying and targeting the genes which control xenobiotic metabolism in fungi can allow researchers to better attribute elements of the xenobiotic degradation mechanism to specific enzymatic activity. With this kind of control, it would be possible to fine-tune the process, upregulating the production of enzymes which have proven directly responsible for xenobiotic degradation.

Metabolic engineering in fungi has already been applied in the model filamentous fungi, *Aspergillus niger*. Generally, the organism is engineered with the intent of producing industrially relevant proteins or chemicals (Poulsen et al. 2005; Schmid et al. 2009; Kuivanen et al. 2015).

Sequenced and annotated genomes are immensely useful, if not required for more nuanced genetic and metabolic engineering (Albertsen et al. 2006; Ghosal et al. 2016). Specialized processes in fungal biodegradation tend to be less genetically conserved than other more basal functions of cellular metabolism (Wisecaver et al. 2014; Wakai et al. 2017). Pinpointing exact sequence data encoding for enzymatic elements of the fungal xenobiotic metabolism is not yet a possibility. Regardless, there is an increasing push for fully or partially sequenced fungal genomes, which may aid in the efforts of geneticists interested in fungal xenobiotic degradation mechanisms (Grigoriev et al. 2014). Fungal genetics are becoming increasingly available in the NCBI BLAST search base, as well as in specific gene libraries.

The Aspergillus Genome Database (AspGD.org) is one example of a library specific to the genus *Aspergillus* and focuses on the genetics of four popular species in academic and industrial research. The four species of fungi featured in the AspGD includes *A. niger*, *A. nidulans*, *A. oryzae*, and *A. fumigatus*. AspGD offers alignment tools, primer design, and links to relevant literature, among other features. The specificity of the AspGD is useful and convenient in its concentrated focus and may offer more carefully curated genomes. However, while the data bank is still available online, some aspects of the sequence data may be dated. The AspGD catalogue has been inactive since 2015; as such, users would be wise to cross-check information from the AspGD with other fungal genetic libraries (Descorps-Declère et al. 2008).

One such library of fungal genetics is the FungiDB (FungiDB.org), a subset of the EuPathDB, which focuses mainly on eukaryote pathogens. FungiDB has combined datasets from a variety of online sources to integrate nearly 100 fungal genomes on one platform. In addition to a gene catalogue, the FungiDB also hosts functional data (proteomics, transcriptomics) and offers tools for gene ontology searches (Basenko et al. 2018). Ontology is further emphasized by the Comparative Fungal Genomics Platform (CFGP). The CFGP offers a localized workspace with features for comparing and organizing genetic and functional data, facilitating trend analysis. For example, the proteomics tool dubbed InterProScan offers protein prediction from sequence data. Additionally, and perhaps more useful for gene inference and manipulation, the CFGP offers multiple sequence alignment by ClustalW, to locate sequence similarities across species. The CFGP library is linked to the NCBI BLAST search base, and its BLASTMatrix feature runs multiple, simultaneous BLAST searches from CFGP datasets. All data mining can be organized and saved to the workspace, making the CFGP a convenient package for layered studies in genomics and functional genetics (Park et al. 2007).

There has been a significant push for data in fungal genetics of late. In efforts with the US Department of Energy Joint Genome Institute, the MycoCosm fungal genetics portal launched a scheme to sequence the genomes of 1000 species of fungi. To help close gaps in knowledge of fungal genetics, users could input their own data or nominate species for genome sequencing (https://genome.jgi.doe.gov/programs/fungi/1000fungalgenomes.jsf) (Grigoriev et al. 2014). This push for sequenced fungal genomes proves fungal genetics to be an emerging and important field with respect to bioengineering. The massive data input in this pursuit may still present challenges in the parsing of these genomes. Most sequenced fungal genes, even in popular model species, remain uncharacterized. While it is relatively straightforward to directly compare DNA and RNA sequence reads, for the information to be more readily accessible, researchers must prioritize processing sequence data to identify protein-coding genes, promotor regions, intron/exon boundaries, etc. through vigorous annotation (Descorps-Declère et al. 2008; Cerqueira et al. 2013; Schmidt-Dannert 2014; de Vries et al. 2017).

## 4.4 Earlier Biotechnologies for Gene Manipulation

Sequenced and annotated genomic data from native fungi with capacity for biodegradation is imperative to the study of microbial metabolism for the goal of understanding and designing species with a competitive advantage for xenobiotic degradation in situ, particularly when the method of gene engineering is highly specific (Ghosal et al. 2016; Thion et al. 2012). Gene manipulation can generally be classified into four different categories, from lowest to highest specificity: mutagenesis (random or site-directed), recombinase-based editing, posttranscriptional gene silencing by RNA interference (RNAi), and endonuclease-based editing.

Traditionally chemical and physical mutagens have been used in fungi to induce base substitutions, UV light and monofunctional alkylating agents (4-nitroquinoline 1-oxide, 4NQO; N-methyl-N′-nitro-N-nitrosoguanidine, MNNG; or ethyl methanesulfonate, EMS) or deletions, ionizing radiation, and bifunctional alkylating agents (diepoxybutane, DEB, or diepoxyoctane, DEO) (reviewed in Talbot 2001).

Ectopic integration of exogenous DNA is frequent in fungi and has proven useful to control the expression of genes for large-scale industrial production processes. In addition, this methodology has allowed the identification of gene function and drug discovery (Li et al. 2017). Since the 1980s, ectopic DNA integration has been used to induce random or quasi-random insertional mutagenesis based on (1) *Agrobacterium tumefaciens*, as in *Aspergillus* (Wang et al. 2014), *Colletotrichum* (Cai et al. 2013) and *Verticillium* (Santhanam 2012), (2) restriction enzymes (restriction enzyme-mediated integration, REMI mutagenesis) (Riggle and Kumamoto 1998; Wang et al. 2007), (3) electroporation (Chakraborty 2015), and also, but less common, (4) particle bombardment (Herzog et al. 1996; Barcellos et al. 1998), or (5) transposon-based DNA delivery (Dufresne and Daboussi 2010).

More precise gene manipulation, gene editing, involves applied biotechnology techniques used to modify the genomes of target organisms by knocking out or replacing specific genes. Gene-editing events are carried out by means of insertions, deletions, or substitutions of DNA sequences and have shown to produce modified organisms with more desirable traits for more than two decades. Unlike transgenesis, gene editing allows changes in the DNA sequence without adding exogenous genes from other organisms. Gene editing makes genetic modification a faster, cheaper, and more accurate method than classical genetics approach (i.e., artificial selection, breeding rare individuals that have desirable phenotypes caused by mutations).

Site-directed mutagenesis and recombinase-mediated gene editing take advantage of the natural ability of the cell to both integrate exogenous DNA and to apply the endogenous recombination system by DNA repair mechanisms. Site-directed mutagenesis uses PCR amplification methods and homologous recombination (HR) to replace/modify the sequence of a specific gene with a copy containing the desired mutation (Stuckey and Storici 2013). In recombinase-mediated editing, site-specific recombinases (i.e., Cre/loxP and Flp/FRT) can knockout or knock-in genes by recognizing 30–40 nucleotide sequences, inducing DNA exchange and enhancing the frequency of HR (Aguiar et al. 2014). In most fungal systems, site-specific recombination frequency is low (<1%) posing a problem for the production of knockout strains (Nakayashiki and Nguyen 2008; Stuckey and Storici 2013).

As an alternative to recombination-based methods, posttranscriptional gene silencing by RNA interference (RNAi) is being successfully used for the manipulation of fungal gene expression (Salame et al. 2011). The expression of one or several genes can be reduced or completely silenced (transcript level can be close to zero in some transformants (Kück and Hoff 2010)) inserting double-stranded RNAs that trigger the degradation of mRNAs. This is particularly useful when the target sequence belongs to a multi-copy gene family or the deletion of the gene is lethal.

The recent development of cost-effective methods for high-throughput DNA sequencing and genome annotation has triggered the rapid development of gene-editing tools based on the action of endonucleases. The capacity of endonucleases to produce double-stranded breaks (DSBs) and break repairs in vivo through DNA damage response pathways has been taken as an opportunity for genetic engineers to edit genomes more precisely by inserting changes at targeted loci. Breaks repaired by nonhomologous end joining (NHEJ) often yield insertions or deletions which can then cause frameshift mutations to knockout gene function. By contrast, breaks repaired by homology-directed repair (HDR) result in a precisely targeted sequence replacement. As such, a template DNA needs to be provided for HDR. These mechanisms for repair include not only single nucleotide substitutions but also single or multiple transgene insertion.

Endonuclease-based editing generally consists of a bipartite system integrated by a customizable DNA-binding domain to recognize virtually any nucleotide sequence in addition to an endonuclease, which uses the recognition site to specifically produce DSBs. Programmable, site-specific nucleases, the so-called molecular scissors, were chosen as Method of the Year in 2011 for being the most promising tool "to knock out or knock in genes, to make allelic mutants, to change gene-regulatory control and to add reporters or epitope tags, all in the endogenous genomic context" (Nature Methods 2012). There are four main classes of endonuclease-mediated tools used for genome editing: meganucleases (MNs), zinc finger nucleases (ZFNs), transcription activator-like effector nucleases (TALENs), and clustered regularly interspaced short palindromic repeats and associated proteins (CRISPR/Cas) (Abdallah et al. 2015).

MNs or homing nucleases were discovered in the 1990s. Since then, hundreds of naturally occurring meganucleases have been described in many different organisms. Although their function is still unclear, MNs are known to recognize and cleave 12–40 bp DNA sequences (most likely unique or nearly so). I-SceI and I-CreI are the most commonly used MNs for genome engineering (Muñoz et al. 2011; Silva et al. 2011). Engineering MNs to modify their recognition site, when possible, is costly and time-consuming compared to other endonuclease-mediated systems.

Synthetic zinc finger proteins have been engineered for nearly two decades, combining different zinc finger domains able to recognize and bind 9–20 bp sequences. Each zinc finger consists of approximately 30 amino acids that can recognize 3 bp motifs. Pairs of zinc finger domains are generally designed and combined to dimeric nuclease FokI to specifically produce DSBs (Davies et al. 2017).

More recently, TALEN, a system of naturally occurring DNA binding proteins derived from transcription activator-like effectors from the plant pathogen *Xanthomonas* sp., has been applied as a technique for creating targeted genetic modifications. TALEN domains are built by tandem conserved repeat modules of 34 amino acids. Each amino acid recognizes a single base pair, but the ones located at positions 12 and 13 are highly variable (repeat variable di-residue, RVD) and crucial for the recognition of a specific nucleotide. Also, gene targeting requires that binding sites have a thymine base in the 5′ end. TALEN domains can be designed to recognize 33–35 bp sequences (reviewed in Abdallah et al. 2015). TALENs form dimers separated by 10–30 bp regions where dimeric FokI can produce DSBs (Guha and Edgell 2017). In comparison with ZFNs, TALEN engineering is simpler, presents higher rates of cleavage activity, and, apart from nucleases, can be combined to numerous effector domains as site-specific recombinases and transcriptional activators (Joung and Sander 2013; Gaj et al. 2013). However, some authors have suggested that hybrid ZFN and TALEN nucleases can increase specificity in multiple cell types (Yan et al. 2013).

By far, the RNA-guided nuclease system of CRISPR/Cas is the method that has shaken the field of gene editing. In 1987 researchers characterized a prokaryotic system for viral defense in *Escherichia coli*, later named for the idiosyncratic region of DNA characterized by clustered regularly interspaced short palindromic repeats

(CRISPR). When first described, it was difficult to foresee the impact CRISPR has in modern genetics (Ishino et al. 1987). The system, with apparent role in DNA repair or gene regulation, was described as a series of short repeats, split up by spacers of similar size (Makarova et al. 2002). Very soon, it was found that spacers are sequences of foreign DNA from phages or plasmids (Mojica et al. 2000). Up to 10% of bacteria and archaea (Burstein et al. 2016) use these spacer sequences to match and recognize invading viral DNA, which is then inactivated by a double-stranded break produced by the CRISPR-associated (Cas) proteins, enzymes with putative nuclease, and helicase domains (Haft et al. 2005). As such, CRISPR is a form of adaptive viral immunity for microorganisms which regularly take in foreign sequences of DNA and RNA (Mojica et al. 2005; Barrangou et al. 2007; Marraffini 2015).

## 4.5 An Introduction to CRISPR/Cas Theory and Methodology

While conventional gene induction techniques have been successful in fungi, the process remains time- and labor-intensive and is considered to be generally inefficient as a means of genetic engineering (Wang et al. 2015). As indicated in the previous section, recent developments in biotechnology introduced a fine-tuned methodology for accessible genetic engineering by the exploitation of a unique element found in bacterial viral defense, CRISPR (Jinek et al. 2012; Doudna and Charpentier 2014). Now widely available is the CRISPR-associated enzyme, Cas9, which has been co-opted as a specific DNA-cutting machine. Cas9 is an RNA-guided endonuclease essential to the microbial adaptive immune system and is presently utilized as a high-precision gene engineering tool in an increasing range of organisms (Jinek et al. 2012; Ran et al. 2013; https://www.broadinstitute.org/what-broad/areas-focus/project-spotlight/crispr-timeline).

There are three types of Cas proteins (I–III) which can be further subdivided into ten subtypes (Makarova et al. 2011). While type I and III systems are complex and need the intervention of multiple Cas proteins to cleave the targeted DNA, type II system (described in *Streptococcus pyogenes*) uses only Cas9 protein (Haft et al. 2005; Makarova et al. 2011, 2013; Chylinski et al. 2013). CRISPR-Cas9 works generally in three steps (reviewed in Karvelis et al. 2013); a first contact with the foreign DNA produces (1) immunization: spacers are acquired and included in CRISPR locus. The presence of recurring viral DNA into the cell induces (2) expression; CRISPR is transcribed into crRNAs containing an individual spacer followed by a process of maturation mediated by a trans-activating CRISPR RNA (tracrRNA), and (3) interference: Cas protein binds crRNA and produces a DSB break at a site 3 base pairs from the protospacer adjacent motif (PAM) site, a two to six base pair DNA sequence immediately downstream the targeted DNA sequence (Shah et al. 2013) (Fig. 4.2).

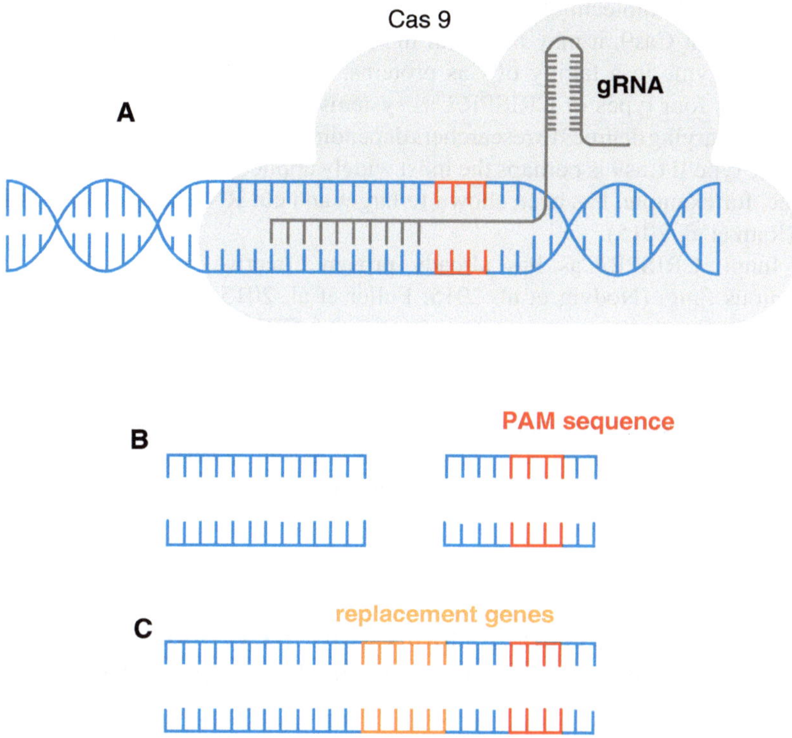


**Fig. 4.2** General representation of genome editing CRISPR/Cas9. (**a**) Target DNA to be edited and guide RNA binds. Cas9 enzyme binds and creates double-stranded break in DNA. (**b**) Errors introduced during repairs of gene disrupter. (**c**) Template used during repair and correct sequence restored

For a general CRISPR/Cas9 gene modification, researchers have tuned the Cas9 nuclease to interact with specific sites of host DNA that correspond to a guide RNA. The guide RNA (gRNA) can be essentially any 20-nucleotide sequence flanked by a protospacer adjacent motif (PAM) site, a string of NGG that acts as a signal for Cas9 to make a cut in the gene of interest. When Cas9 binds to the gRNA, the enzyme is then able to traverse the nuclear membrane, localizing to the target site on the genomic DNA which corresponds to the guide. Cas9 then cleaves the DNA at the target site, and the rapid repair mechanism which follows is innately error prone. As such, what results is a precise mutation in the target gene (Rath et al. 2015).

The Cas9 enzyme can also be modified for more complicated engineering purposes. For example, Cas enzymes can be constructed to include fused activators or repressors to respectively up- or downregulate the target gene (Ran et al. 2013; Ceasar et al. 2016; Graham and Root 2015). The efficacy of these modified Cas enzymes and the relatively abundant presence of PAM sites in the genome have made the CRISPR/Cas system something of a watershed for the potential applicability of biotechnology and genetic engineering. In considering the effectiveness of Cas9, it may be useful in some cases to bear in mind that Cas9 is only one enzyme in a family of Cas proteins. The Cas family is a diverse group which offers four types of CRISPR/Cas systems featuring endonucleases that can be useful in varying degrees to researchers depending on the intent of experimentation. While the type II Cas9 is perhaps the most widely applied variety, the type III Cas10 enzyme, for example, has been shown to target and cut RNA sites as well as DNA sites (Rath et al. 2015).

In fungi, CRISPR/Cas has already proven effective in several species of filamentous fungi (Nødvig et al. 2015; Fuller et al. 2015; Shi et al. 2017; Zheng et al. 2017). Generally, experiments in fungi using CRISPR/Cas systems have been employed to efficiently synthesize chemical products for industry (Kuivanen et al. 2016; Pohl et al. 2016). Though catabolic gene systems in fungi have been of relevance to research in applied bioengineering for some time, in the case of CRISPR/Cas modifications, RNA sequencing events must first be performed due to specificity of the editing system (Fowler and Berka 1991; Descorps-Declère et al. 2008; Kuivanen et al. 2016).

## 4.6 Potential Experiments and Future Directions for Manipulating Fungal Xenobiotic Metabolism

While many of the CRISPR experiments in fungi have been shown to disrupt genes by insertion/deletion and point mutations, there are several possibilities for experiments to apply a degree of metabolic engineering in the degradation of xenobiotics. For example, using an inactive, enzymatically "dead" Cas nuclease, dCas9, researchers may specifically target genes for visualizations or, as mentioned earlier, regulation through attached biomolecules (Qi et al. 2013).

Through annotated genomics and transcriptomics, researchers can identify not only specific genes responsible for encoding proteins involved in xenobiotic metabolism but also the promoter and enhancer regions. A dCas9 enzyme situated on a promotor region could theoretically inspire an uptick in production of PAH-degrading enzymes at the transcription region. An interesting experiment would be to track if an increase in the rate of, for example, certain cytochrome P450 enzymes, a family known to play a role in PAH degradation, results in faster xenobiotic metabolism (Syed et al. 2013). With so little known about fungal xenometabolism on a genomic scale, a considerable experiment would be a mass silencing of genes thought to be involved in degradation. Each gene pinpointed and silenced, through a static, inactivated dCas9, may produce different insights into the mechanistic aspects and genetic elements of PAH degradation. Silencing by inactivation can be more effective considering mortality of the subject, as edited genes may still be transcribed and translated into dysfunctional protein products, causing potential damage to the cell (Watford and Warrington 2017).

To design modified fungi for in situ xenobiotic bioremediation will depend greatly on such preliminary experiments as mentioned above. Viable mutated fungi have shown a capacity to survive laboratory settings, yet the promise of potential in bioremediation through fungi rests on the ability of modified organisms to compete in and detoxify the chemistry of contaminated sites beyond the realm of the controlled experimental backdrop.

## 4.7 Conclusion

The production of multifunctional fungal strains able to efficiently degrade xenobiotic compounds can prove fundamental in efforts to rehabilitate contaminated lands. Genetic engineering experiments to specifically upregulate relevant degradative enzymes can offer an avenue of eliminating contamination in polluted areas. There is fundamental groundwork in basic research that must be executed to apply such experiments in the natural environment. Primarily, available and reliable sequence data is the foundation of genetic engineering in fungi. Secondly, access to efficient gene modification techniques, such as the CRISPR/Cas9 system, shows promise for prospective exploration in fungal xenometabolics. The push for fungal genomes marks that investigations in fungal genetics are a growing field, and this basic data collection is crucial to furthering the potential for manipulation and subsequent augmentation of fungal degradation capacities in real-world applications.

As such, there remains a great deal of research and development in the fields of mycology and genetic engineering for these degradation events using modified fungi to take place. Studies concerning the impact of genetically modified microorganisms on microbial communities and the environment should not be overlooked in the process. Regardless, there is great potential for experiments in fungal genetics and genetic engineering as a means of bioremediation and for understand how to better regulate the potential use of microorganisms in natural ecosystems.

**Acknowledgment** HPS acknowledges the Fulbright Program (PS00247479) for the Open Study/Research Grant. RNP was funded by Ramón y Cajal program (RYC-2011-08653). EA gratefully thanks the Ministry of Economy and Competitiveness (MINECO) and FEDER funds for co-funding the Ramón y Cajal contract (RYC-2013-12481).

## References

Abdallah NA, Prakash CS, McHughen AG (2015) Genome editing for crop improvement: challenges and opportunities. GM Crops Food 6:183–205

Aguiar TQ, Dinis C, Domingues L (2014) Cre-loxP-based system for removal and reuse of selection markers in *Ashbya gossypii* targeted engineering. Fungal Genet Biol 68:1–8

Albertsen A, Ravnskov S, Green H, Jensen DF, Larsen J (2006) Interactions between the external mycelium of the mycorrhizal fungus *Glomus intraradices* and other soil microorganisms as affected by organic matter. Soil Biol Biochem 38:1008–1014

Aranda E (2016) Promising approaches towards biotransformation of polycyclic aromatic hydrocarbons with Ascomycota fungi. Curr Opin Biotechnol 38:1–8

Barcellos FG, Fungaro MH, Furlaneto MC, Lejeune B, Pizziranikleiner AA, de Azevedo JL (1998) Genetic analysis of *Aspergillus nidulans* unstable transformants obtained by the biolistic process. Can J Microbiol 44:1137–1141

Barrangou R, Fremaux C, Deveau H, Richards M, Boyaval P, Moineau S, Romero DA, Horvath P (2007) CRISPR provides acquired resistance against viruses in prokaryotes. Science 315:1709–1712

Basenko EY, Pulman JA, Shanmugasundram A, Harb OS, Crouch K, Starns D, Warrenfeltz S, Aurrecoechea C, Stoeckert CJ, Kissinger JC (2018) FungiDB: an integrated bioinformatic resource for fungi and oomycetes. J Fungi 4:39

Baun A, Ledin A, Reitzel LA, Bjerg PL, Christensen TH (2004) Xenobiotic organic compounds in leachates from ten Danish MSW landfills—chemical analysis and toxicity tests. Water Res 38:3845–3858

Burstein D, Sun CL, Brown CT, Sharon I, Anantharaman K, Probst AJ, Thomas BC, Banfield JF (2016) Major bacterial lineages are essentially devoid of CRISPR-Cas viral defence systems. Nat Commun 7:10613

Cai Z, Li G, Lin C, Shi T, Zhai L, Chen Y, Huang G (2013) Identifying pathogenicity genes in the rubber tree anthracnose fungus *Colletotrichum gloeosporioides* through random insertional mutagenesis. Microbiol Res 168(6):340–350

Ceasar SA, Rajan V, Prykhozhij SV, Berman JN, Ignacimuthu S (2016) Insert, remove or replace: a highly advanced genome editing system using CRISPR/Cas9. BBA-Mol Cel Res 1863:2333–2344

Cerniglia CE (1997) Fungal metabolism of polycyclic aromatic hydrocarbons: past, present and future applications in bioremediation 172. J Ind Microbiol Biotechnol 19:324–333

Cerniglia CE, Sutherland JB (2010) Degradation of polycyclic aromatic hydrocarbons by fungi. In: Timmis K (ed) Handbook of hydrocarbon and lipid microbiology. Springer, Berlin, pp 2079–2110

Cerqueira GC, Arnaud MB, Inglis DO, Skrzypek MS, Binkley G, Simison M, Miyasato SR, Binkley J, Orvis J, Shah P (2013) The *Aspergillus* genome database: multispecies curation and incorporation of RNA-Seq data to improve structural gene annotations. Nucleic Acids Res 42:D705–D710

Chakraborty BN (2015) Electroporation mediated DNA transformation of filamentous fungi. Genet Trans Syst Fungi 1:67–79

Chylinski K, Le Rhun A, Charpentier E. (2013) The tracrRNA and Cas9 families of type II CRISPR-Cas immunity systems. RNA Biology 10(5):726–737

Connell DW (2018) General characteristics of organic compounds which exhibit bioaccumulation. Bioaccumulation of xenobiotic compounds. CRC Press, Boca Raton, pp 47–58

Davies P, Kumar S, Sastry-Dent L (2017) Chapter three – use of zinc-finger nucleases for crop improvement. Prog Mol Biol Transl 149:47–63

de Vries RP, Riley R, Wiebenga A, Aguilar-Osorio G, Amillis S, Uchima CA, Anderluh G, Asadollahi M, Askin M, Barry K, Battaglia E, Bayram Ö, Benocci T, Braus-Stromeyer SA, Caldana C, Cánovas D, Cerqueira GC, Chen F, Chen W, Choi C, Clum A, dos Santos RAC, Damásio ARL, Diallinas G, Emri T, Fekete E, Flipphi M, Freyberg S, Gallo A, Gournas C, Habgood R, Hainaut M, Harispe ML, Henrissat B, Hildén KS, Hope R, Hossain A, Karabika E, Karaffa L, Karányi Z, Kraševec N, Kuo A, Kusch H, LaButti K, Lagendijk EL, Lapidus A, Levasseur A, Lindquist E, Lipzen A, Logrieco AF, MacCabe A, Mäkelä MR, Malavazi I, Melin P, Meyer V, Mielnichuk N, Miskei M, Molnár ÁP, Mulé G, Ngan CY, Orejas M, Orosz E, Ouedraogo JP, Overkamp KM, Park H-S, Perrone G, Piumi F, Punt PJ, Ram AFJ, Ramón A, Rauscher S, Record E, Riaño-Pachón DM, Robert V, Röhrig J, Ruller R, Salamov A,

Salih NS, Samson RA, Sándor E, Sanguinetti M, Schütze T, Sepčić K, Shelest E, Sherlock G, Sophianopoulou V, Squina FM, Sun H, Susca A, Todd RB, Tsang A, Unkles SE, van de Wiele N, van Rossen-Uffink D, Oliveira JVC, Vesth TC, Visser J, Yu J-H, Zhou M, Andersen MR, Archer DB, Baker SE, Benoit I, Brakhage AA, Braus GH, Fischer R, Frisvad JC, Goldman GH, Houbraken J, Oakley B, Pócsi I, Scazzocchio C, Seiboth B, vanKuyk PA, Wortman J, Dyer PS, Grigoriev IV (2017) Comparative genomics reveals high biological diversity and specific adaptations in the industrially and medically important fungal genus *Aspergillus*. Genome Biol 18:28

Descorps-Declère S, Barba M, Labedan B (2008) Matching curated genome databases: a non trivial task. BMC Genomics 9:501

Doudna JA, Charpentier E (2014) Genome editing. The new frontier of genome engineering with CRISPR-Cas9. Science 1258096:346

Dufresne M, Daboussi MJ (2010) Development of impala-based transposon systems for gene tagging in filamentous fungi. Methods Mol Biol 638:41–54

Fowler T, Berka RM (1991) Gene expression systems for filamentous fungi. Curr Opin Biotechnol 2:691–697

Fuchs G, Boll M, Heider J (2011) Microbial degradation of aromatic compounds – from one strategy to four. Nat Rev Microbiol 9:803–816

Fuller KK, Chen S, Loros JJ, Dunlap JC (2015) Development of the CRISPR/Cas9 system for targeted gene disruption in *Aspergillus fumigatus*. Eukaryot Cell: EC 14:00107–00115

Gaj T, Gersbach CA, Barbas CF (2013) ZFN, TALEN and CRISPR/Cas-based methods for genome engineering. Trends Biotechnol 31(7):397–405

Ghosal D, Ghosh S, Dutta TK, Ahn Y (2016) Current state of knowledge in microbial degradation of polycyclic aromatic hydrocarbons (PAHs): a review. Front Microbiol 7:1369

Graham DB, Root DE (2015) Resources for the design of CRISPR gene editing experiments. Genome Biol 16:260

Grigoriev IV, Nikitin R, Haridas S, Kuo A, Ohm R, Otillar R, Riley R, Salamov A, Zhao X, Korzeniewski F, Smirnova T, Nordberg H, Dubchak I, Shabalov I (2014) MycoCosm portal: gearing up for 1000 fungal genomes. Nucleic Acids Res 42:D699–D704

Guha TK, Edgell DR (2017) Applications of alternative nucleases in the age of CRISPR/Cas9. Int J Mol Sci 18:2565

Haft DH, Selengut J, Mongodin EF, Nelson KE (2005) A guild of 45 CRISPR-associated (Cas) protein families and multiple CRISPR/Cas subtypes exist in prokaryotic genomes. PLoS Comput Biol 1:e60

Harms H, Schlosser D, Wick LY (2011) Untapped potential: exploiting fungi in bioremediation of hazardous chemicals. Nat Rev Microbiol 9:177–192

Herzog RW, Daniell H, Singh NK, Lemke PA (1996) A comparative study on the transformation of *Aspergillus nidulans* by microprojectile bombardment of conidia and a more conventional procedure using protoplasts treated with polyethyleneglycol. Appl Microbiol Biotechnol 45:333–337

Holzbaur ELF, Tien M (1988) Structure and regulation of a lignin peroxidase gene from *Phanerochaete chrysosporium*. Biochem Biophys Res Commun 155:626–633

Huarte-Bonnet C, Kumar S, Saparrat MC, Girotti JR, Santana M, Hallsworth JE, Pedrini N (2018) Insights into hydrocarbon assimilation by Eurotialean and Hypocrealean Fungi: roles for CYP52 and CYP53 clans of cytochrome P450 genes. Appl Biochem Biotechnol 184:1047–1060

Iovdijová A, Bencko V (2010) Potential risk of exposure to selected xenobiotic residues and their fate in the food chain-part I. Classification of xenobiotics. Ann Agric Environ Med 17:183–192

Ishino Y, Shinagawa H, Makino K, Amemura M, Nakata A (1987) Nucleotide sequence of the iap gene, responsible for alkaline phosphatase isozyme conversion in *Escherichia coli*, and identification of the gene product. J Bacteriol 169:5429–5433

Jensen KA, Houtman CJ, Ryan ZC, Hammel KE (2001) Pathways for extracellular Fenton chemistry in the brown rot basidiomycete *Gloeophyllum trabeum*. Appl Environ Microbiol 67:2705–2711

Jiang Y, Yang Y, Zhang X (2014) Review on the biodegradation and conversion mechanisms of typical polycyclic aromatic hydrocarbons. Shiyou Xuebao, Shiyou Jiagong/Acta Petrolei Sin (Pet Process Sect) 30:1137–1150

Jinek M, Chylinski K, Fonfara I, Hauer M, Doudna JA, Charpentier E (2012) A programmable dual-RNA–guided DNA endonuclease in adaptive bacterial immunity. Science 337:816

Joung JK, Sander JD (2013) TALENs: a widely applicable technology for targeted genome editing. Nat Rev Mol Cell Biol 14(1):49–55

Karvelis T, Gasiunas G, Miksys A, Barrangou R, Horvath P, Siksnys V (2013) crRNA and tracrRNA guide Cas9-mediated DNA interference in *Streptococcus thermophilus*. RNA Biol 10(5):841–851

Kellner H, Pecyna MJ, Buchhaupt M, Ullrich R, Hofrichter M (2016) Draft genome sequence of the chloroperoxidase-producing fungus *Caldariomyces fumago Woronichin* DSM1256. Genome Announc 4:e00774

Knackmuss H-J (1996) Basic knowledge and perspectives of bioelimination of xenobiotic compounds. J Biotechnol 51:287–295

Knop D, Ben-Ari J, Salame TM, Levinson D, Yarden O, Hadar Y (2014) $Mn^{2+}$-deficiency reveals a key role for the Pleurotus ostreatus versatile peroxidase (VP4) in oxidation of aromatic compounds. Appl Microbiol Biotechnol 98:6795–6804

Kück U, Hoff B (2010) New tools for the genetic manipulation of filamentous fungi. Appl Microbiol Biotechnol 86:51–62

Kuivanen J, Penttilä M, Richard P (2015) Metabolic engineering of the fungal D-galacturonate pathway for L-ascorbic acid production. Microb Cell Factories 14:2

Kuivanen J, Wang Y-MJ, Richard P (2016) Engineering *Aspergillus niger* for galactaric acid production: elimination of galactaric acid catabolism by using RNA sequencing and CRISPR/Cas9. Microb Cell Factories 15:210

Li D, Tang Y, Lin J, Cai W (2017) Methods for genetic transformation of filamentous fungi. Microb Cell Factories 16:168

Liers C, Pecyna MJ, Kellner H, Worrich A, Zorn H, Steffen KT, Hofrichter M, Ullrich R (2013) Substrate oxidation by dye-decolorizing peroxidases (DyPs) from wood- and litter-degrading agaricomycetes compared to other fungal and plant heme-peroxidases. Appl Microbiol Biotechnol 97:5839–5849

Makarova KS, Aravind L, Grishin NV, Rogozin IB, Koonin EV (2002) A DNA repair system specific for thermophilic *Archaea* and bacteria predicted by genomic context analysis. Nucleic Acids Res 30:482–496

Makarova KS, Haft DH, Barrangou R, Brouns SJ, Charpentier E, Horvath P, Moineau S, Mojica FJM, Wolf YI, Yakunin AF, van der Oost J, Koonin EV (2011) Evolution and classification of the CRISPR-Cas systems. Nat Rev Microbiol 9:467–477

Makarova KS, Haft DH, Koonin EV (2013) CRISPR-Cas systems and cas protein families. In: Protein families: relating protein sequence, structure, and function. Wiley, New York, pp 341–381

Marco-Urrea E, García-Romera I, Aranda E (2015) Potential of non-ligninolytic fungi in bioremediation of chlorinated and polycyclic aromatic hydrocarbons. New Biotechnol 32:620–628

Marraffini LA (2015) CRISPR-Cas immunity in prokaryotes. Nature 526:55–61

Martínez AT (2002) Molecular biology and structure-function of lignin-degrading heme peroxidases. Enzym Microb Technol 30:425–444

Mineki S, Suzuki K, Iwata K, Nakajima D, Goto S (2015) Degradation of polyaromatic hydrocarbons by fungi isolated from soil in Japan. Polycycl Aromat Compd 35:120–128

Mojica JM, Díez-Villaseñor C, Soria E, Juez G (2000) Biological significance of a family of regularly spaced repeats in the genomes of archaea, bacteria and mitochondria. Mol Microbiol 36:244–246

Mojica JM, Díez-Villaseñor C, García-Martínez J, Soria E (2005) Intervening sequences of regularly spaced prokaryotic repeats derive from foreign genetic elements. J Mol Evol 60:174–182

Morel M, Meux E, Mathieu Y, Thuillier A, Chibani K, Harvengt L, Jacquot J-P, Gelhaye E (2013) Xenomic networks variability and adaptation traits in wood decaying fungi. Microb Biotechnol 6:248–263

Muñoz IG, Prieto J, Subramanian S, Coloma J, Redondo P, Villate M, Merino N, Marenchino M, D'Abramo M, Gervasio FL, Grizot S, Daboussi F, Smith J, Chion-Sotinel I, Pâques F, Duchateau P, Alibés A, Stricher F, Serrano L, Blanco FJ, Montoya G (2011) Molecular basis of engineered meganuclease targeting of the endogenous human RAG1 locus. Nucleic Acids Res 39(2):729–743

Nakayashiki H, Nguyen QB (2008) RNA interference: roles in fungal biology. Curr Opin Microbiol 11:494–502

Nature Methods (2012) Method of the year 2011. Nat Methods 9:1

Nødvig CS, Nielsen JB, Kogle ME, Mortensen UH (2015) A CRISPR-Cas9 system for genetic engineering of filamentous fungi. PLoS One 10:e0133085

Olicón-Hernández DR, González-López J, Aranda E (2017) Overview on the biochemical potential of filamentous fungi to degrade pharmaceutical compounds. Front Microbiol 8:1792

Park J, Park B, Jung K, Jang S, Yu K, Choi J, Kong S, Park J, Kim S, Kim H (2007) CFGP: a web-based, comparative fungal genomics platform. Nucleic Acids Res 36:D562–D571

Pecyna MJ, Ullrich R, Bittner B, Clemens A, Scheibner K, Schubert R, Hofrichter M (2009) Molecular characterization of aromatic peroxygenase from Agrocybe aegerita. Appl Microbiol Biotechnol 84:885–897

Perna NT, Plunkett G III, Burland V, Mau B, Glasner JD, Rose DJ, Mayhew GF, Evans PS, Gregor J, Kirkpatrick HA (2001) Genome sequence of enterohaemorrhagic *Escherichia coli* O157: H7. Nature 409:529

Pezzella C, Lettera V, Piscitelli A, Giardina P, Sannia G (2013) Transcriptional analysis of *Pleurotus ostreatus* laccase genes. Appl Microbiol Biotechnol 97:705–717

Pohl C, Kiel JA, Driessen AJ, Bovenberg RA, Nygård Y (2016) CRISPR/Cas9 based genome editing of *Penicillium chrysogenum*. ACS Synth Biol 5:754–764

Potin O, Rafin C, Veignie E (2004) Bioremediation of an aged polycyclic aromatic hydrocarbons (PAHs)-contaminated soil by filamentous fungi isolated from the soil. Int Biodeterior Biodegrad 54:45–52

Poulsen BR, Nøhr J, Douthwaite S, Hansen LV, Iversen JJ, Visser J, Ruijter GJ (2005) Increased NADPH concentration obtained by metabolic engineering of the pentose phosphate pathway in *Aspergillus niger*. FEBS J 272:1313–1325

Qi LS, Larson MH, Gilbert LA, Doudna JA, Weissman JS, Arkin AP, Lim WA (2013) Repurposing CRISPR as an RNA-guided platform for sequence-specific control of gene expression. Cell 152:1173–1183

Ran FA, Hsu PD, Wright J, Agarwala V, Scott DA, Zhang F (2013) Genome engineering using the CRISPR-Cas9 system. Nat Protoc 8:2281–2308

Rath D, Amlinger L, Rath A, Lundgren M (2015) The CRISPR-Cas immune system: biology, mechanisms and applications. Biochimie 117:119–128

Riggle PJ, Kumamoto CA (1998) Genetic analysis in fungi using restriction-enzyme-mediated integration. Curr Opin Microbiol 1(4):395–399

Ruiz-Dueñas FJ, Morales M, García E, Miki Y, Martínez MJ, Martínez AT (2008) Substrate oxidation sites in versatile peroxidase and other basidiomycete peroxidases. J Exp Bot 60:441–452

Salame TM, Ziv C, Hadar Y, Yarden O (2011) RNAi as a potential tool for biotechnological applications in fungi. Appl Microbiol Biotechnol 89(3):501–512

Santhanam P (2012) Random insertional mutagenesis in fungal genomes to identify virulence factors. Methods Mol Biol 835:509–517

Schmid J, Stahl U, Meyer V (2009) Genetic and metabolic engineering in filamentous fungi. In: Anke T, Weber D (eds) Physiology and genetics: selected basic and applied aspects. Springer, Berlin, Heidelberg, pp 377–392

Schmidt-Dannert C (2014) NextGen microbial natural products discovery. Microb Biotechnol 8:26–28
Shah SA, Erdmann S, Mojica FJ, Garrett RA (2013) Protospacer recognition motifs: mixed identities and functional diversity. RNA Biol 10(5):891–899
Shi TQ, Liu GN, Ji RY, Shi K, Song P, Ren LJ, Huang H, Ji XJ (2017) CRISPR/Cas9-based genome editing of the filamentous fungi: the state of the art. Appl Microbiol Biotechnol 101(20):7435–7443
Silva G, Poirot L, Galetto R, Smith J, Montoya G, Duchateau P, Pâques F (2011) Meganucleases and other tools for targeted genome engineering: perspectives and challenges for gene therapy. Curr Gene Ther 11(1):11–27
Stuckey S, Storici F (2013) Chapter eight – gene knockouts, in vivo site-directed mutagenesis and other modifications using the delitto perfetto system in *Saccharomyces cerevisiae*. Methods Enzymol 533:103–131
Syed K, Porollo A, Lam YW, Grimmett PE, Yadav JS (2013) CYP63A2, a catalytically versatile fungal P450 monooxygenase capable of oxidizing higher-molecular-weight polycyclic aromatic hydrocarbons, alkylphenols, and alkanes. Appl Environ Microbiol 79:2692–2702
Talbot N (2001) Molecular and cellular biology of filamentous fungi: a practical approach. Oxford University Press, Oxford
Thion C, Cébron A, Beguiristain T, Leyval C (2012) PAH biotransformation and sorption by Fusarium solani and Arthrobacter oxydans isolated from a polluted soil in axenic cultures and mixed co-cultures. Int Biodeterior Biodegrad 68:28–35
Toussaint A, Merlin C, Monchy S, Benotmane MA, Leplae R, Mergeay M, Springael D (2003) The biphenyl-and 4-chlorobiphenyl-catabolic transposon Tn4371, a member of a new family of genomic islands related to IncP and Ti plasmids. Appl Environ Microbiol 69:4837–4845
Tran NH, Urase T, Ngo HH, Hu J, Ong SL (2013) Insight into metabolic and cometabolic activities of autotrophic and heterotrophic microorganisms in the biodegradation of emerging trace organic contaminants. Bioresour Technol 146:721–731
Wakai S, Arazoe T, Ogino C, Kondo A (2017) Future insights in fungal metabolic engineering. Bioresour Technol 245:1314–1326
Wang Y, Guo B, Miao Z, Tang K (2007) Transformation of taxol-producing endophytic fungi by restriction enzyme-mediated integration (REMI). FEMS Microbiol Lett 273(2):253–259
Wang D, He D, Li G, Gao S, LV H, Shan Q, Wang L (2014) An efficient tool for random insertional mutagenesis: *Agrobacterium tumefaciens*-mediated transformation of the filamentous fungus *Aspergillus terreus*. J Microbiol Methods 98:114–118
Wang L, Shao Y, Guan Y, Li L, Wu L, Chen F, Liu M, Chen H, Ma Y, Ma X, Liu M, Li D (2015) Large genomic fragment deletion and functional gene cassette knock-in via Cas9 protein mediated genome editing in one-cell rodent embryos. Sci Rep 5:17517
Watford S, Warrington SJ (2017) Bacterial DNA mutations. StatPearls Publishing, Treasure Island
Wisecaver JH, Slot JC, Rokas A (2014) The evolution of fungal metabolic pathways. PLoS Genet 10:e1004816
Wittich R-M, González B (2016) Editorial overview: environmental biotechnology–quo vadis? Curr Opin Biotechnol 38:viii–viix
Yan W, Smith C, Cheng L (2013) Expanded activity of dimer nucleases by combining ZFN and TALEN for genome editing. Sci Rep 3:2376
Zheng Y-M, Lin F-L, Gao H, Zou G, Zhang J-W, Wang G-Q, Chen G-D, Zhou Z-H, Yao X-S, Hu D (2017) Development of a versatile and conventional technique for gene disruption in filamentous fungi based on CRISPR-Cas9 technology. Sci Rep 7:9250

# Chapter 5
# Phytoremediation and Fungi: An Underexplored Binomial

**Adriana Otero-Blanca, Jorge Luis Folch-Mallol, Verónica Lira-Ruan, María del Rayo Sánchez Carbente, and Ramón Alberto Batista-García**

## Contents



## 5.1 Environmental Pollution: A General Background

Environmental pollution is a worldwide issue that should concern governments and society since it is an ecological problem and a threat to human health. Pollution is defined as the introduction of contaminants into the air, water, and soil that can cause damage to different organisms whose origins are mainly caused by human activities. Some of these activities include refining and distribution of fuel from fossil oil; gas exhaustion from automobiles; domestic, industrial, and agricultural activities; and erroneous deposition of pharmaceutics. The chemicals derived from these activities are often a threat to life and are referred to as xenobiotics (Godheja et al. 2016).

The word "xenobiotic" is derived from two Greek words: *xeno* and *biotic*. In English, the first means strange, unnatural, or different, while the second refers to life. Xenobiotics, also called organic micropollutants, are those organic chemicals

A. Otero-Blanca · V. Lira-Ruan · R. A. Batista-García (✉)
Centro de Investigación en Dinámica Celular-IICBA, Universidad Autónoma del Estado de Morelos, Cuernavaca, Morelos, Mexico
e-mail: rabg@uaem.mx

J. L. Folch-Mallol · M. del Rayo Sánchez Carbente
Centro de Investigación en Biotecnología, Universidad Autónoma del Estado de Morelos, Cuernavaca, Morelos, Mexico

 
R. Prasad, E. Aranda (eds.), *Approaches in Bioremediation*, Nanotechnology in the Life Sciences, https://doi.org/10.1007/978-3-030-02369-0_5

that are not present in the biosphere prior to artificial synthesis or not present in high concentrations without human activities, to which an organism is exposed but which are extrinsic or foreign to its normal metabolism (Sánchez-Avila and Kretzschmar 2017).

Xenobiotics have been classified as classics and emergents. Among the classics are polyaromatic hydrocarbons, cyclic biphenyls, nitroaromatic compounds, aliphatic and aromatic halogenated compounds, triazines, azo dyes, and organic sulfonic acid (Godheja et al. 2016), while examples of emergents are pharmaceutical drugs and their residues, such as analgesics and nonsteroidal antiinflammatory drugs (NSAIDs) like ibuprofen and paracetamol (Joanna et al. 2018). Classic xenobiotics are those whose presence can cause adverse effects, which have been reported for several years, while emergents are those whose toxicity or effects have not been fully comprehended, despite their wide prevalence (Joanna et al. 2018).

Since 1939, with the discovery of the pesticide DDT (dichlorodiphenyltrichloroethane) by chemist Paul Herman Mueller, xenobiotics have been present in the environment. Since 1939, with the discovery of the pesticide DDT (dichlorodiphenyltrichloroethane) by chemist Paul Herman Mueller, xenobiotics have been present in the environment and, more than 600 chemical pesticides have been probed and registered in the USA. Xenobiotics have been reported to cause water pollution, and those that are persistent contribute to soil contamination.

The aquatic enviroments, such as lakes, rivers, seas and groundwater, are strongly affected by organic micropollutants, which are present at nanogram to milligram per liter and can have a negative impact in aquatic organisms (De los Ríos et al. 2016). It has been reported that there are trace amounts of xenobiotics in aquatic environments, such as alkylphenols, polycyclic aromatic hydrocarbons (PAHs), polychlorinated biphenyls, and phthalate esters (Sánchez-Avila et al. 2012; Gao and Wen 2016), among others.

Since xenobiotics are toxic and their rate of degradation is very slow, adverse effects on human and ecological health are possible. Therefore, xenobiotic concentrations in the environment should be maintained at as low level as possible. In consequence, remediation technologies are needed to remove them partially or completely from nature.

## 5.2 Remediation Technologies

The contamination of soil, groundwater, and surface water due to pollutants derived from human activities involving agriculture, mining, industrialization, petroleum extraction, and transport is a global health problem that causes the death of thousands of people around the world. It also causes large ecosystem losses and is linked to global climate change. Almost all human activities produce xenobiotics, which, if not properly treated, can reach noncontaminated areas and extend their deleterious effects. Since 1950, more than 140,000 chemicals and pesticides have been synthetized, and 2500 of them are widely dispersed in the planet and have never been tested for toxicity (Landrigan et al. 2017). Nations have been implementing policies to reduce the production and dispersion of contaminants and have been developing technologies to control pollution. However, some other problems need to be addressed, such as safely removing or destroying the contaminants that have already been produced and dispersed. In the USA, there are 350,000 contaminated sites in need to be cleaned up over the next 30 years. In Europe, there are 3,000,000 potential contaminated sites, of which >8% (~250,000 sites) are highly contaminated (Gong et al. 2016).

Contaminated sites can have prompted the development of remediation technologies, which have been evolving during the last 30 years. The main contaminants of soils and water are heavy metals, petroleum-derived hydrocarbons, aromatic compounds, pesticides, and fertilizers. Different technologies have been created to remove these noxious substances and/or convert them into less toxic forms or totally harmless substances. Contamination sites can have mixtures of different contaminants; thus, it is useful to combine technologies to improve and enhance their remediation capacities. Remediation technologies can be applied directly in the contaminated site: through in situ remediation or, in special facilities to perform the cleaning up, ex situ. In both cases, the cleaning up or removal of the contaminants are achieved through physical, chemical, or biological methods.

Most methods for cleaning soils and ground- and surface water are either physical, chemical, or a combination of both. The first remediation solutions were excavation and confinement of toxic materials, but these are only partial solutions because they do not reduce the concentration of the contaminant but only isolate them. For soil, the simplest remediation technique is the removal of the contaminated soil ex situ by chemical means. Then the contaminated site is filled with non-contaminated soil. This method is applicable only in small areas, and sometimes the soil is only partially removed and mixed with clean soil, which dilutes the contaminant that can be naturally attenuated (Yao et al. 2012). For contaminated water, the simplest method is pump and treat, where the contaminated groundwater is extracted and treated before being reintegrated into the ground. This in situ method is practical and very much used, but it does not completely eliminate the contaminants. The process takes a long time and does not work on contaminants adsorbed into the soil particles. Thus, it is usually combined with other remediation techniques (Khan et al. 2004). Capping is also an in situ technique that isolates more than remediates the contaminated places. A clean layer of unreactive material such as gravel, sand, or rocks is deposited above the contaminated area to cover and isolate it. Different adsorbent materials are added to the cap composition to better prevent the spreading of the xenobiotics to the surrounding soil or water (Gomes et al. 2013).

Air sparging is a useful technology for the remediation of underground water contaminated with volatile organic compounds. It is the injection of air with pressure below the contaminated area; when the air passes through the soil, it can strip the contaminant away or increase the degradation caused by microorganisms that inhabit the soil due to increase in oxygen (Khan et al. 2004; Reddy 2008).

Soil flushing and soil washing use water and solvents to clean soils and groundwater through in situ remediation processes. The nature of the solvents used depends on the contaminant to be removed, semi-volatile organic compounds, petroleum and its derivatives, heavy metals, pesticides, PAHs, and polychlorinated biphenyls. The solution passes through the contaminated area, removing the noxious agents. The solution is recovered and further treated to dispose the contaminants and return the cleaned water (Yao et al. 2012; Lim et al. 2016; Khalid et al. 2017).

Contaminated water is also treated with permeable reactive barriers. The principle is to let the contaminated water pass through a wall constructed with the appropriate material to retain and/or degrade the contaminants. This technique is applicable to groundwater and surface water, and the success of it depends on the material used to construct the barrier. The most common reactive materials are zero-valent iron, carbon-based materials as activated carbon and plant-derived biomass, alkaline-complexing agents, metal oxides, zeolite (Thiruvenkatachari et al. 2008), and more recently carbon-based nanomaterials and graphene in 3D architecture to better adsorb a variety of organic and inorganic water contaminants (Chen et al. 2015a). Moreover, there are reactive permeable barriers that include microorganisms to degrade the contaminants. These are known as permeable reactive bio-barriers.

Among the chemical techniques to clean contaminated soils in situ and ex situ, electrokinetics is one of the most versatile. It has been used to clean soils contaminated with heavy metals, oil and petroleum derivatives, and other organic contaminants. The basic principle consists in the application of a low-intensity current generated by electrodes inserted in the contaminated area in electrolyte wells. The electric current promotes the mobilization of ions and metals by electrophoresis, electromigration, or electroosmosis. This technique has been optimized by the addition of suitable electrolytes to improve its efficiency, such as surfactants or acidic solutions, and in combination with permeable reactive bio-barriers (Gomes et al. 2012; Lim et al. 2016; Mena et al. 2016).

Various soil remediation techniques employ temperature increase to solidify/immobilize, volatilize, or decompose the contaminants.

Vitrification technology is useful to treat soils contaminated with inorganic pollutants such as heavy metals and radionucleids. Electrodes are used to apply electric energy, which promotes increase in temperature. The organic contaminants are volatilized, and the inorganic materials are molten at temperatures between 1000 and 2000 °C. Soil commonly contains silica, which functions as a vitrifying agent; during the cooling of the soil, the materials form a very resistant glass. The vapor produced is collected and treated (Khalid et al. 2017).

Soil contaminated with petroleum derivatives and other industrial solvents can be treated with soil vapor extraction to remove volatile organic compounds. Fresh air is injected into the soil, and a vacuum extraction system is installed to collect the gas, which is further treated. This technique is used in combination with other techniques, like increasing temperature to remove semi-volatile compounds or air sparging (Khan et al. 2004; Lim et al. 2016).

Thermal desorption is a technology used to clean soils contaminated with organic contaminants, oil, and petroleum-derived contaminants and utilizes high (320–560 °C) or low (90–320 °C) temperatures. The vapor produced is collected, separated into water and solvent components, and further used as recycled components or incinerated. The recovered water can be discharged or reused to cool the system (Yao et al. 2012; Gomes et al. 2013).

Despite great research on and the improvement of physical-chemical remediation technologies, there are disadvantages in their use. Some of those applied in situ require the addition of chemical agents to the soil (soil washing and flushing and electrokinetics), which can lead to noncontaminated areas or can alter the chemical composition of soil, rendering it unfit for agriculture or natural preservation. In situ and ex situ technologies require large amounts of energy to function, and some of them, like incineration or thermal adsorption, generate gases that need to be further treated to really eliminate the noxious agents. Thus, alternatives to these remediation techniques have been developed, and the use of living organisms to remove, transform, or degrade xenobiotics is now called bioremediation or biological remediation. The biological agents that are used to perform bioremediation are naturally occurring microorganisms that can degrade or even mineralize many xenobiotic compounds. Most of these treatments are conducted in situ and with techniques that have a much lower cost than physical or chemical remediation. The only disadvantage of bioremediation is that in many cases, the treatment can last for months or even years.

In the next section, we will review in more detail the bioremediation techniques developed so far.

## 5.3 Biological Treatments

As physical and chemical remediation of xenobiotics is costly and most of the time inefficient, bioremediation has attracted attention in recent years. Bioremediation is the treatment of xenobiotic wastes with living organisms or their parts (enzymes, cell walls, secreted polysaccharides, etc.) (Ortiz-Hernández et al. 2011; Hlihor et al. 2017). A huge diversity of organisms has been used to alleviate the effects of xenobiotic compounds in the environment, and thus it seems a promising alternative. Bioremediation is environmentally friendly, low cost, and many times more efficient than physical or chemical remediation since it usually converts the xenobiotic compounds into $CO_2$ and water. Bioremediation may be applied in situ or ex situ and can be achieved through two main techniques: biostimulation and bioaugmentation. In the first case, specific nutrients are provided into the contamination site so specific native microbiota can grow, cometabolizing the xenobiotic compounds. In the second case, native or foreign microorganisms or plants are introduced into the polluted habitat to degrade the pollutants (Alegbeleye et al. 2017). In this situation, especially in ex situ treatments, even transgenic microorganisms can be tested (Balcázar-López et al. 2016). The term bioremediation came as early as 1928, when Gray and Thronton discovered naturally occurring microorganisms that could degrade BTEX (benzene, toluene, ethylbenzene, and xylene) in soil. Here we will discuss three main groups of organisms that have been extensively studied and used for bioremediation.

### 5.3.1 Bacteria

One of earlier reports of a bacterium capable of degrading benzene derivatives (BTEX) was published by Williams and Murray in 1974 (Williams and Murray 1974). Later it was described that this strain of *Pseudomonas putida* carrying the TOL plasmid possessed an enzymatic route to degrade these compounds and use them as carbon source (Worsey and Williams 1975).

Since then, a big number of different bacterial species have been proved to degrade a wide variety of different xenobiotic compounds. Among the most studied are members of the genera *Mycobacterium*, *Pseudomonas*, *Alcanivorax*, *Burkholderia*, *Rhodococcus*, *Aeromonas*, *Arthrobacter*, *Micrococcus*, *Streptomyces*, *Bacillus*, *Sphingomonas*, *Cellulomonas*, *Micrococcus*, *Marinobacter*, *Haemophilus*, *Xanthomonas*, *Acinetobacter*, *Enterobacter*, *Corynebacterium*, etc. These genera can mineralize polycyclic aromatic hydrocarbons (PAHs), pesticides, and azo dyes and remove or change the redox state of heavy metals.

Although the degradation of many xenobiotics via oxidases has been better studied in fungi, ligninolytic-like enzymes have also been found in bacteria. Enzymes called yellow laccases have been described, but they show a lower redox potential toward their substrates than true fungal blue laccases (Valderrama et al. 2003; Riva 2006). Bacteria typically do not show peroxidase activity; however, another kind of peroxidase activity (by proteins nonhomologous to the fungal ones), known as dye-decolorizing peroxidases, has been described and studied (Van Bloois et al. 2010). In general, these enzymes are unspecific oxidases that can activate many xenobiotic compounds through the production of free radicals for their further mineralization or polymerization, rendering them nonbioavailable. Also, enzymes involved in oxidative stress such as catalases and superoxide dismutases have been involved in PAH degradation (de Gonzalo et al. 2016).

For pesticides such as organophosphate or carbamate compounds (ethyl or methyl parathion, coumaphos, carbofuran, carbaryl, etc.), several hydrolytic enzymes have been described (Singh et al. 2017). Although these compounds may also be oxidized, the hydrolytic route via phosphotriesterases is preferable since it will not produce toxic intermediate quinones (Ortiz-Hernández et al. 2011).

Heavy metals, on the other hand, cannot be degraded. The means for their bioremediation consists primarily in adsorption into the cell wall, compartmentalization in vacuoles or other organelles in eukaryotes, or changing their redox state to less soluble forms and thus making them less bioavailable. Usually, for these xenobiotic elements, plants are the best option, but a striking example of biostimulation is the capacity of heavy metals to transform into less toxic forms with the use of *Geobacter sulfurreducens* (Hernández-Eligio et al. 2017). In this system, acetate is pumped into the ground that is polluted by heavy metals (this has been especially efficient for manganese, uranium, or chromium) so that the anaerobe *G. sulfurreducens* can "respire" this substrate and transfer electrons to metals, converting them into less soluble forms that make them unavailable. The striking feature here is that, through this process, an electrical current is generated and conducted through specialized pili, and so this system is also being studied as a bioenergy-generating alternative.

### 5.3.2 Fungi

Fungi have been one of the most promising bioresources for bioremediation (especially white-rot fungi, pertaining to the *Basidiomycota*) since they produce extracellularly a plethora of oxidative and hydrolytic enzymes (Baldrian 2008). Fungi are eukaryotes that grow saprotrophically, forming hyphae, tube-like long cells that can form tight mats called mycelia. One of the advantages in using fungi for bioremediation in situ is the ability to cover and penetrate large surface areas by hyphae. Fungi have evolved to use complex carbon sources since they are sessile organisms and can decompose lignin, cellulose, and hemicellulose, which have different recalcitrant structures. Particularly, lignin, an amorphous phenolic polymer, shows some structural and chemical similarities to PAHs, organophosphate aromatic pesticides, and industrial dyes.

The most widely studied species belongs to the division *Basidiomycota*, although a large number of other species have been studied. Among the most studied genera, we can find *Trametes*, *Phanerochaete*, *Pleurotus*, *Bjerkandera*, *Coriollopsis* (*Basidiomycota*), and several different *Aspergilli*, *Trichoderma*, and *Fusarium* (*Ascomycota*). Although there are reports on yeasts (*Rhodotorula* spp., *Candida* spp., and *Yarrowia* spp.), these have been less studied.

Yeasts, ascomycetes and basidiomycetes have been also studied for the production of these aforementioned oxidases: laccases, peroxidases, lytic polysaccharide monooxygenases, and the intracellular monooxygenase cytochrome P450. Activities for all these are quite unspecific and thus can use as substrates large amounts of different compounds, such as PAHs, pesticides, endocrine disruptors, dyes, and even explosives such as trinitrotoluene. It is worth to note that the oxidases (especially laccase and peroxidases) produced by fungi have a much larger redox potential than their bacterial counterparts, thus being more efficient in xenobiotic degradation (Singh et al. 2017).

Other fungal enzymes involved in the mineralization of xenobiotic compounds are glucose oxidase, aryl alcohol oxidase, quinone oxidoreductase, and cellobiose dehydrogenase (Leonowicz et al. 1999).

Fungi can also be used for bioremediation of soil and water polluted by heavy metals (Abbas et al. 2014). Several mechanisms are used by these organisms that give an advantage over bacteria to control heavy-metal-polluted environments. The main mechanisms for heavy metal control is probably adsorption into the cell wall (which presents many carbonyl groups to which heavy metals can bind), and this is also true for bacteria and algae. These organisms can also conjugate heavy metals in several organic molecules such as glutathione or organic acids. However, fungi are eukaryotes that have cell compartments, and, in many cases, compartmentalization of the metals in different organelles (vacuole, endoplasmic reticulum, or the Golgi apparatus) is achieved to contend with the toxicity caused by the metals. Furthermore, fungi express specific proteins called metallothioneins, which are cysteine-rich proteins that can also "trap" these xenobiotics in the Golgi apparatus (Siddiquee et al. 2015).

In conclusion, bacteria are probably the most studied and used microorganisms for bioremediation of xenobiotic compounds, but fungi are promising bioremediator agents since they grow fast, can encompass large areas (e.g., be placed in large filters), and produce powerful and robust extracellular enzymes (Prasad 2017, 2018). Furthermore, in many cases, they can withstand harsher conditions than bacteria.

### *5.3.3 Phytoremediation*

Plants are also very versatile organisms because they are sessile and thus have evolved to withstand a plethora of environmental challenges than other organisms can evade by running away from biological, chemical, or physical threats.

Bioremediation with plants has been mainly applied to alleviate pollution by heavy metals, but it has also proven useful with other kinds of xenobiotics, such as PAHs, pesticides, dyes, etc. (Rasmussen and Olsen 2004; Lin et al. 2005; Dixit et al. 2015; Tripathi et al. 2016). Plants use several mechanisms for the bioremediation of different compounds: phytovolatilization, phytostabilization, phytodegradation, and rhizodegradation (Arslan et al. 2017). The mechanism applies mainly to volatile compounds such as BTEX, which consists of the plant absorb such pollutants and transferring them to the air through stomata. However, this only takes the pollutant from one place to another, so no real remediation occurs.

Phytostabilization refers to the immobilization of the pollutants in the plant. Mostly, the toxic compounds are adsorbed into the roots, or by the root system, and its faith is then decided, depending on whether they can be degraded or just accumulated. In general, plants are able to absorb easily nonpolar contaminants such as PAHs or pesticides. The lipid content of the plant tissue is crucial for the plant to be able to absorb these kinds of xenobiotics (Goodin and Webber 1995).

Phytodegradation of organic pollutants has been poorly studied, although it is controversial. However, we must not forget that plants produce peroxidases and laccases that could act as a starting point for the degradation of PAHs, pesticides, and dyes (Günther et al. 1996).

On the other hand, rhizodegradation is the best studied and most successful mechanism. It involves the participation of plant-associated microorganisms that live in close vicinity with plant roots. Even if the microbes can many times degrade pollutants by themselves, the rhizospheric environment greatly enhances its efficiency. In some scenarios, microbes cannot mineralize xenobiotics without the plant counterpart (Grosser et al. 1995; Chen et al. 2015b; Hong et al. 2015).

In the rhizosphere, plants offer nutrients and residency to the microbial communities that live in the vicinity. As payback, some of these microbes often promote plant growth and develop tolerance to abiotic stress and resistance to phytopathogens for the plant. There are two main classes of rhizospheric interactions: surface root colonization and endophytic microorganisms penetrating the plant tissues and residing within the plant (Compant et al. 2010). The beneficial effect of microorganisms to the plant is given by the fact that microbes contribute to nutrient uptake by fixing nitrogen; producing siderophores (which can provide Fe); some excrete organic acids that dissolve insoluble minerals containing phosphate, calcium, and other essential nutrients; the production of plant hormones such as auxins and gibberellins; and enzymes that degrade ethylene, a hormone that stops plant growing in stressful situations. These interactions also induce both induced systemic resistance (ISR) and systemic acquired resistance (SAR), which protects the plant against pathogens (fungi, bacteria, and viruses) through the expression of proteins such as defense- and pathogen-related proteins (often chitinases and glucanases that degrade the cell wall of the pathogen). Finally, pH changes in the soil and other factors induce the response to oxidative stress, which protects the plant against drought, heat, cold, and other abiotic stresses (Brotman et al. 2013; Pelagio-Flores et al. 2017). Furthermore, as mentioned before, fungi produce a whole set of extracellular enzymes that are able to degrade or transform persistent organic pollutants, so the combination of plant and fungi for bioremediation looks very promising.

Among the most studied endophytic interactions is that of *Trichoderma* spp. with a whole different set of plants, from *Arabidopsis thaliana* to agroeconomical important plants such as cucumber, tomato, and beans (Salas-Marina et al. 2011; Dildey et al. 2016). Other fungi that also have been widely studied are those that form arbuscular mycorrhizal symbiosis. These microorganisms can also penetrate the plant tissues and help nurture the plant, especially in the acquisition of nitrogen and phosphate.

Many groups are now studying the potential of the plant-microbe associations to implement bioremediation strategies that have been found to be very promising. For example, up to 49% of hydrocarbon from petrol was removed in a consortium formed by oat (*Avena sativa*), *Acinetobacter* sp. (bacteria), and *Rhizophagus intraradices* (formerly *Glomus intraradices*) (an arbuscular mycorrhizal fungus) (Xun et al. 2015). Other cases of success were achieved by using common bean (*Phaseolus vulgaris*) mixed with N-fixing bacteria *Rhizobium* sp. to remove the pesticide atrazine from agricultural soils (Madariaga-Navarrete et al. 2017). Salami and coworkers (Salami et al. 2017) also obtained good results from using *Trichoderma harzianum* and the AMF fungus *Funneliformis mosseae* (formerly *Glomus mosseae*) for the treatment of *Capsicum annum* L. plants irrigated with water from a mining site.

In conclusion, bioremediation is a potentially effective strategy for the remediation of soils and waters polluted with xenobiotics. It is an emerging low-cost technology that still has to be explored but that has already proven to be efficient since very high percentages of persistent organic pollutants or heavy metals have been shown to be removed.

## 5.4 Phytoremediation and Fungi: Cases of Study and Perspectives

Mycorrhizoremediation has been identified as an enhanced form of phytoremediation. Fungi are very important organisms in the colonization of different plants. It has been suggested that glomalean fungi are essential in the colonization of briophyte-like land plants (Simon et al. 1993). Arbuscular mycorrhizal fungi are one of the oldest fungi in nature, and they establish symbiotic relationships with approximately 80–90% of land plants (Brundrett 2002). These fungi when associated with plants, improve the plants' nutrition, root development, productivity, and health. The fungus–plant interaction is the base of the mycophytore mediation.

Specifically, AMF fungi have been largely studied with respect to their ability to stabilize and sorb heavy metals in soils and also during their interaction with plants (Khan 2006). They produce glomalin, a related soil protein (GRSP) with properties to chelate metals in different environments. Some authors have proposed that glomalins immobilize heavy metals like a filter in the soil/hypha interface, and then they facilitate metal internalization into mycorrhizas. Thus, in association with roots, AMF fungi are crucial to phytostabilize heavy metals in soils. Even though there are many studies describing the biotechnological potential of AMF to develop phytoremediation systems (Khan 2006), few works critically analyze the molecular mechanisms involved in the phytostabilization of heavy metal by the AMF plant system. This represents a challenge for further study and at the same time should be considered a perspective in the field. The full understanding of these mechanisms will allow a better comprehension of the ecological role of glomalean fungi and, in consequence, the improvement of new phytoremediation systems. AMF also plays an important role in the aggregation of soil particles, because of which they produce glycol-soil-proteinaceous substances (glomalins) (Rillig et al. 2003; Sharma et al. 2017). These compounds have been poorly studied, and only little is known about their potential to bind heavy metals. Another unknown aspect is glomalin interaction with plants and the microbiota present in the soil. Further studies are required to establish the specific physiological and ecological roles of these proteins. But glycol-soil-proteinaceous substances are unique proteins secreted by AMF involved in the mycoremediation of heavy metals. Surely, other proteins with an important role in the sorption and sequestration of toxic elements would exist, not only metals but also organic xenobiotics.

Many studies to explore the potential applications of phytoremediation of heavy metals have been addressed with AMF. Greenhouse pot experiments were conducted using *Glomus versiforme* to determine its contribution to cadmium hyperaccumulation by *Solanum nigrum*. The *G. versiforme* inoculation has a significant effect on the extractable Cd concentrations and also enhanced acid phosphatase activity and phosphorous acquisition. This fungus supported the high growth of *S. nigrum* (Liu et al. 2015). Other *Glomus* species have exhibited good potentials for phytoremediation. *Rhizophagus intraradices* (previously known as *Glomus intrara-*

*dices*) enhanced zinc adsorption during its interaction with tobacco plants (Audet and Charest 2006), and *Funneliformis mosseae* (previously known as *Glomus mosseae*) allowed high removal of lead in association with vetiver roots (Wong et al. 2007; Punamiya et al. 2010). *Claroideoglomus etunicatum* (previously known as *Glomus etunicatum*) showed capabilities to reduce Ni concentration in the roots of *Sorghum vulgare* and enhanced plant growth, contributing to nickel phytostabilization (Amir et al. 2013). In general terms, mycorrhizal fungi represent a great friendly alternative for phytoremediation.

There are many works evaluating the biotechnological potential of mycorrhizal fungi for phytoremediation, but only few studies test these potentialities at a greenhouse scale. The major challenge is to transfer the laboratory scale results to the field, where other microbial communities might be playing a role in the process. Thus, mycophytoremediation deserves further efforts to generate novel knowledge to accelerate the application at real-time scale of those results generated at laboratory level and, therefore, attain improvement in mycorrhizal fungi biotechnology (Khan 2006).

Endophytic fungi also represent good candidates for phytoremediation. They play a crucial role in several processes such as organic and inorganic transformations, element cycling, rock and mineral biotransformations, bioweathering, mycogenic mineral formation, fungal–clay interactions, metal-fungal interactions, as well as organic compound–fungal interactions. These elements suggest that endophytic fungi could improve phytoremediation efficiencies (Deng and Cao 2017). This is another area with big opportunities to generate novel knowledge related to mycophytoremediation because has exhibited a discreet advance in recent years. As Deng and Cao pointed out, the role of endophytic fungi in phytoremediation has been poorly studied (Deng and Cao 2017). More attention should be concentrated on this topic to describe in detail the physiological properties of endophytic fungi and their potentialities in phytoremediation. In this context, yeasts have been scarcely studied with few reports related to their effects on phytoremediation of heavy metals. However, some works describing the use of endophytic filamentous fungi in the phytoremediation of organic pollutants have been published. For example, the endophyte *Ceratobasidium stevensii* enhanced the degradation of phenolic acids when it was inoculated to watermelon plants. At the same time, it promoted plant growth, helped grow more stems, and enhanced leaf length (Xiao et al. 2010). Another study evaluated the infection of endophytic fungi in the grass species *Festuca arundinacea* and *Festuca pratensis* during the phytoremediation of hydrocarbons in an aged-contaminated soil. It was demonstrated that the endophytic fungi significantly contributed to the total degradation of petroleum hydrocarbon, as well as root formation in the grasses (Soleimani et al. 2010).

Fungi could also act as endophytic microorganisms since they can colonize roots. Endophytic microbes, especially fungi, establish intimate and symbiotic relationships with plant, and these interactions could contribute to the efficiency of phytoremediation. Thus, the mycoremediation implementation at field level needs a full understanding of the endophytic fungi–microbiota–plant interaction (Prasad 2017, 2018).

As petroleum-derivate-polluted soils are very common in many countries, fungi have also been used for phytoremediation of diesel at laboratory levels. It has been proved that diesel is a potent inhibitor of growth for a wide range of plants. Some plants exhibit better growth in the presence of diesel when they are inoculated with AMF (Alarcón et al. 2006; Joner and Leyval 2003; Hernández-Ortega et al. 2012). During the interaction between *Melilotus albus* and *Glomus* Zac-19 in a soil contaminated with diesel, *M. albus* showed better growth, and diesel was significantly degraded when *Glomus* Zac-19 was used in the experiment (Hernández-Ortega et al. 2012). It was also demonstrated that *Glomus* Zac-19 significantly reduced diesel toxicity on the plants because of the fungus-enhanced plant biomass, nutrient sequestration, and the total antioxidant activity involved (Hernández-Ortega et al. 2012).

Another AMF has been investigated for the phytoremediation of cadmium and organic pollutants, such as decabromodiphenyl ether (BDE-209) (Li et al. 2018). *Funneliformis mosseae*– and *Rhizophagus intraradices*–*Solanum nigrum* interactions revealed an improvement of shoot biomass and the cadmium contents in shoots in comparison with uninoculated plants (Li et al. 2018). Thus, fungi are ideal for phytoremediation since they can coremove both organic and inorganic pollutants. Additional efforts are necessary in the investigation of fungi-based phytoremediation of soil cocontaminated with heavy metals and organic compounds since studies on this have been scarce and poli-polluted environments are more frequently found. These studies will find new insights in order to establish new/novel cost-effective, efficient, and environmentally friendly mycophytoremediation strategies for the removal of multiclass pollutants.

Interested in the development of mycophytoremediation technologies, Mohsenzadeh et al. (2010) studied some fungal–plant interactions in petroleum-polluted soils. Seven plants showed tolerance and growth in the presence of petroleum (Mohsenzadeh et al. 2010). They were *Alhaji cameleron*, *Amranthus retroflexus*, *Convolvulus arvensis*, *Chrozophora hierosolymitana*, *Noea mucronata*, *Poa* sp., and *Polygonum aviculare*. Eleven fungi isolated from these plants, some endophytic fungi between them, tolerated 1% (v/v) of petroleum, while *Fusarium* species resisted 10% (v/v). The study demonstrated that plants of *P. aviculare* inoculated with *F. acuminatum*, *F. equiseti*, *F. reticulatum*, and *F. oxysporum* significantly alleviate the petroleum pollution in the soil in comparison to plant and fungi separately (Mohsenzadeh et al. 2010). While huge attention has been given to the investigation of fungi as a promising degrader agent, little attention is noted in analyzing the role and biotechnological potentialities in phytoremediation of fungal-plant systems in the bioremediation of hydrocarbons. On the other hand, no mycorrhizal fungi have received poor attention. Further researches and efforts should be considered to facilitate new knowledge related to mycophytoremediation using free-living fungi.

In conclusion, AMF exhibits an excellent potential in the phytoremediation of metals and organic pollutants. Figure 5.1 (Rajtor and Piotrowska-Seget 2016) shows the benefits derived from these types of fungi since they enhance nutrient and water acquisition, promote plant growth, facilitate the stabilization and aggre-

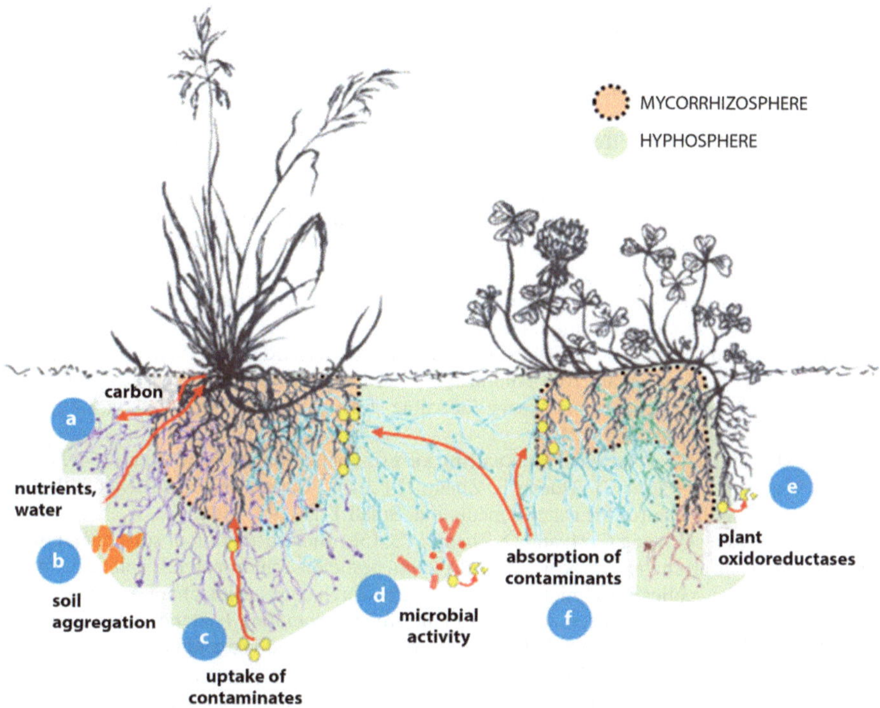


**Fig. 5.1** AMF and its potential in phytoremediation. Fungal growth (**a**) facilitates the acquisition of nutrients and water, (**b**) improves the stabilization and aggregation of soil particles, (**c**) promotes plant growth, (**d**) participates in carbon degradation using oxidoreductases and a huge battery of degrading enzymes, (**e**) changes the profile of root exudations and the microbiota composition, (**e**) increases the adsorption area in roots since it enhances the lipid content in the root system

gation of soil particles, allow the immobilization of organic pollutants, contribute to carbon degradation, stimulate root exudates and the proliferation of the rhizomicrobiota increasing enzymatic activities with a crucial role in the degradation of organic contaminants, and protect plants from oxidative stress (Rajtor and Piotrowska-Seget 2016).

## References

Abbas S, Ismail I, Mostafa T, Abbas S (2014) Biosorption of heavy metals: review. J Chem Sci Technol 3:74–102

Alarcón A, Delgadillo-Martínez J, Franco-Ramírez A, Davies FT, Ferrera-Cerrato R (2006) Influence of two polycyclic aromatic hydrocarbons on spore germination, and phytoremediation potential of Gigaspora margarita-Echynochloa polystachya symbiosis in benzo[a]pyrene-polluted substrate. Rev Int Contam Ambient 22:39–47

Alegbeleye OO, Opeolu BO, Jackson VA (2017) Polycyclic aromatic hydrocarbons: a critical review of environmental occurrence and bioremediation. Environ Manag 60:758–783

Amir H, Lagrange A, Hassaïne N, Cavaloc Y (2013) Arbuscular mycorrhizal fungi from New Caledonian ultramafic soils improve tolerance to nickel of endemic plant species. Mycorrhiza 23:585–595

Arslan M, Imran A, Khan QM, Afzal M (2017) Plant–bacteria partnerships for the remediation of persistent organic pollutants. Environ Sci Pollut Res 24:4322–4336

Audet P, Charest C (2006) Effects of AM colonization on "wild tobacco" plants grown in zinc-contaminated soil. Mycorrhiza 16:277–283

Balcázar-López E, Méndez-Lorenzo LH, Batista-García RA, Esquivel-Naranjo U, Ayala M, Kumar VV, Savary O, Cabana H, Herrera-Estrella A, Folch-Mallol JL (2016) Xenobiotic compounds degradation by heterologous expression of a *Trametes sanguineus* laccase in *Trichoderma atroviride*. PLoS One 11:e0147997

Baldrian P (2008) Wood-inhabiting ligninolytic basidiomycetes in soils: ecology and constraints for applicability in bioremediation. Fungal Ecol 1:4–12

Brotman Y, Landau U, Cuadros-Inostroza Á, Takayuki T, Fernie AR, Chet I, Viterbo A, Willmitzer L (2013) *Trichoderma*-plant root colonization: escaping early plant defense responses and activation of the antioxidant machinery for saline stress tolerance. PLoS Pathog 9 http://doi.org/101371/journalppat1003221

Brundrett MC (2002) Coevolution of roots and mycorrhizas of land plants. New Phytol 154:275–304

Chen B, Ma Q, Tan C, Lim TT, Huang L, Zhang H (2015a) Carbon-based sorbents with three-dimensional architectures for water remediation. Small 11:3319–3336

Chen M, Xu P, Zeng G, Yang C, Huang D, Zhang J (2015b) Bioremediation of soils contaminated with polycyclic aromatic hydrocarbons, petroleum, pesticides, chlorophenols and heavy metals by composting: applications, microbes and future research needs. Biotechnol Adv 33:745–755

Compant S, Clément C, Sessitsch A (2010) Plant growth-promoting bacteria in the rhizo- and endosphere of plants: their role, colonization, mechanisms involved and prospects for utilization. Soil Biol Biochem 42:669–678

de Gonzalo G, Colpa DI, Habib MHM, Fraaije MW (2016) Bacterial enzymes involved in lignin degradation. J Biotechnol 236:110–119

De los Ríos A, Echavarri-Erasun B, Lacorte S, Sánchez-Ávila J, De Jonge M, Blust R, Orbea A, Juanes JA, Cajaraville MP (2016) Relationships between lines of evidence of pollution in estuarine areas: linking contaminant levels with biomarker responses in mussels and with structure of macroinvertebrate benthic communities. Mar Environ Res 121:49–63

Deng Z, Cao L (2017) Fungal endophytes and their interactions with plants in phytoremediation: a review. Chemosphere 168:1100–1106

Dildey ODF, Broetto L, Rissato BB, Gonçalves-Trevisoli EDV, Coltro-Roncato S, Dal'Maso EG, Webler TFB (2016) *Trichoderma*-bean interaction: defense enzymes activity and endophytism. Afr J Agric Res 11(43):4286–4292

Dixit R, Wasiullah Malaviya D, Pandiyan K, Singh UB, Sahu A, Shukla R, Singh BP, Rai JP, Sharma PK, Lade H, Paul D (2015) Bioremediation of heavy metals from soil and aquatic environment: an overview of principles and criteria of fundamental processes. Sustainability 7:2189–2212

Gao DW, Wen ZD (2016) Phthalate esters in the environment: a critical review of their occurrence, biodegradation, and removal during wastewater treatment processes. Sci Total Environ 541:986–1001

Godheja J, Shekhar SK, Siddiqui SA, Modi DR (2016) Xenobiotic compounds present in soil and water: a review on remediation strategies. J Environ Anal Toxicol 6:5

Gomes HI, Dias-Ferreira C, Ribeiro AB (2012) Electrokinetic remediation of organochlorines in soil: enhancement techniques and integration with other remediation technologies. Chemosphere 87:1077–1090

Gomes HI, Dias-Ferreira C, Ribeiro AB (2013) Overview of in situ and ex situ remediation technologies for PCB-contaminated soils and sediments and obstacles for full-scale application. Sci Total Environ 445–446:237–260

Gong Y, Tang J, Zhao D (2016) Application of iron sulfide particles for groundwater and soil remediation: a review. Water Res 89:309–320
Goodin JD, Webber MD (1995) Persistence and fate of anthracene and benzo(a)pyrene in municipal sludge treated soil. J Environ Qual 24:271–278
Grosser R, Warshawsky D, Vestal R (1995) Mineralization of polycyclic and N-heterocyclic aromatic compounds in hydrocarbon-contaminated soils. Environ Toxicol Chem 14:375–382
Günther T, Dornberger U, Fritsche W (1996) Effects of ryegrass on biodegradation of hydrocarbons in soil. Chemosphere 33(2):203–215
Hernández-Eligio A, Andrade Á, Soto L, Morett E, Juárez K (2017) The unphosphorylated form of the PilR two-component system regulates pilA gene expression in *Geobacter sulfurreducens*. Environ Sci Pollut Res 24:25693–25701
Hernández-Ortega HA, Alarcón A, Ferrera-Cerrato R, Zavaleta-Mancera HA, López-Delgado HA, Mendoza-López MR (2012) Arbuscular mycorrhizal fungi on growth, nutrient status, and total antioxidant activity of *Melilotus albus* during phytoremediation of a diesel-contaminated substrate. J Environ Manag 95:S319–S324
Hlihor RM, Gavrilescu M, Tavares T, Fvier L, Olivieri G (2017) Editorial bioremediation: an overview on current practices, advances, and new perspectives in environmental pollution treatment Hindawi. Biomed Res Int 2:3–5
Hong Y, Liao D, Chen J, Khan S, Su J, Li H (2015) A comprehensive study of the impact of polycyclic aromatic hydrocarbons (PAHs) contamination on salt marsh plants *Spartina alterniflora*: implication for plant-microbe interactions in phytoremediation. Environ Sci Pollut Res 22:7071–7081
Joanna Ż, Pi A, Marchlewicz A, Hupert-kocurek K, Wojcieszy D (2018) Organic micropollutants paracetamol and ibuprofen – toxicity, biodegradation, and genetic background of their utilization by bacteria. Environ Sci Pollut Res Int 25:21498–21524
Joner E, Leyval C (2003) Phytoremediation of organic pollutants using mycorrhizal plants: a new aspect of rhizosphere interactions. Agronomie EDP Sci 23(5-6):495–502
Khalid S, Shahid M, Niazi NK, Murtaza B, Bibi I, Dumat C (2017) A comparison of technologies for remediation of heavy metal contaminated soils. J Geochem Explor 182:247–268
Khan AG (2006) Mycorrhizoremediation – an enhanced form of phytoremediation. J Zhejiang Univ Sci B 7:503–514
Khan FI, Husain T, Hejazi R (2004) An overview and analysis of site remediation technologies. J Environ Manag 71:95–122
Landrigan PJ, Fuller R, Acosta NJR, Adeyi O, Arnold R, Basu N, Baldé AB, Bertollini R, Bose-O'Reilly S, Boufford JI, Breysse PN, Chiles T, Mahidol C, Coll-Seck AM, Cropper ML, Fobil J, Fuster V, Greenstone M, Haines A, Hanrahan D, Hunter D, Khare M, Krupnick A, Lanphear B, Lohani B, Martin K, Mathiasen KV, McTeer MA, Murray CJL, Ndahimananjara JD, Perera F, Potočnik J, Preker AS, Ramesh J, Rockström J, Salinas C, Samson LD, Sandilya K, Sly PD, Smith KR, Steiner A, Stewart RB, Suk WA, van Schayck OCP, Yadama GN, Yumkella K, Zhong M (2017) The lancet commission on pollution and health. Lancet 391(3–9):462–512
Leonowicz A, Matuszewska A, Luterek J, Ziegenhagen D, Wojta-Wasilewska M, Cho N-S, Hofrichter M, Rogalski J, Wasilewska W (1999) Biodegradation of lignin by white rot fungi. Fungal Genet Biol 27:175–185
Li H, Li X, Xiang L, Zhao HM, Li YW, Cai QY, Zhu L, Mo CH, Wong MH (2018) Phytoremediation of soil co-contaminated with Cd and BDE-209 using hyperaccumulator enhanced by AM fungi and surfactant. Sci Total Environ 613–614:447–455
Lim MW, Von Lau E, Poh PE (2016) A comprehensive guide of remediation technologies for oil contaminated soil – present works and future directions. Mar Pollut Bull 109:14–45
Lin CH, Lerch RN, Garrett HE, George MF (2005) Incorporating forage grasses in riparian buffers for bioremediation of atrazine, isoxaflutole and nitrate in Missouri. Agrofor Syst 63:91–99
Liu H, Yuan M, Tan S, Yang X, Lan Z, Jiang Q, Ye Z, Jing Y (2015) Enhancement of arbuscular mycorrhizal fungus (*Glomus versiforme*) on the growth and Cd uptake by Cd-hyperaccumulator *Solanum nigrum*. Appl Soil Ecol 89:44–49

Madariaga-Navarrete A, Rodríguez-Pastrana BR, Villagómez-Ibarra JR, Acevedo-Sandoval OA, Perry G, Islas-Pelcastre M (2017) Bioremediation model for atrazine contaminated agricultural soils using phytoremediation (using *Phaseolus vulgaris* L) and a locally adapted microbial consortium. J Environ Sci Heal B Pestic Food Contam Agric Wastes 52:367–375

Mena E, Villaseñor J, Rodrigo MA, Cañizares P (2016) Electrokinetic remediation of soil polluted with insoluble organics using biological permeable reactive barriers: effect of periodic polarity reversal and voltage gradient. Chem Eng J 299:30–36

Mohsenzadeh F, Nasseri S, Mesdaghinia A, Nabizadeh R, Zafari D, Khodakaramian G, Chehregani A (2010) Phytoremediation of petroleum-polluted soils: application of *Polygonum aviculare* and its root-associated (penetrated) fungal strains for bioremediation of petroleum-polluted soils. Ecotoxicol Environ Saf 73:613–619

Ortiz-Hernández ML, Sánchez-Salinas E, Olvera-Velona A, Folch-Mallo JL (2011) Pesticides in the environment: impacts and their biodegradation as a strategy for residues treatment. In: Pesticides, Intech Open, p 22

Pelagio-Flores R, Esparza-Reynoso S, Garnica-Vergara A, López-Bucio J, Herrera-Estrella A (2017) *Trichoderma*-induced acidification is an early trigger for changes in Arabidopsis root growth and determines fungal phytostimulation. Front Plant Sci 8:1–13

Prasad R (2017) Mycoremediation and environmental sustainability, vol 1. Springer International Publishing, Cham, Switzerland https://doi.org/10.1007/978-3-319-68957-9

Prasad R (2018) Mycoremediation and environmental sustainability, vol 2. Springer International Publishing, Cham, Switzerland https://www.springer.com/us/book/9783319773858

Punamiya P, Datta R, Sarkar D, Barber S, Patel M, Das P (2010) Symbiotic role of *Glomus mosseae* in phytoextraction of lead in vetiver grass [*Chrysopogon zizanioides* (L)]. J Hazard Mater 177:465–474

Rajtor M, Piotrowska-Seget Z (2016) Prospects for arbuscular mycorrhizal fungi (AMF) to assist in phytoremediation of soil hydrocarbon contaminants. Chemosphere 162:105–116

Rasmussen G, Olsen RA (2004) Sorption and biological removal of creosote-contaminants from groundwater in soil/sand vegetated with orchard grass (*Dactylis glomerata*). Adv Environ Res 8:313–327

Reddy KR (2008) Physical and chemical groundwater remediation technologies. In: Overexploitation contam shar groundw resour. Springer, Dordrecht, pp 257–274

Rillig MC, Ramsey PW, Morris S, Paul EA (2003) Glomalin, an arbuscular-mycorrhizal fungal soil protein, responds to land-use change. Plant Soil 253:293–299

Riva S (2006) Laccases: blue enzymes for green chemistry. Trends Biotechnol 24:219–226

Salami AO, Opadiran AE, Idowu OO (2017) Bioremediation potentials of *Trichoderma harzianum* and *Glomus mosseae* on the growth of *Capsicum annum* L grown on soil irrigated with water from mining site. Int J Biosci Agric Technol 8(9):64–72

Salas-Marina MA, Silva-Flores MA, Uresti-Rivera EE, Castro-Longoria E, Herrera-Estrella A, Casas-Flores S (2011) Colonization of *Arabidopsis* roots by *Trichoderma atroviride* promotes growth and enhances systemic disease resistance through jasmonic acid/ethylene and salicylic acid pathways. Eur J Plant Pathol 131:15–26

Sánchez-Avila JI, Kretzschmar T (2017) Simultaneous determination of polycyclic aromatic hydrocarbons, alkylphenols, phthalate esters and polychlorinated biphenyls in environmental waters based on headspace–solid phase microextraction followed by gas chromatography–tandem mass spectrometry. J Environ Anal Chem 4:11

Sánchez-Avila J, Tauler R, Lacorte S (2012) Organic micropollutants in coastal waters from NW Mediterranean Sea: sources distribution and potential risk. Environ Int 46:50–62

Sharma S, Prasad R, Varma A, Sharma AK (2017) Glycoprotein associated with *Funneliformis coronatum, Gigaspora margarita* and *Acaulospora scrobiculata* suppress the plant pathogens in vitro. Asian Journal of Plant Pathology https://doi.org/10.3923/ajppaj.2017

Siddiquee S, Rovina K, Al AS, Naher L, Suryani S, Chaikaew P (2015) Microbial & biochemical technology heavy metal contaminants removal from wastewater using the potential filamentous fungi biomass: a review. Microbiol Biochem Technol 7:384–393

Simon L, Bousquet J, Levesque RC, Lalonde M (1993) ©19 9 3 Nature Publishing Group Nature 363:67–69

Singh P, Jain R, Srivastava N, Borthakur A, Pal DB, Singh R, Madhav S, Srivastava P, Tiwary D, Mishra PK (2017) Current and emerging trends in bioremediation of petrochemical waste: a review. Crit Rev Environ Sci Technol 47:155–201

Soleimani M, Afyuni M, Hajabbasi MA, Nourbakhsh F, Sabzalian MR, Christensen JH (2010) Phytoremediation of an aged petroleum contaminated soil using endophyte infected and non-infected grasses. Chemosphere 81:1084–1090

Thiruvenkatachari R, Vigneswaran S, Naidu R (2008) Permeable reactive barrier for groundwater remediation. J Ind Eng Chem 14:145–156

Tripathi V, Edrisi SA, O'Donovan A, Gupta VK, Abhilash PC (2016) Bioremediation for fueling the biobased economy. Trends Biotechnol 34:775–777

Valderrama B, Oliver P, Medrano-Soto A, Vazquez-Duhalt R (2003) Evolutionary and structural diversity of fungal laccases. Antonie Van Leeuwenhoek 84:289–299

Van Bloois E, Torres Pazmiño DE, Winter RT, Fraaije MW (2010) A robust and extracellular heme-containing peroxidase from *Thermobifida fusca* as prototype of a bacterial peroxidase superfamily. Appl Microbiol Biotechnol 86:1419–1430

Williams PA, Murray K (1974) Metabolism of benzoate and the methylbenzoates by mt-2: evidence for the existence of a TOL plasmid metabolism of benzoate and the methylbenzoates by *Pseudomonas putida* (arvilla) mt-2: evidence for the existence of a TOL plasmid. J Bacteriol 120:416–423

Wong CC, Wu SC, Kuek C, Khan AG, Wong MH (2007) The role of mycorrhizae associated with vetiver grown in Pb-/ Zn-contaminated soils: greenhouse study. Restor Ecol 15:60–67

Worsey MJ, Williams PA (1975) Metabolism of toluene and xylenes by *Pseudomonas putida* (arvilla) mt-2: evidence for a new function of the TOL plasmid. J Bacteriol 124:7–13

Xiao X, Luo S, Zeng G, Wei W, Wan Y, Chen L, Guo H, Cao Z, Yang L, Chen J, Xi Q (2010) Biosorption of cadmium by endophytic fungus (EF) Microsphaeropsis sp LSE10 isolated from cadmium hyperaccumulator *Solanum nigrum* L. Bioresour Technol 101:1668–1674

Xun F, Xie B, Liu S, Guo C (2015) Effect of plant growth-promoting bacteria (PGPR) and arbuscular mycorrhizal fungi (AMF) inoculation on oats in saline-alkali soil contaminated by petroleum to enhance phytoremediation. Environ Sci Pollut Res 22:598–608

Yao Z, Li J, Xie H, Yu C (2012) Review on remediation technologies of soil contaminated by heavy metals. Procedia Environ Sci 16:722–729

# Chapter 6
# Bioremediation of Polythenes and Plastics: A Microbial Approach

**Shubha Rani Sharma**

## Contents



## 6.1 Introduction

Human beings have always endeavored to make life easier. They have devised ways and means to have a comfortable and luxurious life. They have tried to maintain everything around us by packaging like food packaging, water packaging, etc., and for this purpose, they have used plastics and polythenes. Plastics can be defined as polymer that has the capacity of being shaped into any form with the aid of high temperature and pressure. Polythene comprises of polymer of the ethylene monomers that is the major nondegradable solid waste produced from daily use. Plastics being typically inert and resistant to attack by microbes continue to exist for years. Exploiting the additional properties of plastics like low density and toughness, we can mold plastics into different types of products. Due to their durability and low cost, synthetic polymers are widely used, but discarding of these materials has cropped up as a major issue in the area of solid waste management causing main threat to environmental pollution.

---

S. R. Sharma (✉)
Department of Bio-engineering, Birla Institute of Technology, Mesra, Ranchi, India
e-mail: srsharma@bitmesra.ac.in

 
R. Prasad, E. Aranda (eds.), *Approaches in Bioremediation*, Nanotechnology in the Life Sciences, https://doi.org/10.1007/978-3-030-02369-0_6

Now, the use of polythenes and plastics has been so profuse that it has become a serious problem as to how to decompose them. Bakelite was invented by Leo Baekeland in 1907 as the first fully synthetic plastic. As the plastics are non-biodegradable, they have become a matter of serious concern for the proper disposal of these plastics and polythenes. The industrial and urban wastes produced by the human activities are alarmingly increasing the environmental contamination. The use of plants to convert the poisonous compounds into nonpoisonous forms is known as phytoremediation (Ghosh and Singh 2005). The use of plants in bioremediation for cleanup of contaminated environment dates back about 300 years. Polythene or polyethylene which is commonly used in our day-to-day life like grocery bags, carry bags, sachets, etc. is the product of polymerization of ethylene. The general formula for polythene is $(C_2H_4)_nH_2$.

The various classes of polyethylene used are low-density polyethylene, medium-density polyethylene, high-density polyethylene, and very low-density polyethylene, out of which the low-density polyethylene is widely used for common purposes like carry bags used in grocery, packaging materials, electrical casing, etc. Plastic containers in laboratory and kitchen are very versatile as they are acid resistant. Now the versatility of the plastic which makes it so useful in every scenario has become a headache for the environmentalists as the profuse use of the versatile material in every field is resulting into heaping and piling of non-biodegradable material in the environment causing pollution. Now the time has come when we should ponder over the concerns related to plastic and polythene pollution. The physical and chemical methods of degradation of plastics may be very cumbersome and may result into production of more toxic substances. So we should go for an alternative method of plastic degradation which is safe and easy. The answer to this problem is the microbial degradation of plastics that is the use of microbial enzymes and bioactive compounds for the degradation. There are different types of plastics found in our environment. Each type of plastic has unique type of properties as well as find use in different areas. Some of them are enumerated in the Table 6.1.

## 6.2 Environmental Effects of Plastic Pollution

Plastics being very less degradable can accumulate in the environment, making it polluted. If at all they are partially degradable, their chemical by-products are more toxic, producing adverse effects. Plastic pollution in present scenario is one of the biggest environmental problems. Nowadays we have become so much used to plastics and polythenes that it seems that life without plastics is not possible. Although plastics are very helpful to us in daily life, if we come down to review the after effects of plastic pollution, we would rather quit using plastics and polythenes. As the population grows, the quantity of plastic refuse accumulates in the world causing plastic pollution (Song et al. 2009). The toxic pollutants in the plastics have adverse effect on environment. The decomposition of plastics takes a very long time say thousands of years causing serious damage to the environment. The effect of

**Table 6.1** Types of plastics, properties, and their uses

| No | Types | Name | Properties | Uses |
|---|---|---|---|---|
| 1 | PET | Polyethylene terephthalate | Absorbs odors and flavors from food and drinks that are stored in them, high heat resistant, extremely effective moisture barrier, shatterproof | Beverage bottles, medicine jars, rope, clothing, and carpet fiber |
| 2 | HDPE | High-density polyethylene | Excellent moisture barrier properties, chemical resistance | Containers for milk, motor oil, shampoos and conditioners, soap bottles, detergents, and bleaches |
| 3 | PVC | Polyvinyl chloride | Excellent transparency, hard, rigid, long-term stability, brittle, waxy surface | Plumbing pipes, credit cards, carpet backings, window or door frames, synthetic leather products |
| 4 | LDPE | Low-density polyethylene | Tough, flexible, good transparency, low melting point | Cling film, sandwich bags, squeezable bottles, and plastic grocery bags |
| 5 | PP | Polypropylene | Excellent chemical resistance, high melting point, hard but flexible, strong, translucent | Lunch boxes, margarine containers, yogurt pots, syrup bottles, prescription bottles, plastic bottle caps |
| 6 | PS | Polystyrene | Clear to opaque, glassy surface, rigid or foamed, brittle, high clarity | Disposable coffee cups, plastic food boxes, plastic cutlery, and packing foam |
| 7 | Others | Polycarbonate and polylactide | Layered or multi-material mixed polymers | Baby bottles, compact discs, and medical storage containers |

sunlight on plastic is immense which leads to photodecomposition resulting in trash and toxin-like chemicals, and these are the main cause of soil and water pollution by plastics. Plastic, synthetic polymer consisting of carbon, hydrogen, silicon, oxygen, chloride, and nitrogen, is manufactured from oil, coal, natural gas, etc. Since plastics are extremely stable and robust, they are extensively used. There are various types of plastics, e.g., polyethylene (PE), polystyrene (PS), polyvinyl chloride (PVC), nylons, polyethylene terephthalate (PET), polypropylene (PP), and polyurethane (PUR). These polymers accumulate in the surroundings as there is no efficient method for degradation or method for safe disposal, thus posing constantly aggravating threat to the flora and fauna.

A new study suggests that as of the 1950s, over 9 billion tons of plastic has been produced, and most of them have been converted into scrap which permanently persists in our environment. Around 6300 Mt of plastic waste had been produced till 2018, out of which only 9% have been recycled, 12% were incinerated, and the residual 79% have their fate in landfills thrown into the environment (Geyer et al. 2017). According to the Global Industry Analysts, plastic consumption throughout the world has been approximated roughly equal to 260 million tons in 2008, and it was anticipated to be over 300 tons in 2015. These plastics end up in the environment as a permanent disaster producing threat to wildlife and humans. Worldwide utilization of plastic material in 2015 was expected to be 45 kg/person. The highest consumption of plastic material per person for the year 2015 was expected to be in NAFTA region which was 139 kg/person followed by Western Europe where 136 kg use of plastic per person was observed. The lowest consumption of 16 kg/person was found in the Middle East and African region. If existing production and the inadequate waste management trends persist, then by 2050, we will have a burden of approximately 12,000 Mt of plastic waste. By the data given by Smithers Rapra, worldwide plastic consumption hiked during the period 2007–2012, reaching to nearly 19.6 million tons. Global specialty plastic utilization has been predicted to upsurge in the period 2012–2017 to reach 25.3 million tons. Worldwide, about 380 million tons of plastic will be produced till 2018. Over 5 million tons of plastic is consumed each year in the UK only, out of which one-quarter is only recycled and the remaining are directed to landfills. There are studies which announce that the bodies of 90% of seabirds are found to contain plastic waste. Now that people have understood the havoc caused by the accumulation of the non-biodegradable plastics in the environment, major steps are being taken to trim down the aftereffects of plastic pollution, by dropping plastic consumption and enhancing plastic recycling.

There are many hazards of plastic pollution. Let's ponder on some major aspects of this plastic pollution. Garbage dumps and landfills containing plastic refuse are one of the major problems which let the pollutants to enter the earth affecting the flora and fauna as well as groundwater for long terms. The nets used for commercial fishing are usually made of plastic. When these nets are kept underwater for the purpose of fishing for long hours, the toxins are released affecting the water and fish of the area. When the water organisms are intoxicated with the poison in the plastic ingested by them, it results into greater problem for larger organisms, which make the higher part of the food chain, thus affecting the whole food chain and web. Plastics in the landfills interact with water forming hazardous toxins which in their turn seep into the earth affecting the water quality of groundwater.

Plastics can affect animals that may come in the surrounding area and might choke them to death. When plastic is burnt in open air, it releases toxic substances leading to environmental pollution. The humans and animals are forced to inhale the polluted air which results in health hazards and can cause respiratory troubles. Due to exposure to sunlight, rainfall, and other ecological conditions, the plastics disposed in the ocean decompose at a faster rate, releasing toxic chemicals such as bisphenol A (Teuten et al. 2009). Plastic bits are taken in by filter-feeding organisms and planktons in the ocean which in turn are taken up by the higher animals leading to bioaccumulation of these toxins in the food chain and food web. The most recognized undesirable environmental effects of polyethylene films are when they are swallowed by wild animals. The most general types of plastic pollution in oceans are the polystyrene, plastic pellets, plastic bags, and food containers. Chlorinated plastics release harmful chemicals which can then seep into groundwater causing serious injury to the species that consume this water. Different types of plastics continuously accumulate in the landfills.

In the manufacture of plastics, some of the chemicals that are used are easily absorbed by human beings through skin absorption and cause different types of diseases like dermatitis, etc. There are innumerable side effects of plastics and polythenes which we are not even aware of. Now the most important need of the hour is the alternative to plastics and the biodegradation of plastic heaps which have already accumulated and clogged our environment causing serious threats. Most of the countries restrict the use of incineration process for disposal of wastes as plastics are the major constituent. As an alternative, uncontrolled burning and landfilling are used to get rid of the plastics. Now the open burning can pose various health hazards. Some of the organic pollutants like furans and dioxins are produced as the by-product of polyvinyl chloride (PVC) plastic burning. Various immune diseases and lung disorders are produced in the burning of polyethylene, polyurethane, polyvinyl chloride, and polystyrene.

## 6.3 Microbial Role in Biodegradation of Plastics

Some of the living organisms are instrumental in the degradation of plastics, and, as a result, the degraded stuff is returned back to the environment. Both aerobic and anaerobic microbes processes are employed for degradation of plastics. Methane, which is considered as the greenhouse gas, is released in the landfills, resulting into global warming. Fungal degradation can proceed strictly under aerobic conditions, whereas bacterial degradation can occur both in aerobic and anaerobic conditions (Kumar et al. 2011). The polymer decomposes and produces chemicals which act as nutrients and support the microbial growth. Biological degradation is the process of biological conversion of organic compounds into some other specific product by the use of microbes (Restrepo-Florez et al. 2014).

Biodegradation is the practice in which organic substances are broken down by living organisms. It is anticipated that biodegradation is one of the chief methods for the release of most of the chemicals into the atmosphere. The ability of microorganisms to control degradation through physical, chemical, or enzymatic action can be aptly named as biodegradation. Here the microbes use the components in the plastics as carbon and energy source for their growth. Cracking, erosion, discoloration, and phase separation are the different methods by which plastics are degraded. Organic materials can be degraded aerobically with oxygen or anaerobically without oxygen. Microorganisms play a major role in the degradation of both synthetic and natural polymers which is depicted in Fig. 6.1. The microbes can biodegrade the polymers but not use it directly as a nutrient. The decomposition of plastics is considered to be an extremely slow process which is first started by the natural forces and effects like temperature, pH, and ultraviolet rays, and then the microbes adopt a special strategy to completely decompose the remains of the plastics.

In order to degrade the plastics, we can employ different types of methods like chemical, thermal, photooxidation, and biodegradation. Depending on the type of polymer, different time periods are witnessed to degrade them. Different types of

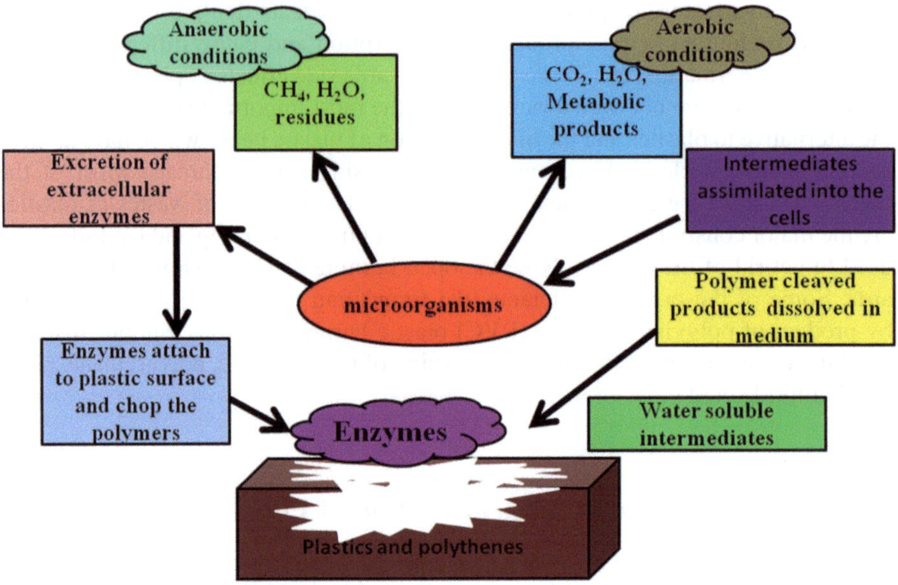


**Fig. 6.1** Mechanism of plastic degradation

research reveal that microorganisms can be used for the biodegradation and enormous genera of bacteria (including Actinomycetes) and fungi have proved to degrade plastics (Mahdiyah and Mukti 2013). Usually, microorganisms biodegrade plastic very slowly, and also certain specific plastics cannot be degraded by them (Singh and Gupta 2014). Biodegradable plastics are especially fabricated so that they are ready to be degraded under environmental conditions; thus new avenues were targeted for waste management. Polyhydroxyalkanoates (PHAs) are a bioplastic produced by many of the microbes. It is considered very safe and nontoxic. These can be biodegraded. PHAs are completely biodegradable and biocompatible. They resemble to the petrochemical-based traditional plastics, and they are available in a variety of polymers.

## 6.4 Bacteria Involved in Biodegradation of Plastics and Polythenes

Microorganisms, chiefly bacteria, are responsible for the biodegradation of substances (Shah and Fariha 2008). As bacteria remineralize the organic carbon in the wastes, they can be considered as saprophytic scavengers. The hydrocarbons which are the major components of plastics are degraded by bacteria which are specialized for the purpose. *Ideonella sakaiensis*, a bacterium from the genus *Ideonella* and family *Comamonadaceae*, was found to be capable of breaking down PET (polyethylene terephthalate) plastic which was a by-product from a plastic recycling plant in Sakai, Japan (Yoshida et al. 2016). The bacterium gets nourished solely on PET and uses enzymes to break them. PETase is the most important enzyme secreted by *I. sakaiensis* 201-F6, a plastic-feeding bacterium. PETase degrades PET into mono(2-hydroxyethyl) terephthalic acid (MHET) which is then further broken down into terephthalic acid and ethylene glycol. The final products are used as food source by the bacterium (Austin et al. 2018).

Chemically different kinds of plastic bags HL (plastic bags containing nano-additives from the Netherlands) and VN1 (plastic bags with additives sold at supermarkets in Vietnam) were found to be degraded by a new thermophilic bacterial strain *Bacillus* sp. BCBT21 (Dang et al. 2018). This species that was isolated from agricultural waste undergoing composting in Vietnam was found to produce extracellular hydrolase enzymes including lipase, CMCase, xylanase, chitinase, and protease (Dang et al. 2018).

Microbes responsible for degradation of plastics were isolated from two different soil sample plastic strips. The isolates were identified as *Streptococcus* sp., *Pseudomonas* sp., and *Bacillus* sp. for bacteria and *Aspergillus* sp. and *Fusarium* sp. for fungi. The isolates were tested for their biodegradative ability, and after about 30 days, it was observed that 23% degradation was by bacterial sp. and 44% by fungal sp. (Vignesh et al. 2016).

The decomposed vulcanized natural rubber and synthetic poly(cis-1,4-isoprene) are used by the bacterial strains, i.e., *Streptomyces coelicolor* 1A and *Pseudomonas citronellolis*. Five Gram-positive and two Gram-negative bacteria and eight fungal species of *Aspergillus* were found associated with the degrading plastics. The chief species were *Streptococcus*, *Staphylococcus*, *Micrococcus* (Gram positive), *Moraxella*, and *Pseudomonas* (Gram negative) and two species of fungi (*Aspergillus glaucus* and *A. niger*). Shaker cultures were employed to check the efficiency of the microbial species in degradation of plastics and polythene. The 20.54% of polythene and 8.16% of plastics were found to be degraded by *Pseudomonas* species, degraded in 1-month period. *A. glaucus* among the fungal species was found to degrade 28.80% of polythene and 7.26% of plastics in 1-month period (Kathiresan 2003). The bacterium *Alcaligenes faecalis* produces polycaprolactone depolymerase (Oda et al. 1997) which degraded polycaprolactone (PCL) which is a biopolymer. Biodegradation of plastics was found to be efficiently done by fungal species like *Phanerochaete* and bacterial species *Streptomyces* (Lee et al. 1991). This was the first report indicating bacterial degradation by oxidization of polyethylenes in pure culture. As an electron acceptor, the aerobic bacteria use oxygen and decompose the organic chemicals finally into $CO_2$ and water.

It has been found in the recent studies that nitrate, sulfate, iron, manganese, and carbon dioxide are utilized as their electron acceptors by some anaerobic microbes and thus decompose the organic chemicals into smaller compounds (Datta et al. 1998). *Serratia marcescens marcescens* is found to degrade polyethylene plastic films like linear low-density polyethylene (LLDPE) which is used in the packaged water. A novel process of dissolving technique for plastics was used causing an increase in the surface area as well as creating an interface with the environment that may be the cause of microbial growth (Odusanya et al. 2013). Wax worms, or Indian meal moths, were found to be capable of degrading polyethylene films. *Enterobacter asburiae* YT1 and *Bacillus* sp. YP1 were the two bacterial strains capable of degrading PE, isolated from this worm's gut (Yang et al. 2014).

## 6.5 Fungi Involved in Plastic Biodegradation

The fungus *Penicillium simplicissimum* YK was found to utilize the integral polyethylene as the carbon source and grew better on agar plates containing irradiated polythene (which contained carbonyl groups) than intact polyethylene having no carbonyl groups. It was also observed that hyphae of the fungus were more efficient in degrading intact polyethylene than the spores (Yamada-Onodera et al. 2001). PCL (polycaprolactone) is a biodegradable polymer, which is degraded by microorganisms aerobically as well as anaerobically. Fungi like *Penicillium* and *Aspergillus* are very active in degrading PCL. A report was furnished where they showed that the *Aspergillus* strain ST-01 degraded PCL when incubated at 50 °C for 6 days. It took 12 days for the *Penicillium* species strain 26-1 to degrade PCL (Bhardwaj et al. 2013). *Penicillium funiculosum* and *Aspergillus flavus* were shown to produce enzymes which cleave the PCL. Lipases and esterases are the enzymes which are involved in the degradation of PCL. Organic acid esters like dioctyl adipate (DOA) and dioctyl phthalate (DOP) existing in plasticized polyvinyl chloride (pPVC) make it susceptible to microbial attack (Webb et al. 1999).

Colonization of fungus and biodegradation of pPVC under in situ and ex situ conditions were investigated by Webb and his colleagues in 1999. They reported that 25 and 40 weeks were required for *Aureobasidium pullulans* attached itself on pPVC. It was demonstrated that *A. pullulans* can use pure pPVC as source of carbon. It produces extracellular esterase by degrading DOA. This suggested that fungus *A. pullulans* is trusted to be essential in degrading pPVC. Both fungus and bacteria are found to be instrumental in degrading polyester polyurethane (PUR). The endophytic fungi can degrade synthetic polymer polyester polyurethane (PUR) (Russell et al. 2011). In both aerobic and anaerobic conditions, *Pestalotiopsis microspora* was the fungus which was growing on PUR by using it as sole source of carbon. It was shown that serine hydrolase from the fungus was instrumental in degrading PUR. Here we must make a mention that the endophytes can be considered as the microorganisms which live in the plant tissues and actively participate in decomposition of the plant after the host plant dies.

In 2003, Kathiresan conducted experiments and concluded that the microbes obtained from mangrove soil were also slowly able to degrade plastics. Thus, the work revealed that many plastic-degrading microbes can be isolated from mangrove soil. Moreover, mangrove soil from the Niger Delta was also studied for plastic-degrading microbe population. Two *Aspergillus* species were isolated which were studied for degradation of low-density polyethylene (LDPE) (Pramila Vijaya Ramesh, 2012), and some of the fungal isolates have been found to degrade polyethylene sheets (high-density polyethylene (HDPE) and low-density polyethylene, LDPE). *Penicillium oxalicum* NS4 (KU559906) and *Penicillium chrysogenum* NS10 (KU559907) are the two potential strains which have been witnessed to have the plastic-degrading potential. The degradation of the polythenes like HDPE and LDPE films was found to be done by the unique enzymatic activities in the two fungal strains like *Penicillium oxalicum* NS4 (KU559906) and *Penicillium chrysogenum* NS10 (KU559907) (Ojha et al. 2017). Here prerequisites like prior oxidation or other chemical treatments are not required in the process of degradation. Endophytic fungi were researched for their capability to degrade the synthetic polymer polyester polyurethane (PUR) (Álvarez-Barragán et al. 2016). Under aerobic as well as anaerobic conditions, *Pestalotiopsis microspora* was found to grow on PUR as the sole carbon source. When the molecular characterization of the organism was done, it was found that a particular enzyme, known as serine hydrolase, is accountable for degradation of PUR (Russell et al. 2011). Fungi were found to be outstanding in degrading of polymers in soil as they produce a plethora of enzymes such as glucosidase, cutinase, amylase, lipase, esterase, cellulase, pectinase, and hemicellulase. Fungal strains like *Aspergillus niger* and *A. japonicus* were found to be effective in polythene carry bags of low-density polyethylene (LDPE) degradation (Raaman et al. 2012). As fungi are able to produce an array of enzymes such as glucosidase, cutinase, amylase, lipase, esterase, cellulase, pectinase, and hemicellulase, they would be very instrumental in degrading of polymers (Anastasi et al. 2013).

A fungus with the potential to destroy non-biodegradable plastics was found by the scientists at the Chinese Academy of Sciences' Kunming Institute of Botany. Waste plastics were degraded in a week's time by the fungus. *Aspergillus tubingensis* was a soil fungus which could happily live on the plastic surface by secreting enzymes (Khan et al. 2017). The fungi isolated from El-Sharqia soil, Egypt, were found to produce protease, esterase, and lipase followed by those isolated from Ismailia soil. The fungi with high esterase activity were recognized as *Monascus ruber*, *Monascus sanguineus*, and *Monascus* sp. Maximum esterase was found to be produced by *M. ruber* followed by *M. sanguineus*. *Monascus* sp. isolated from El-Sharqia was the most efficient isolate in degradation of polyurethane to Impranil DLN. Among the ten fungal strains that were isolated from Red Sea water, Jeddah, Saudi Arabia, *Aspergillus* and *Penicillium* species were found to biodegrade the low-density polyethylene, like polyethylene films and powder. The highest percentage in reduction of polyethylene weight was exhibited by *Penicillium* sp. (Alshehrei 2017). Two fungal strains isolated from municipal landfill area were tested for the biodegradation of low-density polyethylene. The degrading ability of the two fungal strains exhibited promising degradation pattern as analyzed by colonization studies, SEM and Sturm test analysis. The fungi were recognized as *Mucor circinelloides* and *Aspergillus flavus*.

## 6.6 Factors Involved in Biodegradation of Plastics

The biodegradation of plastics by bacteria and fungi takes place at different rates depending upon the soil conditions as well as the types and constitution of the plastics. The different factors that direct the biodegradation process of plastics depend mostly on the type of organism involved in the degradation, the properties of the polymer, and the type of pretreatment given to the polymer.

There are many factors responsible for plastic degradation by microbes which are enumerated in Fig. 6.2. The physicochemical parameters that decide the plastic degradation can be enumerated as the melting temperature, molecular weight, surface area, hydrophilic and hydrophobic nature of polymer, crystallinity, chemical structure, etc. Chemical and physical properties of plastics play important role in their biodegradation. The polymers with side chains are less degradable when compared to polymers with no side chains. Along with the high molecular weight, the morphology, the melting temperature, and the crystal property play a crucial role in the degree of biodegradation of plastics. The amorphous polymer is degraded easily in contrast to crystalline polymer.

Polymers with high melting temperatures are difficult to biodegrade. Polycaprolactone with higher molecular weight was degraded slowly by *Rhizopus delemar* lipase (endo-cleavage type) than that with low molecular weight (Tokiwa and Suzuki 1978). The various factors affecting the biodegradation of plastics and polythenes may be enumerated as composition that defines its chemical and physical properties; concentration; abiotic factors like temperature, salinity, presence of water, etc.; and biotic factors which include the composition of microbial community.

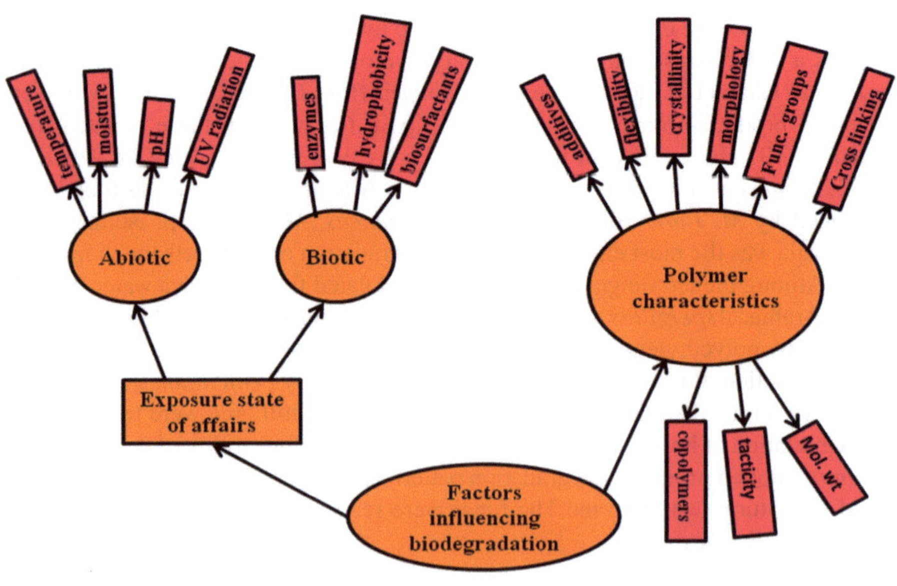


**Fig. 6.2** Various factors affecting the biodegradation of plastics and polythenes

The knowledge composition of the hydrocarbons is very crucial for the rate of degradation of the plastics and polythenes (Gajendiran et al. 2016). The hydrocarbons present in the plastics and polythenes are of different types and have been placed in the decreasing order of vulnerability to biodegradation. The n-alkanes are the most susceptible ones for biodegradation by the microbes than the branched alkanes followed by low molecular weight aromatics and the cyclic alkanes, with high molecular weight aromatics and polar compounds being extremely resistant. Nevertheless the order of biodegradation can vary under some conditions depending upon the composition of microbial community or the abiotic factors. Nutrients are very important ingredients for successful biodegradation of hydrocarbon pollutants especially nitrogen, phosphorus, and in some cases iron. The rate of biodegradation of the plastics follows the Michaelis-Menten kinetics (Chinaglia et al. 2018).

The microbial populations can degrade the substrates depending upon concentration of the hydrocarbons. The microbial degradation of high molecular weight hydrocarbons, such as long (>C12) alkanes with solubility less than 0.01 mg/l, occurs at rates that exceed the rates of their dissolution and are a function of the hydrocarbon surface area available for emulsification or physical attachment by cells and therefore do not display the dependence on concentration. At lower temperatures, the speed of plastic biodegradation generally decreases (Tokiwa et al. 2009). This may be accounted due to the decrease in enzymatic activity at lower temperatures. The rates of hydrocarbon metabolism are increased by higher temperatures, the optimum range being 30–40 °C. The oxidation of the substrate by oxygenases is the prerequisite in the biodegradation process of plastics by bacteria and fungi. Scarcity of nitrogen and phosphorus inhibits microbial plastic degradation in the environment, so for the process of biodegradation of plastics, the nitrogen and phosphorus concentration should be high in the environment (Ong et al. 2017).

The presence of salinity is inversely related with the biodegradation of plastics as the high salinity decreases the existence of microbes (Le Borgne et al. 2008). It was witnessed that rates of hydrocarbon consumption start to reduce when the salinity is between 3.3% and 28.4%. Crude oil-degrading *Streptomyces albiaxialis* and an n-alkane (C10–C30)-degrading member of the *Halobacterium* group however are capable of biodegrading plastics at high salinity. The effect of pressure on biodegradation of hydrocarbons is also observed (Schedler et al. 2014). It was seen that microorganisms degrade the plastic at the bottom of the sea or ocean extremely slowly, and so they continue to stay there for years and years altogether.

Most favorable rates of biodegradation of hydrocarbons are witnessed at 30–90% of water availability (Sudhakar et al. 2007). Thus the biodegradation of hydrocarbon in terrestrial ecosystems may be restricted by the amount of available water. The bacterial degradation is increased in the moist conditions, thus increasing the evolution of gas. Most of the bacteria and fungi involved in biodegradation prefer neutral pH (Siddique et al. 2002), and most of the times, the fungi are seen to be more tolerant to acidic conditions. Bacteria and fungi are the pioneers in biodegradation of hydrocarbons in the environment, and although the algae and protozoa both belong to the community of microorganisms of aquatic and terrestrial ecosystems, their role in hydrocarbon biodegradation is still a mystery.

## 6.7 Different Steps of Plastic Degradation by Microorganisms

The polymer compounds are broken down by the microorganisms into simpler forms by the help of biochemical reactions. Biodegradation of polymer is carried by digestion by microbial enzymes leading to decrease in molecular weight, loss of properties like mechanical strength, etc. Microorganisms produce catalytic enzymes for the purpose of biodegradation (Nigam 2013). This proves to be very instrumental in environmental waste management. Different enzymatic reactions of the microbial enzymes and breaking of bonds result into biodegradation. Sequential degradation steps are involved in the biodegradation process like bio-fragmentation by enzymatic cleavage and assimilation of the products by microorganisms and mineralization which involves formation of oxidized metabolites that is shown in Fig. 6.3. Both aerobic and anaerobic conditions are essential for mineralization of polymers. $CO_2$ and $H_2O$ are the products of aerobic oxidation, while $CH_4$, $CO_2$, and $H_2O$ are produced under anaerobic conditions (Singh and Sharma 2008).

Different steps involved in the plastic biodegradation process are:

1. *Biodeterioration*: Decomposing microbial communities are accountable for the physical and chemical degradation which results in modification of different properties of plastics like mechanical, physical, and chemical properties. Most of the times, abiotic parameters are useful either as a synergistic factor or to initiate

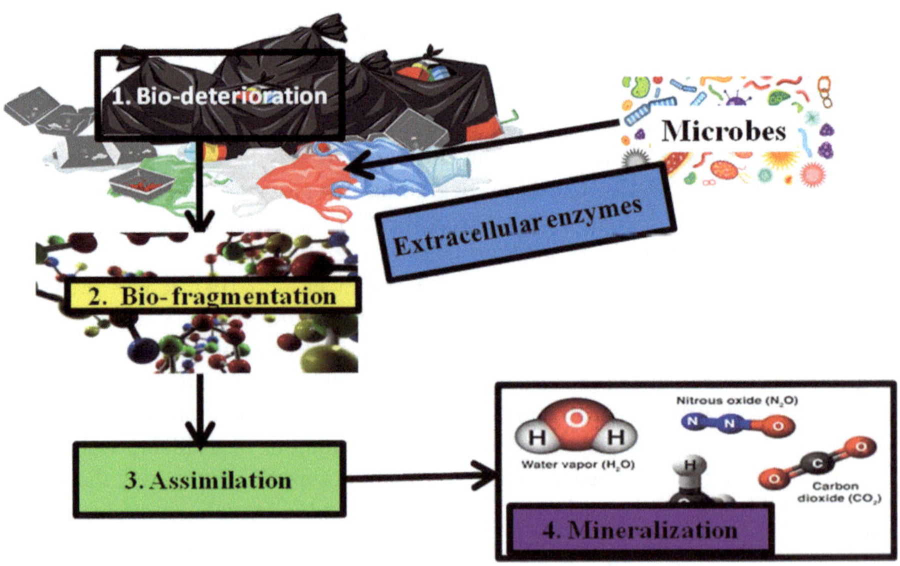


**Fig. 6.3** Different steps involved in plastic biodegradation

the biodegradation process and deteriorate the polymeric structure (Helbling et al. 2006; Ipekoglu et al. 2007). The formation of a microbial biofilm on the surface aggravates biodeterioration.

2. *Bio-fragmentation*: It is the catalytic events where enzymes secreted by microorganisms cleave the plastics into monomers. Microorganisms secrete extracellular enzymes which catalyze reactions at the margins of the plastic polymer. The bacteria that are capable of degrading the plastics generally contain oxygenase enzyme which can add oxygen to a long carbon chain making it susceptible to cleavage. Post-formation of carboxylic groups, the lipases and esterases result in further degradation of the substrates.
3. *Assimilation*: Only those monomers which can be assimilated by the microbes, which are able to cross the cell wall. Those monomers which use particular carriers to cross the cell wall are able to get assimilated; others are left behind as they are. The assimilation produces a number of secondary metabolites which is carried out of cell to be utilized for further degradation. The absolute degradation of primary and secondary metabolites results in production of $CO_2$, $N_2$, $CH_4$, $H_2O$, etc. which is better known as mineralization of metabolites.
4. *Mineralization*: The final step in the process of biodegradation is the absolute degradation of molecules of the products of biodegradation that results in release of $CO_2$, $N_2$, $CH_4$, $H_2O$, etc.

## 6.8 Enzymes Involved in Biodegradation of Plastics and Polythenes

The biodegradation of polyethylene requires the set of microbial enzymes capable of degrading lignin in plant cells, which is a heterogeneous cross-linked phenolic polymer (Restrepo-Flórez et al. 2014). These microbial enzymes comprise of laccases, manganese peroxidase, and lignin peroxidases. It was found that, a copper-binding thermozyme, laccase obtained from *Rhodococcus ruber* C208, was very effective in degradation of the UV-treated polyethylene films in the presence of copper (Sivan et al. 2006). A laccase from *Trametes versicolor* in the presence of 1-hydroxybenzotriazole was instrumental in oxidation of non-phenolic substrates to degrade polyethylene membrane (Fujisawa et al. 2001).

The key enzyme involved in the degradation of a high molecular weight PE membrane is named as manganese peroxidase (MnP) which has been obtained from the white rot fungus *Phanerochaete chrysosporium* ME-446 and isolate IZU-154 (Iiyoshi et al. 1998). Extracellular laccases as well as MnP produced from *Bacillus cereus* were found to degrade UV-irradiated polyethylene (Sowmya et al. 2014). Seventy percent of a preoxidized high molecular weight PE was found to be degraded by lignin peroxidase and MnP that were produced by *Phanerochaete chrysosporium* MTCC-787 (Mukherjee and Kundu 2014).

Terminal or subterminal oxidation of hydrocarbon oligomers by alkane hydroxylases can catalyze the degradation (Rojo 2010). Alkane hydrolases, alkane monooxygenase, rubredoxin, and rubredoxin reductase from *Pseudomonas aeruginosa* E7 were very instrumental in degrading polyethylenes (Jeon and Kim 2015). It has been witnessed that whole cells rather than the isolated enzymes have more potential for biodegradation of plastic and polythenes. A number of lipases, esterases, and cutinases from fungal and actinomycetes species hydrolyze amorphous PET and modify the surface of PET films and fibers (Zimmermann and Billig 2011). Polyethylene terephthalate (PET) fibers were seen to be partially degraded by carboxylesterases obtained from *Bacillus licheniformis*, *Bacillus subtilis*, and *Thermobifida fusca* (Barth et al. 2016). The most active fungal polyester hydrolase is cutinase HiC, which is thermostable and is obtained from *Thermomyces* (formerly *Humicola*) *insolens* (Ronkvist et al. 2009).

There is a plethora of microbial enzymes for the degradation of the plastic which are reflected in Fig. 6.4. Some examples are alkane hydroxylase obtained from *Pseudomonas aeruginosa* for biodegradation of polyethylene (Jeon and Kim 2015), lipases (Bhardwaj et al. 2012), laccase (*Rhodococcus ruber*), lignin- and manganese-dependent peroxidase (LiP and MnP), serine hydrolases, esterases, polyurethanases, heme peroxidase (lignin peroxidase), protease, etc. The enzyme such as the oxygenase increases the solubility of plastic polymer, rendering them more readily degradable by bacteria. Microbial lipases and esterases specifically attack carboxylic groups and amine group's endopeptidases. The humification of various phenolic substances produced from the decomposition of lignin is carried by the

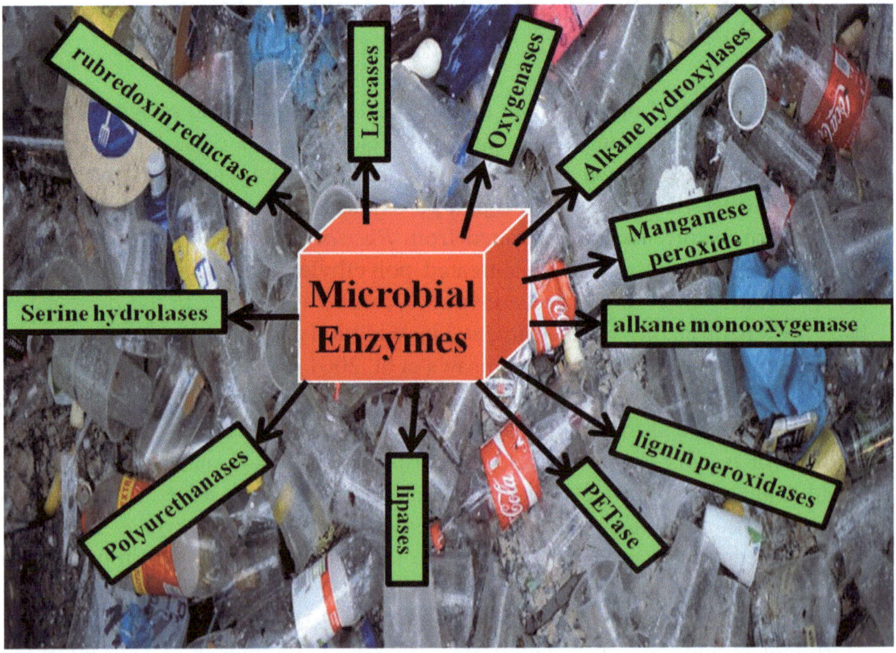


**Fig. 6.4** Microbial enzymes used in biodegradation of plastics and polythenes

oxidoreductases. Different kinds of enzymes like laccase, manganese peroxidase, and lignin peroxidase are the main weapons of the fungal population that they use for biodegradation of the plastics and polythenes. The main extracellular oxidoreductase enzyme is the chief enzyme which helps the fungi in the process of biodegradation. The monooxygenases oxidize substrates ranging from alkanes to complex endogenous molecules such as steroids and fatty acids.

## 6.9 Conclusion

Synthetic or man-made plastics can be considered as xenobiotics. They are very extensively used materials in our day-to-day life and are omnipresent. They have become a matter of ponderance as they clog the various sites of the environment. They are so recalcitrant and nondegradable that it has become a burning issue as to how to get rid of them. They are causing a great environmental havoc, and there is an urgent need to address the problem of accumulation of the plastics and polythenes in the nature. The different methods of degradation of the synthetic polymers have been researched on, be it physical methods, chemical methods, as well as microbial methods. Out of these we found that the participation of the microbes has come up with wonderful results for biodegradation of plastics and polymers.

*Pseudomonas* which is the member of metabolically active species shows special performance in the biodegradation of synthetic polymers due to their capability to degrade as well as metabolize the synthetic substrates. Microbial species isolated from different ecological areas have been witnessed to degrade polyethylene, polyethylene terephthalate, polypropylene, polyethylene glycol, polystyrene, polyurethane, polyethylene succinate, polyvinyl alcohol, and polyvinyl chloride. Spotlight of research should be on the safe disposal of most common constituent of the plastics and polythenes like polyethylene, polypropylene, polyurethane, polystyrene, etc. These polymers are also long-lasting as well as the most durable plastics. Considering these problems, different types of strategies have been suggested to come to the rescue of the problem of safe waste disposal. The most efficient technique to overcome the problem is the biodegradation of plastic by enzymes obtained from the microbes. Lots and lots of research activities have been started on a large scale be it at international level or national level to overcome the problem of pollution that is being caused by the plastics.

The mechanism of biodegradation of the synthetic products by the microbial enzymes has been studied and employed, and more research in this field is being carried. The study of molecular mechanisms involved in the process of biodegradation like bio-fragmentation, bio-assimilation, and bio-mineralization is still in their infancy, and thus a lot of investigations are still on their way to explore these phenomenons.

There is a diverse population of microbes involved in the process of biodegradation which are very efficient in producing different types of enzymes for the biodegradation of recalcitrant substrates. *R. ruber*, *C. thermocellum*, *P. aeruginosa*, *P.*

*stutzeri*, *S. badius*, *S. setonii*, *C. acidovorans*, etc. some of them are the chief bacterial spp. related with the biodegradation of polymers. *P. aeruginosa* is one of the most studied microorganisms for polymer. Low-density polyethylenes are degraded very efficiently by *P. aeruginosa* CA9. Hydrolyzing enzymes produced by *Rhodococcus ruber* have been shown to degrade polyethylene by formation of a biofilm. Microbial peroxidases have also been found to enhance the polyethylene biodegradation by introducing in vitro thermal and photochemical mineralization.

Hence, further future studies on the screening of effective microbial strains which produce an array of hydrolytic enzymes are essential to eliminate the plastic pollution from the environment. The future prospects for a cleaner and healthy environment will lie on the product of only biodegradable products in the time to come, but our focus should be to get rid of the pile of tough plastics and polymers that have previously been designed and produced by exploring new enzymes and new microorganisms.

## References


Alshehrei F (2017) Biodegradation of low density polyethylene by fungi isolated from Red Sea water. Int J Curr Microbiol App Sci 6(8):1703–1709

Álvarez-Barragán J, Domínguez-Malfavón L, Vargas-Suárez M, González-Hernández R et al (2016) Biodegradative activities of selected environmental fungi on a polyester polyurethane varnish and polyether polyurethane foams. Appl Environ Microbiol 82:5225–5235

Anastasi A, Tigini V, Varese GC (2013) The bioremediation potential of different ecophysiological groups of fungi, Fungi as Bioremediators, soil biology, vol 32. Springer, Berlin

Austin HP, Allen MD, Donohoe BS, Rorrer NA, Kearns FL, Silveira RL, Pollard BC, Dominick G, Duman R, El Omari K, Mykhaylyk V, Wagner A, Michener WE, Amore A, Skaf MS, Crowley MF, Thorne AW, Johnson CW, Woodcock HL, McGeehan JE, Beckham GT (2018) Characterization and engineering of a plastic-degrading aromatic polyesterase. Proc Natl Acad Sci U S A 115:E4350–E4357

Barth M, Honak A, Oeser T, Wei R, Belisario-Ferrari MR, Then J et al (2016) A dual enzyme system composed of a polyester hydrolase and a carboxylesterase enhances the biocatalytic degradation of polyethylene terephthalate films. Biotechnol J 11:1082–1087

Bhardwaj H, Gupta R, Tiwari A (2012) Microbial population associated with plastic degradation. Sci Rep 5:272–274

Bhardwaj H, Gupta R, Tiwari A (2013) Communities of microbial enzymes associated with biodegradation of plastics. J Polym Environ 21:575–579

Dang TCH et al (2018) Plastic degradation by thermophilic Bacillus sp. BCBT21 isolated from composting agricultural residual in Vietnam. Adv Nat Sci: Nanosci Nanotechnol 9(1):015014. https://doi.org/10.1088/2043-6254/aaabaf

Chinaglia S, Maurizio T, Degli-Innocenti F (2018) Biodegradation rate of biodegradable plastics at molecular level. Polym Degrad Stab 147:237–244

Datta PK, Mishra K, Kumar MNVR (1998) Popular plastics and packaging. Mahindra Publishers, New Delhi, p 73

Fujisawa M, Hirai H, Nishida T (2001) Degradation of polyethylene and nylon-66 by the laccase-mediator system. J Polym Environ 9:103–108

Gajendiran A, Krishnamoorthy S, Abraham J (2016) Microbial degradation of low-density polyethylene (LDPE) by *Aspergillus clavatus* strain JASK1 isolated from landfill soil. 3 Biotech 6(1):52

Geyer R, Jambeck JR, Law KL (2017) Production, use, and fate of all plastics ever made. Sci Adv 3:e1700782
Ghosh M, Singh SP (2005) A review on phytoremediation of heavy metals and utilization of its byproducts. Appl Ecol Environ Res 3(1):1–18
Helbling C, Abanilla M, Lee L, Karbhari VM (2006) Issues of variability and durability under synergistic exposure conditions related to advanced polymer composites in civil infrastructure. Compos Part A Appl Sci Manuf 37(8):1102–1110
Iiyoshi Y, Tsutsumi Y, Nishida T (1998) Polyethylene degradation by lignin-degrading fungi and manganese peroxidase. J Wood Sci 44(3):222–229
Ipekoglu B, Böke H, Cizer O (2007) Assessment of material use in relation to climate in historical buildings. Build Environ 42:970–978
Jeon HJ, Kim MN (2015) Functional analysis of alkane hydroxylase system derived from Pseudomonas aeruginosa E7 for low molecular weight polyethylene biodegradation. Int Biodeterior Biodegrad 103:141–146
Kathiresan K (2003) Polythene and plastics-degrading microbes from the mangrove soil. Rev Biol Trop 51:3–4
Khan S, Nadir S, Shah ZU, Shah AA, Karunarathna SC, Xu J, Khan A, Munir S, Hasan F (2017) Biodegradation of polyester polyurethane by Aspergillus tubingensis. Environ Pollut 225:469–480
Kumar AA, Karthick K, Arumugam KP (2011) Biodegradable polymers and its applications. Int J Biosci Biochem Bioinforma 1(3):173–176
Le Borgne S, Paniagua D, Vazquez-Duhalt R (2008) Biodegradation of organic pollutants by halophilic bacteria and archaea. J Mol Microbiol Biotechnol 15:74–92
Lee B, Pometto AL, Fratzke A, Bailey TB (1991) Biodegradation of degradable plastic polyethylene by phanerochaete and streptomyces species. Appl Environ Microbiol 57:678–685
Mahdiyah D, Mukti BH (2013) Isolation of polyethylene plastic degrading-bacteria. Biosci Int 2:29–32
Mukherjee S, Kundu PP (2014) Alkaline fungal degradation of oxidized polyethylene in black liquor: studies on the effect of lignin peroxidases and manganese peroxidases. J Appl Polym Sci 131:40738
Nigam PS (2013) Microbial enzymes with special characteristics for biotechnological applications. Biomol Ther 3(3):597–611
Oda Y, Oida N, Urakami T, Tonomura K (1997) Polycaprolactone depolymerase produced by the bacterium *Alcaligenes faecalis*. FEMS Microbiol Lett 152:339–343
Odusanya SA, Nkwogu JV, Alu N, Udo GE, Ajao JA, Osinkolu GA, Uzomah AC (2013) Preliminary studies on microbial degradation of plastics used in packaging potable water in Nigeria. Niger Food J 31(2):63–72
Ojha N, Pradhan N, Singh S, Barla A, Shrivastava A, Khatua P, Rai V, Bose S (2017) Evaluation of HDPE and LDPE degradation by fungus, implemented by statistical optimization. Sci Rep 7:39515
Ong SY, Chee JY, Sudesh K (2017) Degradation of Polyhydroxyalkanoate (PHA): a review. J SibFU 10(2):211
Pramila R, Vijaya Ramesh K (2012) Biodegradation of low density polyethylene (LDPE) by fungi isolated from municipal landfill area. J Microbiol Biotechnol Res 1(4):131–136
Raaman N, Rajitha N, Jayshree A, Raman J (2012) Biodegradation of plastic by *Aspergillus* spp. isolated from polythene polluted sites around Chennai. J Acad Ind Res 1:313–316
Restrepo-Florez JM, Bassi A, Thompson MR (2014) Microbial degradation and deterioration of polyethylene – a review. Int Biodeterior Biodegrad 88:83–90
Rojo F (2010) Enzymes for aerobic degradation of alkanes. In: Handbook of hydrocarbon and lipid microbiology, 2, 781–797, https://doi.org/10.1007/978-3-540-77587-4_59
Ronkvist ASM, Xie W, Lu W, Gross RA (2009) Cutinase-catalyzed hydrolysis of poly(ethylene terephthalate). Macromolecules 42:5128–5138

Russell JR, Huang J, Anand P, Kucera K, Sandoval AG, Dantzler KW, Hickman D, Jee J, Kimovec FM, Koppstein D, Marks DH, Mittermiller PA, Nunez SJ, Santiago M, Townes MA, Vishnevetsky M, Williams NE, Vargas MP, Boulanger LA, Bascom-Slack C, Strobel SA (2011) Biodegradation of polyester polyurethane by endophytic fungi. Appl Environ Microbiol 77:6076–6084

Schedler M, Hiessl R, Valladares Juárez AG, Gust G, Müller R (2014) Effect of high pressure on hydrocarbon-degrading bacteria. AMB Express 4:77

Shah AA, Fariha H (2008) Biological degradation of plastics: a comprehensive review. Biotechnol Adv 26:246–265

Siddique T, Okeke BC, Arshad M, Frankenberger WT (2002) Temperature and pH effects on biodegradation of hexachlorocyclohexane isomers in water and a soil slurry. J Agric Food Chem 50:5070–5076

Singh J, Gupta K (2014) Screening and identification of low density polyethylene (LDPE) degrading soil fungi isolated from polythene polluted sites around Gwalior city (MP). Int Curr Microbiol Appl Sci 3:443–448

Singh B, Sharma N (2008) Mechanistic implications of plastic degradation. Polym Degrad Stab 93:561–584

Sivan A, Szanto M, Pavlov V (2006) Biofilm development of the polyethylene-degrading bacterium *Rhodococcus ruber*. Appl Microbiol Biotechnol 72:346–352

Song JH, Murphy RJ, Narayan R, Davies GBH (2009) Biodegradable and compostable alternatives to conventional plastics. Philos Trans R Soc Biol 364:2127–2139

Sowmya HV, Ramalingappa MK, Thippeswamy B (2014) Biodegradation of polyethylene by *Bacillus cereus*. Adv Polym Sci Technol-Int J 4:28–32

Sudhakar M, Trishul A, Doble M, Suresh Kumar K, Syed Jahan S et al (2007) Biofouling and biodegradation of polyolefins in ocean waters. Polym Degrad Stab 92:1743–1752

Teuten EL, Saquing JM, Knappe DRU, Barlaz MA, Jonsson S, Björn A et al (2009) Transport and release of chemicals from plastics to the environment and to wildlife. Philos Trans R Soc Lond Ser B Biol Sci 364(1526):2027–2045

Tokiwa Y, Suzuki T (1978) Hydrolysis of polyesters by *Rhizopus delemar* lipase. Agric Biol Chem 42:1071–1072

Tokiwa Y, Calabia BP, Ugwu CU, Aiba S (2009) Biodegradability of plastics. Int J Mol Sci 10(9):3722–3742

Vignesh R, Deepika RC, Manigandan P, Janani R (2016) Screening of plastic degrading microbes from various dumped soil samples. Int Res J Eng Tech 3(4):2493–2498

Webb JS, Van der Mei HC, Nixon M, Eastwood IM, Greenhalgh M, Read SJ, Robson GD, Handley PS (1999) Plasticizers increase adhesion of the deteriogenic fungus *Aureobasidium pullulans* to PVC. Appl Environ Microbiol 65:3575–3581

Yamada-Onodera K, Mukumoto H, Katsuyaya Y, Saiganji A, Tani Y (2001) Degradation of polyethylene by a fungus, *Penicillium simplicissimum* YK. Polym Degrad Stab 72:323–327

Yang J, Yang Y, Wu W-M, Zhao J, Jiang L (2014) Evidence of polyethylene biodegradation by bacterial strains from the guts of plastic-eating waxworms. Environ Sci Technol 48:13776–13784

Yoshida S, Hiraga K, Talehana T, Taniguchi I, Yamaji H, Maeda Y, Toyohara K, Miyamoto K, Kimura Y, Oda K (2016) A bacterium that degrades and assimilates poly(ethylene terephthalate). Science 351(6278):1196–1199

Zimmermann W, Billig S (2011) Enzymes for the biofunctionalization of poly(ethylene terephthalate). Adv Biochem Eng Biotechnol 125:97–120

# Chapter 7
# Microbial Dynamics During the Bioremediation of Petroleum Hydrocarbon-Contaminated Soils Through Biostimulation: An Overview

**José A. Siles and Mercedes García-Sánchez**

## Contents



## 7.1 Introduction

Petroleum (crude oil) is one of the most important sources of the world's energy (Kotowicz et al. 2010). Human activities such as heating, transportation, and power production are dependent on petroleum-derivative fuels (Safdari et al. 2018). Petroleum also represents an important raw material for numerous industries (Varjani 2017). To meet these demands, the global crude oil production in 2017 was 92.65 million barrels per day, and it is estimated that 100 million barrels of oil per day will be needed up to 2021 (dos Santos and Maranho 2018).

J. A. Siles (✉)
Department of Plant and Microbial Biology, University of California, Berkeley, Berkeley, CA, USA
e-mail: jose.siles@berkeley.edu

M. García-Sánchez
UMR Eco&Sols, Ecologie fonctionelle and Biogéochimie des sols and Agro-écosystémes, Montpellier, France



R. Prasad, E. Aranda (eds.), *Approaches in Bioremediation*, Nanotechnology in the Life Sciences, https://doi.org/10.1007/978-3-030-02369-0_7

Petroleum is a naturally occurring, oily, and flammable liquid, which is formed from the organic decomposition of ancient plants and animals under conditions of high temperatures and pressures and is present at varying depths in the subsoil of the Earth (dos Santos and Maranho 2018). It is mainly a mixture of hydrocarbons, although other elements such as sulfur, oxygen, and nitrogen are also present. The compounds found in petroleum can be classified into four different fractions: (i) saturates (aliphatics), (ii) aromatics, (iii) resins, and (iv) asphaltenes (Chandra et al. 2013; Varjani 2017) (Fig. 7.1). Saturates are defined as hydrocarbons without double bonds and represent the highest percentage of crude oil constituents. They are categorized according to their chemical structures into alkanes and cycloalkanes (Abbasian et al. 2015). Aromatics are hydrocarbon molecules that have one or several aromatic rings usually substituted with different alkyl groups and are classified as (i) monocyclic aromatic hydrocarbons, i.e., benzene, toluene, ethylbenzene, and xylene (BTEX), and (ii) polycyclic aromatic hydrocarbons (PAHs), e.g., naphthalene, phenanthrene, anthracene, pyrene, or benzo[a]pyrene (Varjani 2017) (Fig. 7.1). Resins and asphaltenes contain non-hydrocarbon polar compounds and are characterized by the presence of very complex and mostly unknown C structures (Fig. 7.1). Resins consist of compounds containing nitrogen, sulfur, and oxygen that are dissolved in oil, while asphaltenes are large and complex molecules that are colloidally dispersed in oil (Nelyubov et al. 2017). The exact chemical composition of crude oil varies from one location to another (Chandra et al. 2013).

The current extensive exploitation of petroleum and its derivatives (mainly gasoline, creosote, and diesel) is having a negative effect on the environment (Hussain et al. 2018). Many places around the world are being contaminated by petroleum hydrocarbons (PHCs) as a consequence of industrial runoffs, effluent releases, and

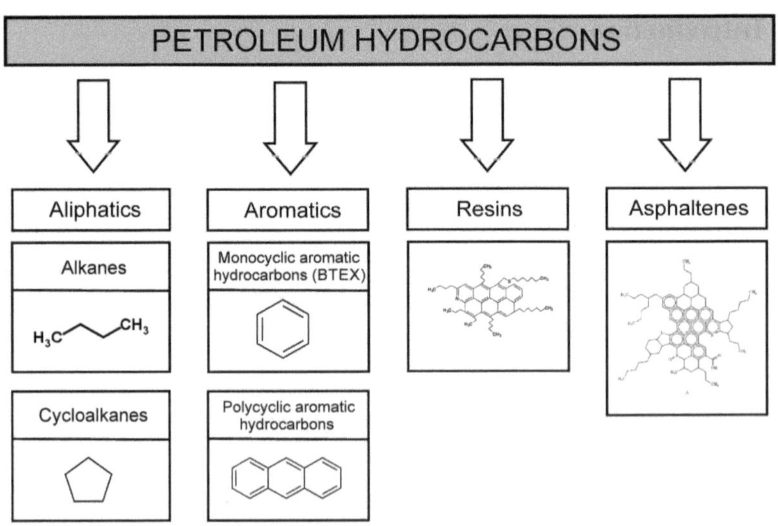


**Fig. 7.1** Schematic representation of the four different fractions of compounds that can be found in crude oil (petroleum)

accident spills that occur during the three major segments of activities that comprise the petroleum industry: (i) activities of exploration and production, (ii) labors of refining and marketing, and (iii) activities related to the transport of crude oil and other petroleum-derived products (Brassington et al. 2007; Gkorezis et al. 2016; Yavari et al. 2015). PHC contamination is happening at several environmental compartments: soils (Kim et al. 2018; Siles and Margesin 2018), groundwaters (Guerin et al. 2002; Moussavi et al. 2011), shores (Kankara et al. 2016; Zhu et al. 2017), sediments (Frysinger et al. 2003; Zhang et al. 2015; Acosta-González and Marqués 2016), and oceans (Bagby et al. 2017; Nicolaus et al. 2017). The magnitude of worldwide contamination by petroleum-derived products is such that PHCs are considered the most widespread class of organic contaminants worldwide (Brassington et al. 2007). PHCs have been classified as priority pollutants since they have a negative impact on quality services of ecosystems, animal life, and human health (Cole 2018). In humans, prolonged exposure to PHCs can produce damages on the central nervous system, dysfunction of the respiratory system, disruption of the endocrine system, and even cancer (Gkorezis et al. 2016).

Soil is an ecosystem especially vulnerable to PHC contamination. After an oil product spill, PHCs differently behave through the soil profile according to their nature (chemical composition) and abundance (volume of the spill), the soil structure, and the soil organic matter composition as well as the chemical processes that may occur (Cocârţă et al. 2017). Soluble PHCs are the most susceptible to changes due to biodegradation, volatilization, and filtration and usually reach groundwater rapidly (Mena et al. 2016). On the contrary, the PHCs that are most absorbed and adsorbed by the organic matter in the soil are the most resistant to losses or alterations by other processes. They alter soil chemical, physical, and biological properties for longer periods of time and can also form an impermeable coating at the soil surface, which prevents water circulation in soil and gas exchange between the soil and air, causing serious alterations in the status and growth of plants (Streche et al. 2018). Ultimately, extensive soil pollution with TPHs causes infertile soils (Hussain et al. 2018). The development of effective strategies for the decontamination of PHC-polluted soils is thus critical. However, the issue of soils contaminated with PHCs and their remediation are among the most complex tasks in the environmental protection field in terms of financial and organizational aspects (Streche et al. 2018).

Thermal (e.g., incineration, thermal desorption, and microwave frequency heating), chemical (e.g., chemical oxidation and electrokinetics), and physicochemical (e.g., ultrasound, flotation, and extraction by solvent or by steam) technologies have been developed and applied to remediate soils contaminated by PHCs (dos Santos and Maranho 2018). Nevertheless, these approaches require heavy machinery, are not economically viable, and can involve significant site disturbances (Shahi et al. 2016a). On the other hand, the environmental consequences of removing PHCs from soils with some of these methods may result in massive air pollution (Chandra et al. 2013). As an alternative, bioremediation has shown to be an efficient, cost-effective, versatile, and environmentally sound approach to clean up PHC-polluted soils (Chikere et al. 2012; Bordoloi and Boruah 2018).

## 7.2 Bioremediation

Bioremediation is defined as "the use of biologically mediated processes to detoxify, degrade or transform pollutants to an innocuous state" (Gkorezis et al. 2016). This approach takes advantage of the capacity of many microorganisms to use hydrocarbons as carbon and energy source (biodegradation), transforming or mineralizing these pollutants into less harmful or nonhazardous substances, which are then integrated into natural biogeochemical cycles (Margesin 2013; Siles and Margesin 2018).

The intrinsic bioremediation potential of a soil ecosystem is known as natural attenuation (Azubuike et al. 2016). Microbial PHC degradation in soil (or, in other words, effectiveness of the bioremediation process) can be limited for three groups of factors (Adams et al. 2015): (i) characteristics of the indigenous microbial community (in terms of taxonomy, gene regulation and expression, metabolic diversity, tolerance to metals and other toxic xenobiotics, substrate uptake or adherence mechanisms, chemotaxis, and biofilm formation); (ii) environmental conditions (i.e., nutrient availability, terminal electron acceptors, salinity, pressure, temperature, pH, and moisture); and (iii) chemical nature and physicochemical properties of the PHCs (i.e., solubility, concentration, hydrophobicity, volatility, and molecular mass) (Gkorezis et al. 2016). To overcome the limitations that bioremediation-based strategies may undergo, bioaugmentation (i.e., inoculating soil with exogenous or endogenous hydrocarbon-degrading microorganisms) and biostimulation (i.e., addition of the appropriate nutrients and/or electron acceptors to stimulate the degradation capacity of the indigenous soil microbial community) approaches have been developed (Wu et al. 2016). Different treatments of biostimulation and/or bioaugmentation have been extensively applied for bioremediation of soils from an ample range of environments using laboratory (ex situ) or in situ approaches (Margesin and Schinner 1999; Adams et al. 2015; Azubuike et al. 2016; Kuppusamy et al. 2017; Varjani 2017). The effectivity of biostimulation and bioaugmentation treatments seems to be case-specific and varies with the kind of nutrients and inoculants that are applied (Suja et al. 2014; Wu et al. 2016). In general, biostimulation is regarded as the best option to speed up the treatment of oligotrophic soils, while bioaugmentation has been recognized as the most suitable strategy to recover soils with poor microbial communities in PHC degraders (Kuppusamy et al. 2017). However, several works comparing removal rates of PHCs during the simultaneous bioremediation of different kinds of PHC-polluted soils via biostimulation and bioaugmentation have shown better results for biostimulation at the long term (Qiao et al. 2014; Masy et al. 2016; Wu et al. 2016; Polyak et al. 2018). These results have been explained as the result of the scarce survival of the inoculated microorganisms during the biostimulation treatment since they have to adapt to the prevailing environmental conditions, to compete with indigenous microbiota, and to survive to predators (Adams et al. 2015; Azubuike et al. 2016; Mapelli et al. 2017). In biostimulation strategies, pollutant degradation is undertaken by already present microorganisms (native) in soil that are well suited to the existing conditions as well as properly spatially distributed over the soil matrix (Adams et al. 2015). In fact, biostimulation has also regarded as a useful approach for the identification of suitable microorganisms to be used for bioaugmentation purposes. Bioremediation of PHC-contaminated soils via biostimulation is the object of study in this chapter.

## 7.3 Biostimulation

PHCs are mainly carbon and hydrogen compounds, as previously explained. Therefore, the contamination of a soil with an oil-derived product provokes an imbalance in the C:N (carbon/nitrogen) ratio, which limits the availability of essential nutrients for microbial growth and activity (Vandera and Koukkou 2017). The lack of a balanced C:N:P (carbon/nitrogen/phosphorus) ratio constrains most of the bioremediation processes that involve autochthonous microorganisms (Safdari et al. 2018). Consequently, the existing biostimulation strategies are based on the soil addition of inorganic nutrients (e.g., inorganic fertilizers rich in N and P) or different organic materials (e.g., animal manure, domestic sewage, rice straw biochar, crop residues, and different types of composts varying in composition and degree of stabilization) to improve the biodegradation potential of the indigenous microorganisms (Gkorezis et al. 2016). When compost is applied to soil, the microbes added with it may also be involved in the PHC degradation. In this case, soil compost amendment for bioremediation purposes is considered as a simultaneously biostimulation and bioaugmentation approach (Kastner and Miltner 2016). Other biostimulation strategies involve the application of components such as biosurfactants and electron acceptors (e.g., $O_2$, chelated Fe (III), nitrates, or sulfate) (Gkorezis et al. 2016; Shahi et al. 2016a).

During the biostimulation of a soil, it is important to add the appropriate levels of nutrients. The excessive fertilization or amendment can lead to ammonia toxicity, soil eutrophication, and the excessive multiplication of r-strategist microorganisms, which may not be hydrocarbon degraders (Yavari et al. 2015; Brzeszcz et al. 2016). Theoretically, 150 mg of N and 30 mg of P are required to degrade 1 g of hydrocarbons. Consequently, a C:N:P ratio of 100:5:1 has been regarded as optimum for the biostimulation of PHC-contaminated soils (Shahi et al. 2016a). However, the appropriate C:N:P ratio needs to be identified for each contaminated site as each petroleum-derived product has different properties and each contaminated site is influenced by different environmental factors. For example, Turgay et al. (2010) successfully biostimulated (applying leonardite) soil from a crude oil-contaminated site regulating the C:N:P ratio to 100:15:1, while Qin et al. (2013) cleaned up petroleum-contaminated soil through the application of rice straw biochar by using a C:N:P ratio of 100:10:1.

## 7.4 Microbial Dynamics During the Biostimulation of Petroleum Hydrocarbon-Contaminated Soils

During the bioremediation of PHC-polluted soils via biostimulation, the study of the mediating microbial communities is important since the data obtained from these surveys may be useful to identify the specific microbial characteristics that lead to enhanced decontamination rates (Siles and Margesin 2018). These data can be further used to optimize the existing bioremediation strategies and to develop new ones (Stenuit et al. 2008; Pal et al. 2017). In soil, groups of bacteria, archaea, fungi, and algae are involved in PHC degradation (Militon et al. 2010). However, the number of works simultaneously studying the characteristics of microbial communities of all three microbial domains involved in bioremediation is scarce. Therefore, in the present chapter, we aim at giving an overview of the dynamics of soil microbial communities, including bacterial and archaeal domains as well as fungal kingdom, in terms of (i) activity, (ii) abundance, and (iii) taxonomic composition during the bioremediation of PHC-contaminated soils through biostimulation.

### 7.4.1 *Activity of Soil Microbial Communities*

Soil microbial enzyme activities play an important role in nutrient cycling and can be sensitive indicators of environmental pollution (Polyak et al. 2018). In this way, microbial enzyme activities are routinely measured during processes of soil bioremediation. Potential (i.e., under optimum laboratory conditions of temperature and substrate concentration) enzyme activities such as dehydrogenase, catalase, lipase, and urease have been measured (using mostly photometric methods) in short- and long-term biostimulation experiments of PHC-polluted soils using different kinds of fertilizers and amendments (Margesin et al. 2007; Xu and Lu 2010; Silva-Castro et al. 2013; Fan et al. 2014; Kaczyńska et al. 2015; Wu et al. 2016; Liu et al. 2018; Polyak et al. 2018). In general, biostimulation treatments lead to an increase in microbial enzyme production due to the stimulation of the activity of autochthonous microorganisms, being these increments more evident at the beginning of the bioremediation experiments. Higher values for soil enzyme activities such as dehydrogenase and catalase in concomitance with decreasing PHC contents reflect the clear participation of native microbial communities in PHC degradation (Maila and Cloete 2005). Liu et al. (2018) detected that dehydrogenase activity had a positive correlation with the rapid increase in PHC removal rate during the biostimulation with aged refuse from landfills of a petroleum-contaminated soil. The incorporation of nutrients supports thus hydrocarbon biodegradation. However, very high doses of nutrients may not result in an increment in microbial enzyme activities (Margesin et al. 2007). On the other hand, the monitoring of extracellular enzymes like β-glucosidase and phosphatases (acid or alkaline) has been regarded as useful when the biostimulation treatment includes application of compost (Bastida et al. 2016).

Microbial respiration is another parameter that is usually measured during the biostimulation of contaminated soils. Changes in $CO_2$ production may indirectly reflect the microbial breakdown of hydrocarbons. In general, increasing microbial respiration rates and enzyme activities are concomitant with decreasing PHC contents in soil during biostimulation treatments (Polyak et al. 2018).

### *7.4.2 Abundance of Soil Microbial Communities*

Methodological approaches such as phospholipid fatty acids (PLFA) and quantitative PCR (qPCR) analyses have been used to monitor the relative size of bacterial, archaeal, and fungal communities during the bioremediation of PHC-contaminated soils via biostimulation. Alternatively, qPCR has also been used to quantify key genes in PHC degradation, which gives useful information regarding the relative size of the microbial population potentially conducting bioremediation process.

Soil bacterial biomass usually increases after the application of biostimulation treatments since increased levels of nutrients stimulate the bacterial growth, especially that of r-strategists (Siles and Margesin 2018). The response of bacterial communities to the nutrient input is immediate. For example, Margesin et al. (2007) reported a significant increase in PLFA-based bacterial abundances 7 days after the application of inorganic NPK (nitrogen-, phosphorus-, and potassium-based) fertilization during the bioremediation of a soil artificially contaminated with several concentrations of diesel oil. Over time, the relative size of bacterial communities in biostimulation treatments remains higher than that of the control treatment (natural attenuation) (Mair et al. 2013). This is especially evident when the applied amounts of nutrients keep a balanced C:N:P ratio over the process. For instance, Han et al. (2017) found that bacterial 16S rRNA gene copy number after 90 days of biostimulation of a PAH-contaminated soil using three different types of wastes was significantly higher than that of unamended soil. However, some works have not obtained a significant correlation between increased bacterial abundances and PHC removal rates (Ka et al. 2001). Siles and Margesin (2018) suggested that improved decontaminated rates could be explained to a higher extent by the stimulation of the activities of the indigenous soil microbes than by the enhancement of microbial abundances. In this way, the quantification of specific bacterial functional genes (e.g., *alkB* (encoding for alkane monooxygenase), *phnAc* (naphthalene dioxygenase large subunit), *nah* (naphthalene dioxygenase), or PAH-RHD$_{\alpha}$ (PAH-ring hydroxylating dioxygenase alpha subunit)) has shown to be a more informative approach to know the abundance dynamics of bacteria involved in PHC degradation (Shahi et al. 2016a). One of the genes most commonly used as functional marker is *alkB* since alkane monooxygenase is a key enzyme in bacterial alkane aerobic degradation. This enzyme acts in a chain reaction with electron carriers to reduce the alkane to an alcohol, which subsequently enters bacterial β-oxidation pathway (Vandera and Koukkou 2017). Han et al. (2017), in the aforementioned biostimulation study, reported that the three types of wastes used (wheat stalk, mushroom cultivation substrate waste, and cow manure) significantly increased the abundances of the PAH-degrading genes tested (*pdo1* (encoding pyrene dioxygenase) and *nah*). Interestingly, Masy et al. (2016), comparing biostimulation and bioaugmentation methods for the bioremediation of a diesel-polluted clay-rich soil, calculated the ratio between *alkB* and 16S rRNA genes. This can be an interesting approach to know the dynamics of the abundances of the bacterial populations actively involved in PHC degradation in relation to the abundance of the entire bacterial community.

Unlike bacteria, the abundance of archaeal communities does not increase after the addition of nutrients in biostimulation treatments, as demonstrated by Röling et al. (2004) and de Jesus et al. (2015). On the other hand, Siles and Margesin (2018) noticed a decrease in archaeal abundances in the presence of fertilization. These findings have been explained as a consequence of the adaptation of archaea to oligotrophic nutrient conditions and their poor competitive ability in respect to other microbial groups that are favored by nutrient addition (Karlsson et al. 2012).

In the case of fungal communities, Covino et al. (2016b) documented increased fungal abundances (as determined by ergosterol measurement) during the bioremediation of a long-term oil-contaminated soil through biostimulation for 30 and 60 days using a lignocellulosic mixture. Accordingly, Mair et al. (2013), during the bioremediation of PHC-polluted soil from an Alpine former military site by using three types of biostimulation treatments (NPK-based fertilization and two commercial products, Inipol and Terramend), observed a significant increment of the fungal PLFA biomarker 18:2ω6,9 in all the fertilized treatments in comparison with the unfertilized soil after 15 and 30 weeks of experiment. Therefore, the beneficial effect of biostimulation treatments on the relative size of soil fungal communities as a consequence of the increment of nutrient levels seems clear. However, Siles and Margesin (2018) have recently argued that the shifts in fungal abundance during bioremediation experiments could be driven by factors different of nutrient amounts, e.g., incubation temperature or other soil physicochemical and environmental factors.

### 7.4.3 Taxonomic Composition of Soil Microbial Communities

The remediation of PHC-contaminated soils through biostimulation approaches has been intensively studied during the last decades. However, the knowledge about the impact of these methods on the autochthonous soil microbial communities in terms of taxonomic composition is scarcely known. High-throughput sequencing techniques have revolutionized the study of the soil microbes and can provide a thorough explanation of a site microbial ecology, as explained by Czaplicki et al. (2016). Therefore, their use in bioremediation experiments allows us to discover the composition of microbial communities involved in PHC degradation and to understand why specific bioremediation strategies fail. Here, we provide an overview of the main groups of soil bacterial, archaeal, and fungal communities potentially involved in PHC degradation (and their dynamics) during the bioremediation of polluted soils via biostimulation by using pyrosequencing or Illumina technologies.

#### 7.4.3.1 Bacterial Communities

Several pyrosequencing- and Illumina-based studies have described the diversity of bacterial communities during the biostimulation of PHC-polluted soils from diverse geographical locations by using different kinds of stimulating agents (Table 7.1). These works along with those investigating how the presence of PHC contamination impacts the soil bacterial community composition, via high-throughput sequencing techniques and chemical monitoring of PHC concentration, have shown to be useful approaches for the identification of groups of bacteria involved in PHC degradation (natural attenuation) (Sutton et al. 2013; Zhou et al. 2017). Bacteria have been described as the most active agents in PHC degradation (Varjani 2017).

There are three possible ways for PHC degradation: (i) phototrophic, anoxygenic; (ii) chemotrophic, aerobic; and (iii) chemotrophic, anaerobic. Chemical reactions such as oxidation, reduction, hydroxylation, and dehydrogenation are common for both aerobic and anaerobic pathways of microbial PHC degradation (Varjani 2017). In the works listed in Table 7.1, biostimulation was mainly conducted under aerobic conditions.

All the proteobacterial classes, except *Epsilonproteobacteria*, have shown to respond to the supplementation of nutrients in biostimulation treatments (Vandera and Koukkou 2017). In the case of the works reviewed in Table 7.1, *Gammaproteobacteria* and *Alphaproteobacteria* seemed to have an important role on decontamination processes. A number of works have reported the active involvement of *Gammaproteobacteria* in the degradation of PHCs, and the "gamma shift" is a well-known phenomenon that occurs after the PHC contamination of soil (when the contaminant acts as a source of nutrient for the degraders) or under conditions of nutrient oversupply (Militon et al. 2010; Dong et al. 2015). Within *Gammaproteobacteria*, *Pseudomonas* spp. have been recognized as effective alkane and aromatic hydrocarbon degraders in petroleum-contaminated soils, and their relative abundances significantly correlate with high degradation rates (Qin et al. 2013). For example, *Pseudomonas putida* and *Pseudomonas fluorescens* have the ability to adapt to many different hydrocarbons not solely with catabolic enzymes but also with metabolic regulation (Kulkarni et al. 2012). According to Table 7.1, *Pseudomonas* spp. were actively involved in the recovery of some of the polluted soils studied. This genus is widely distributed in soil, and many *Pseudomonas* species are able to adapt to variable nutritional (both oligotrophic and copiotrophic) environments and pH conditions (Siles et al. 2018). This would explain the dynamics of *Pseudomonas* spp. during biostimulation processes. The significant

**Table 7.1** Key studies describing changes in the composition of bacterial communities during the bioremediation of petroleum hydrocarbon-contaminated soils via biostimulation by using pyrosequencing or Illumina techniques

| Geographical location | Type of contamination | Initial petroleum hydrocarbon concentration (mg $kg^{-1}$) | Biostimulation treatment | Duration of the biostimulation experiment (days) | Bacterial groups involved in petroleum hydrocarbon degradation | Reference |
|---|---|---|---|---|---|---|
| Salisbury, NC, USA | PAHs | 295 (only PAHs) | Inorganic NPK fertilization | 534 | *Alphaproteobacteria*, *Gammaproteobacteria* | Singleton et al. (2013) |
| Barcelona, Spain | Creosote | 2815 | Lignocellulosic substrate (wheat straw and wheat bran) | 120 | *Alphaproteobacteria*, *Gammaproteobacteria* | Lladó et al. (2014) |
| Belgium | Diesel and heating fuel | Ranging from 6117 to 196 | Oxygen | 100 | *Betaproteobacteria* | Masy et al. (2016) |
| Hangzhou, China | PAHs | 1.3 (only PAHs) | Rhamnolipids, Tween 80, and sodium dodecyl benzenesulfonate | 90 | *Pseudomonas* (*Gammaproteobacteria*), *Bacillus* (*Firmicutes*), *Sphingomonas* (*Alphaproteobacteria*) | Wang et al. (2016) |
| Gela, Italy | Oil-derived products | 10,200 | Lignocellulosic mixture (wheat straw and poplar wood) | 60 | *Alphaproteobacteria*, *Actinobacteria* | Covino et al. (2016b) |
| Turkey | Crude oil | Not shown | Urea ($(NH_2)_2CO$) and $KH_2PO_4$ | 49 | *Proteobacteria* (*Pseudomonas*), *Firmicutes*, *Bacteroidetes* | Shahi et al. (2016b) |
| Welsberg-Taisten, Italy | Diesel oil | 6220 | Inorganic NPK fertilization | 105 | *Gammaproteobacteria*, *Bacteroidia* (*Bacteroidetes*) | Siles and Margesin (2018) |
| Saskatchewan, Canada | Oil-derived products | 5196 | NPK fertilization and humate amendments | 260 | *Alphaproteobacteria*, *Gammaproteobacteria* | Kim et al. (2018) |

involvement of plasmids in PHC degradation is another characteristic of *Pseudomonas* spp. In fact, many of these plasmid-encoding pathways for hydrocarbon degradation have been characterized (Vasudevan et al. 2007). Due to these characteristics, multiple species of the genus *Pseudomonas* are used for bioaugmentation purposes (Lim et al. 2016). Other gammaproteobacterial genera such as *Acinetobacter*, *Marinobacter*, *Stenotrophomonas*, and *Vibrio* are also reported as PHC degraders (Varjani 2017).

In the studies presented in Table 7.1, *Sphingomonas* spp., belonging to the class *Alphaproteobacteria*, were shown to positively respond to the stimulating agents and to be involved in PHC degradation. Members of the genus *Sphingomonas* are frequently isolated from PHC-contaminated soils, suggesting that the *Sphingomonas* are probably key members of natural PHC-degrading microbial consortia. For instance, *Sphingomonas* sp. strain HXN-200 could use $C_5$–$C_{16}$ as a sole carbon source (van Beilen et al. 2006), while *Sphingomonas paucimobilis* DSM 1098 could grow on n-decane and n-hexadecane (Vomberg and Klinner 2000). Additionally, Zheng et al. (2018) discovered that one species belonging to the genus *Sphingomonas* could degrade $C_9$–$C_{30}$. Genetic analysis of several PAH-degrading pathways in *Sphingomonas* strains revealed the presence of a unique group of genes for aromatic compound degradation that were distantly related to those in other genera (Zheng et al. 2018).

Biostimulation treatments enrich soil with *Betaproteobacteria* according to the studies of Table 7.1. Relative abundances of this bacterial group have positively related to degradation in nutrient-rich soils, while a negative correlation of its abundances with degradation has been reported in the case of unamended soils. This strongly suggests that the effectiveness of *Betaproteobacteria* as PHC degraders may be related to the sufficient availability of nutrients (Vandera and Koukkou 2017).

Besides *Proteobacteria*, other bacterial groups like *Bacteroidetes* have demonstrated to have a significant role in PHC degradation during biostimulation, as shown by Shahi et al. (2016b) and Siles and Margesin (2018). Additionally, Gandolfi et al. (2010) found that the compost amendment of a soil contaminated with PAHs caused a complete change of predominant bacterial community composition from *Alphaproteobacteria* and *Gammaproteobacteria* to *Bacteroidetes* and *Firmicutes*. Interestingly, at the end of this experiment, *Bacteroidetes* became predominant in the microbial community, which was concomitant with an effective recovery of the polluted soil. The use of vermicompost from olive mill wastes for bioremediation of PAH-contaminated soils proved to be effective.

#### 7.4.3.2 Archaeal Communities

Members of the archaeal phylum *Euryarchaeota* have shown to commonly inhabit hydrocarbon-rich environments (Joshi et al. 2014; Jeanbille et al. 2016). Siles and Margesin (2018), by using Illumina amplicon sequencing, reported that *Thaumarchaeota* and *Euryarchaeota* dominated archaeal community in a diesel oil-contaminated soil from a former military site. Nevertheless, they did not find statistical evidences of the involvement of any archaeal group in the biostimulation of that soil via inorganic NPK fertilization. In this line, Kim et al. (2018) found a very low proportion of archaeal sequences in the Illumina-based analysis of the prokaryotic community mediating the bioremediation of nutrient-amended PHC-polluted soils using a pilot-scale biopile field experiment. The apparent nonsignificant role of archaeal communities in the aforementioned studies could be a consequence of the aerobic conditions that characterized both processes since archaea play a noteworthy role on hydrocarbon degradation only under anaerobiosis. Under these conditions, methanogenesis becomes a common process through the activity of methanogenic archaea in syntrophic association with specific bacterial groups due to the metabolic dependence of methanogenic organisms on simple compounds such as $O_2$, $H_2$, acetate, methanol, or formate (Das and Kazy 2014; Fowler et al. 2016). Sutton et al. (2013), investigating (pyrosequencing-based analysis) the prokaryotic community inhabiting a site with a diesel contamination history of 40 years, observed that archaea belonging to the genus *Methanosaeta* (phylum *Euryarchaeota*) in association with bacteria classified at class level as *Anaerolineae* (phylum *Chloroflexi*) were involved in the bioremediation of this site via natural attenuation through the anaerobic degradation of oil-related compounds.

As compared to bacteria, only poor attention is currently paid to the study of archaeal communities during the cleaning-up of PHC-polluted soils, independently of the bioremediation process considered. This is reflected in the scarce number of works studying archaeal communities during bioremediation experiments.

#### 7.4.3.3 Fungal Communities

The application of amendments has been proposed to promote the proliferation of specific fungal groups, which might be involved in overall pollutant degradation at a given PHC-contaminated site. The ability to secrete catabolic enzymes involved in PHC degradation such as laccases and/or peroxidases has been described in fungi belonging to *Ascomycota* phylum (Prasad 2017, 2018). For instance, certain species such as *Lasiodiplodia theobromae*, *Trichoderma asperellum*, and *Aspergillus fumigatus* were able to degrade compounds derived from PHCs (Zafra et al. 2014).

Wheat straw has proved to be one of the most suitable materials to stimulate the biodegradation capabilities of soil fungi in biostimulation processes (Lladó et al. 2014). This complex material may induce some specific catabolic enzymes (i.e., laccases and/or peroxidases) as well as lead to an increase in both nutrient availability and oxygen mass transfers (Federici et al. 2012). Stella et al. (2017) observed a significant increment in the abundance of the genera *Agrocybe*, *Sphaerobolus*, *Pluteus*, and *Cryptococcus* (*Basidiomycota*) when a similar lignocellulosic material was applied to PHC-contaminated soil. The populations of fungi belonging to *Ascomycota* phylum (e.g., *Penicillium* spp. and *Stachybotrys* spp.) were also significantly enhanced using this material. Interestingly, this finding was also in line with previous evidences indicating the ability of *Penicillium* species such as *P. chrysogenum*,

*P. purpurescens*, *P. digitatum*, and *P. aurantiogriseum* (*Ascomycota*) to degrade certain PHC compounds (Tigini et al. 2009; Mouhamadou et al. 2013). Cebron et al. (2015) reported a predominance of *Ascomycota* after soil addition of wood sawdust. This study described the proliferation of *Chaetomium* and *Neurospora* genera, revealing the potential abilities of these fungi in the degradation of cellulose as they possess an efficient cellobiose dehydrogenase activity (Harreither et al. 2011). On the other hand, a significant increment in the relative abundance of the genera *Fusarium* and *Scedosporium* (*Ascomycota*) was reported by adding lignocellulosic materials to contaminated soils (Lladó et al. 2014). This finding is in line with previous studies reporting the ability of these fungi to degrade hydrocarbon compounds through laccase activity (Wu and Nian 2014; Zafra et al. 2014). Other studies have been focused on the analysis of the fungal composition during the co-composting of materials in a PHC-contaminated site (Covino et al. 2016a). *Ascomycota* has shown to be the most abundant phylum during the first phase of co-composting of materials such as fresh mushroom substrate. The analysis of the fungal composition at order level revealed a predominance of *Chaetothyriales* and *Helotiales*, which are described as filamentous black yeasts with the ability to assimilate small aromatic hydrocarbons (Prenafeta-Boldu et al. 2006; Zhao et al. 2010; Thion et al. 2012; Dogen et al. 2013). However, the co-composting of materials such as fresh grass cutting provoked a significant proliferation of the *Saccharomycetales* order during the first and late phases. Authors concluded that the abundance of *Saccharomycetales* might be associated with the excess of grass hydrolysates or with an acidification as the result of organic acid release and/or the $CO_2/O_2$ pressure. In any case, it has been described the ability of yeast to produce extracellular peroxidases that are competent in the degradation of hydrocarbon compounds (Yang et al. 2013). On the other hand, the success of the biostimulation approaches might be limited by some soil factors such as the temperature, among others. A study conducted by Siles and Margesin (2018) has shown that the interaction between the inorganic fertilization and the incubation temperature may significantly influence fungal richness and diversity (Shannon index) during soil biostimulation. However, this finding was not concomitant with the PHC degradation, as previously reported (Covino et al. 2016a). Authors explained this finding as the result of the interaction of other physicochemical parameters than PHC content in soils, indicating that these geochemical factors would be influencing the fungal community structure.

In the case of fungi, biostimulation studies have been regarded as useful to identify fungal PHC degraders. A recent pyrosequencing-based study of fungal communities in three types of polluted soils subjected to bioaugmentation using *Pleurotus ostreatus* and *Irpex lacteus* has demonstrated the ability of the two selected fungi to compete with the autochthonous mycobiota (Stella et al. 2017). In addition, this study showed that fungal community was dominated by *Basidiomycota*, *Ascomycota* (divisions), and *Mucoromycotina* subdivision. Interestingly, a significant enhancement in *Basidiomycota* fungi was found compared with *Ascomycota* and *Mucoromycotina*, probably as the result of the bioaugmentation treatment. However, the sequence abundance for basidiomycetes dramatically decreased during the last weeks of this experiment with a concomitant increase in unidentified fungal sequences. This finding is not surprising; soil does not represent a natural habitat for these fungi, as they are wood-decay fungi. Therefore, it is not unexpected that fungi used in bioaugmentation are unable to compete with the autochthonous soil mycobiota. To answer questions about the role of the native microbiota, fungi used for bioaugmented procedures were examined in sterile and non-sterile soils. Fungi showed different behaviors in sterile soils in comparison to non-sterile soils, which indicates that fungi did not have adequate ability to compete with the native microbes for a space in the soil's ecology (Andersson et al. 2000). Previous studies have revealed a marked antagonist effect of the indigenous microbial populations during augmentation processes using wood-decay fungi (Lladó et al. 2013; García-Sánchez et al. 2018). In a survey conducted by Zafra et al. (2016), the impact of a microbial PHC-degrading consortium composed of four fungal species (*Aspergillus flavus* H6, *Aspergillus nomius* H7, *Rhizomucor variabilis* H9, and *Trichoderma asperellum* H15) and five bacterial trains (*Klebsiella pneumoniae* B1, *Bacillus cereus* B4, *P. aeruginosa* B6, *Klebsiella* sp. B10, and *Stenotrophomonas maltophilia* B14) on fungal taxonomic composition was evaluated. Fungal populations remained stable over time, being *Ascomycota* the most abundant phylum. *Pezizomycotina* dominated among *Ascomycota*. All fungal strains used in this consortium belonged to *Pezizomycotina* group, which indicates a better adaptability of fungi belonging to *Ascomycota*, in comparison to those classified as *Basidiomycota*, to perform augmentation approaches.

## 7.5 Conclusions and Final Remarks

Biostimulation has proved to be an efficient approach for the decontamination of PHC-contaminated soils. Most of the biostimulation treatments yield final rates of PHC removal ranging from 70% to 90%. The remaining 10 ± 30% consists of hydrocarbons that are structurally less available for biodegradation due to their very limited bioavailability and recalcitrance (Polyak et al. 2018). The successful decontamination rates obtained via biostimulation approaches are concomitant with deep shifts in the mediating soil bacterial, archaeal, and fungal communities in terms of activity, abundance, and composition. In general, the nutrients and electron acceptors added with biostimulation improve soil microbial activity, increase the total abundance of bacteria and fungi, and promote the selective proliferation of bacterial, archaeal, and fungal PHC degraders. In the future, further functional studies (metagenomic- or metatranscriptomic-based) are needed during treatments of bioremediation to decipher the characteristics of the microbial communities in charge of PHC degradation in terms of activity.

# References

Abbasian F, Lockington R, Mallavarapu M, Naidu R (2015) A comprehensive review of aliphatic hydrocarbon biodegradation by bacteria. Appl Biochem Biotechnol 176(3):670–699

Acosta-González A, Marqués S (2016) Bacterial diversity in oil-polluted marine coastal sediments. Curr Opin Biotechnol 38:24–32

Adams GO, Fufeyin PT, Okoro SE, Ehinomen I (2015) Bioremediation, biostimulation and bioaugmentation: a review. Int J Environ Bioremediat Biodegrad 3(1):28–39

Andersson BE, Welinder L, Olsson PA, Olsson S, Henrysson T (2000) Growth of inoculated white-rot fungi and their interactions with the bacterial community in soil contaminated with polycyclic aromatic hydrocarbons, as measured by phospholipid fatty acids. Bioresour Technol 73(1):29–36

Azubuike CC, Chikere CB, Okpokwasili GC (2016) Bioremediation techniques–classification based on site of application: principles, advantages, limitations and prospects. World J Microb Biot 32(11):180

Bagby SC, Reddy CM, Aeppli C, Fisher GB, Valentine DL (2017) Persistence and biodegradation of oil at the ocean floor following deepwater horizon. Proc Natl Acad Sci U S A 114(1):E9–E18

Bastida F, Jehmlich N, Lima K, Morris BEL, Richnow HH, Hernandez T, von Bergen M, Garcia C (2016) The ecological and physiological responses of the microbial community from a semi-arid soil to hydrocarbon contamination and its bioremediation using compost amendment. J Proteomics 135:162–169

Bordoloi J, Boruah HPD (2018) Analysis of recent patenting activities in the field of bioremediation of petroleum hydrocarbon pollutants present in the environment. Recent Pat Biotechnol 12(1):3–20

Brassington KJ, Hough RL, Paton GI, Semple KT, Risdon GC, Crossley J, Hay I, Askari K, Pollard SJT (2007) Weathered hydrocarbon wastes: a risk management primer. Crit Rev Environ Sci Technol 37(3):199–232

Brzeszcz J, Steliga T, Kapusta P, Turkiewicz A, Kaszycki P (2016) R-strategist versus K-strategist for the application in bioremediation of hydrocarbon-contaminated soils. Int Biodeterior Biodegrad 106:41–52

Cebron A, Beguiristain T, Bongoua-Devisme J, Denonfoux J, Faure P, Lorgeoux C, Ouvrard S, Parisot N, Peyret P, Leyval C (2015) Impact of clay mineral, wood sawdust or root organic matter on the bacterial and fungal community structures in two aged PAH-contaminated soils. Environ Sci Pollut Res Int 22(18):13724–13738

Chandra S, Sharma R, Singh K, Sharma A (2013) Application of bioremediation technology in the environment contaminated with petroleum hydrocarbon. Ann Microbiol 63(2):417–431

Chikere CB, Surridge K, Okpokwasili GC, Cloete TE (2012) Dynamics of indigenous bacterial communities associated with crude oil degradation in soil microcosms during nutrient-enhanced bioremediation. Waste Manag Res 30(3):225–236

Cocârţă DM, Stoian MA, Karademir A (2017) Crude oil contaminated sites: evaluation by using risk assessment approach. Sustainability 9(8):1365

Cole GM (2018) Assessment and remediation of petroleum contaminated sites. CRC Press, Boca Raton

Covino S, Fabianova T, Kresinova Z, Cvancarova M, Burianova E, Filipova A, Voriskova J, Baldrian P, Cajthaml T (2016a) Polycyclic aromatic hydrocarbons degradation and microbial community shifts during co-composting of creosote-treated wood. J Hazard Mater 301:17–26

Covino S, Stella T, D'Annibale A, Lladó S, Baldrian P, Čvančarová M, Cajthaml T, Petruccioli M (2016b) Comparative assessment of fungal augmentation treatments of a fine-textured and historically oil-contaminated soil. Sci Total Environ 566–567:250–259

Czaplicki LM, Cooper E, Ferguson PL, Stapleton HM, Vilgalys R, Gunsch CK (2016) A new perspective on sustainable soil remediation–case study suggests novel fungal genera could facilitate in situ biodegradation of hazardous contaminants. Remediat J 26(2):59–72

Das R, Kazy SK (2014) Microbial diversity, community composition and metabolic potential in hydrocarbon contaminated oily sludge: prospects for in situ bioremediation. Environ Sci Pollut Res 21(12):7369–7389

de Jesus HE, Peixoto RS, Cury JC, van Elsas JD, Rosado AS (2015) Evaluation of soil bioremediation techniques in an aged diesel spill at the Antarctic Peninsula. Appl Microbiol Biotechnol 99(24):10815–10827

Dogen A, Ilkit M, de Hoog GS (2013) Black yeast habitat choices and species spectrum on high altitude creosote-treated railway ties. Fungal Biol 117(10):692–696

Dong Y, Lang Z, Kong X, Lu D, Liu Z (2015) Kinetic and multidimensional profiling of accelerated degradation of oil sludge by biostimulation. Environ Sci Process Impact 17(4):763–774

dos Santos JJ, Maranho LT (2018) Rhizospheric microorganisms as a solution for the recovery of soils contaminated by petroleum: a review. J Environ Manag 210:104–113

Fan MY, Xie RJ, Qin G (2014) Bioremediation of petroleum-contaminated soil by a combined system of biostimulation-bioaugmentation with yeast. Environ Technol 35(4):391–399

Federici E, Giubilei MA, Covino S, Zanaroli G, Fava F, D'Annibale A, Petruccioli M (2012) Addition of maize stalks and soybean oil to a historically PCB-contaminated soil: effect on degradation performance and indigenous microbiota. New Biotechnol 30(1):69–79

Fowler SJ, Toth CRA, Gieg LM (2016) Community structure in methanogenic enrichments provides insight into syntrophic interactions in hydrocarbon-impacted environments. Front Microbiol 7:562

Frysinger GS, Gaines RB, Xu L, Reddy CM (2003) Resolving the unresolved complex mixture in petroleum-contaminated sediments. Environ Sci Technol 37(8):1653–1662

Gandolfi I, Sicolo M, Franzetti A, Fontanarosa E, Santagostino A, Bestetti G (2010) Influence of compost amendment on microbial community and ecotoxicity of hydrocarbon-contaminated soils. Bioresour Technol 101(2):568–575

García-Sánchez M, Košnář Z, Mercl F, Aranda E, Tlustoš P (2018) A comparative study to evaluate natural attenuation, mycoaugmentation, phytoremediation, and microbial-assisted phytoremediation strategies for the bioremediation of an aged PAH-polluted soil. Ecotox Environ Safe 147:165–174

Gkorezis P, Daghio M, Franzetti A, Van Hamme JD, Sillen W, Vangronsveld J (2016) The interaction between plants and bacteria in the remediation of petroleum hydrocarbons: an environmental perspective. Front Microbiol 7:1836

Guerin TF, Horner S, McGovern T, Davey B (2002) An application of permeable reactive barrier technology to petroleum hydrocarbon contaminated groundwater. Water Res 36(1):15–24

Han X, Hu H, Shi X, Zhang L, He J (2017) Effects of different agricultural wastes on the dissipation of PAHs and the PAH-degrading genes in a PAH-contaminated soil. Chemosphere 172:286–293

Harreither W, Sygmund C, Augustin M, Narciso M, Rabinovich ML, Gorton L, Haltrich D, Ludwig R (2011) Catalytic properties and classification of cellobiose dehydrogenases from ascomycetes. Appl Environ Microbiol 77(5):1804–1815

Hussain I, Puschenreiter M, Gerhard S, Schöftner P, Yousaf S, Wang A, Syed JH, Reichenauer TG (2018) Rhizoremediation of petroleum hydrocarbon-contaminated soils: improvement opportunities and field applications. Environ Exp Bot 147:202–219

Jeanbille M, Gury J, Duran R, Tronczynski J, Agogué H, Ben Saïd O, Ghiglione J-F, Auguet J-C (2016) Response of core microbial consortia to chronic hydrocarbon contaminations in coastal sediment habitats. Front Microbiol 7:1637

Joshi MN, Dhebar SV, Dhebar SV, Bhargava P, Pandit A, Patel RP, Saxena A, Bagatharia SB (2014) Metagenomics of petroleum muck: revealing microbial diversity and depicting microbial syntrophy. Arch Microbiol 196(8):531–544

Ka JO, Yu Z, Mohn WW (2001) Monitoring the size and metabolic activity of the bacterial community during biostimulation of fuel-contaminated soil using competitive PCR and RT-PCR. Microb Ecol 42(3):267–273

Kaczyńska G, Borowik A, Wyszkowska J (2015) Soil dehydrogenases as an indicator of contamination of the environment with petroleum products. Water Air Soil Pollut 226(11):372

Kankara R, Arockiaraj S, Prabhu K (2016) Environmental sensitivity mapping and risk assessment for oil spill along the Chennai Coast in India. Mar Pollut Bull 106(1–2):95–103

Karlsson AE, Johansson T, Bengtson P (2012) Archaeal abundance in relation to root and fungal exudation rates. FEMS Microbiol Ecol 80(2):305–311

Kastner M, Miltner A (2016) Application of compost for effective bioremediation of organic contaminants and pollutants in soil. Appl Microbiol Biotechnol 100(8):3433–3449

Kim J, Lee AH, Chang W (2018) Enhanced bioremediation of nutrient-amended, petroleum hydrocarbon-contaminated soils over a cold-climate winter: the rate and extent of hydrocarbon biodegradation and microbial response in a pilot-scale biopile subjected to natural seasonal freeze-thaw temperatures. Sci Total Environ 612:903–913

Kotowicz J, Chmielniak T, Janusz-Szymańska K (2010) The influence of membrane $CO_2$ separation on the efficiency of a coal-fired power plant. Energy 35(2):841–850

Kulkarni SV, Palande AS, Deshpande MV (2012) Bioremediation of petroleum hydrocarbons in soils. In: Satyanarayana T, Johri BN (eds) Microorganisms in environmental management: microbes and environment. Springer Netherlands, Dordrecht, pp 589–606

Kuppusamy S, Thavamani P, Venkateswarlu K, Lee YB, Naidu R, Megharaj M (2017) Remediation approaches for polycyclic aromatic hydrocarbons (PAHs) contaminated soils: technological constraints, emerging trends and future directions. Chemosphere 168:944–968

Lim MW, Lau EV, Poh PE (2016) A comprehensive guide of remediation technologies for oil contaminated soil – present works and future directions. Mar Pollut Bull 109(1):14–45

Liu Q, Li Q, Wang N, Liu D, Zan L, Chang L, Gou X, Wang P (2018) Bioremediation of petroleum-contaminated soil using aged refuse from landfills. Waste Manag 77:576–585

Lladó S, Gràcia E, Solanas AM, Viñas M (2013) Fungal and bacterial microbial community assessment during bioremediation assays in an aged creosote-polluted soil. Soil Biol Biochem 67:114–123

Lladó S, Covino S, Solanas AM, Petruccioli M, D'Annibale A, Viñas M (2014) Pyrosequencing reveals the effect of mobilizing agents and lignocellulosic substrate amendment on microbial community composition in a real industrial PAH-polluted soil. J Hazard Mater 283:35–43

Maila MP, Cloete TE (2005) The use of biological activities to monitor the removal of fuel contaminants – perspective for monitoring hydrocarbon contamination: a review. Int Biodeterior Biodegrad 55(1):1–8

Mair J, Schinner F, Margesin R (2013) A feasibility study on the bioremediation of hydrocarbon-contaminated soil from an Alpine former military site: effects of temperature and biostimulation. Cold Reg Sci Technol 96:122–128

Mapelli F, Scoma A, Michoud G, Aulenta F, Boon N, Borin S, Kalogerakis N, Daffonchio D (2017) Biotechnologies for marine oil spill cleanup: indissoluble ties with microorganisms. Trends Biotechnol 35(9):860–870

Margesin R (2013) Bioremediation and biodegradation of hydrocarbons by cold-adapted yeasts. In: Buzzini P, Margesin R (eds) Cold-adapted yeasts: biodiversity, adaptation strategies and biotechnological significance. Springer, Berlin, pp 465–480

Margesin R, Schinner F (1999) Biological decontamination of oil spills in cold environments. J Chem Technol Biotechnol 74(5):381–389

Margesin R, Hammerle M, Tscherko D (2007) Microbial activity and community composition during bioremediation of diesel-oil-contaminated soil: effects of hydrocarbon concentration, fertilizers, and incubation time. Microb Ecol 53(2):259–269

Masy T, Demanèche S, Tromme O, Thonart P, Jacques P, Hiligsmann S, Vogel TM (2016) Hydrocarbon biostimulation and bioaugmentation in organic carbon and clay-rich soils. Soil Biol Biochem 99:66–74

Mena E, Villaseñor J, Rodrigo MA, Cañizares P (2016) Electrokinetic remediation of soil polluted with insoluble organics using biological permeable reactive barriers: effect of periodic polarity reversal and voltage gradient. Chem Eng J 299:30–36

Militon C, Boucher D, Vachelard C, Perchet G, Barra V, Troquet J, Peyretaillade E, Peyret P (2010) Bacterial community changes during bioremediation of aliphatic hydrocarbon-contaminated soil. FEMS Microbiol Ecol 74(3):669–681

Mouhamadou B, Faure M, Sage L, Marcais J, Souard F, Geremia RA (2013) Potential of autochthonous fungal strains isolated from contaminated soils for degradation of polychlorinated biphenyls. Fungal Biol 117(4):268–274

Moussavi G, Khosravi R, Farzadkia M (2011) Removal of petroleum hydrocarbons from contaminated groundwater using an electrocoagulation process: batch and continuous experiments. Desalination 278(1):288–294

Nelyubov DV, Semikhina LP, Vazhenin DA, Merkul'ev IA (2017) Influence of resins and asphaltenes on the structural and rheological properties of petroleum disperse systems. Petrol Chem 57(3):203–208

Nicolaus EEM, Wright SR, Barry J, Bolam TPC, Ghareeb K, Ghaloom M, Al-Kanderi N, Harley BFM, Le Quesne WJF, Devlin MJ, Lyons BP (2017) Spatial and temporal analysis of the risks posed by total petroleum hydrocarbon and trace element contaminants in coastal waters of Kuwait. Mar Pollut Bull 120(1):422–427

Pal S, Kundu A, Banerjee TD, Mohapatra B, Roy A, Manna R, Sar P, Kazy SK (2017) Genome analysis of crude oil degrading *Franconibacter pulveris* strain DJ34 revealed its genetic basis for hydrocarbon degradation and survival in oil contaminated environment. Genomics 109(5–6):374–382

Polyak YM, Bakina LG, Chugunova MV, Mayachkina NV, Gerasimov AO, Bure VM (2018) Effect of remediation strategies on biological activity of oil-contaminated soil – a field study. Int Biodeterior Biodegrad 126:57–68

Prasad R (2017) Mycoremediation and environmental sustainability, vol 1. Springer International Publishing, Cham, Switzerland https://doi.org/10.1007/978-3-319-68957-9

Prasad R (2018) Mycoremediation and environmental sustainability, vol 2. Springer International Publishing, Cham, Switzerland https://www.springer.com/us/book/9783319773858

Prenafeta-Boldu FX, Summerbell R, Sybren de Hoog G (2006) Fungi growing on aromatic hydrocarbons: biotechnology's unexpected encounter with biohazard? FEMS Microbiol Rev 30(1):109–130

Qiao J, Zhang C, Luo S, Chen W (2014) Bioremediation of highly contaminated oilfield soil: bioaugmentation for enhancing aromatic compounds removal. Front Environ Sci Eng 8(2):293–304

Qin G, Gong D, Fan M-Y (2013) Bioremediation of petroleum-contaminated soil by biostimulation amended with biochar. Int Biodeterior Biodegrad 85(Supplement C):150–155

Röling WFM, Couto de Brito IR, Swannell RPJ, Head IM (2004) Response of archaeal communities in beach sediments to spilled oil and bioremediation. Appl Environ Microbiol 70(5):2614–2620

Safdari MS, Kariminia HR, Rahmati M, Fazlollahi F, Polasko A, Mahendra S, Wilding WV, Fletcher TH (2018) Development of bioreactors for comparative study of natural attenuation, biostimulation, and bioaugmentation of petroleum-hydrocarbon contaminated soil. J Hazard Mater 342:270–278

Shahi A, Aydin S, Ince B, Ince O (2016a) Evaluation of microbial population and functional genes during the bioremediation of petroleum-contaminated soil as an effective monitoring approach. Ecotoxicol Environ Saf 125:153–160

Shahi A, Aydin S, Ince B, Ince O (2016b) Reconstruction of bacterial community structure and variation for enhanced petroleum hydrocarbons degradation through biostimulation of oil contaminated soil. Chem Eng J 306:60–66

Siles JA, Margesin R (2018) Insights into microbial communities mediating the bioremediation of hydrocarbon-contaminated soil from an Alpine former military site. Appl Microbiol Biotechnol 102(10):4409–4421

Siles JA, Öhlinger B, Cajthaml T, Kistler E, Margesin R (2018) Characterization of soil bacterial, archaeal and fungal communities inhabiting archaeological human-impacted layers at Monte Iato settlement (Sicily, Italy). Sci Rep 8(1):1903

Silva-Castro GA, Rodelas B, Perucha C, Laguna J, González-López J, Calvo C (2013) Bioremediation of diesel-polluted soil using biostimulation as post-treatment after oxidation with Fenton-like reagents: assays in a pilot plant. Sci Total Environ 445–446:347–355

Singleton DR, Jones MD, Richardson SD, Aitken MD (2013) Pyrosequence analyses of bacterial communities during simulated in situ bioremediation of polycyclic aromatic hydrocarbon-contaminated soil. Appl Microbiol Biotechnol 97(18):8381–8391

Stella T, Covino S, Čvančarová M, Filipová A, Petruccioli M, D'Annibale A, Cajthaml T (2017) Bioremediation of long-term PCB-contaminated soil by white-rot fungi. J Hazard Mater 324:701–710

Stenuit B, Eyers L, Schuler L, Agathos SN, George I (2008) Emerging high-throughput approaches to analyze bioremediation of sites contaminated with hazardous and/or recalcitrant wastes. Biotechnol Adv 26(6):561–575

Streche C, Cocârţă DM, Istrate I-A, Badea AA (2018) Decontamination of petroleum-contaminated soils using the electrochemical technique: remediation degree and energy consumption. Sci Rep 8(1):3272

Suja F, Rahim F, Taha MR, Hambali N, Rizal Razali M, Khalid A, Hamzah A (2014) Effects of local microbial bioaugmentation and biostimulation on the bioremediation of total petroleum hydrocarbons (TPH) in crude oil contaminated soil based on laboratory and field observations. Int Biodeterior Biodegrad 90(Supplement C):115–122

Sutton NB, Maphosa F, Morillo JA, Abu Al-Soud W, Langenhoff AA, Grotenhuis T, Rijnaarts HH, Smidt H (2013) Impact of long-term diesel contamination on soil microbial community structure. Appl Environ Microbiol 79(2):619–630

Thion C, Cebron A, Beguiristain T, Leyval C (2012) Long-term in situ dynamics of the fungal communities in a multi-contaminated soil are mainly driven by plants. FEMS Microbiol Ecol 82(1):169–181

Tigini V, Prigione V, Di Toro S, Fava F, Varese GC (2009) Isolation and characterisation of polychlorinated biphenyl (PCB) degrading fungi from a historically contaminated soil. Microb Cell Factories 8:5

Turgay OC, Erdogan EE, Karaca A (2010) Effect of humic deposit (leonardite) on degradation of semi-volatile and heavy hydrocarbons and soil quality in crude-oil-contaminated soil. Environ Monit Assess 170(1–4):45–58

van Beilen JB, Funhoff EG, van Loon A, Just A, Kaysser L, Bouza M, Holtackers R, Rothlisberger M, Li Z, Witholt B (2006) Cytochrome P450 alkane hydroxylases of the CYP153 family are common in alkane-degrading eubacteria lacking integral membrane alkane hydroxylases. Appl Environ Microbiol 72(1):59–65

Vandera E, Koukkou AI (2017) Bacterial community response to hydrocarbon contamination in soils and marine sediments: a critical review of case studies. In: Cravo-Laureau C, Cagnon C, Lauga B, Duran R (eds) Microbial ecotoxicology. Springer International Publishing, Cham, pp 185–226

Varjani SJ (2017) Microbial degradation of petroleum hydrocarbons. Bioresour Technol 223(Supplement C):277–286

Vasudevan N, Bharathi S, Arulazhagan P (2007) Role of plasmid in the degradation of petroleum hydrocarbon by *Pseudomonas fluorescens* NS1. J Environ Sci Health A Tox Hazard Subst Environ Eng 42(8):1141–1146

Vomberg A, Klinner U (2000) Distribution of alkB genes within n-alkane-degrading bacteria. J Appl Microbiol 89(2):339–348

Wang L, Li F, Zhan Y, Zhu L (2016) Shifts in microbial community structure during in situ surfactant-enhanced bioremediation of polycyclic aromatic hydrocarbon-contaminated soil. Environ Sci Pollut Res 23(14):14451–14461

Wu Y-R, Nian D-L (2014) Production optimization and molecular structure characterization of a newly isolated novel laccase from *Fusarium solani* MAS2, an anthracene-degrading fungus. Int Biodeterior Biodegrad 86:382–389

Wu M, Dick WA, Li W, Wang X, Yang Q, Wang T, Xu L, Zhang M, Chen L (2016) Bioaugmentation and biostimulation of hydrocarbon degradation and the microbial community in a petroleum-contaminated soil. Int Biodeterior Biodegrad 107:158–164

Xu Y, Lu M (2010) Bioremediation of crude oil-contaminated soil: comparison of different biostimulation and bioaugmentation treatments. J Hazard Mater 183(1–3):395–401

Yang Q, Zhang H, Li X, Wang Z, Xu Y, Ren S, Chen X, Xu Y, Hao H, Wang H (2013) Extracellular enzyme production and phylogenetic distribution of yeasts in wastewater treatment systems. Bioresour Technol 129:264–273

Yavari S, Malakahmad A, Sapari NB (2015) A review on phytoremediation of crude oil spills. Water Air Soil Pollut 226(8):18

Zafra G, Absalón ÁE, Cuevas MDC, Cortés-Espinosa DV (2014) Isolation and selection of a highly tolerant microbial consortium with potential for PAH biodegradation from heavy crude oil-contaminated soils. Water Air Soil Pollut 225(2):1826

Zafra G, Taylor TD, Absalón AE, Cortés-Espinosa DV (2016) Comparative metagenomic analysis of PAH degradation in soil by a mixed microbial consortium. J Hazard Mater 318:702–710

Zhang Z, Lo IMC, Yan DYS (2015) An integrated bioremediation process for petroleum hydrocarbons removal and odor mitigation from contaminated marine sediment. Water Res 83:21–30

Zhao J, Zeng J, de Hoog GS, Attili-Angelis D, Prenafeta-Boldu FX (2010) Isolation and identification of black yeasts by enrichment on atmospheres of monoaromatic hydrocarbons. Microb Ecol 60(1):149–156

Zheng J, Feng JQ, Zhou L, Mbadinga SM, Gu JD, Mu BZ (2018) Characterization of bacterial composition and diversity in a long-term petroleum contaminated soil and isolation of high-efficiency alkane-degrading strains using an improved medium. World J Microbiol Biotechnol 34(2):34

Zhou Z-F, Wang M-X, Zuo X-H, Yao Y-H (2017) Comparative investigation of bacterial, fungal, and archaeal community structures in soils in a typical oilfield in Jianghan, China. Arch Environ Contam Toxicol 72(1):65–77

Zhu H, Taylor AA, Astor SR, Terry N (2017) Enhancing salt grass germination and growth in a saline soil contaminated with petroleum hydrocarbons. Plant Soil 412(1):189–199

# Chapter 8
# Microalgae-Bacteria Consortia for the Removal of Phenolic Compounds from Industrial Wastewaters

**Paula Maza-Márquez, Alejandro González-Martínez, Belén Juárez-Jiménez, Belén Rodelas, and Jesús González-López**

## Contents



P. Maza-Márquez (✉) · A. González-Martínez · B. Juárez-Jiménez · B. Rodelas
J. González-López
Department of Microbiology, Faculty of Pharmacy, and Institute of Water Research, University of Granada, Granada, Spain
e-mail: paulamaza@ugr.es; agon@ugr.es; belenjj@ugr.es; mrodelas@ugr.es; jgl@ugr.es

 
R. Prasad, E. Aranda (eds.), *Approaches in Bioremediation*, Nanotechnology in the Life Sciences, https://doi.org/10.1007/978-3-030-02369-0_8

## 8.1 Phenolic Compounds (PCs): Definition, Occurrence in the Environment, Sources, and Toxicity for Living Organisms

Phenolic compounds (PCs), also termed phenolics, are substances which possess an aromatic ring bearing one (phenol) or more (polyphenol) hydroxyl substituents (Lattanzio et al. 2006). Phenol ($C_6H_5OH$) is the basic compound of the family with the simplest structure, consisting of a single hydroxyl group bond to a benzene ring, but the range of derivatives of phenol is wide, going from monocyclic structures (simple phenols) to complex polymers comprising a large number of phenol units. Examples of some common mono- and polycyclic structures are shown in Fig. 8.1. Several ways for the classification of PCs have been proposed, being the one devised by Harborne and Simmonds (1964) one of the most acknowledged, which divides PCs into 20 main groups according to their number of carbon atoms (Vermerris and Nicholson 2008). Other classifications are based on the number of phenol units in the molecule, their distribution in nature, or location in plants (Anku et al. 2017).

PCs are a very large and diverse family of molecules, occurring naturally mostly as secondary metabolites of plants, which accumulate in their roots, stems, leaves,

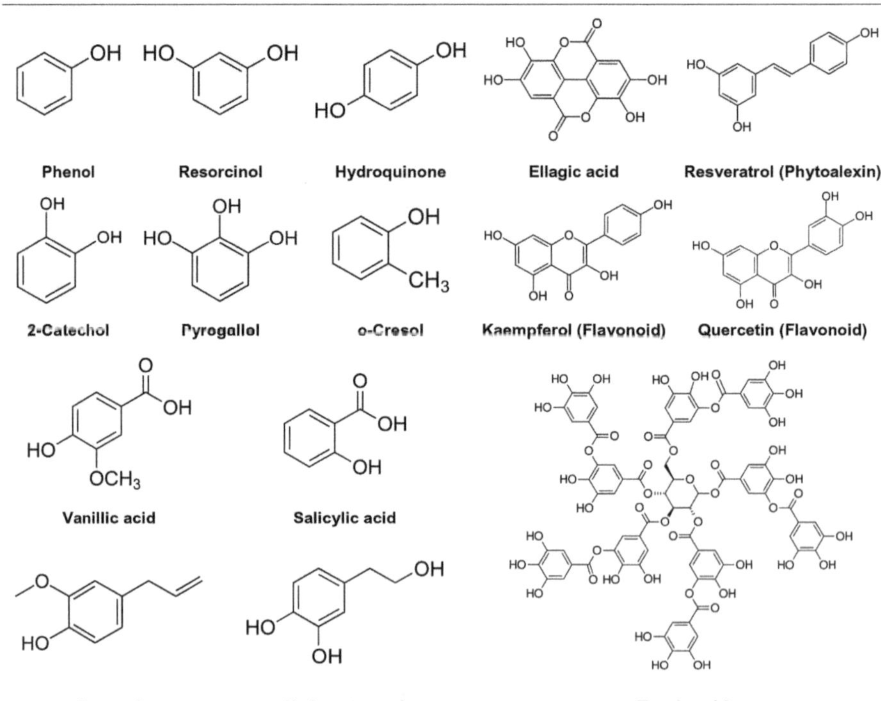


**Fig. 8.1** Examples of chemical structures of simple and polycyclic phenolic compounds

and fruits, where they fulfill important biological functions. Polymers like lignin and tannins contribute to the structural integrity of plants, while smaller phenolic molecules like flavonoids are involved in plant growth regulation, provide protection from biotic and abiotic stresses, and act as chemotactic and signaling agents in mutualistic and pathogenic plant-microbial interactions (Lattanzio et al. 2006; Bhattacharya et al. 2010). Many vegetal pigments, dyes, and flavors are also polyphenolic molecules. An increasing number of plant polyphenols occurring in foods are regarded for their positive effects on human health, being included as active ingredients of pharmaceutical, cosmetic, and dietary supplement formulations (Vermerris and Nicholson 2008; Sharma 2014).

Many of the simple phenols are highly or moderately soluble in water (Sobiesak 2017) and thus easily distribute in the aquatic media. Alongside the natural sources of PCs, PCs of anthropogenic origin occurring in industrial, agricultural, or municipal wastes also enter the environment (Anku et al. 2017). Currently, >10 tons of phenol are yearly produced worldwide (Plotkin 2016), which are mostly used as raw material for a variety of synthetic and semisynthetic compounds. Vast amounts of PCs are generated from coking plants, oil and gas refineries, chemical industries (manufacturing of explosives, dyes, paints, papers, pesticides, plastics, pharmaceuticals, resins, textile fibers), metal smelting, tannery, olive oil, and other food industries (Busca et al. 2008; Al-Khalid and El Naas 2012, Anku et al. 2017). Polyphenolic insecticides, herbicides, and other pesticides used in agriculture reach water bodies by runoff, and many phenol derivatives are also included in the formulations of pharmaceutical drugs, personal care products, and domestic disinfectants (Anku et al. 2017). Consequently, PCs of anthropogenic origin are often found in wastewaters, effluents of wastewater treatment plants, and landfill leachates, from which they reach underground waters (Zhong et al. 2012). Table 8.1 summarizes the concentrations of PCs reported in various types of wastewaters, which are highly variable, ranging from less than 1 mg/L to several g/L.

**Table 8.1** Ranges of concentration of phenolic compounds (PCs) reported for various types of wastewaters

| Activity | Range of concentration (mg/L) | References |
|---|---|---|
| Coal processing | 9–6800 | Busca et al. (2008) |
| Coal gasification | 850–950 | Ji et al. (2015) |
| Coking plants | 28–3900 | Busca et al. (2008) |
| Manufacture of petrochemicals | 2.8–1220 | Busca et al. (2008) |
| Municipal sewage | 0.02–0.77 | Folke and Lund (1983) and Zhong et al. (2012) |
| Olive oil production | 0.5–80000 | Niaounakis and Halvadakis (2006) and Rahmanian et al. (2014) |
| Oil refineries | 6–500 | Busca et al. (2008) |
| Pharmaceutical plants, wood products, paints, papermaking | 0.1–1600 | Busca et al. (2008) |
| Plastics | 600–2000 | Jusoh and Razali (2008) |

Several PCs, mostly of anthropogenic origin, are regarded as major pollutants of aquatic systems. The toxicity levels of PCs usually are in the 9–25 mg/L range for humans and aquatic life (Cordova-Villegas et al. 2016), but harmful effects to living organisms are reported even at concentrations lower than 1 mg/L (Duan et al. 2018). Phenol and some of its chloro- and nitro-derivatives are highly corrosive chemicals. In humans, they cause acute toxicity, irritating the eyes and respiratory tract and causing skin necrosis after exposition, being absorbed fast through the skin and damaging the liver, kidneys, and other internal organs (Michałowicz and Duda 2007). Long-term toxicity due to mutagenic, carcinogenic, neurotoxic, immunotoxic, or endocrine-disrupting effects has been reported for a significant number of PCs (Michałowicz and Duda 2007; Busca et al. 2008). Chlorophenols are one of the largest and widespread groups of phenol derivatives, particularly abundant in wastewaters from chemical, metallurgic, pharmaceutical, pesticide, and textile industries. Carcinogenic effects have been reported for several chlorophenols, and significant quantities have been detected in drinking water, vegetables, and poultry products for human consumption (Michałowicz and Duda 2007). Alkylphenols, such as nonylphenol and octylphenol ethoxylates, are a major group of nonionic surfactants, broadly included in industrial cleaning formulations. These compounds enter wastewater treatment facilities, where they undergo microbial biotransformation generating several toxic and bioaccumulative metabolites, some of which are structural analogues of estrogens which act as endocrine disruptors in animals and humans, and may also affect the nervous and immune systems (Priac et al. 2017; Acir and Guenther 2018). In this context, the international regulatory bodies have established strict discharge limits for phenol and its derivatives (Cordova-Villegas et al. 2016). Both the US Environmental Protection Agency (EPA) and the European Union include several PCs, mostly chlorophenols, nitrophenols, and alkylphenols, on their lists of priority pollutants in the field of water policy (EU 2013; EPA 2014).

## 8.2 Overview of the Strategies for the Removal of PCs from Wastewaters

In order to minimize the negative impact of PCs on the natural environment, a wide range of approaches for their removal from industrial, agricultural, or municipal wastewaters have been investigated to date. These include physicochemical strategies, biological methods, or a combination of both.

### *8.2.1 Physicochemical Methods*

The physicochemical methods comprise a diverse range of treatments, from simple precipitation using flocculants and coagulants (Shen 2002), to more advanced treatments like separation by distillation, adsorption, solvent extraction, ultra- or nanofiltration, reverse osmosis, chemical or electrochemical oxidation, advanced oxidation processes (i.e., Fenton, UV/$H_2O_2$, wet-air oxidation, ozonation, photocatalysis), combustion/pyrolysis, and biochemical abatement by enzymes (Busca et al. 2008; Liotta et al. 2009; Rahmanian et al. 2014; Cordova-Villegas et al. 2016; Anku et al. 2017). Some of the aforementioned approaches are also focused on the recovery of valuable plant-derived compounds in wastes like those from the olive oil industry (Rahmanian et al. 2014; Araújo et al. 2015), where the concentration of phenolics may largely exceed 1000 mg/L (Table 8.1). However, these methods are generally characterized as expensive, due to the cost of reactives and special equipment. In addition, they sometimes display low efficiency or generate hazardous by-products (Busca et al. 2008; Krastanov et al. 2013). A summary of the drawbacks reported for some of the physicochemical approaches applied to the removal of phenolics in wastewaters is provided in Table 8.2.

**Table 8.2** Disadvantages of the application of some physicochemical methods to the removal of phenolic compounds (PCs) from industrial wastewaters

| Method | Disadvantages | References |
|---|---|---|
| Adsorption in activated carbon | Adsorbent cannot be regenerated due to molecules adsorbing irreversibly. Costly disposal of potentially hazardous spent adsorbent | Alexander et al. (2012), Priac et al. (2017), and Terzyk (2007) |
| Nanofiltration, reverse osmosis | High operating and maintenance cost<br>Fouling issues reduce membrane life span | Cui et al. (2016) and Priac et al. (2017) |
| Biochemical abatement with enzymes (laccase) | High cost of enzymes and difficulties for their supply at large scale<br>Low rate of reaction<br>Limitations of the operation conditions (temperature and pH) | Xiong et al. (2017) |
| Advanced oxidation processes (AOPs) | General drawbacks:<br>Pretreatment necessary to remove suspended solids, radical scavengers, and competing ions<br>Generation of oxidation-refractory compounds<br>High cost of reactives and/or equipment | Alexander et al. (2012) and Cañizares et al. (2009) |
| Fenton | Involves storage and handling of $H_2O_2$<br>Tight control of pH required<br>Generation of additional toxic by-products | Alexander et al. (2012), Liotta et al. (2009), and Throop (1975) |
| Ozonation | Slow reaction<br>Incomplete degradation<br>Possible generation of intermediates<br>High energy cost | Liotta et al. (2009) and Priac et al. (2017) |
| Electrolytic oxidation | High cost of equipment<br>Generation of additional toxic by-products | Radjenovic and Sedlak (2015) |
| Photocatalysis | Acid pH (6.0) required for optimal degradation<br>Possible generation of by-products<br>High cost | Das et al. (2005) and Priac et al. (2017) |

The selection of the most suitable physicochemical approach is determined by many factors and should be reviewed for each particular case, since the efficiency of these technologies is highly variable depending on the volume of wastewaters to be treated, the concentration and nature of the specific PCs, and the co-occurrence of other contaminants (Busca et al. 2008; Cañizares et al. 2009). Combination of biological and physicochemical methods often offers an interesting and cost-effective alternative for many industrial wastewaters (reviewed by Oller et al. 2011).

### *8.2.2 Biological Treatments*

The suitability of biological treatments for the removal of pollutants from household and industrial wastes has been discussed since the 1950s (Heukelekian 1956). Currently, these approaches constitute the most attractive technologies for such purpose. Compared to physicochemical methods, biological treatments are regarded as cost-saving, environmentally compatible, and reliable, are easy to manage and minimize the generation of by-products (Paraskeva and Diamadopoulos 2006; Al-Khalid and El Naas 2012; Krastanov et al. 2013; Ji et al. 2015; Azubuike et al. 2016). Living organisms may contribute to the removal of organic and inorganic pollutants from their surrounding media by several mechanisms. Pollutants can undergo passive extracellular biosorption or active uptake and subsequent intracellular bioaccumulation (Xiong et al. 2017). The transformation of an organic pollutant to a less toxic or nontoxic intermediate form is termed detoxification (Suthersan 1996). Biodegradation implies the biologically catalyzed reduction of the complexity of a chemical compound, while the complete conversion of the organic structure to inorganic molecules is called mineralization (Joutey et al. 2013). Biodegradation can be mediated by both extracellular and intracellular enzymes (Xiong et al. 2017). Many microorganisms biodegrade/mineralize organic pollutants to use them as carbon and/or energy source (Kanekar et al. 1998; Basha et al. 2010).

On a general basis, phenolic molecules are regarded as recalcitrant toward biodegradation. Smaller, man-made phenol derivatives are the most difficult to degrade because of their aromatic nature and their toxicity to living organisms at low concentrations (Al-Khalid and El Naas 2012; Krastanov et al. 2013). The breakage of the aromatic ring, which has a high thermodynamic stability, requires the introduction of hydroxyl radicals as a first step before the molecules can be degraded by different metabolic pathways (Gibson and Subrawahian 1984). Chlorinated and nitro-derivatives are particularly resistant to biodegradation, since Cl- and $NO_2$-groups reduce the electron density on the aromatic rings due to an inductive effect, making them less prone to enzymatic cleavage (Nikel et al. 2013; Arora et al. 2018). Large natural polymers like lignin are hard to decompose due to their particular molecular architecture (Ruiz-Dueñas and Martínez 2009). Nevertheless, many bacteria, archaea, fungi, and algae are described able to biodegrade phenolics in the environment. Efficient biodegradation can be performed either by single species of microorganisms or microbial consortia. Several studies reported that the indigenous microbes inhabiting the polluted wastewater sources or other contaminated sites

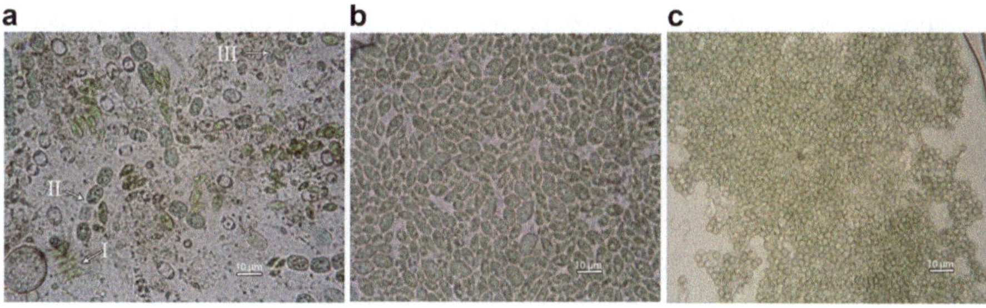


**Fig. 8.2** Microscopic images (×1000 magnification) of the indigenous microalgae mixed community (**a**) and the major microalgal strains (**b** and **c**) in olive washing wastewater (OWW) samples. Complex morphological types found in the OWW are shown in (**a**). (**b**) *Tetradesmus obliquus* (=*Scenedesmus obliquus*) and (**c**) *Chlorella vulgaris* isolates, growing in pure cultures on Rodríguez-López medium. (Reprinted from Maza-Márquez et al. (2014), with permission from Elsevier)

more often hold the ability to tolerate and effectively degrade PCs (Al-Khalid and El Naas 2012; Krastanov et al. 2013; Rucká et al. 2017). As an example, Fig. 8.2 displays microscopic images of two eukaryotic microalgae isolated from olive washing wastewater (OWW), a by-product of the olive oil manufacturing industry, which were able to tolerate up to 150 mg/L of phenols (Maza-Márquez et al. 2014).

In the design of a biotreatment process, the selection of the microorganisms largely depends on the nature of the target pollutants, the particular metabolic ability of the microorganisms, and the conditions in which the specific effluent is more suitably treated, i.e., aerobic or anaerobic (Paraskeva and Diamadopoulos 2006). An important feature for the assessment of the microbial strains is the biodegradation rate, which varies greatly, even among different strains from the same species (Krastanov et al. 2013). Changes in the efficiency of biodegradation may occur when mixtures of different pollutants are combined in the same waste. For instance, phenol inhibited the degradation of benzene and naphthalene by microorganisms from an oil refinery wastewater (Meyer et al. 1984). Addition of phenol hampers 4-chlorophenol biological removal in bioaugmented soils, while sodium benzoate enhances it (Nowak and Mrozik 2018). The ecological interactions among different species may also be considered; this topic will be discussed later in this chapter (Sect. 8.6.1.2). These facts highlight the difficulties to foresee the fate in the environment of those pollutants sourced from complex wastes, as well as the success of full-scale biotreatment technologies.

## 8.3 Removal of PCs by Bacteria and Archaea

The biodegradation of phenol and its derivatives by bacteria have been thoroughly investigated for over a century. Microorganisms able to degrade phenol were firstly isolated and described at the beginning of the twentieth century (*Bacillus hexacarbovorum*, Störmer 1908), and to date a vast number of bacterial strains able to transform, biodegrade, or mineralize phenol and its derivatives have been characterized physiologically, genetically, or both. Only representative examples have been summarized in Table 8.3; more comprehensive reviews have been published elsewhere (Field and Sierra-Álvarez 2008; Al-Khalid and El Naas 2012; Levén et al. 2012).

**Table 8.3** A review of Bacteria and Archaea strains described in the literature for their ability to remove phenolic compounds

| Microorganisms | Experimental conditions | Summary of results | References |
|---|---|---|---|
| **Bacteria, pure isolates** | **Aerobic culture** | | |
| *Acinetobacter baumannii* | Synthetic minimal medium added with phenol (125–1000 mg/L) as sole carbon source. Batch cultures | Phenol (1000 mg/L) was fully removed after 48 h of incubation | Prasad et al. (2010) |
| *Bacillus brevis* | Synthetic mineral medium added with phenol (500–1750 mg/L) as sole carbon source. Batch cultures | Removal of phenol was maximum at pH 8, 34 °C and 0.5% (v/v) inoculum, without any co-substrate. Phenol (1750 mg/L) was fully removed after 120 h | Arutchelvan et al. (2006) |
| *Bacillus* sp. Cro 3.2 (thermophilic strain) | Synthetic mineral medium added with different concentrations of phenol as sole carbon source. Stirred-tank bioreactor | Phenol was tolerated up to a concentration of 0.1% (w/v). Enzymes of the *meta*-pathway of degradation were detected | Ali et al. (1998) |
| *Delftia tsuruhatensis* BM90 | Synthetic medium. Repeated batch cultures. Cells immobilized on macroporous cellulose | A mixture of 20 phenolic compounds (500 mg/L) was 90% removed after every 24 h of repeated incubations | Juárez-Jiménez et al. (2012) |
| *Pseudomonas aeruginosa* SZH16 | Synthetic mineral medium added with phenol (200 mg/L) as sole carbon source<br>Phenol-spiked soil (100 mg/kg soil), inoculated with $10^8$ CFU per 20 g | In mineral medium, phenol was 100% removed after 18 h, and 85% was removed from spiked soil after 15 days. In soil samples, phytotoxic effects on corn seedlings were concomitantly reduced | Wang et al. (2011) |
| *Pseudomonas aeruginosa* MTCC 4997 | Synthetic medium added with phenol at different concentrations as sole carbon source. Packed-bed bioreactor, cells immobilized on several matrices | Cells immobilized in polyurethane foam removed phenol up to 1500 mg/L after 12.5 h | Kotresha and Vidyasagar (2017) |

(continued)

**Table 8.3** (continued)

| Microorganisms | Experimental conditions | Summary of results | References |
|---|---|---|---|
| *Pseudomonas putida* ATCC 17484 | Synthetic medium added with 1000 mg/L phenol as sole carbon source. Stirred tank bioreactor (STB) and fluidized bed bioreactor (FBB), cells immobilized on alginate beads, continuous operation | Phenol was 100% removed in the STB after 10 days (HRT 2–4 days). Phenol was 100% removed in the FBB after 12 days (HRT 4–0.25 days). Phenol was >90% degraded in both systems at loads over 4 g/day FBB provided better control of operation | González et al. (2001) |
| *Pseudomonas stutzeri* OX1 | Synthetic mineral medium added with phenol (450 mg/L) as sole carbon source. Batch cultures. Internal-loop airlift bioreactor continuously fed with 190 mg/L phenol | In batch cultures, phenol was >80% removed in 140 h. In the airlift bioreactor, phenol conversion was >95% | Viggiani et al. (2006) |
| **Bacteria, pure isolates** | **Anoxic/anaerobic culture** | | |
| *Cryptanaerobacter phenolicus* ATCC BAA-820T | Synthetic medium, added with phenol (3.5 mM) and 4-hydroxy benzoate (4OHB) (6.5 mM). Incubated anaerobically ($H_2/CO_2/N_2$ 10:10:80) | Requires phenol or 4OHB for growth. Carboxylates phenol into 4OHB and dehydroxylates 4OHB into benzoate, producing energy that is conserved for growth | Juteau et al. (2005) |
| *Desulfobacterium anilini* AK1 | Isolated from a sediment-enriched culture-degrading PCs. Batch cultures, minimal salt medium with sulfate | Removes phenol, 4-hydroxy-phenylacetate, and 2-fluorophenol, among other organic compounds. 800 μM phenol was removed within 6 days and an additional 1500 μM after 4 days | Ahn et al. (2009) |
| *Syntrophorabdus aromaticivorans* UI$^{T}$ | Isolated from a methanogenic isophthalate-degrading enrichment culture (syntrophic co-culture with *Methanospirillum hungatei*) | Grows on phenol under methanogenic conditions. Phenol (4.64 μM) was 31% removed after 100 days. Besides phenol, this strain removes *p*-cresol and other aromatic compounds | Qiu et al. (2008) |

(continued)

**Table 8.3** (continued)

| Microorganisms | Experimental conditions | Summary of results | References |
|---|---|---|---|
| *Thauera aromatica* K 172$^T$ | Isolated from enriched cultures of anaerobic sewage sludge. Batch cultures, 1 mM phenolic compounds as sole carbon source, 5 mM nitrate, anaerobic conditions | Strain K172$^T$ used phenol, *p*-cresol, phenylacetate, 4-hydroxyphenil acetate, and other aromatic compounds as sole carbon and energy sources, under denitrifying conditions | Tschech and Fuchs (1987) and Anders et al. (1995) |
| **Archaea** | **Aerobic culture** | | |
| *Halobacterium* spp. DSM 11147 | Synthetic media, 25% NaCl, added with trichlorophenols, 1–50 μM. Batch cultures. Incubation at 40 °C | Cells become sequentially adapted to increasing concentrations of trichlorophenols. The adapted strain completely removed trichlorophenols (50 μM) after 7–8 days of incubation | Oesterhelt et al. (1998) |
| *Natrialba* sp. C21 | Synthetic media, 25% NaCl, added with phenol, 3% (v/v) as sole carbon source. Incubation at 40 °C, pH = 7.0 | The strain grew with phenol as sole carbon and energy source | Khemili-Talbi et al. (2015) |
| *Sulfolobus solfataricus* 98/2 | Synthetic mineral medium added with phenol (51–745 mg/L) a sole carbon source. Batch cultures, pH 3.2, incubated at 80 °C | After adaptation, the strain grew and removed 100% phenol (51–745 mg/L) after 35–146 h of incubation | Christen et al. (2012) |
| Unidentified archaeon A235 | Synthetic media, 20% NaCl, added with phenol (50–200 mg/L) as sole carbon source. Batch cultures | After adaptation, the strain removed 20–80% of phenol within 15 days. Optimal degradation rates were measured at 100 mg/L phenol, 37 °C, and pH = 7.5 | Acikgoz and Ozcan (2016) |
| **Bacterial consortia** | **Aerobic culture** | | |
| *Acinetobacter* sp. AG22 + *Alcaligenes* sp. AG21, AG23 | Synthetic medium added with 2-chlorophenol, phenol, or *m*-cresol (100, 50, and 50 mg/L, respectively) as sole carbon source. Batch cultures | Strains AG22 and AG23 removed >99% phenol (50 mg/L) after 11 and 20 h, respectively. Strain AG21 removed >99% 2-chlorophenol (100 mg/L) after 27 h. Strain AG23 removed >99% *m*-cresol (50 mg/L) after 8 h. The mixture of the three PCs was completely removed after 38 h by the bacterial consortium | Gallegos et al. (2003) |

(continued)

**Table 8.3** (continued)

| Microorganisms | Experimental conditions | Summary of results | References |
|---|---|---|---|
| *Alcaligenes odorans* + *Bacillus subtilis* + *Corynebacterium propinquum* + *Pseudomonas aeruginosa* | Strains isolated from petroleum refinery wastewater. Batch cultures in raw wastewater (3.71–1.20 mg/L phenol) | A consortium of the four strains reduced phenol and oil content of the petroleum refinery wastewater (70–85%) after 10 days of incubation | Singh et al. (2013) |
| *Comamonas* sp. PG-08 + *Propioniferax*-like PG-02 | Batch mono- and co-cultures with 250 mg/L phenol as sole carbon source. Bioaugmentation in sequencing batch reactors (SBR) inoculated with activated sludge, fed synthetic wastewater added with phenol (500 mg/L), at a rate of 1.5 g phenol/L day | In batch cultures, PG-02 + PG-08 removed phenol faster (29 h) than each strain in monoculture (53 h and >80 h, respectively). Bioaugmentation of the SBR with both strains improved phenol removal and aerobic granulation, compared to non-bioaugmented or monoculture experiments. Complete removal was achieved after 2 days of operation | Jiang et al. (2006) |
| *Pantoea agglomerans* PM15 + *Raoultella terrigena* PM3 | Batch cultures in non-sterile olive washing wastewater (OWW) | Both strains grew on agar media amended with 150 mg/L of a 1:1:1 mixture of caffeic, *p*-OH-benzoic, and coumaric acids. The consortium reduced PCs, COD, $BOD_5$, and color of OWW (93, 89, 91, and 62%, respectively) after 6 days of incubation | Maza-Márquez et al. (2013) |
| *Sphingomonas* + *Rhizobiaceae*-related α-proteobacteria | Olive washing wastewater (OWW), polyphenols content: 104.6 mg/L. Laboratory-scale fixed biofilm bioreactor. Indigenous bacteria of OWW identified by TGGE | COD and $BOD_5$ were reduced (90% and 85%, respectively) in continuous operation under aeration (0.1 L/min) and water inlet flow of 1.3 L/day. Polyphenolic content was 50% reduced with OWW flows of 1.3 and 2.3 L/day | Pozo et al. (2007) |

(continued)

**Table 8.3** (continued)

| Microorganisms | Experimental conditions | Summary of results | References |
|---|---|---|---|
| Unidentified mixed culture, nine strains | Culture enriched from hydrocarbon-polluted soil using mineral media added with alkylphenols (1 mM each) as sole carbon source. Isolated strains were used at equal ratios to construct the consortium. Batch cultures | The mixed culture removed phenol, cresol isomers (*o*-, *m*-, *p*-), 2-ethylphenol, and xylenol isomers (2,5- and 3,4-dimethylphenol). Phenol was 100% removed after 10 h incubation, while 2,5-dimethylphenol degradation required 150 h | Acuña-Arguelles et al. (2003) |
| **Bacterial consortia** | **Anoxic culture** | | |
| *Alcaligenes faecalis* n1 + *Enterobacter* sp. o1 | Synthetic mineral medium added with phenol (up to 600 mg/L) and other PCs (100 mg/L) as sole carbon source. Batch cultures | The consortium was capable of removing phenol and other PCs under anoxic conditions using $NO_3^-$ as electron acceptor | Thomas et al. (2002) |
| *Azoarcus* spp. + *Microbulbifer* spp. + *Pseudomonas* spp. + *Thauera* spp. | Synthetic coke-oven wastewater. Laboratory-scale activated-sludge anoxic-aerobic bioreactor, 21–23 °C, pH 7.5–7.9 | A sample from the anoxic tank was taken and added with [$^{13}C_6$] phenol (91 mg/L) and nitrate and incubated anaerobically. Anaerobic phenol removal was accompanied by a substantial reduction in nitrate concentrations. The members of the phenol-assimilating consortia were identified by RNA-stable isotope probing (RNA-SIP) | Sueoka et al. (2009) |

Since PCs are widespread in nature, microorganisms with the ability to use them as a carbon and/or energy source are found in many different habitats. Biodegradation of phenolics is achieved both under aerobic and anaerobic conditions (Nešvera et al. 2015) and in extreme environments subjected to low or elevated temperatures, acidic or alkaline pH, high salt concentrations, or high hydrostatic pressure (Margesin and Schiner 2001). Many of the described isolates have been retrieved from soils and aquatic media, where they contribute to the decay of lignin and other naturally sourced PCs. As an example, phenol is broken down by soil and water microbiota in 2–5 and 9 days, respectively (Busca et al. 2008). However, native soil or aquatic microbiotas lack enough attenuation capacity for those PCs generated by anthropogenic contamination; hence, treatment of phenolic wastewaters is required before their discharge to the environment.

Most of the bacterial isolates described in the literature as degraders of PCs have been affiliated to the genus *Pseudomonas*, but Gram-negative and Gram-positive strains identified as *Acinetobacter*, *Alcaligenes*, *Arthrobacter*, *Bacillus*, *Burkholderia*, *Cupriavidus*, *Paenibacillus*, *Rhodococcus*, *Sphingomonas*, and several other genera have been as well reported (Basha et al. 2010; Al-Khalid and El Naas 2012; Krastanov et al. 2013). Nonetheless, it must be taken into consideration that a large part of the available studies was mostly focused on the degradative ability of the isolates rather than on its systematics, and the identification of the strains is frequently missing or relies on morphological and phenotypical traits only; thus, taxonomic affiliations are sometimes ambiguous.

Aerobic bacteria transform phenol into nontoxic intermediate compounds that enter the tricarboxylic acid cycle, through *ortho-* or *meta-*pathways of degradation (Krastanov et al. 2013). Both pathways start with the conversion of phenol to catechol by monohydroxylation of the $C_6$ ring at the *o*-position, a step catalyzed by the enzyme phenol hydroxylase (Al Khalid and El Naas 2012). Phenol hydroxylase is a monooxygenase, an enzyme which inserts an atom of molecular oxygen into its substrate, reducing the other oxygen atom to $H_2O$ using a hydrogen donor. This enzyme can catalyze the hydroxylation of phenol-, hydroxyl-, amino-, halogen-, or methyl-substituted phenols (Krastanov et al. 2013). Alkylphenols of short and medium chain size are also converted by phenol hydroxylase to alkylcatechols (Nešvera et al. 2015). The aromatic ring of catechol is further opened by the enzymes catechol 1,2-dioxygenase (*ortho*-pathway), leading to the formation of succinyl-CoA and acetyl CoA, or catechol 2,3-dioxygenase (*meta*-pathway), leading to the formation of pyruvate and acetaldehyde (Al Khalid and El Naas 2012). The *ortho*-cleavage is also commonly known as the β-ketoadipate pathway, since β-ketoadipate is a key intermediate in the route (Harwood and Parales 1996). Chlorophenol and (chloromethyl)phenol hydroxylases (monooxygenases) are also used by aerobic bacteria to convert chlorophenols to the respective chlorinated catechols or chloromethylcatechols (Nešvera et al. 2015), which are further cleaved by modified *ortho-* or *meta-*pathways (Arora et al. 2014). Long-chain alkylphenols, bisphenols, nitrophenols, and chloronitrophenols require specific peripheral degradation pathways (thoroughly reviewed by Arora et al. 2014, 2018; Nešvera et al. 2015).

Anaerobic biodegradation of phenol and related compounds has been reported for several isolates of strict anaerobes (iron-reducing, sulfate-reducing, fermentative, and acetogenic bacteria) and facultative anaerobes (denitrifiers) (Qiu et al. 2008; Schmeling and Fuchs 2009, Table 8.3). Historically, aerobic processes for biological treatment of PCs were preferred mostly due to their low cost; however, in the last decades, the development and optimization of anaerobic treatment strategies for phenolic wastewaters have received considerable attention (reviewed by Veeresh et al. 2005). The anaerobic approaches display some advantages over other biological operations, since these systems tolerate high organic loading rates and are more resilient to fluctuating loads of PCs, produce lower volumes of sludge, and offer the chance of valorization through methane generation (Rosenkranz et al.

2013). Nevertheless, it must be taken into consideration that the inhibition of microorganisms by PCs, particularly the acetoclastic methanogens, may affect the anaerobic digestion processes and negatively impact the performance of both organic matter degradation and methane production (Franchi et al. 2018). In addition, the digestate often becomes unsuitable for its valorization as a fertilizer in agricultural soils, due to the occurrence of hazardous pollutants in the sludge (Levén et al. 2012).

Under methanogenic conditions, the mineralization of phenols proceeds through different pathways and requires a consortium of various microorganisms: fermenters, homoacetogens, syntrophs, and acetoclastic/hydrogenotrophic methanogens (Lv et al. 2014; Kato et al. 2015). Two alternative degradation pathways for phenol to acetate are known so far: by cleavage of the aromatic ring into caproate or via 4-hydroxybenzoate and the benzoyl-CoA pathway (Levén et al. 2012). Anaerobic degradation of phenol has been reported at either mesophilic (37 °C) or thermophilic temperatures (55 °C); however, most of the known phenol-degrading isolates and consortia characterized under methanogenic conditions are mesophilic (Levén et al. 2012). Temperature is acknowledged as a main factor influencing community structure and population shifts in anaerobic bioreactors (Calderón et al. 2013). On a general basis, removal of PCs is more efficient in the mesophilic digestors (Levén et al. 2012), which usually bear more diverse microbial communities (Calderón et al. 2013; Kim et al. 2018).

The ability to transform phenolics and other aromatic compounds has been described in a limited number of isolates of Archaea from extreme habitats. Extremely halophilic Archaea require >20% (w/v) NaCl for growth (Le Borgne et al. 2008). *Haloferax* sp. D1227, isolated from a petroleum-contaminated soil, was the first reported archaeon able to degrade 3-phenyl propionic acid, among other aromatic compounds (Emerson et al. 1994). Strains from the genera *Halobacterium*, *Haloharcula*, and *Haloferax* are described able to become adapted to concentrations up to 1 mM of trichlorophenols and remove them from culture media (Table 8.3), by mechanisms not fully elucidated. Evidences of the involvement of the *meta*-cleavage pathway in the degradation of PCs by an unidentified halophilic archaeal isolate were revealed by Acikgoz and Özcan (2016). Some strains of the hyperthermophilic and acidophilic archaeon *Sulfolobus solfataricus* able to remove phenol have been also described (Christen et al. 2012). These extremophilic Archaea have an interesting potential for the bioremediation of organic pollutants in hypersaline, acidic, or high-temperature environments. The process of selection of halophilic archaeal strains adapted to PCs has been patented (Oesterhelt et al. 1998).

A wide variety of technological approaches have been tested in order to take advantage of the biodegradation capacities of Bacteria and Archaea for the removal of PCs from wastewaters. Description of such approaches and discussion on their particular advantages and drawbacks are out of the scope of this chapter; interested readers can refer to previously published reviews on the topic (Basha et al. 2010; Al-Khalid and El Naas 2012).

## 8.4 Removal of PCs by Fungi

Fungi are cosmopolitan, heterotrophic, and metabolically versatile microorganisms, well known for their ability to use complex carbohydrates, heteropolymers, aromatic hydrocarbons, and other recalcitrant molecules, either as sole carbon sources or in co-metabolism (Harms et al. 2011; Aranda 2016). Fungi express a wide array of extra- and intracellular oxidative, hydrolytic, and conjugative enzymes. The involvement of fungi in the bioremediation of aromatic compounds which are poorly catabolized by prokaryotic microorganisms is of particular interest for the development of more efficient biotreatment technologies; nonetheless, these approaches remain rather unexploited, compared to the efforts based on the use of bacteria.

Many studies have reported the ability of fungal strains to degrade aromatic compounds, including PCs. Most of the well-characterized pollutant degraders belong to the phyla *Ascomycota* and *Basidiomycota* (Harms et al. 2011). Both yeast and filamentous fungi of the genera *Aspergillus*, *Candida*, *Coprinus*, *Fusarium*, *Geotrichum*, *Penicillium*, *Phanerochaete*, *Pleurotus*, *Trametes*, and *Trichosporon* have been described as able to either use phenol as the major carbon and energy source or biotransform it (Al Khalid and El Naas 2012; Krastanov et al. 2013). Examples available in the literature are summarized in Table 8.4. Many of the approaches based on the application of yeast cultures for the removal of PCs are referred to *Candida tropicalis* isolates (Galíndez-Mayer et al. 2008). White rot fungi, which are in great part responsible of the decay of lignin in nature, are particularly resistant to high concentrations of organic pollutants and display the ability to degrade chlorophenols (Al Khalid and El Naas 2012).

The biodegradation potential of fungi is mostly related to their rich arsenal of fairly non-specific extracellular oxidoreductase enzymes, such as laccases (phenoloxidases) and peroxidases, able to degrade a broad spectrum of aromatic substrates (reviewed by Harms et al. 2011; Kües 2015; Martínková et al. 2016). Intracellular fungal monooxygenases such as cytochrome p-450 oxidases and tyrosinases also lack substrate specificity (Harms et al. 2011). Fungal laccases and tyrosinases biodegrade a wide range of PCs (Martínková et al. 2016). These enzymes are particularly attractive to be used in purified immobilized form for the biochemical abatement of pollutants, since they cleave the aromatic rings in an $O_2$-dependent manner and, unlike peroxidases, do not require the addition of $H_2O_2$ (Kües 2015; Martínková et al. 2016). Laccases are also produced by a few genera of bacteria (mostly Gram-positive). These enzymes display some interesting properties, such as stability at high temperatures and alkaline pH; however, their biotechnological potential has been not thoroughly explored (Chauhan et al. 2017).

Degradation pathways specific for phenol and its derivatives have been also revealed in fungi, mainly in yeast species (Stoilova et al. 2007). Phenol hydroxylases have been isolated and purified from several yeasts such as *Cutaneotrichosporon cutaneum* (formerly named *Trichosporon cutaneum*) and *Candida tropicalis* and identified by genome sequencing in the filamentous fungus *Aspergillus fumigatus*

**Table 8.4** A review of fungal strains described in the literature for their ability to remove phenolic compounds

| Fungi | Experimental conditions | Summary of results | References |
|---|---|---|---|
| **Pure isolates** | **Aerobic culture** | | |
| *Aspergillus awamori* NRRL 3112 | Synthetic medium with PCs as sole carbon source (phenol 0.3–1.0 g/L; catechol, 2,4-dichlorophenol, and 2,6-dimethoxyphenol, 1.0–3.0 g/L) | Fully removes phenol (1 g/L) after 8 days; catechol (3 g/L) after 5 days; 2,4-dichlorophenol (2 g/L) after 5 days; and 2,6-dimethoxyphenol (1 g/L) after 7 days | Stoilova et al. (2006) |
| | | Removes binary mixtures of the aforementioned PCs | Stoilova et al. (2007) |
| *Aspergillus* sp. LEBM2 | Synthetic wastewater containing glucose (500 mg/L) and phenol (250–500 mg/L). Batch cultures | Complete removal of phenol was achieved after 72 or 96 h, at concentrations of 250 and 500 mg/L, respectively (removal rate of 7.71 mg/L h at 500 mg/L phenol) | Passos et al. (2010) |
| [Candida] oregonensis $B02_1$, *Candida subhashii* $A01_1$, *Nadsonia starkeyi-henricii* (=*Schizoblastosporion starkeyi-henricii*) $L01_2$ | Synthetic mineral salt medium added with phenol (500–1000 mg/L) as sole carbon source.<br>Batch cultures, 18 °C | The three yeasts were able to remove phenol at 500 and 750 mg/L after 2 days. Strains $A01_1$ and $L01_2$ removed 1000 mg/L phenol after 2 days | Filipowicz et al. (2017) |
| *Candida tropicalis* 3118, 3556, *Candida lipolytica* 3472 | Water added with different concentrations of phenol, 2,6 dimethylphenol, $\alpha$-naphtol, *o*-cresol, *m*-cresol, *o*-chlorophenol, *p*-nitrophenol. Batch cultures | The three strains removed 90% phenol after 48 h. Strain 3556 metabolized >95% and 70% phenol at 2 and 4 g/L, respectively, after 16 h incubation. At a concentration of 1 g/L, the three strains were able to remove most PCs tested | Varma and Gaikwad (2008) |
| *Candida tropicalis* CC1 | Synthetic basal medium added with phenol (1500–2500 mg/L) and 4-chlorophenol (100 mg/L). Bubble-column continuous fluidized bed reactor, cells immobilized onto granular activated carbon | >98% removal efficiency in the range of volumetric loading rates of 4.1 mg/L h (4-chlorophenol) and 55 mg/L h (phenol) | Galíndez-Mayer et al. (2008) |

(continued)

| | | | |
|---|---|---|---|
| *Fusarium* sp. HJ01 | Synthetic mineral medium added with 420 mg/L phenol | Grows with phenol as sole carbon source, complete removal was achieved after 8 days | Cai et al. (2007) |
| *Mucor* sp. AF2, *Rhizopus* sp. AF1 | Water added with phenol (25–100 mg/L). Batch cultures, incubated for 25 days | Both strains removed >50% phenol in monoculture at all concentrations tested. Maximal removal (86%) was recorded for *Rhizopus* sp. at 100 mg/L phenol | Al-Fawwaz et al. (2016) |
| *Pleurotus ostreatus* LGAM P113, LGAM P115 | Diluted and sterilized olive oil mill wastewater (OMW) with variable concentrations of PCs (150–4500 mg/L). Batch cultures | Removal efficiency of PCs ranged 80–50%, being higher when OMW was diluted. Laccase was the only ligninolytic enzyme detected | Aggelis et al. (2003) |
| *Trametes versicolor* CCT4521 | Diluted olive washing wastewater (OWW), total PCs = 277 mg/L. Bubble column bioreactor continuously operated | 89% reduction of the content of PCs after 192 h operation, with an optimal dilution rate of 0.225 $m^3$ of OWW treated daily per $m^3$ of bioreactor | Cerrone et al. (2011) |
| **Fungal consortia** | | | |
| *Aspergillus awamori* NRRL 3112 + *Cutaneotrichosporon cutaneum* (=*Trichosporon cutaneum*) R57 | Synthetic medium added with phenol (1000–1500 mg/L) and nine other different PCs (100–300 mg/L) as sole carbon sources. Batch culture, cells adapted and immobilized on polyamide beads | The consortium *Aspergillum*/*Cutaneotrichosporon* at 1:4 ratio removed 1000 mg/L phenol faster than either strain in monoculture (48 and 24 h, respectively). Phenol, bisphenol A, and 2,4 dichlorophenol (300 mg/L) were fully removed after 16 h | Yordanova et al. (2013) |

(Krastanov et al. 2013). In both yeasts and filamentous fungi, phenol is metabolized by an *ortho*-degradative pathway. Genes encoding key enzymes of this route have been identified in several fungal genera, i.e., *Aspergillus*, *Neurospora*, *Rhodotorula*, and *Trichosporon* (Krastanov et al. 2013). However, both 1,2-dioxygenase and catechol 2,3-dioxygenase activities have been detected in free cell extracts obtained from *Fusarium* sp. HJ01 grown on phenol, suggesting that this strain can cleave catechol by either *ortho*- or *meta*-pathways (Cai et al. 2007).

## 8.5 Removal of PCs by Microalgae

The non-taxonomic term "microalgae" refers to microscopic, unicellular photosynthetic organisms inhabiting freshwater, marine waters, or other environments and encompasses a taxonomically miscellaneous group, which includes >50000 species of eukaryotic algae and prokaryotic cyanobacteria (Cuéllar-Bermúdez et al. 2017; Osanai et al. 2017). Microalgae inhabit almost all ecological niches and are fast-growing organisms, contributing globally to the mitigation of greenhouse gas emissions by assimilating $CO_2$ (Asha et al. 2011). The majority of microalgae are mixotrophs, being capable of combining photoautotrophy and heterotrophy (Liká and Papadakis 2009), which provides them with great flexibility to survive and proliferate even under extreme conditions, thus making them choice candidates for the bioremoval of contaminants in wastewaters (Xiong et al. 2017).

The use of microalgae for wastewater treatment in oxidation ponds was first proposed in the 1950s (Oswald and Gotaas 1957). This approach offers several environmental and economic advantages: the involved organisms use a cheap, abundant, and renewable energy source (sunlight) and are easy to grow (Liká and Papadakis 2009). Microalgae uptake $NH_4^+$ and $NO_3^-$ as N sources and contribute to P removal by enhancing precipitation of phosphates (by generating an alkaline pH through $CO_2$ assimilation) and also by luxury uptake of P in intracellular granules (Larsdotter 2006; Abdel-Raouf et al. 2012). Algal cells display a high capacity to adsorb most of the heavy metal ions occurring in municipal wastewaters and reduce the numbers of fecal coliforms by the generation of $O_2$ and the increase of pH (Oswald 2003). Last but not least, the algal biomass generated can be valorized for a number of uses, such as animal feed, biodiesel or $H_2$ production, and the recovery of molecules valuable for cosmetics, nutraceutical, and pharmaceutical products (Oswald 2003; Larsdotter 2006; Liká and Papadakis 2009; Rizwan et al. 2018). In particular, the use of microalgae for the production of biodiesel has focused considerable attention in the past decades, since some species are able to accumulate hydrocarbons up to 30–70% of their dry weight, and their oil content is 1000-fold that of soybeans grown on the same area of land (Bitog et al. 2011). It is postulated that a "zero-waste" model could be implemented through the utilization of wastewater as nutrient source for microalgae followed by the subsequent valorization of the produced biomass (Xiong et al. 2017).

The term phycoremediation was coined to refer to the removal of organic and inorganic compounds (carbon, nitrogen, or phosphorus), metals, and emerging contaminants in wastewater by microalgae (algae and cyanobacteria) (Cuéllar-Bermudez et al. 2017). Regarding the bioremediation of organic pollutants, over 30 species of microalgae of the prokaryotic phylum *Cyanobacteria* and the eukaryotic phyla *Bacillariophyta* (diatoms), *Euglenida*, and *Chlorophyta* (green algae) have been investigated (Xiong et al. 2017). As earlier stated in this chapter, the bioremoval of phenols by either bacteria or fungi has been thoroughly investigated for over a century, but phycoremediation-based approaches for the biological treatment of wastes containing PCs or other organic pollutants have only been proposed and developed more recently (Semple et al. 1999; Safonova et al. 2004; Ghasemi et al. 2011). Cyanobacteria and eukaryotic microalgae strains have been characterized for their ability to biotransform PCs; representative publications are summarized in Table 8.5. Several of the available studies reported the ability of single strains or consortia of microalgae to remove single phenols, chlorophenols, nitrophenols, alkylphenols, and bisphenol A, among other PCs, from synthetic media in batch cultures, and consequently several microalgae have been tested for the bioremediation of complex effluents such as those of the coal gasification, olive oil, paper, and petroleum refinery industries (Tarlan et al. 2002; Pinto et al. 2003; Das et al. 2015; Di Caprio et al. 2015; Dayana Priyadharshini and Bakthavatsalam 2017a, b; Di Caprio et al. 2018).

To date, the metabolic degradation pathways of phenolic molecules by microalgae have been seldom investigated thoroughly. By definition, microalgae are obligate aerobes; hence, the degradation of organic compounds is thought to always require $O_2$ to proceed. Some species of microalgae are capable of heterotrophic growth using aromatic compounds as sole carbon and energy sources (Semple et al. 1999). Semple and Cain (1996) characterized an obligatory aerobic *meta*-pathway for the cleavage of phenol in the eukaryotic freshwater algae *Ochromonas danica* under non-photosynthetic conditions. These authors used radiolabeled phenol to demonstrate that the aromatic compound was mineralized to $CO_2$ (65%) and $^{14}C$ was incorporated to the algal cell molecules. In contrast, Wurster et al. (2003) identified an *ortho*-fission pathway for the degradation of phenol in the cyanobacterial strain *Synechococcus* sp. PCC7002 cultivated in the dark, but this strain was not able to assimilate the organic carbon. *Auxenochlorella pyrenoidosa* (formerly named *Chlorella pyrenoidosa*) degrades phenol by both *ortho*- and *meta*-pathways (Das et al. 2015). The degradation pathway of bisphenol A by a strain of *Desmodesmus* sp. was characterized recently by Wang et al. (2017), who identified metabolites generated from oxidative hydroxylation, glycosylation, and oxidative cleavage.

Polyphenol oxidase and laccase enzymes were first characterized in a phenol-degrading strain of the marine cyanobacterium *Leptolyngbya valderiana* (formerly named *Phormidium valderianum*) (Shashirekha et al. 1997). In a more recent study, laccase activity was tested in 35 isolates belonging to different genera of the *Cyanobacteria*, of which 29 (>80%) gave positive results ranging 4.68–66.74 U/mL, suggesting that this is a fairly widespread trait among the prokaryotic microalgae (Afreen et al. 2017). Laccase activities of two species of green eukaryotic microalgae (*Tetracystis aeria* and *Chlamydomonas moewusii*) able to enzymatically

**Table 8.5** A review of microalgae strains and consortia described in the literature for their ability to remove phenolic compounds

| Microorganisms | Experimental conditions | Summary of results | References |
|---|---|---|---|
| **Freshwater algae, pure isolates** | **Phototrophic culture (light)** | | |
| *Chlorolobion braunii* (=*Ankistrodesmus braunii*) CCAP 202.7a, *Scenedesmus quadricauda* UTEX 76 | Synthetic medium added with 400 mg/L of 9 PCs commonly present in olive oil mill wastewater (OMW): tyrosol, hydroxytyrosol, catechol, 4-hydroxybenzoic, ferulic, *p*-coumaric, synapic, caffeic, and vanillic acids. Batch cultures | In the synthetic medium, each of the strains was able to remove >70% of most of the assayed PCs, after 5 days of incubation | Pinto et al. (2002) |
| *Auxenochlorella pyrenoidosa* (=*Chlorella pyrenoidosa*) | Synthetic medium added with phenol (25–200 mg/L) as sole carbon source. Also cultivated in raw refinery wastewater (23.33 mg/L phenols) | Removed 81.56% phenol within 3 days and 100% after 6 days, at a concentration of 200 mg/L. 38.32% phenol in refinery wastewater was removed after 7 days | Das et al. (2015) |
| *Auxenochlorella pyrenoidosa* (=*Chlorella pyrenoidosa*) | Phenolic effluent of a coal gasification plant (total PCs, 1475 mg/L). Effluent used at 20, 40, 60, and 80%, inoculated with 1 g wet biomass/L. Batch cultures | >90% phenols were removed for effluent concentrations up to 60%, after 7 days of incubation | Dayana Priyadharshini and Bakthavatsalam (2017a) |
| | Undiluted effluent. Batch culture, supplemented or not with nutrients (K and urea) and inoculated with 4 g wet biomass/L | Phenols >95% removed after 7 days when nutrients were supplemented. 46% average reduction of phenolics in the absence of nutrients. Organic carbon was used for growth | Dayana Priyadharshini and Bakthavatsalam (2017b) |

(continued)

**Table 8.5** (continued)

| Microorganisms | Experimental conditions | Summary of results | References |
|---|---|---|---|
| *Carteria inversa* NIES 422, *Chlamydomonas fasciata* NIES 437, *Chlorella sorokiniana* IAM C-21, *Scenedesmus vacuolatus* (=*Chlorella fusca* var. *vacuolata*) IAM C-28 | Synthetic media added with *o*-, *m*-, and *p*-nitrophenol, 2,4-dinitrophenol, 2,4,6-trinitrophenol, bisphenol A, *p*-chlorophenol, 2,4-dichlorophenol (40 μM). Removal of nitrophenols was tested in batch cultures, removal of chlorophenols in a small photobioreactor | The four strains degraded 20–57% of 2,4 dinitrophenol after 72-h incubation. *S. vacuolatus* removed 60–100% of all tested mono- and di-nitrophenols and bisphenol A, after 5-day incubation. This strain also removed chlorophenols to a lesser extent | Hirooka et al. (2003) |
| | Synthetic media added with bisphenol A (10–160 μM) | *S. vacuolatus* removed >90% of bisphenol A (up to 80 μM), after 7 days of incubation, either under light or dark conditions | Hirooka et al. (2005) |
| *Chlorella vulgaris*, *Chlorella miniata* WW1, *Chlorella* sp. 1uoai, *Chlorella* sp. 2f5aia | Synthetic medium, added with nonylphenol (1 mg/L). Batch cultures | The four strains removed nonylphenol. *C. vulgaris* had the highest efficiency (99% removal and >80% degradation, after 168-h incubation) | Gao et al. (2011) |
| *Chlorella* sp. | Synthetic medium added with phenol up to 1000 mg/L. Batch cultures. Strain was subjected to adaptive laboratory evolution | The adapted strain fully removed 500 mg/L after 7 days. Up to 700 mg/L phenol were 86% removed after 8 days | Wang et al. (2016) |
| *Chlorella* sp., *Tetradesmus obliquus* (=*Scenedesmus obliquus*) | Synthetic medium, several PCs added as sole carbon source at different concentrations. Batch cultures | Phenol (1000 mg/L) was removed by both algae. *Chlorella* sp. converted 2,4-dimethyl phenol (1000 mg/L) into dimethyl benzenediol and partially dechlorinated 2-chlorophenol (200 mg/L). *Tetradesmus* removed 2,4-dinitrophenol (190 mg/L) after adaptation | Klekner and Kosaric (1992) |

(continued)

**Table 8.5** (continued)

| Microorganisms | Experimental conditions | Summary of results | References |
|---|---|---|---|
| *Desmodesmus* sp. WR1 | Synthetic medium added with bisphenol A (1–13.5 mg/L). Batch cultures | 57%, 25%, 18%, and 26% removal of bisphenol A were achieved after 10 days of incubation for each initial concentration (1, 3, 5.5, and 13.5 mg/L, respectively) | Wang et al. (2017) |
| *Scenedesmus* sp. | Unsterilized olive oil mill wastewater (OMW), 9% (v/v)<br>Batch culture, four different media tested | Best removal of phenol (22%) achieved in BG11 medium | Di Caprio et al. (2015) |
| *Tribonema minus* SAG 880-3 | Synthetic medium added with 250–1000 mg/L phenol. Batch cultures | Phenol removed efficiently up to a concentration of 700 mg/L | Cheng et al. (2017) |
| Unidentified strain VT1 | Synthetic medium added with 0–40 mg/L pentachlorophenol. Batch cultures. Radiolabeled pentachlorophenol ($^{14}C$) was used to test assimilation/mineralization | The strain mineralized up to 13.8% of radiolabeled substrate to $^{14}CO_2$ after 11 days of incubation, and $^{14}C$ was incorporated to cell molecules. Mineralization was further enhanced adding glucose | Tikoo et al. (1997) |
| **Freshwater algae, pure isolates** | **Phototrophic/ heterotrophic culture (light/dark)** | | |
| *Chlorolobion braunii* (=*Ankistrodesmus braunii*) CCAP 202.7a, *Scenedesmus quadricauda* UTEX 76 | Olive oil mill wastewater (OMW), PC content =1000–15,000 mg/L. Used 1:2 and 1:10 diluted. Free or immobilized cells (alginate beads), incubated both in the light and under dark conditions | Higher removal of phenols (48–55%) from OMW (1000 mg/L) achieved with the 1:10 dilution, in dark conditions, for both strains, with non-immobilized cultures | Pinto et al. (2003) |
| *Scenedesmus* sp. | Tap water-based medium supplied with olive oil mill wastewater (OMW), total phenols 4880 mg/L. Batch supply, 9% OMW (v/v) added initially; fed-batch supply, 1 mL/100 mL added each 24 h; two-stage supply, 9% OMW (v/v) added after 10 day of incubation, incubated in the dark afterwards | 55% phenol removal was achieved by the fed-batch strategy for phenol concentrations <100 mg/L and by the two-stage strategy when the heterotrophic stage lasted >8–10 days | Di Caprio et al. (2018) |

(continued)

**Table 8.5** (continued)

| Microorganisms | Experimental conditions | Summary of results | References |
|---|---|---|---|
| **Freshwater algae, pure isolates** | **Heterotrophic culture (dark)** | | |
| *Ochromonas danica* CCAP 933/2B | Synthetic media added phenol or *p*-cresol (up to 1 mM) as sole carbon substrates. Radiolabeled phenol ($^{14}C$) was used to test assimilation/ mineralization | Grew heterotrophically on phenol and *p*-cresol and removed them after 3 days of incubation. Tolerated up to 4 mM phenol. Radiolabeled phenol was 65% mineralized to $CO_2$ and $^{14}C$ incorporated to cell molecules | Semple and Cain (1996) |
| | Synthetic medium added with mixtures of phenol plus cresols or xylenol (500 μM) | Also grows heterotrophically on binary mixtures of phenol and other PCs | Semple (1998) |
| **Marine algae, pure isolates** | **Phototrophic culture (light)** | | |
| *Skeletonema costatum* CCMP 1332 (diatom) | Synthetic medium added with 6 mg/L 2,4 dichlorophenol. Batch cultures | 2,4-Dichlorophenol was 19% additionally removed by the strain, compared to uninoculated controls | Yang et al. (2002) |
| *Tetraselmis marina* CCMP 898 | Artificial seawater added with phenol, *o*-, *m*-, and *p*-chlorophenol (20 mg/L). Batch cultures | Phenol was not removed, while chlorophenols were removed up to 43% after 10 days of incubation. 65% *p*-chlorophenol (20 mg/L) was removed after 10 days of incubation with 1 g/L $NaHCO_3$ | Petroutsos et al. (2007) |
| *Thalassiosira* sp. HP9101 (diatom) | Artificial seawater added with benzoate, catechol, phenol, or protocatechuate (1 mM). Batch cultures | Growth was observed in the presence of 0.25 mM phenol and 1 mM benzoate. Catechol and protocatechuic acids inhibited growth. Phenol-supplemented *Thalassiosira* sp. was also capable of dechlorinating chlorophenols | Lovell et al. (2002) |

(continued)

Table 8.5 (continued)

| Microorganisms | Experimental conditions | Summary of results | References |
|---|---|---|---|
| **Freshwater *Cyanobacteria*, pure isolates** | **Phototrophic culture (light)** | | |
| *Anabaena cylindrica* NIES19, *Trichormus variabilis* (=*Anabaena variabilis*) NIES23, *Microcystis aeruginosa* NIES 44 | Synthetic media added with bisphenol A, *o*- *m*-, and *p*-nitrophenol, 2,4-dinitrophenol, 2,4,6-trinitrophenol, *p*-chlorophenol, 2,4-dichlorophenol (40 μM). Batch cultures (nitrophenols and bisphenol A) and lab-scale photobioreactor (chlorophenols) | The 3 strains degraded 20–86% 2,4 dinitrophenol after 72-h incubation. *T. variabilis* removed 100% *o*- and *m*-nitrophenol, 95% 2,4 dinitrophenol, and 51% 2,4,6-trinitrophenol, after 5-day incubation. Chlorophenols were also removed at a lesser extent | Hirooka et al. (2003) |
| *Arthrospira maxima* (=*Spirulina maxima*) | Synthetic medium, added with phenol (1000 mg/L) as sole carbon substrate. Batch cultures | Removed 100% phenol after 6-day incubation | Klekner and Kosaric (1992) |
| *Arthrospira maxima* (=*Spirulina maxima*) | Synthetic medium, added with phenol (50–1000 mg/L) as sole carbon substrate. Batch cultures. Isotopic labeling ($^{13}$C phenol) used to test assimilation/mineralization | *A. maxima* cells grew on phenol up to 400 mg/L. 97.5% of phenol was removed within 24 h. $^{13}$C-phenol-carbon was assimilated to biomass and mineralized to $CO_2$ | Lee et al. (2015) |
| **Marine *Cyanobacteria*, pure isolates** | **Phototrophic culture (light)** | | |
| *Leptolyngbya valderiana* (=*Phormidium valderianum*) BDU 30501 | Synthetic medium added with phenol (25–100 mg/L). Batch cultures | The strain tolerated up to 50 mg/L phenol and removed 38 mg/L after 7 days of incubation | Shashirekha et al. (1997) |
| **Microalgae consortia** | **Phototrophic culture (light)** | | |
| *Trichormus variabilis* (=*Anabaena variabilis*) NIES23 + *Anabaena cylindrica* NIES19 | Synthetic media added with 2,4-dinitrophenol (5–150 μM) | *A. variabilis* removed 100% 2,4-dinitrophenol (5–40 μM) and >90% at concentrations of 80–150 μM after 5 days of incubation. 2-amino-4-nitrophenol was accumulated in the medium. *A. variabilis* + *A. cylindrica* 100% removed 2,4-dinitrophenol (40 μM), and low concentrations of 2-amino-4-nitrophenol were detected | Hirooka et al. (2006) |

(continued)

**Table 8.5** (continued)

| Microorganisms | Experimental conditions | Summary of results | References |
|---|---|---|---|
| *Auxenochlorella pyrenoidosa* (= *Chlorella pyrenoidosa*) + *Chlorella vulgaris* | Synthetic medium, added with *p*-nitrophenol (10–50 mg/L) | The consortium (3:1 ratio) removed up to 50 mg/L *p*-nitrophenol after 3 days of incubation, faster than either strain in monoculture | Lima et al. (2003) |
| | Synthetic medium, added with *p*-chlorophenol (25–150 mg/L) | The consortium removed 150 mg/L *p*-chlorophenol after 9 days of incubation | Lima et al. (2004) |

transform phenolic micropollutants such as bisphenol A, nonylphenol, or triclosan have been also characterized recently (Otto et al. 2015). The microalgal enzymes were more efficient at alkaline pH, while the optimal pH for fungal laccases is more often acidic, which is an interesting feature regarding their applications in bioremediation (Otto et al. 2015).

## 8.6 The Potential of Microalgae-Bacteria Consortia for the Removal/Biodegradation of PCs from Industrial Wastewaters

Eukaryotic microalgae, cyanobacteria, and non-phototrophic bacteria are ubiquitous and colonize all habitats on Earth, including extreme environments (Ramanan et al. 2016). In such mixed microbial communities, they interact together playing essential roles for many ecosystems processes. In agreement with this long-term coexistence through evolution, microalgae-bacteria interactions comprise nearly the whole spectra of symbiotic relationships and are thought to be universal in the environment (Ramanan et al. 2016).

Consortia of microalgae and bacteria are regarded as highly efficient for nutrient removal and detoxification of both inorganic and organic pollutants in wastewaters, compared to the use of the individual microorganisms (Xiong et al. 2017). The use of microalgal-bacteria consortia for wastewater treatment processes provides advantages at several levels. Through oxygenic photosynthesis, microalgae generate the $O_2$ required for the aerobic degradation of organic molecules by heterotrophic bacteria (Borde et al. 2003; Muñoz et al. 2003; Min et al. 2011; Subashchandrabose et al. 2013); consequently, the implementation of this technology is cost-saving compared to conventional activated sludge treatments, since little or even no energy is required to supply aeration (Muñoz et al. 2005). Another advantage of photosynthetic oxygenation is the uptake by microalgae of the $CO_2$ released during the bacterial aerobic mineralization of the organic substrates, which contributes to the prevention of greenhouse gas emissions in wastewater treatment plants (Muñoz and

Guieysse 2006; Essam et al. 2014). *Chlorophyta*, the green algae, display up to 50 times better solar energy absorbing efficiency in addition to $CO_2$ fixation compared to that of terrestrial plants, and certain species are able to grow under very high $CO_2$ concentrations (up to 80%), although optimal concentrations are much lower (Bitog et al. 2011). Finally, bacteria often require long periods of adaption to degrade specific organic molecules, while many microalgae are described for their ability to fast transform such compounds, thus accelerating its degradation in consortium (Semple et al.1999).

In that context, several studies have reported that photosynthetic oxygenation supports the aerobic treatment of toxic wastewaters containing PCs and other organic pollutants, such as acetonitrile, azo compounds, benzopyrene, black oil, crude oil, naphtalene, phenanthrene, phenol, or salicylate (Borde et al. 2003; Safonova et al. 2004; Muñoz et al. 2005; Muñoz and Guieysse 2006; Tang et al. 2010; Subashchandrabose et al. 2013; Essam et al. 2014; Maza-Márquez et al. 2014; Mahdavi et al. 2015; Ryu et al. 2017). The results of these studies together with those summarized in Sect. 8.5 demonstrate the potential of microalgae to contribute to biotreatment or bioremediation strategies for PCs, either by their direct involvement on the transformation/degradation of the pollutant molecules, or by indirectly enhancing the degradation potential of the coexistent bacterial communities (Semple et al. 1999).

## 8.6.1 Selection of Microalgae and Bacteria for the Construction of Consortia Able to Remove PCs

### 8.6.1.1 Adaptation of Strains to the Target Pollutants

PCs are toxic for many living organisms, including plentiful microalgae and bacteria. Thus, the first important factor to consider is the toxicity of the target PCs in the wastewater to be treated. Microalgae are more likely to suffer inhibition during the treatment of phenolic effluents under real, full-scale conditions, since they tend to be more sensitive than bacteria to changes in the concentration of the hazardous organic pollutants (Borde et al. 2003), and are influenced by other toxics concomitantly occurring in the industrial wastewaters; in some cases, pretreatment of the industrial effluents is required to remove such compounds and prevent algal inhibition (Muñoz and Guieysse 2006; Essam et al. 2014).

There are two main approaches for the implementation of microalgae-bacteria consortia for the treatment of wastewaters contaminated with organic pollutants: (a) natural colonization and (b) artificial inoculation (Xiong et al. 2017). Natural colonization involves the selection of consortia comprised by native pollutant-removing microorganisms of the wastewater to be treated, while artificial inoculation relies on the construction of consortia based on the combination of strains already characterized for their removal capacity of the target pollutants. Several studies have reported that phenolic-contaminated wastewaters and soils carry already adapted microbial populations, which are commonly isolated by selective enrichment strategies (Tikoo et al. 1997; Borde et al. 2003; Lima et al. 2003, 2004; Maza-Márquez et al. 2013, 2014). However, in order to achieve optimal pollutant removal rates by means of biological treatment, adaptation of the indigenous microorganisms is often needed. The simplest adaptation procedures may consist of prolonged incubations in the presence of the target pollutant or sequential cultivation increasing its concentration stepwise (Klekner and Kosaric 1992; Oesterhelt et al. 1998; Arutchelvan et al. 2006; Christen et al. 2012; Yordanova et al. 2013; Maza-Márquez et al. 2014; Acikgoz and Ozcan 2016). More sophisticated adaptive laboratory evolution experiments involve the continuous cultivation of the organisms under the pollutant pressure over multiple generations, either by repeated batch cultures (during which the medium is periodically withdrawn and substituted with fresh medium), or in continuous culture systems (i.e., chemostats) (Juárez-Jiménez et al. 2012; Wang et al. 2016). In the case of microalgae, the adaptation mechanisms to extreme conditions, either due to high concentration of organic compounds or physical factors (salt, light radiation, etc.), have been explained by genetic changes, caused by spontaneous mutations or physiological adaptation; however, investigations at the molecular level are still too scarce to properly understand the role of stress factors in the improvement of the removal efficiency of organic pollutants by these organisms (Xiong et al. 2017).

#### 8.6.1.2 Ecological Microalgae-Bacteria Interactions

Beyond the tolerance to the target pollutants and the efficiency for their removal, an essential factor to consider for the building of the microbial consortia is to explore the compatibility between the selected species/strains. In the particular case of microalgae-bacteria consortia, balanced exchange of $CO_2$ and $O_2$ is essential for an optimal performance. $CO_2$ is the major carbon source for photosynthetic microalgae; however, at too high concentrations, it generates a low pH, which could be inhibitory to some microalgal strains (Wang et al. 2012). The amount of $CO_2$ required for microalgae growth varies according to the species and is also dependent on the particular configuration of the cultivation system (Bitog et al. 2011). On the other hand, accumulation of $O_2$ produced by photosynthesis must be avoided, since at high levels it induces photooxidative damage in microalgae (Muñoz and Guieysse 2006). Therefore, maintaining both $CO_2$ and $O_2$ in the optimal concentration ranges is important for the stable removal efficiency of pollutants by microalgae-bacteria consortia.

According to Maza-Márquez et al. (2014), microalgae/bacteria cell ratios constitute an essential factor for the optimization of consortia for industrial applications. When two different microalgae/bacteria ratios were tested ($10^6/10^3$ and $10^{12}/10^3$ cells/mL) in batch cultures, the rate of removal of PCs was improved at higher microalgae numbers, due to the increased availability of $O_2$ in the culture for the bacterial partners (Fig. 8.3). In that line, earlier work by Guieysse et al. (2002) already highlighted the importance of the optimization of the inoculation strategy and initial composition of the microalgae-bacteria consortia aimed for pollutants'

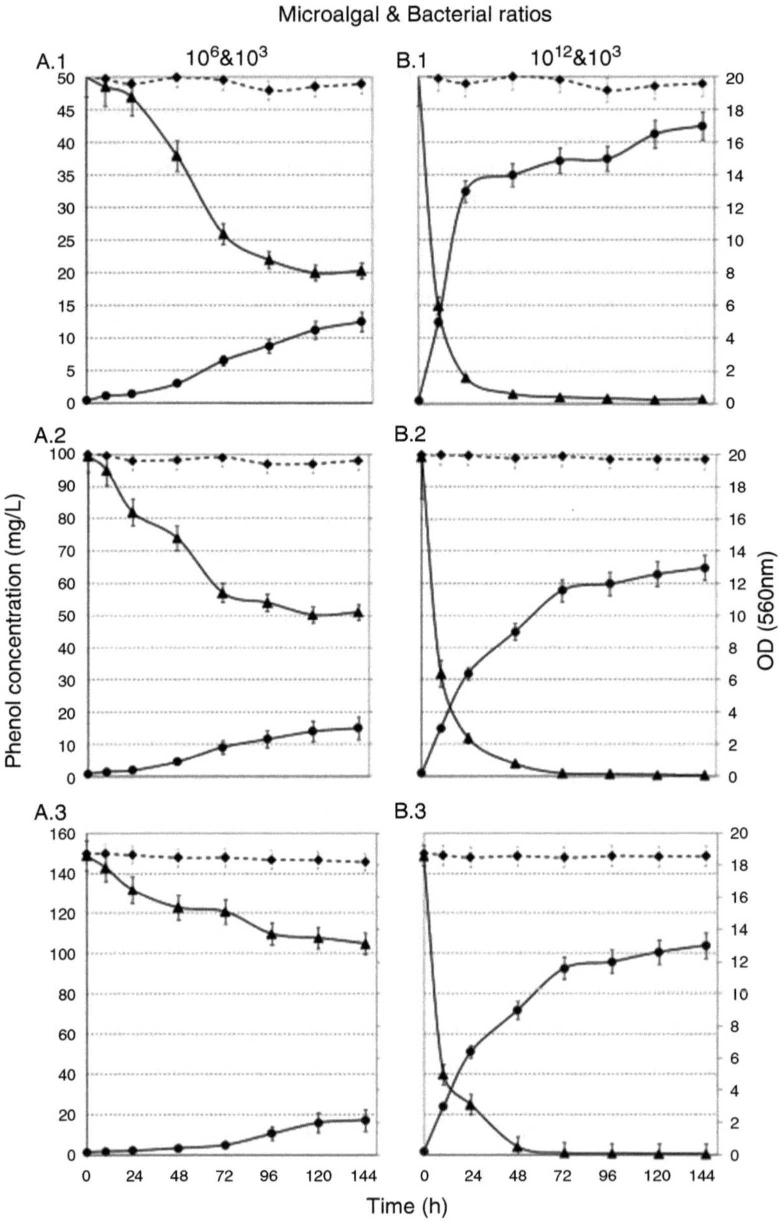


**Fig. 8.3** Phenolic compounds' concentration (▲) and consortium growth (●) of inoculated (continuous lines) and un-inoculated (dotted lines) cultures containing synthetic olive washing wastewater (OWW) with different amounts of polyphenols (A.1, B1 = 50 mg/L; A.2, B.2 = 100 mg/L; A.3, B.3 = 150 mg/L) and microalgae/bacteria cell ratios (A.1, A.2, A.3 $=10^6/10^3$; B.1, B.2, B.3 $=10^{12}/10^3$, respectively). (Reprinted from Maza-Márquez et al. (2014), with permission from Elsevier)

removal. Other authors concluded that the level of dissolved $O_2$ concentration measured in microalgae-bacterial mixed cultures is a good indicator to monitor the efficiency of the removal of organic pollutants in complex wastewaters (Muñoz et al. 2003, 2004; Essam et al. 2006).

Positive interactions other than the mutual dependence for $CO_2$ and $O_2$ have been also described in microalgae-bacteria consortia. Bacteria promote the growth of microalgae by the synthesis and excretion of organic factors such as auxins and vitamins $B_1$ (thiamine) and $B_{12}$ (cobalamin), and in marine habitats they supply algae with iron through production of photosensitive siderophores (Kazamia et al. 2012). Concurrently, bacteria can uptake organic carbon substrates (protein, lipids, carbohydrates) secreted by algae and colonize their cell surfaces in search of a stable habitat (Grossart and Simon 2007; Kazamia et al. 2012; Brodie et al. 2017; Xiong et al. 2017). These synergistic interactions enhance the removal efficiency of biotreatment systems; however, negative symbiotic interactions may also occur which should be taken into consideration. Some strains of bacteria secrete antibiotics such as streptomycin which are toxic for microalgae (Harras et al. 1985). Bacteria are also reported to induce the cell lysis of eukaryotic algae (Wang et al. 2010). This algalytic effect of bacteria has been related to the enzymatic degradation of their cell walls by glucosidases, chitinases, and cellulases (Ramanan et al. 2016). On the other hand, algae may alkalinize the medium pH up to 9–10 due to their autotrophic metabolism, inhibiting or slowing down the growth of many bacteria (since they are mostly neutrophilic), and may also inhibit bacterial growth by increasing in excess temperature or dissolved $O_2$ concentration or by secreting bactericidal substances (Muñoz and Guieysse 2006).

Due to their larger cell size, eukaryotic algae have longer generation times than heterotrophic bacteria, which may outcompete them for limiting nutrients such as N sources (Grossart and Simon 2007), leading in turn to a limitation of $O_2$ availability for bacteria which may hamper the efficiency of pollutants' removal in biotreatment systems. For this reason, fast-growing microalgae with fast $O_2$-releasing rates and high $CO_2$-sinking ability are preferred (Muñoz and Guieysse 2006; Xiong et al. 2017). In Table 8.6, different types of interactions previously characterized between microalgae and bacteria or fungi are summarized. Mutualistic interactions among eukaryotic green algae and *Alphaproteobacteria* of the *Rhizobiaceae*, well known for their growth-promoting effects in plants, have been very often described and seem to be widespread in nature (Kim et al. 2014).

### *8.6.2 Cultivation of Microalgae and Microalgae-Bacteria Consortia. Photobioreactors (PBRs)*

As previously stated in Sect. 8.5 of this chapter, cultivation of microalgae for different purposes including wastewater treatment and biomass production was initiated in the 1950s. The success of such cultivation techniques strongly relies on their

**Table 8.6** Different types of interactions of microalgae with bacteria and fungi

| Microorganisms involved | Type of interaction | Reference |
|---|---|---|
| *Chlorella vulgaris* (algae)<br>*Azospirillum brasilense* (bacteria) | **Mutualism**<br>*A. brasilense* promotes *C. vulgaris* growth by producing indole-3-acetic acid | González and Bashan (2000) and De-Bashan et al. (2008) |
| *Chlorella vulgaris* (algae)<br>*Bacillus pumilus* (bacteria) | **Commensalism**<br>*C. vulgaris* growth is promoted by N supplied by *B. pumilus* through $N_2$ fixation | Hernández et al. (2009) |
| *Chlorella vulgaris* (algae)<br>*Microbacterium* sp. (bacteria)<br>*Exophiala* sp. (fungi) | **Competition**<br>*Microbacterium* competed with algae for some essential nutrients resulting in a reduction in the growth rate of algae<br>**Antagonism**<br>The *Exophiala* strain was algicidal | Cho et al. (2015) |
| *Chlorella vulgaris* (algae)<br>*Rhizobium* sp. (bacteria) | **Mutualism**<br>*Rhizobium* promotes algal growth, and both organisms increase their growth rates in co-culture | Kim et al. (2014) |
| *Lobomonas rostrata* (algae)<br>*Mesorhizobium loti* (bacteria) | **Mutualistic facultative symbiosis**<br>Exchange of photosynthate for vitamin $B_{12}$ | Kazamia et al. (2012) |
| *Chlorella sorokiniana* (algae)<br>*Ralstonia pickettii* (bacteria)<br>*Sphingomonas* sp. (bacteria)<br>*Microbacterium trichotecenolyticum* (bacteria)<br>*Micrococcus luteus* (bacteria)<br>*Acremonium* sp. (fungus) | **Mutualism and commensalism**<br>*Microbacterium* and *Acremonium* promoted the growth of *C. sorokiniana*<br>*Acremonium* and *Ralstoni*a lived directly adhered to the *Chlorella* cell surface | Watanabe et al. (2005) |
| *Microcystis aeruginosa* (*Cyanobacteria*)<br>*Achromobacter* sp. (bacteria) | **Antagonism**<br>*Achromobacter* lyses *Microcystis* cells by extracellular thermostable factors | Wang et al. (2010) |

design and performance. Cultivation systems which incorporate a natural or artificial light source to allow the growth of phototrophic microorganisms are termed photobioreactors (PBRs) (Chang et al. 2017). PBRs must be designed to fulfil the following characteristics: even illumination and reduced shading, efficient use of light, efficient mixing, control over the reaction conditions, low hydrodynamic stress on the photosynthetic organisms, fast mass transfer of $CO_2$ and $O_2$, and good scalability (Muñoz and Guieysse 2006; Gupta et al. 2015).

#### 8.6.2.1 Types of PBRs

On a general basis, there are two main types of PBRs available: open and closed bioreactors. Examples of several types of such designs are depicted in Fig. 8.4. Detailed descriptions and discussion on the advantages and limitations of either type of PBRs for microalgae cultivation were thoroughly reviewed elsewhere (Bitog et al. 2011; Wang et al. 2012; Gupta et al. 2015; Chang et al. 2017). In brief, open systems are easier to build and operate but have several important drawbacks such as large space requirements, significant loss of water by evaporation, limited control over environmental parameters (i.e., temperature and light intensity), diffusion of $CO_2$ to the atmosphere, nutrient limitation, and ineffective homogenization resulting in poor mass transfer rates (Gupta et al. 2015; Chang et al. 2017). In contrast, closed PBRs have large surface/volume ratios, provide better penetration of light, minimize the evaporation of water, and offer more control over external parameters, resulting in much higher photosynthetic efficiencies and biomass productivities (Wang et al. 2012), but on the other hand, they are expensive to construct and more difficult to design, scale up, and operate (Larsdotter 2006; Muñoz and Guieysse 2006).

Classic open bioreactors are typically large and shallow open ponds lacking internal mixing (Muñoz and Guieysse 2006) or circular ponds where mixing is pro-

**Fig. 8.4** Photobioreactors (PBRs) used for the growth of microalgae: (**a**) open PBR, raceway pond, (**b**) closed PBR, flat-plate type, (**c**) closed PBR, inclined tubular type, and (**d**) closed PBR, horizontal/continuous tubular type. (Reprinted from Bitog et al. (2011), with permission from Elsevier)

vided by a rotating arm (Chang et al. 2017). Raceway ponds, in which cells and water are circulated continuously around a racetrack aided by paddle wheels (Fig. 8.4a), are the most widespread reactors, since they are easily scaled up and offer flexibility of use (Larsdotter 2006; Chang et al. 2017). These designs were improved in the 1980s by Oswald, who developed the so-called high-rate algal ponds (HRAP), less than 1-m deep and continuously mixed by gentle stirring to achieve 100% aerobic volume (Larsdotter 2006).

Closed systems are composed of transparent containers of different shapes through which cultures are flown. Stirred tanks are the simplest type of closed PBRs, consisting of cylindrical vessels where mixing is provided by mechanical agitation (Gupta et al. 2015). Flat-plate PBRs (Fig. 8.4b) are formed by units of orthohedric shape consisting of a narrow frame closed at both sides by transparent plates, designed to hold in very thin layers of culture (Gupta et al. 2015). This kind of PBRs offers a very high surface/volume ratio and excellent light distribution, allowing high photosynthetic efficiencies (Eriksen 2008; Chang et al. 2017), but displays some limitations for full-scale applications, such as low dissolved $O_2$ concentrations, difficulties for their cleaning, or poor control of temperature (Gupta et al. 2015).

In tubular PBRs, the culture flows through a tubing system made of a highly transparent material (most commonly plexiglass or polyethylene) and is recirculated using mechanical pumps or by means of aeration (Wang et al. 2012; Chang et al. 2017). Tubular PBRs can be arranged vertically (Fig. 8.4c), horizontally (Fig. 8.4d), or coiled. The vertical systems are reported to provide higher productivities and photosynthetic efficiencies (de Vree et al. 2015). Among vertical PBRs, bubble column and airlift PBRs are differentiated on the basis of the mode of liquid flow used (Chang et al. 2017). Both types are widely used for industrial applications because of their simple construction and operation (Bitog et al. 2011). Vertical tubular PBRs are most suitable for outdoor uses (Gupta et al. 2015) and have been tested for full-scale urban or industrial wastewater treatment (Gouveia et al. 2016; Maza-Márquez et al. 2017b).

#### 8.6.2.2 Optimization of Operating Conditions for the Removal of PCs by Microalgae-Bacteria Consortia in PBRs

Light, availability of N and P nutrients, pH, temperature, and mixing are important variables for the growth of either microalgae or their bacterial partners in PBR systems; thus, the optimization and control of these factors are essential for the efficiency of microalgae-bacteria-based biological treatments.

Light is the most important parameter for the cultivation of photosynthetic organisms, and both its intensity and spectral quality are limiting factors for the growth of microalgae (Gupta et al. 2015; Chang et al. 2017). The photosynthetically active radiation (PAR) lies within the 400–700 nm range (Wang et al. 2012). Light may be provided artificially by electric lamps either external or internal to the PBR, by taking advantage of sunlight in open or closed PBRs installed outdoors, or a combination of both (Bitog et al. 2011). Either continuous illumination or a photoperiod can be applied. Sun as a light source has the advantage of reducing cost but is subjected to fluctuations. Besides, only 50% of sunlight falls within the PAR, while light radiance outside the PAR contributes to temperature rise, and the ultraviolet wavelengths are lethal for microorganisms (Huang et al. 2017).

In open ponds, microalgae cells tend themselves to stay near the surface by floating mechanisms, such as the accumulation of fats or production of extracellular polymers (Larsdotter 2006). In the design of closed PBRs, the transparent tubing or flat plates are shaped to provide narrow light paths, which improve the efficiency of the utilization of light by the cultures (Gupta et al. 2015). Providing good light penetration allows achieving high cell densities and productivity; however, light intensity saturation must be also avoided. Algal activity is increased by light intensities up to 200–400 $\mu E/m^2$ s (Muñoz and Guieysse 2006), but at intensities only slightly above the optimal range, the photosynthetic apparatus becomes saturated, and photo-inhibition leading to oxidative cell damage occurs (Bitog et al. 2011; Chang et al. 2017). Light saturation is more likely to happen in outdoor systems using uncontrolled sunlight (Muñoz and Guieysse 2006). Another important factor is the control of cell density, since at high densities mutual shading occurs and light availability decreases (Larsdotter 2006). Wastewaters with dark color are also limiting for the growth of microalgae because the color interacts with the transmittance of light into the cultures; for this purpose, strains able to grow and remove pollutants under limiting light conditions must be selected, and pretreatment of wastewater to reduce color may be required (Cuéllar-Bermudez et al. 2017).

The availability of N and P nutrients in the PBR influent is a key factor, and the N/P ratio must be adjusted depending on the algal species (Cuéllar-Bermúdez et al. 2017). The preferred N source for microalgae is ammonia, although nitrate, nitrite, and urea can be also assimilated, and some *Cyanobacteria* (i.e. *Anabaena* and *Nostoc* spp.) can fix $N_2$ (Larsdotter 2006; Subashchandrabose et al. 2011). High ammonia concentrations inhibit microalgae growth. As previously stated in Sect. 8.5, depletion of ammonium, nitrate, and phosphate by microalgae contributes to efficient nutrient removal from wastewaters (Cai et al. 2013; Vasconcelos-Fernandes et al. 2015).

pH affects many biological processes and strongly governs the growth of both microalgae and bacteria. In PBRs, pH influences the solubility of $CO_2$ and nutrients such as phosphates (Chang et al. 2017). As previously mentioned in Sect. 8.5, microalgae tend to increase the pH of their surrounding media through $CO_2$ assimilation. Freshwater microalgae often tolerate a wide range of pH, while marine microalgae are more sensitive to changes in this physical factor and prefer pH values in the 8.2–8.7 range (Huang et al. 2017). At pH > 10, complete inhibition of bacteria has been observed in open-culture systems (Muñoz and Guieysse 2006). At high pH levels, the availability of $CO_2$ may also become limiting for the growth of microalgae (Bitog et al. 2011).

Temperature also strongly influences the growth rates and survival of microorganisms. Most microalgae grow optimally between 16 and 27 °C (Bitog et al. 2011) and adapt well to temperatures below their optimum but in contrast have low tolerance for increases in temperature, which fast reduce their growth rates (Larsdotter 2006; Muñoz and Guieysse 2006). Cold-adapted microalgae are capable to remove pollutants from wastewaters at temperatures <15 °C (Muñoz and Guieysse 2006); however, under cold climate conditions, it must be taken into account that inhibition by light intensity is enhanced at low temperatures (Larsdotter 2006). In outdoor PBRs, large diurnal and seasonal temperature fluctuations can occur which significantly influence microalgae cell growth and efficiency (Chang et al. 2017). Finally, a substantial part of the light energy is converted to heat by the photosynthetic organisms, which may lead to undesirably high temperatures negatively influencing bacterial growth (Larsdotter 2006).

Efficient mixing is required for microalgae-bacteria consortia growing in PBR systems in order to keep cultures in suspension, provide an even distribution of nutrients, avoid thermal stratification, and enhance mass transfer of $CO_2$ to the microalgae and effective removal of dissolved $O_2$ by heterotrophs preventing its accumulation (Gupta et al. 2015; Chang et al. 2017). In vertical column or tubular systems, if the path of tubing is wide enough, mixing allows the mobility of algal cells among the lightened surface and darker internal zones, inducing alternative light/dark cycles (Gupta et al. 2015; Chang et al. 2017). It has been reported that the photosynthetic efficiency of microalgae under intermittent illumination is higher than under continuous illumination (Liao et al. 2014). Mixing can be provided by mechanical elements such as paddle wheels or stirrers, circulation by peristaltic pumps, or aeration. Excess mixing must be avoided, since hydrodynamic stress can be created inhibiting cell metabolic activity and damaging cells (Chang et al. 2017).

### 8.6.3 Removal of PCs by Microalgae-Bacteria Consortia in PBRs

The successful removal of PCs from wastewaters by microalgae-bacteria consortia has been described in several studies, although most of the available results were obtained from experiments in laboratory-scale batch cultures or continuously operated small PBRs, most often using synthetic media or artificial wastewater rather than real industrial effluents.

Borde et al. (2003) reported for the first time the photosynthetically enhanced biodegradation of PCs and other aromatic compounds by microalgae-bacteria consortia under laboratory conditions (batch cultures in mineral media, incubated under an Ar atmosphere). In their study, *Chlorella sorokiniana* 211/8k supported growth and >85% removal of salicylate or phenol by strains of *Cupriavidus basilensis* (formerly named *Ralstonia basilensis*) and *Acinetobacter haemolyticus*, respectively, which were incubated under continuous illumination for 93 h. The same *C. basilensis* strain was shown able to remove salicylate in consortia with other microalgal strains: *Chlorella vulgaris* H 1987, a *Tetradesmus obliquus* (formerly named *Scenedesmus obliquus*) strain, an unidentified wild-type Bolivian algal strain, and a strain of the cyanobacteria *Microcystis aeruginosa* (Muñoz et al. 2003). The later experiments were carried out in batch cultures under an inert atmosphere using synthetic mineral media added with sodium salicylate (800 mg/L) and inoculated with binary combinations of *C. basilensis* and each microalgal strain. *C. sorokiniana* and the unidentified Bolivian isolate yielded the highest salicylate removal rates (18.5 ± 2.4 and 17.5 ± 1.2 mg/L h, respectively). These authors demonstrated that salicylate removal was totally supported by photosynthetic oxygenation, since the compound was not photodegraded or assimilated by the microalgae, and the *C. basilensis* strain was known able to remove salicylate under aerobic conditions only (Guieysse et al. 2002).

Further studies corroborated the efficiency to remove salicylate of the *C. sorokiniana* 211/8k + *C. basilensis* consortium, using different types of continuously operated laboratory-scale PBRs. Muñoz et al. (2004) inoculated a stirred-tank reactor (600 mL) with *C. sorokiniana* 211/8k + *C. basilensis* at initial concentrations of 51 mg dw/L and 1.2 mg dw/L, respectively, with an inlet sodium salicylate concentration of 1000 mg/L, an agitation rate of 300 rpm, and variable hydraulic retention times (HRTs), temperatures, and light inputs. Under optimized conditions (300 μE/$m^2$ s, 30 °C, and applying biomass recirculation), sodium salicylate was removed at a maximum constant rate of 87 mg/L h, corresponding to an estimated oxygenation capacity of 77 mg $O_2$/L h, and complete salicylate removal was achieved at HRTs of 0.9–1.5 days. Six different PBR configurations were further tested for the removal of biological oxygen demand at 5 days ($BOD_5$) and salicylate from synthetic industrial wastewater, also evaluating the growth of the consortium as attached biomass on suspended carriers or forming biofilms onto the PBR walls (Muñoz et al. 2009). Salicylate removal rates ranged 23–100% and 74–100% in flat-plate and tubular PBRs, respectively, and attachment of biomass to the PBR walls was found to improve removal stability.

Ryu et al. (2017) investigated the ability for the removal of PCs from coke wastewater of a consortium formed by the microalgae *Scenedesmus quadricauda* AG10003 and the indigenous bacterial community of an activated sludge sewage treatment plant in Korea. Filtered coke factory effluent containing 429.0 ± 9.2 mg/L PCs was used either undiluted or 80%, 60%, 40%, and 20% diluted for batch cultures, which were inoculated with 1:1 biomass of *S. quadricauda* and activated sludge. Under illumination, the microalgae-bacteria consortia removed 100% phenol at all dilutions of coke wastewater, after 94 h of incubation.

To the best of the authors' knowledge, there is only one published study available combining microalgae and fungi for the removal of PCs (Al-Fawwaz et al. 2016). *Desmodesmus* sp. AF1 + *Rhizopus* sp. AF1 were cocultivated in a synthetic medium added with phenol (25–100 mg/L). After 25 days of incubation, 95% removal of phenol at the lowest concentration was achieved by the consortium, compared to 70–66% removal by each of the isolates in monoculture.

A consortium formed by *Chlorella vulgaris* MM1 and *Pseudomonas* MT1 was tested in several studies for its ability to remove mixtures of PCs from artificial wastewaters using laboratory-scale continuous PBRs. This consortium was able to completely remove mixtures of phenol and pyridine (up to 4.6 and 4.4 mM, respectively) with an HRT of 2.7 days (Essam et al. 2013) and mixtures of salicylate and phenol (6.4 mM and 2.0 mM, respectively) with an HRT of 4 days (Essam et al. 2014). In the latter study, thiocyanate (1 mM) and cyanide (0.72 mM) were also added to check their influence on PC removal. The mixture of salicylate, phenol, and thiocyanate was 100% removed; however, after cyanide addition, all pollutants accumulated, due to the toxicity of cyanide for the algae.

Few reports are available in the literature attempting the scale-up of microalgae-bacteria-based treatments in PBRs aimed for real industrial wastewaters rich in PCs. Essam et al. (2006) designed and evaluated the application of a PBR to biologically treat the wastewater of a coking factory in Egypt. A consortium was constructed comprising two microorganisms isolated from the wastewater aeration tank of the coke factory: a *Chlorella vulgaris* strain and *Alcaligenes* sp. TW1, which was able to use phenol (up to 1200 mg/L) and other PCs as the sole source of carbon. The wastewater contained phenol (127 mg/L), 4-chlorophenol (14.2 mg/L), and 2,6-dichlorophenol (6.2 mg/L). Phenol removal by the consortium in batch tests was compared for both artificial and real wastewaters. Complete phenol removal from the artificial wastewater required light and inoculation with both bacteria and algae. While the consortium removed 100% phenol from the artificial wastewater after 5 days, the removal efficiency in the real wastewater dropped to 16% after 7 days, due to the occurrence of substances other than PCs which were inhibitory for algae. Pretreatment of the real wastewater with activated carbon or UV radiation allowed the subsequent nearly complete removal of phenol by the consortium. Removal of phenol was also optimized in a continuously fed 600-mL conical glass photobioreactor using artificial wastewater (325 mg/L phenol) with an HRT of 3.6 days and by adding $NaHCO_3$ (8 g/L) to enhance algae growth. A consortia of the same *Chlorella vulgaris* strain plus an acclimated combo of a phenol-degrading *Alcaligenes* sp. and a *p*-nitrophenol-degrading *Arthrobacter* sp. were tested for the simultaneous removal of phenol (100 mg/L) and *p*-nitrophenol (50 mg/L) in batch tests, but photosynthetic oxygenation was not enough to support the removal of the mixture of PCs due to the toxicity of *p*-nitrophenol for microalgae (Essam et al. 2007). Photocatalytic pretreatment of the mixture during 60 h removed 51% phenol and 62% *p*-nitrophenol in the medium, allowing the subsequent biological removal of the remaining contaminants.

A consortium of two microalgae (*Chlorella vulgaris* 1 and *Tetradesmus obliquus* 2, formerly named *Scenedesmus obliquus*) and two bacterial strains (*Pantoea agglomerans* PM15 and *Raoultella terrigena* PM3) was designed aiming for the removal of PCs from OWW. Olive oil production starts with the washing of the olives to eliminate vegetable residues and soil particles; the water used for this purpose must be drinkable water since olive oil is directed to human consumption. The resulting wastewater is called OWW and is characterized for a moderate content of PCs, around 10–100 times lower than the more concentrated residue of olive oil milling (olive oil mill wastewater, OMW) (Pozo et al. 2007). In a first study, the microbial consortium was tested on synthetic OWW added with 50, 100, or 150 mg/L of a 1:1:1 mixture of caffeic, *p*-OH-benzoic, and coumaric acids as the only carbon source in laboratory batch cultures (Maza-Márquez et al. 2014). The microalgae

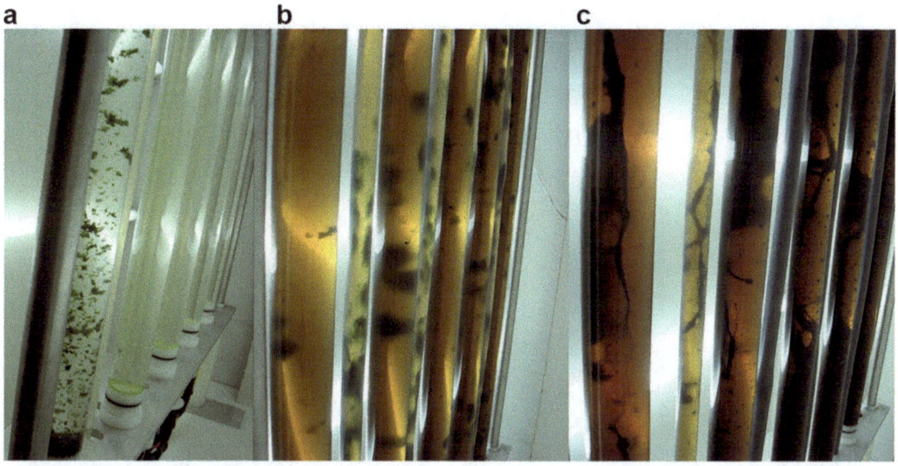


**Fig. 8.5** Development of a microbial biofilm in a laboratory-scale tubular PBR treating olive washing wastewater (OWW). (**a**) Microbial biofilm in the PBR fed with synthetic medium at time 0. (**b**) Microbial biofilm in the PBR fed with real OWW operated at 5 days HRT. (**c**) Microbial biofilm in the PBR fed with real OWW operate at 3 days HRT. (Reprinted by permission from Springer Nature, Maza-Márquez et al. (2017a))

grew with PCs up to a concentration of 150 mg/L, and the four-strain consortium removed 99% of phenolic compounds after 48 h of incubation. In further work (Maza-Márquez et al. 2017a), a laboratory-scale tubular PBR (14.5 L) fed with real OWW was inoculated with the consortium using *C. vulgaris* and *T. obliquus* at $10^{12}$ cells/mL and *P. agglomerans* and *R. terrigena* at $10^3$ CFU/mL. The microalgae-bacteria consortium was able to colonize the abiotic surface of the PBR walls forming stable biofilms throughout the whole experimental period (Fig. 8.5). Total content of PCs, chemical oxygen demand (COD), $BOD_5$, turbidity, and color were 90.3 ± 11.4%, 80.7 ± 9.7%, 97.8 ± 12.7%, 82.9 ± 8.4%, and 83.3 ± 10.4% removed, respectively, using a HRT of 3 days.

In a more recent study, Maza-Márquez et al. (2017b) investigated the feasibility of treating real OWW (total content of PCs = 274.0 ± 4.5 mg/L) in a full-scale tubular PBR composed of eight modules with ten tubes each, able to treat 1374–3120 L/day of the industrial effluent of a small olive oil factory in South Spain, testing HRTs ranging from 2 to 5 days (Fig. 8.6a). The PBR was constructed outdoors and was subjected to intense sunlight radiation, high environmental temperature, and variability of pH and concentrations of PCs and other pollutants in the OWW to be treated. Real OWW entered the system after a pretreatment with activated carbon. After being initially filled with 60% diluted OWW, the PBR was operated in recirculation for a week to allow for the development of a consortium of the indigenous microbiota in the wastewater, before the PBR was fed raw OWW in continuous flow. The culture was circulated by peristaltic pumps, and no external aeration was supplied. The PBR achieved average removal efficiencies of 94.84 ± 0.55%, 85.86

A.

**Fig. 8.6** (**a**) Experimental design of the vertical tubular photobioreactor (PBR) installed at Nuestra Señora de los Desamparados, Puente Genil, Córdoba, Spain. (**b**) Color removal succession throughout the PBR treatment steps. OWW: olive washing wastewater sample, PBR: effluent of the PBR, UF membrane: effluent of the ultrafiltration membrane modules, NF modules: effluent of the nanofiltration membrane modules. (Pictures by P. Maza-Márquez (unpublished))

± 1.24%, 99.12 ± 0.17%, 95.86 ± 0.98%, and 87.24 ± 0.91% for phenols, COD, $BOD_5$, turbidity, and color of the OWW, respectively, with little influence of HRT.

As a complement to the aforementioned study by Maza-Márquez et al. (2017b), ultrafiltration and nanofiltration membrane modules were tested as tertiary treatment for the PBR effluent treatment, which significantly enhanced the efficiency of the removal of color (up to 99%, Fig. 8.6b), generating water suitable for direct reuse (Maza-Márquez, unpublished data).

Maza-Márquez et al. (2017b) used Illumina sequencing (Figs. 8.7 and 8.8) to investigate the community diversity of the stable consortium in the full-scale PBR, revealing that it was composed by associations of green algae (*Sphaeropleales* and *Chlorellales*), cyanobacteria (*Hapalosiphon* and *Acarychloris*), and bacteria (*Silanimonas*, *Ahrensia*, *Azotobacter*, and *Nitrobacter*). It is worth noting that the

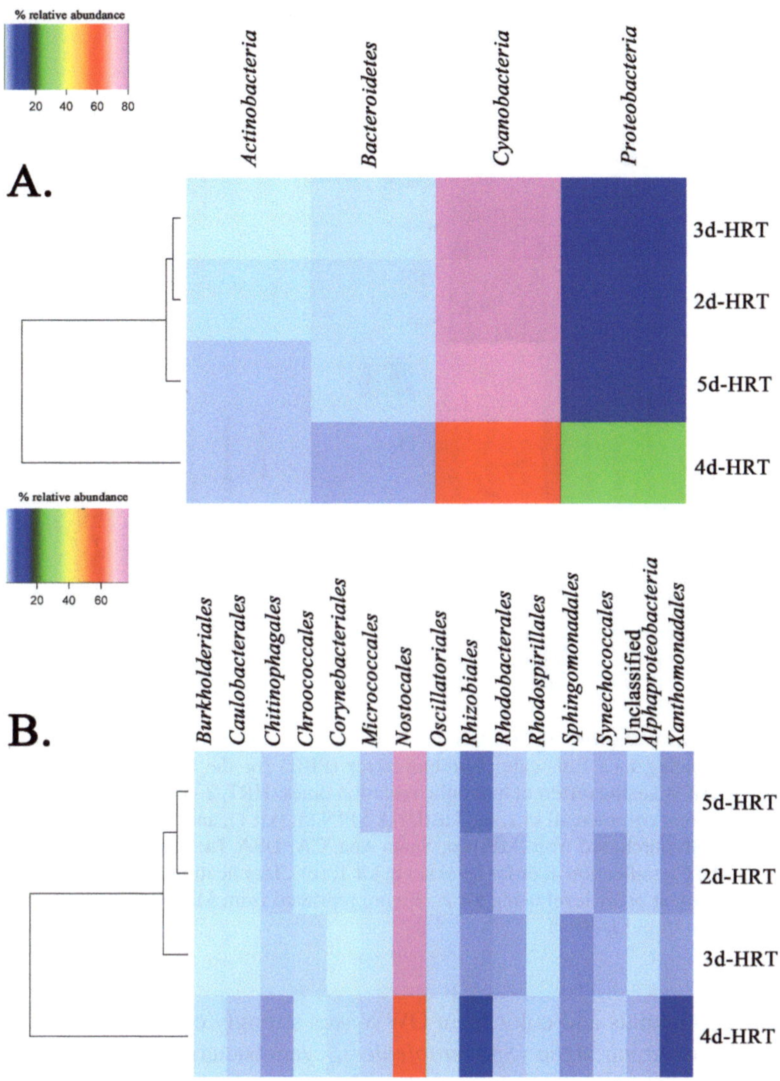


**Fig. 8.7** Heat maps showing the relative abundances of bacterial phylotypes detected by Illumina-sequencing in a full-scale photobioreactor (PBR) for the treatment of olive washing wastewater (OWW) operated under different hydraulic retention times (HRT, 2–5 days). DNA was extracted using the FastDNA SPIN kit, purchased from MP-Bio, Santa Ana, CA, USA. (**a**) Classification at Phylum level. (**b**) Classification at order level. (Reprinted from Maza-Márquez et al. (2017b), with permission from Elsevier)

selection of the nucleic acid extraction protocol was essential for the accurate characterization of the microbial community by molecular methods, since the output of two different commercial DNA extraction kits was compared in this study and one of them failed to recover DNA from the branched cyanobacteria of the *Nostocales*, which were actually dominating the bacterial community (Fig. 8.7). The removal of

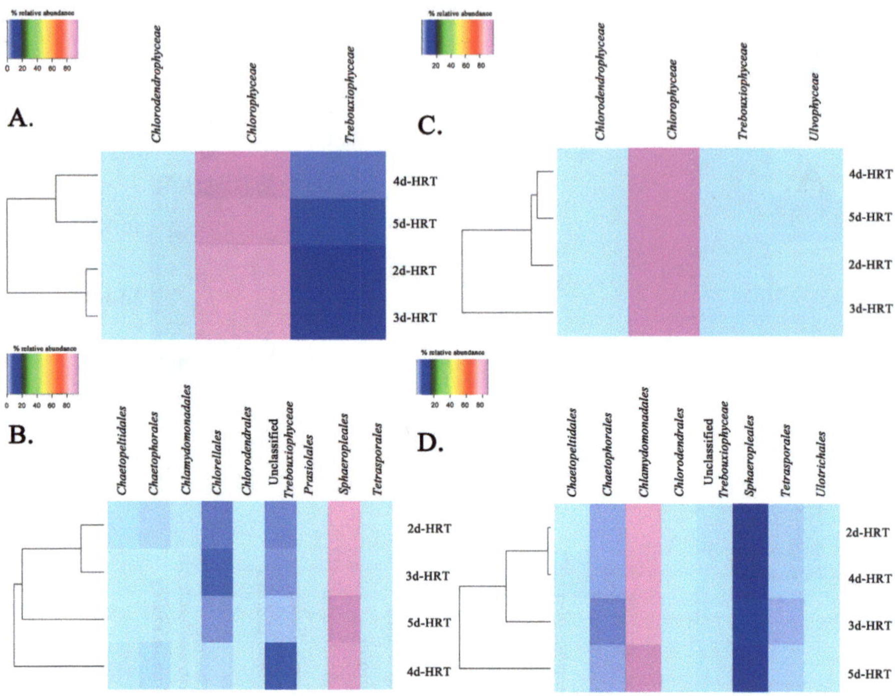


**Fig. 8.8** Heat maps showing the relative abundances of phylotypes of *Chlorophyta* detected by Illumina-sequencing in a full-scale photobioreactor (PBR) for the treatment of olive washing wastewater (OWW) using different hydraulic retention times (HRT, 2–5 days). DNA was extracted using two different commercial kits: the FastDNA SPIN kit (kit 1), and the FastDNA-SPIN kit for Soil (kit 2), both purchased from MP-Bio, Santa Ana, CA, USA. (**a**) Classification at class level using kit 1. (**b**) Classification at order level using kit 1. (**c**) Classification at class level using kit 2. (**d**) Classification at order level using kit 2. (Figure reprinted from Maza-Márquez et al. (2017b), with permission from Elsevier)

phenolic compounds and color from OWW was strongly correlated to the relative abundances of green algae (*Sphaeropleales*), cyanobacteria (*Hapalosiphon*), and *Proteobacteria* (*Rhodopseudomonas*, *Azotobacter*), indicating the relevance of the interplay among these groups for the efficiency of the biological treatment.

## 8.7 Conclusions and Future Prospects

Extensive research has been carried out in the last century to understand the microbial mechanisms for the removal of phenol and its derivatives and exploit these capabilities for the design and application of biotreatment strategies for polluted industrial wastewaters. Biotreatments relying on microalgae-bacteria consortia display several advantages for the removal of PCs; however, to date, only a few studies have attempted the implementation of the PBR technology to real-scale applications. It must be taken into consideration that most of the existing PBR designs and configurations, including those commercially available, have been conceived for microalgae biomass production (often in monocultures) or wastewater treatment by microalgae only. Thus, exploitation of microalgae-bacteria consortia for wastewater treatment is currently somehow limited by the scarce development of cultivation systems specifically aimed for such purpose. In this sense, new designs for the cultivation of microalgae-bacteria consortia with wastewater as a substrate for growth have been explored in recent years, such as hybrid bioreactors supporting cultivation of both heterotrophic and autotrophic microorganisms (Wu et al. 2011) or aerobic algal-bacterial granular systems (Zhang et al. 2018). The need for complementing the biological treatment with pre- or post-physicochemical treatments to enable better efficiencies of PC removal or allow for the direct reuse of the treated water is also challenging for the economic feasibility of the technology.

It is also worth noting that, to date, little is known about the structure and diversity of the microalgae-bacteria mixed communities developed in PBRs aimed for pollutant removal from wastewaters. Massive parallel sequencing technologies providing thousands to millions of DNA sequence reads in a single reaction have been widely used for further in-depth analysis of microbial diversity in many natural and engineered habitats; however, few studies are currently available taking advantage of these tools for the characterization of mixed algae-bacteria communities, particularly those developed in PBRs aimed for wastewater treatment (i.e., Maza-Márquez et al. 2017b; Zhang et al. 2018). In this sense, multivariate analyses provide strong tools to reveal the links between removal efficiencies of pollutants and the occurrence/relative abundance of specific microbial populations in those bioremediation processes driven by mixed communities, significantly helping to interpret the large and complex datasets derived from massive parallel sequencing platforms.

## References

Abdel-Raouf N, Al-Homaidan AA, Ibraheem IBM (2012) Microalgae and wastewater treatment. Saudi J Biol Sci 19(3):257–275

Acikgoz E, Ozcan B (2016) Phenol biodegradation by halophilic archaea. Int Biodeterior Biodegradation 107:140–146

Acir IH, Guenther K (2018) Endocrine-disrupting metabolites of alkylphenol ethoxylates – a critical review of analytical methods, environmental occurrences, toxicity, and regulation. Sci Total Environ 635:1530–1546

Acuña-Argüelles ME, Olguin-Lora P, Razo-Flores E (2003) Toxicity and kinetic parameters of the aerobic biodegradation of the phenol and alkylphenols by a mixed culture. Biotechnol Lett 25(7):599

Afreen A, Bano F, Ahmad N, Fatma T (2017) Screening and optimization of laccase from cyanobacteria with its potential in decolorization of anthraquinonic dye Remazol Brilliant Blue R. Biocatal Agric Biotechnol 10:403–410

Aggelis G, Iconomou D, Christou M, Bokas D, Kotzailias S, Christou G, Tsagou V, Papanikolaou S (2003) Phenolic removal in a model olive oil mill wastewater using *Pleurotus ostreatus* in bioreactor cultures and biological evaluation of the process. Water Res 37(16):3987–3904

Ahn YB, Chae JC, Zylstra GJ, Häggblom MM (2009) Degradation of phenol via phenylphosphate and carboxylation to 4-hydroxybenzoate by a newly isolated strain of the sulfate-reducing bacterium *Desulfobacterium anilini*. Appl Environ Microbiol 75(13):4248–4253

Alexander JT, Hai FI, Al-Aboud TM (2012) Chemical coagulation-based processes for trace organic contaminant removal: current state and future potential. J Environ Manag 111:195–207

Al-Fawwaz AT, Jacob JH, Al-Wahishe TE (2016) Bioremoval capacity of phenol by green micro-algal and fungal species isolated from dry environment. Int J Sci Technol Res 5(8):155–160

Ali S, Lafuente RL, Cowan DA (1998) Meta-pathway degradation of phenolics by thermophilic *Bacilli*. Enzym Microb Technol 23(7–8):462–468

Al-Khalid T, El Naas MH (2012) Aerobic biodegradation of phenols: a comprehensive review. Crit Rev Environ Sci Technol 42:1631–1690

Anders HJ, Kaetzke A, Kämpfer P, Ludwig W, Fuchs G (1995) Taxonomic position of aromatic-degrading denitrifying *Pseudomonas* strains K 172 and KB 740 and their description as new members of the genera *Thauera*, as *Thauera aromatica* sp. Nov., and *Azoarcus*, as *Azoarcus evansii* sp. nov., respectively, members of the beta subclass of the Proteobacteria. Int J Syst Bacteriol 45(2):327–333

Anku WW, Mamo MA, Govender PP (2017) Phenolic compounds in water: sources, reactivity, toxicity and treatment methods. In: Soto-Hernández M, Palma-Tenango M, García-Mateos MR (eds) Phenolic compounds-natural sources, importance and applications, chapter 17. InTech Open, Croatia, pp 420–443

Aranda E (2016) Promising approaches towards biotransformation of polycyclic aromatic hydrocarbons with *Ascomycota*. Curr Opin Biotechnol 38:1–8

Araújo M, Pimentel FB, Alves RC, Oliveira MBPP (2015) Phenolic compounds from olive mill wastes: health effects, analytical approach and application as food antioxidants. Trends Food Sci Technol 45(2):200–211

Arora PK, Srivastava A, Singh VP (2014) Bacterial degradation of nitrophenols and their derivatives. J Hazard Mater 266:42–59

Arora PK, Srivastava A, Garg SK, Singh VP (2018) Recent advances in degradation of chloronitrophenols. Bioresour Technol 250:902–909

Arutchelvan V, Kanakasabai V, Elangovan R, Nagarajan S, Muralikrishnan V (2006) Kinetics of high strength phenol degradation using *Bacillus brevis*. J Hazard Mater B 129(1–3):216–222

Asha P, Nira KS, Ashok P, Edgard G, Datta M (2011) Cyanobacteria and microalgae: a positive prospect for biofuels. Bioresour Technol 102(22):10163–10172

Azubuike CC, Chikere CB, Okpokwasili GC (2016) Bioremediation techniques–classification based on site of application: principles, advantages, limitations and prospects. World J Microbiol Biotechnol 32(11):180

Basha KM, Rajendran A, Thangavelu V (2010) Recent advances in the biodegradation of phenol: a review. Asian J Exp Biol Sci 1(2):219–234

Bhattacharya A, Sood P, Citovsky V (2010) The roles of plant phenolics in defence and communication during *Agrobacterium* and *Rhizobium* infection. Mol Plant Pathol 11(5):705–719

Bitog JP, Lee IB, Lee CG, Kim KS, Hwang HS, Hong SW, Seo IH, Kwon KS, Mostafa E (2011) Application of computational fluid dynamics for modeling and designing photobioreactors for microalgae production: a review. Comput Electron Agric 76(2):131–147

Borde X, Guieysse B, Delgado O, Muñoz R, Hatti-Kaul R, Nugier-Chauvin C, Patin H, Mattiasson B (2003) Synergistic relationships in algal-bacterial microcosms for the treatment of aromatic pollutants. Bioresour Technol 86(3):293–300

Brodie J, Ball SG, Bouget FY, Chan CX, De Clerck O, Cock JM, Gachon C, Grossman AR, Mock T, Raven JA, Saha M, Smith AG, Vardi A, Yoon HS, Bhattacharya D (2017) Biotic interactions as drivers of algal origin and evolution. New Phytol 216(3):670–681

Busca G, Berardinelli S, Resini C, Arrighi L (2008) Technologies for the removal of phenol from fluid streams: a short review of recent developments. J Hazard Mater 160(2–3):174–183

Cai W, Li J, Zhang Z (2007) The characteristics and mechanisms of phenol biodegradation by *Fusarium* sp. J Hazard Mater 148:38–42

Cai T, Park SY, Li Y (2013) Nutrient recovery from wastewater streams by microalgae: status and prospects. Renew Sust Energ Rev 19:360–369
Calderón K, González-Martínez A, Gómez-Silván C, Osorio F, Rodelas B, González-López J (2013) Archaeal diversity in biofilm technologies applied to treat urban and industrial wastewater: recent advances and future prospects. Int J Mol Sci 14:18572–18598
Cañizares P, Paz R, Sáez C, Rodrigo MA (2009) Costs of the electrochemical oxidation of wastewaters: a comparison with ozonation and Fenton oxidation processes. J Environ Manag 90(1):410–420
Cerrone F, Barghini P, Pesciaroli C, Fenice M (2011) Efficient removal of pollutants from olive washing wastewater in bubble-column bioreactor by *Trametes versicolor*. Chemosphere 84(2):254–259
Chang JS, Show PL, Ling TC, Chen CY, Ho SH, Tan CHD, Nagarajan D, Phong WN (2017) Photobioreactors. In: Larroche C, Sanroman MA, Du G, Pandey A (eds) Current developments in biotechnology and bioengineering: bioprocess, Bioreactors and Controls. Elsevier, Amsterdam, pp 313–352
Chauhan PS, Goradia B, Saxena A (2017) Bacterial laccase: recent update on production, properties and industrial applications. 3 Biotech 7(5):323
Cheng T, Zhang W, Zhang W, Yuan G, Wang H, Liu T (2017) An oleaginous filamentous microalgae *Tribonema minus* exhibits high removing potential of industrial phenol contaminants. Bioresour Technol 238:749–754
Cho DH, Ramanan R, Heo J, Lee J, Kim BH, Oh HM, Kim HS (2015) Enhancing microalgal biomass productivity by engineering a microalgal-bacterial community. Bioresour Technol 175:578–585
Christen P, Vega A, Casalot L, Simon G, Auria R (2012) Kinetics of aerobic phenol biodegradation by the acidophilic and hyper thermophilic archaeon *Sulfolobus solfataricus* 98/2. Biochem Eng J 62:56–61
Cordova-Villegas L, Mashhadi N, Chen M, Mukherjee D, Taylor KE, Biswas N (2016) A short review of techniques for phenol removal from wastewater. Curr Pollut Rep 2:157–167
Cuéllar-Bermúdez SP, Alemán-Nava GS, Chandra R, Garcia-Perez JS, Contreras-Angulo JR, Markou G, Muylaert K, Rittmann BE, Parra-Saldivar R (2017) Nutrients utilization and contaminants removal. A review of two approaches of algae and cyanobacteria in wastewater. Algal Res 24(Part B):438–449
Cui Y, Liu XY, Chung TS, Weber M, Staudt C, Maletzko C (2016) Removal of organic micropollutants (phenol, aniline and nitrobenzene) via forward osmosis (FO) process: evaluation of FO as an alternative method to reverse osmosis (RO). Water Res 91:104–114
Das DP, Parida K, De BR (2005) Photo-oxidation of phenol over titania pillared zirconium phosphate and titanium phosphate. J Mol Catal A Chem 240(1–2):1–6
Das B, Mandal TK, Patra S (2015) A comprehensive study on *Chlorella pyrenoidosa* for phenol degradation and its potential applicability as biodiesel feedstock and animal feed. Appl Biochem Biotechnol 176(5):1382–1401
Dayana Priyadharshini S, Bakthavatsalam AK (2017a) Phycoremediation of phenolic effluent of a coal gasification plant by *Chlorella pyrenoidosa*. Process Saf Environ Prot 111:31–39
Dayana Priyadharshini S, Bakthavatsalam AK (2017b) Effect of nutrients on *Chlorella pyrenoidosa* for treatment of phenolic effluent of coal gasification plant. Environ Sci Pollut Res 24(15):13594–13603
de Vree JH, Bosma R, Janssen M, Barbosa MJ, Wijffels RH (2015) Comparison of four outdoor pilot-scale photobioreactors. Biotechnol Biofuels 8:215
De-Bashan LE, Antoun H, Bashan Y (2008) Involvement of indole-3-acetic acid produced by the growth-promoting bacterium *Azospirillum* spp. in promoting growth of *Chlorella vulgaris*. J Phycol 44(4):938–947
Di Caprio F, Altimari P, Pagnanelli F (2015) Integrated biomass production and biodegradation of olive mill wastewater by cultivation of *Scenedesmus* sp. Algal Res 9:306–311

Di Caprio F, Altimari P, Pagnanelli F (2018) Integrated microalgae biomass production and olive mill wastewater biodegradation: optimization of the wastewater supply strategy. Chem Eng J 349:539–546

Duan W, Meng F, Cui H, Lin Y, Wang G, Wu J (2018) Ecotoxicity of phenol and cresols to aquatic organisms: a review. Ecotoxicol Environ Saf 15(157):441–456

Emerson D, Chauchan S, Oriel P, Breznak JA (1994) *Haloferax* sp. D1227, a halophilic archaeon capable of growth on aromatic compounds. Arch Microbiol 161(6):445–452

Environmental Protection Agency, EPA (2014) Toxic and priority pollutants under the Clean Water Act, 40 CFR Part 423, Appendix A. Available at: https://19january2017snapshot.epa.gov/sites/production/files/2015-09/documents/priority-pollutant-list-epa.pdf

Eriksen NT (2008) The technology of microalgal culturing. Biotechnol Lett 30(9):1525–1536

Essam T, Amin MA, ElTayeb O, Mattiasson B, Guieysse B (2006) Biological treatment of industrial wastes in a photobioreactor. Water Sci Technol 53(11):117–125

Essam T, Amin M, ElTayeb O, Mattiasson B, Guieysse B (2007) Solar-based detoxification of phenol and *p*-nitrophenol by sequential $TiO_2$ photocatalysis and photosynthetically aerated biological treatment. Water Res 41(8):1697–1704

Essam T, ElRakaiby M, Hashem A (2013) Photosynthetic based algal/bacterial combined treatment of mixtures of organic pollutants and $CO_2$ mitigation in a continuous photobioreactor. World J Microbiol Biotechnol 29(6):969–974

Essam T, El Rakaiby M, Agha A (2014) Remediation of the effect of adding cyanides on an algal/bacterial treatment of a mixture of organic pollutants in a continuous photobioreactor. Biotechnol Lett 36(9):1773–1781

European Union, EU (2013) Council Directive 2013/39/EU amending directives 2000/60/EC and 2008/105/EC as regards priority substances in the field of water policy. Off J Eur Union L226

Field JA, Sierra-Álvarez R (2008) Microbial degradation of chlorinated phenols. Rev Environ Sci Biotechnol 7(3):211–241

Filipowicz N, Momotko M, Boczkaj G, Pawlikowski T, Wanarska M, Cieśliński H (2017) Isolation and characterization of phenol-degrading psychrotolerant yeasts. Water Air Soil Pollut 228(6):210

Folke J, Lund U (1983) Occurrence of low- and high-chlorinated phenols in municipal sewage before and after passing through biological treatment plants. J Chromatogr 279:189–198

Franchi O, Bovio P, Ortega-Martínez E, Rosenkranz F, Chamy R (2018) Active and total microbial community dynamics and the role of functional genes *bamA* and *mcrA* during anaerobic digestion of phenol and *p*-cresol. Bioresour Technol 264:290–297

Galíndez-Mayer J, Ramón-Gallegos J, Ruiz-Ordaz N, Juárez-Ramírez C, Salmerón-Alcocer A, Poggi-Varaldo HM (2008) Phenol and 4-chlorophenol biodegradation by yeast *Candida tropicalis* in a fluidized bed reactor. Biochem Eng J 38(2):147–157

Gallegos A, Fortunato MS, Foglia J, Rossi S, Gemini V, Gomez L, Gomez CE, Higa LE, Korol SE (2003) Biodegradation and detoxification of phenolic compounds by pure and mixed indigenous cultures in aerobic reactors. Int Biodeterior Biodegradation 52(4):261–267

Gao QT, Wong YS, Tam NFY (2011) Removal and biodegradation of nonylphenol by different *Chlorella* species. Mar Pollut Bull 63(5–12):445–451

Ghasemi Y, Rasoul-Amini S, Fotooh-Abadi E (2011) Review: the biotransformation, biodegradation and bioremediation of organic compounds by microalgae. J Phycol 47(5):969–980

Gibson DT, Subrawahian V (1984) Microbial degradation of aromatic hydrocarbons. In: Gibson DT (ed) Microbial degradation of organic compounds. Marcel Dekker, New York, pp 181–252

González LE, Bashan Y (2000) Increased growth of the microalga *Chlorella vulgaris* when coimmobilized and cocultured in alginate beads with the plant-growth-promoting bacterium *Azospirillum brasilense*. Appl Environ Microbiol 66(4):1527–1531

González G, Herrera MG, García MT, Peña MM (2001) Biodegradation of phenol in a continuous process: comparative study of stirred tank and fluidized-bed bioreactors. Bioresour Technol 76(3):245–251

Gouveia L, Graça S, Sousa C, Ambrosano L, Ribeiro B, Botrel EP, Neto PC, Ferreira AF, Silva CM (2016) Microalgae biomass production using wastewater: treatment and costs: scale-up considerations. Algal Res 16:167–176
Grossart HP, Simon M (2007) Interactions of planktonic algae and bacteria: effects on algal growth and organic matter dynamics. Aquat Microb Ecol 47(2):163–176
Guieysse B, Borde X, Muñoz R, Hatti-Kaul R, Nugier-Chauvin C, Patin H, Mattiasson B (2002) Influence of the initial composition of algal-bacterial microcosms on the degradation of salicylate in a fed-batch culture. Biotechnol Lett 24(7):531–538
Gupta PL, Lee SM, Choi HJ (2015) A mini review: photobioreactors for large scale algal cultivation. World J Microbiol Biotechnol 31(9):1409–1417
Harborne JB, Simmonds NW (1964) Biochemistry of phenolic compounds. Academic, London
Harms H, Schlosser D, Wick LY (2011) Untapped potential: exploiting fungi in bioremediation of hazardous chemicals. Nat Rev Microbiol 9(3):177–192
Harrass MC, Kindig AC, Taub FB (1985) Responses of blue-green and green algae to streptomycin in unialgal and paired culture. Aquat Toxicol 6(1):1–11
Harwood CS, Parales RE (1996) The β-ketoadipate pathway and the biology of self-identity. Ann Rev Microbiol 50:553–590
Hernández JP, de-Bashan L, Johana-Rodriguez D, Rodriguez Y, Bashan Y (2009) Growth promotion of the freshwater microalga *Chlorella vulgaris* by the nitrogen-fixing, plant growth-promoting bacterium *Bacillus pumilus* from arid zone soils. Eur J Soil Biol 45(1):88–93
Heukelekian H (1956) Purifying chemically polluted waters. Ind Eng Chem 48(9):1403
Hirooka T, Akiyama Y, Tsuji N, Nakamura T, Nagase H, Hirata K, Miyamoto K (2003) Removal of hazardous phenols by microalgae under photoautotrophic conditions. J Biosci Bioeng 95(2):200–203
Hirooka T, Nagase H, Uchida K, Hiroshige Y, Ehara Y, Nishikawa J, Nishihara T, Miyamoto HZ (2005) Biodegradation of bisphenol A and disappearance of its estrogenic activity by the green alga *Chlorella fusca* var. *vacuolata*. Environ Toxicol Chem 24(8):1896–1901
Hirooka T, Nagase H, Hirata K, Miyamoto K (2006) Degradation of 2,4-dinitrophenol by a mixed culture of photoautotrophic microorganisms. Biochem Eng J 29(1):157–162
Huang Q, Jiang F, Wang L, Yang C (2017) Design of photobioreactors for mass cultivation of photosynthetic organisms. Engineering 3:318–329
Ji Q, Tabassum S, Yu G, Chua C, Zhang Z (2015) A high efficiency biological system for treatment of coal gasification wastewater – a key in-depth technological research. RSC Adv 5(50):40402–40413
Jiang HL, Tay JH, Maszenan AM, Tay STL (2006) Enhanced phenol biodegradation and aerobic granulation by two coaggregating bacterial strains. Environ Sci Technol 40(19):6137–6142
Joutey NT, Bahafid W, Sayel H, El Ghachtouli N (2013) Biodegradation: involved microorganisms and genetically engineered microorganisms. In: Chamy R (ed) Biodegradation: life of science, chapter 11. In Tech Open, Croatia, pp 289–320
Juárez-Jiménez B, Reboleiro-Rivas P, González-López J, Pesciaroli C, Barghini P, Fenice M (2012) Immobilization of *Delftia tsuruhatensis* in macro-porous cellulose and biodegradation of phenolic compounds in repeated batch process. J Biotechnol 157(1):148–153
Jusoh N, Razali F (2008) Microbial consortia from residential wastewater for bioremediation of phenol in a chemostat. J Teknol 48:51–60
Juteau P, Côté V, Duckett MF, Beaudet R, Lépine F, Villemur R, Bisaillon JG (2005) *Cryptanaerobacter phenolicus* gen. nov., sp nov., an anaerobe that transforms phenol into benzoate via 4-hydroxybenzoate. Int J Syst Evol Microbiol 55(P1):245–250
Kanekar PP, Sarneik SS, Kelkar AS (1998) Bioremediation of phenol by alkaliphilic bacteria isolated form alkaline Lake of Lonar, India. J Appl Microbiol 85(Suppl 1):128S–133S
Kato S, Chino K, Kamimura N, Masai E, Yumoto I, Kamagata Y (2015) Methanogenic degradation of lignin-derived monoaromatic compounds by microbial enrichments from rice paddy field soil. Sci Rep 5:14295

Kazamia E, Czesnick H, Nguyen TT, Croft MT, Sherwood E, Sasso S, Hodson SJ, Warren MJ, Smith AG (2012) Mutualistic interactions between vitamin $B_{12}$-dependent algae and heterotrophic bacteria exhibit regulation. Environ Microbiol 14(6):1466–1476

Khemili-Talbi S, Kebbouche-Gana S, Akmoussi-Toumi S, Angar Y, Gana ML (2015) Isolation of an extremely halophilic arhaeon *Natrialba* sp. C21 able to degrade aromatic compounds and to produce stable biosurfactant at high salinity. Extremophiles 19(6):1109–1120

Kim BH, Ramanan R, Cho DH, Oh HM, Kim H (2014) Role of *Rhizobium*, a plant growth promoting bacterium, in enhancing algal biomass through mutualistic interaction. Biomass Bioenergy 69:95–105

Kim E, Lee J, Han G, Hwang S (2018) Comprehensive analysis of microbial communities in full-scale mesophilic and thermophilic anaerobic digesters treating food waste-recycling wastewater. Bioresour Technol 259:442–450

Klekner V, Kosaric N (1992) Degradation of phenols by algae. Environ Technol 13:493–501

Kotresha D, Vidyasagar GM (2017) Phenol degradation in a packed bed reactor by immobilized cells of *Pseudomonas aeruginosa* MTCC 4997. Biocatal Agric Biotechnol 10:386–389

Krastanov A, Alexieva Z, Yemendzhiev H (2013) Microbial degradation of phenol and phenolic derivatives. Eng Life Sci 13(1):76–87

Kües U (2015) Fungal enzymes for environmental management. Curr Opin Biotechnol 33:268–278

Larsdotter K (2006) Wastewater treatment with microalgae – a literature review. Vatten 62:31–38

Lattanzio V, Lattanzio VMT, Cardinali A (2006) Role of phenolics in the resistance mechanisms of plants against fungal pathogens and insects. In: Imperato E (ed) Phytochemistry: advances in research. Research Signpost, Trivandrum, pp 23–67

Le Borgne S, Paniagua D, Vazquez-Duhalt R (2008) Biodegradation of organic pollutants by halophilic bacteria and archaea. J Mol Microbiol Biotechnol 15(2–3):74–92

Lee HC, Lee M, Den W (2015) *Spirulina maxima* for phenol removal: study on its tolerance, biodegradability and phenol-carbon assimilability. Water Air Soil Pollut 226:395

Levén L, Nyberg K, Schnürer A (2012) Conversion of phenols during anaerobic digestion of organic solid waste: a review of important microorganisms and impact of temperature. J Environ Manag 95:99–103

Liao Q, Li L, Chen R, Zhu X (2014) A novel photobioreactor generating the light/dark cycle to improve microalgae cultivation. Bioresour Technol 161:186–191

Liká K, Papadakis IA (2009) Modeling the biodegradation of phenolic compounds by microalgae. J Sea Res 62:135–146

Lima SAC, Castro PML, Morais R (2003) Biodegradation of *p*-nitrophenol by microalgae. J Appl Phycol 15(2–3):137–142

Lima SAC, Filomena M, Raposo J, Castro PML, Morasis RM (2004) Biodegradation of *p*-chlorophenol by microalgae consortium. Water Res 38(1):97–102

Liotta LF, Gruttadauria M, Di Carlo G, Perrini G, Librando V (2009) Heterogeneous catalytic degradation of phenolic substrates: catalysts activity. J Hazard Mater 162(2–3):588–606

Lovell CH, Eriksen NT, Lewitus AJ, Chen YP (2002) Resistance of the marine diatom *Thalassiosira* sp. to toxicity of phenolic compounds. Mar Ecol Prog Ser 229:11–18

Lv Y, Chen Y, Song W, Hu Y (2014) Enhanced selection of micro-aerobic pentachlorophenol degrading granular sludge. J Hazard Mater 280:134–142

Mahdavi H, Prasad V, Liu Y, Ulrich AC (2015) *In situ* biodegradation of naphthenic acids in oil sands tailings pond water using indigenous algae-bacteria consortium. Bioresour Technol 187:97–105

Margesin R, Schiner F (2001) Biodegradation and bioremediation of hydrocarbons in extreme environments. Appl Microbiol Biotechnol 56(5–6):650–663

Martínková L, Kotik M, Marková E, Homolka L (2016) Biodegradation of phenolic compounds by *Basidiomycota* and its phenol oxidases: a review. Chemosphere 149:373–382

Maza-Márquez P, Martínez-Toledo MV, González-López J, Rodelas B, Juárez-Jiménez B, Fenice M (2013) Biodegradation of olive washing wastewater pollutants by highly efficient phenol-degrading strains selected from adapted bacterial community. Int Biodeterior Biodegradation 82:192–198

Maza-Márquez P, Martínez-Toledo MV, Fenice M, Andrade L, Laserrot A, González-López J (2014) Biotreatment of olive washing wastewater by a selected microalgal-bacterial consortium. Int Biodeterior Biodegradation 88:69–76
Maza-Márquez P, González-Martínez A, Martínez-Toledo MV, Fenice M, Laserrot A, González-López J (2017a) Biotreatment of industrial olive washing waster by synergetic association microalgal-bacterial consortia in a photobioreactor. Environ Sci Pollut Res Int 24:527–538
Maza-Márquez P, González-Martínez A, Rodelas B, González-López J (2017b) Full-scale photobioreactor for biotreatment of olive washing water: structure and diversity of the microalgae-bacteria consortium. Bioresour Technol 238:389–398
Meyer JS, Marcus MD, Bergman HL (1984) Inhibitory interactions of aromatic organics during microbial degradation. Environ Sci Technol 3(4):583–587
Michałowicz J, Duda W (2007) Phenols – sources and toxicity. Pol J Environ Stud 16(3):347–362
Min M, Wang L, Li Y, Mohr M, Hu B, Zhou W, Chen P, Ruan R (2011) Cultivating *Chlorella* sp. in a pilot-scale photobioreactor using centrate wastewater for microalgae biomass production and wastewater nutrient removal. Appl Biochem Biotechnol 165(1):123–137
Muñoz R, Guieysse B (2006) Algal-bacterial processes for the treatment of hazardous contaminants: a review. Water Res 40(15):2799–2815
Muñoz R, Köllner C, Guieysse B, Mattiasson B (2003) Salicylate biodegradation by various algal–bacterial consortia under photosynthetic oxygenation. Biotechnol Lett 25(22):1905–1911
Muñoz R, Köllner C, Guieysse B, Mattiasson B (2004) Photosynthetically oxygenated salicylate biodegradation in a continuous stirred tank photobioreactor. Biotechnol Bioeng 87(6):797–803
Muñoz R, Jacinto MSA, Guieysse B, Mattiasson B (2005) Combined carbon and nitrogen removal from acetonitrile using algal-bacterial reactors. Appl Microbiol Biotechnol 67(5):609–707
Muñoz R, Köllner C, Guieysse B (2009) Biofilm photobioreactors for the treatment of industrial waste-waters. J Hazard Mater 161(1):29–34
Nešvera J, Rucká L, Pátek M (2015) Catabolism of phenol and its derivatives in bacteria: genes, their regulation, and use in the biodegradation of toxic pollutants. Adv Appl Microbiol 93:107–160
Niaounakis M, Halvadakis C (2006) Olive processing waste management, Waste Management Series, vol 5. Elsevier, London
Nikel PI, Pérez-Pantoja D, de Lorenzo V (2013) Why are chlorinated pollutants so difficult to degrade aerobically? Redox stress limits 1,3-dichloprop-1-ene metabolism by *Pseudomonas pavonaceae*. Philos Trans R Soc B 368(1616):20120377
Nowak A, Mrozik A (2018) Degradation of 4-chlorophenol and microbial diversity in soil inoculated with single *Pseudomonas* sp. CF600 and *Stenotrophomonas maltophilia* KB2. J Environ Manag 215:216–229
Oesterhelt D, Patzelt H, Kesler B (1998) Decomposition of halogenated hydrocarbons by halophilic bacteria. Patent DE19639894
Oller I, Malato S, Sánchez-Pérez JA (2011) Combination of advanced oxidation processes and biological treatments for wastewater decontamination – a review. Sci Total Environ 409:4141–4166
Osanai T, Park YI, Nakamura Y (2017) Editorial: biotechnology of microalgae, based on molecular biology and biochemistry of eukaryotic algae and cyanobacteria. Front Microbiol 8:118
Oswald WJ (2003) My sixty years in applied algollogy. J Appl Phycol 15:99–106
Oswald WJ, Gotaas HB (1957) Photosynthesis in sewage treatment. Trans Am Soc Civil Eng 122(1):73–97
Otto B, Beuchel C, Liers C, Reisser W, Harms H, Schlosser D (2015) Laccase-like enzyme activities from chlorophycean green algae with potential for bioconversion of phenolic pollutants. FEMS Microbiol Lett 362(11):pii:fnv072
Paraskeva P, Diamadopoulos E (2006) Technologies for olive mill wastewater (OMW) treatment: a review. J Chem Technol Biotechnol 81(9):1475–1485
Passos CT, Michelon M, Burkert JFM, Kalil SJ, Burkert CAV (2010) Biodegradation of phenol by free and encapsulated cells of a new *Aspergillus* sp. isolated from a contaminated site in southern Brazil. Afr J Biotechnol 9(40):6716–6720

Petroutsos D, Katapodis P, Christakopoulos P, Kekos D (2007) Removal of *p*-chlorophenol by the marine microalga *Tetraselmis marina*. J Appl Phycol 19(5):485–490

Pinto G, Pollio A, Previtera L, Temussi F (2002) Biodegradation of phenols by microalgae. Biotechnol Lett 24(24):2047–2051

Pinto G, Pollio A, Previtera L, Stanzione M, Temussi F (2003) Removal of low molecular weight phenols from olive oil mill wastewater using microalgae. Biotechnol Lett 25(19):1657–1659

Plotkin JS (2016) What's new in phenol production? ASC News. Available at: https://www.acs.org/content/acs/en/pressroom/cutting-edge-chemistry/what-s-new-in-phenol-production-.html

Pozo C, Rodelas B, Martínez-Toledo MV, Vílchez R, González-López J (2007) Removal of organic load from olive washing water by an aerated submerged biofilter and profiling of the bacterial community involved in the process. J Microbiol Biotechnol 17(5):784–791

Prasad SBC, Babu RS, Chakrapani R, Ramachandra R, Rao CSV (2010) Kinetics of high concentrated phenol biodegradation by *Acinetobacter baumannii*. Int J Biotechnol Biochem 6(4):609–615

Priac A, Morin-Crini N, Druart C, Gavoille S, Bradu C, Lagarrigue C, Torri G, Winterton P, Crini G (2017) Alkylphenol and alkylphenol polyethoxylates in water and wastewater: a review of options for their elimination. Arab J Chem 10(Supplement 2):S3749–S3773

Qiu YL, Hanada S, Ohashi A, Harada H, Kamagata Y, Sekiguchi Y (2008) *Syntrophorhabdus aromaticivorans* gen. nov., sp. nov., the first cultured anaerobe capable of degrading phenol to acetate in obligate syntrophic associations with a hydrogenotrophic methanogen. Appl Environ Microbiol 74(7):2051–2058

Radjenovic J, Sedlak DL (2015) Challenges and opportunities for electrochemical processes as next-generation technologies for the treatment of contaminated water. Environ Sci Technol 49(19):11292–11302

Rahmanian N, Jafari SM, Galanakis CM (2014) Recovery and removal of phenolic compounds from olive mill wastewater. J Am Oil Chem Soc 91(1):1–18

Ramanan R, Kim BH, Cho DH, Oh HM, Kim HS (2016) Algae-bacteria interactions: evolution, ecology and emerging applications. Biotechnol Adv 34(1):14–29

Rizwan M, Mujtaba G, Memon SA, Leed K, Rashide N (2018) Exploring the potential of microalgae for new biotechnology applications and beyond: a review. Renew Sustain Energy Rev 92:394–404

Rosenkranz F, Cabrol L, Carballa M, Donoso-Bravo A, Cruz L, Ruiz-Filippi G, Chamy R, Lema JM (2013) Relationship between phenol degradation efficiency and microbial community structure in an anaerobic SBR. Water Res 47(17):6739–6749

Rucká L, Nešvera J, Pátek M (2017) Biodegradation of phenol and its derivatives by engineered bacteria: current knowledge and perspectives. World J Microbiol Biotechnol 33(9):174

Ruiz-Dueñas FJ, Martínez AT (2009) Microbial degradation of lignin: how a bulky recalcitrant polymer is efficiently recycled in nature and how we can take advantage of this. Microb Biotechnol 2009 2(2):164–177

Ryu BG, Kim J, Han JI, Yang JW (2017) Feasibility of using a microalgal-bacterial consortium for treatment of toxic coke wastewater with concomitant production of microbial lipids. Bioresour Technol 225:58–66

Safonova E, Kvitko KV, Iankevitch MI, Surgko LF, Afti IA, Reisser W (2004) Biotreatment of industrial wastewater by selected algal-bacterial consortia. Eng Life Sci 4(4):347–353

Schmeling S, Fuchs G (2009) Anaerobic metabolism of phenol in proteobacteria and further studies of phenylphosphate carboxylase. Arch Microbiol 191(12):869–878

Semple KT (1998) Heterotrophic growth on phenolic mixtures by *Ochromonas danica*. Res Microbiol 149(1):65–72

Semple KT, Cain RB (1996) Biodegradation of phenols by the alga *Ochromonas danica*. Appl Environ Microbiol 62(4):1265–1273

Semple KT, Cain RB, Schmidt S (1999) Biodegradation of aromatic compounds by microalgae. FEMS Microbiol Lett 170(2):291–300

Sharma R (2014) Polyphenols in health and disease. In: Watson RR, Preedy VR, Zibadi S (eds) Polyphenols in human health and disease, vol 1. Elsevier, Oxford, pp 757–777

Shashirekha S, Uma L, Subramanian G (1997) Phenol degradation by the marine cyanobacterium *Phormidium valderianum* BDU 30501. J Ind Microbiol Biotechnol 19(2):130–133
Shen YH (2002) Removal of phenol from water by adsorption–flocculation using organobentonite. Water Res 36(5):1107–1114
Singh A, Kumar V, Srivastana JN (2013) Assessment of bioremediation of oil and phenol contents in refinery waste water via bacterial consortium. J Pet Environ Biotechnol 4:3
Sobiesak M (2017) Chemical structure of phenols and its consequence for sorption processes. In: Soto-Hernández M, Palma-Tenango M, García-Mateos MR (eds) Phenolic compounds-natural sources, importance and applications, chapter 1. InTech Open, Croatia, pp 3–26
Stoilova I, Krastanov A, Stanchev V, Daniel D, Gerginova M, Alexieva Z (2006) Biodegradation of high amounts of phenol, catechol, 2,4-dichlorophenol and 2,6-dimethoxyphenol by *Aspergillus awamori* cells. Enzym Microb Technol 39(5):1036–1041
Stoilova I, Krastanov A, Yanakieva I, Kratchanova M, Yemendjiev H (2007) Biodegradation of mixed phenolic compounds by *Aspergillus awamori* NRRL 3112. Int Biodeterior Biodegradation 60(4):342–346
Störmer K (1908) Ueber die Wirkung des Schwefelkohlenstoffs und ähnlicher Stoffe auf den Boden. Zentralbl Bakteriol Parasitenkd Infektionskr Hyg Abt 2 Naturwiss 20:282–286
Subashchandrabose SR, Ramakrishnan B, Megharaj M, Venkateswarlu K, Naidu R (2011) Consortia of cyanobacteria/microalgae and bacteria: biotechnological potential. Biotechnol Adv 29:896–907
Subashchandrabose SR, Ramakrishnan B, Megharaj M, Venkateswarlu K, Naidu R (2013) Mixotrophic cyanobacteria and microalgae as distinctive biological agents for organic pollutant degradation. Environ Int 51:59–72
Sueoka K, Satoh H, Onuki M, Mino T (2009) Microorganisms involved in anaerobic phenol degradation in the treatment of synthetic coke-oven wastewater detected by RNA stable-isotope probing. FEMS Microbiol Lett 291(2):169–174
Suthersan S (1996) Remediation engineering: design concepts, Geraghty & Miller Environmental Sciences and Engineering Series. Lewis Publishers, CRC Press, Boca Raton
Tang X, He LY, Tao XQ, Dang Z, Guo CL, Lu GN, Yi XY (2010) Construction of an artificial microalgal-bacterial consortium that efficiently degrades crude oil. J Hazard Mater 181(1–3):1158–1162
Tarlan E, Dilek FB, Yetis U (2002) Effectiveness of algae in the treatment of a wood-based pulp and paper industry wastewater. Bioresour Technol 84(1):1–5
Terzyk AP (2007) The impact of carbon surface chemical composition on the adsorption of phenol determined at the real oxic and anoxic conditions. Appl Surf Sci 253(13):5752–5755
Thomas S, Sarfaraz S, Misharaz LC, Iyengar L (2002) Degradation of phenol and phenolic compounds by a defined denitrifying bacterial culture. World J Microbiol Biotechnol 18(1):57–63
Throop WM (1975) Alternative methods of phenol wastewater control. J Hazard Mater 1(4):319–329
Tikoo V, Scragg AH, Shales SW (1997) Degradation of pentachlorophenol by microalgae. J Chem Technol Biotechnol 68(4):425–431
Tschech A, Fuchs G (1987) Anaerobic degradation of phenol by pure cultures of newly isolated denitrifying pseudomonads. Arch Microbiol 148(3):213–217
Varma RJ, Gaikwad BG (2008) Rapid and high biodegradation of phenols catalyzed by *Candida tropicalis* NCIM 3556 cells. Enzym Microb Technol 43(6):431–435
Vasconcelos-Fernandes T, Shrestha R, Sui Y, Papini G, Zeeman G, Vet LEM, Wijffels RH, Lamers P (2015) Closing domestic nutrient cycles using microalgae. Environ Sci Technol 49(20):12450–12456
Veeresh GS, Kumar P, Mehrotra I (2005) Treatment of phenol and cresols in upflow anaerobic sludge blanket (UASB) process: a review. Water Res 39(1):154–170
Vermerris W, Nicholson R (2008) Phenolic compound biochemistry. Springer Science + Business Media B.V., Dordrecht

Viggiani A, Olivieri G, Siani L, Di Donato A, Marzocchella A, Salatino P, Barbieri P, Galli E (2006) An airlift biofilm reactor for the biodegradation of phenol by *Pseudomonas stutzeri* OX1. J Biotechnol 123(4):464–477

Wang H, Liu L, Liu ZP, Qin S (2010) Investigations of the characteristics and mode of action of an algalytic bacterium isolated from Tai Lake. J Appl Phycol 22(4):473–478

Wang Y, Song J, Zhao W, He X, Chen J, Xiao M (2011) In situ degradation of phenol and promotion of plant growth in contaminated environments by a single *Pseudomonas aeruginosa* strain. J Hazard Mater 192(1):354–360

Wang B, Lan CQ, Horsman M (2012) Closed photobioreactors for production of microalgal biomasses. Biotechnol Adv 30(4):904–912

Wang L, Xue C, Wang L, Zhao Q, Wei W, Sun Y (2016) Strain improvement of *Chlorella* sp. for phenol biodegradation by adaptive laboratory evolution. Bioresour Technol 205:264–268

Wang R, Diao P, Chen Q, Wu H, Xu N, Duan S (2017) Identification of novel pathways for biodegradation of bisphenol A by the green alga *Desmodesmus* sp. WR1, combined with mechanistic analysis at the transcriptome level. Chem Eng J 321:424–431

Watanabe K, Takihana H, Aoyagi H, Hanada S, Watanabe Y, Ohmura N, Saiki H, Tanaka H (2005) Symbiotic association in *Chlorella* culture. FEMS Microbiol Ecol 51(2):187–196

Wu Y, Hu Z, Yang L, Graham B, Kerr PG (2011) The removal of nutrients from non-point source wastewater by a hybrid bioreactor. Bioresour Technol 102(3):2419–2426

Wurster M, Mundt S, Hammer E, Schauer F, Lindequist U (2003) Extracellular degradation of phenol by the cyanobacterium *Synechococcus* PCC 7002. J Appl Phycol 15(2–3):171–176

Xiong JQ, Kurade MB, Jeon BH (2017) Can microalgae remove pharmaceutical contaminants from water? Trends Biotechnol 36(1):30–44

Yang S, Wu RSS, Kong RYC (2002) Biodegradation and enzymatic responses in the marine diatom *Skeletonema costatum* upon exposure to 2,4-dichlorophenol. Aquat Toxicol 59(3–4):191–200

Yordanova G, Godjevargova T, Nenkova R, Ivanova D (2013) Biodegradation of phenol and phenolic derivatives by a mixture of immobilized cells of *Aspergillus awamori* and *Trichosporon cutaneum*. Biotechnol Biotechnol Equip 27(2):3681–3688

Zhang B, Lens PNL, Shi W, Zhang R, Zhang Z, Guo Y, Bao X, Cui F (2018) Enhancement of aerobic granulation and nutrient removal by an algal–bacterial consortium in a lab-scale photobioreactor. Chem Eng J 334:2373–2382

Zhong W, Wang D, Xu X (2012) Phenol removal efficiencies of sewage treatment processes and ecological risks associated with phenols in effluents. J Hazard Mater 217–218:286–292

# Chapter 9
# Fungal Technology Applied to Distillery Effluent Treatment

**Luciana Melisa Del Gobbo and Verónica L. Colin**

## Contents



## 9.1 Sugarcane Vinasse: Physicochemical Characteristics and Conventional Management Practices

The Northwest of Argentina has numerous sugar industries, and about 99.5% of the total sugar production is concentrated in the provinces of Tucumán, Salta, and Jujuy. Most of the sugar mills coexist with distilleries (integrated sugarcane factories) and utilize the molasses or cane juice from cane sugar manufacturing as the starting material for bioethanol production. The liquid fraction generated from rectification and distillation operations of bioethanol, known as vinasse, is a dark-colored acid effluent with an unpleasant odor. Vinasse itself is not a hazardous waste (EPA 2016), but because of its complex composition, it is considered potentially dangerous. Certainly, distilleries are one of the most polluting industries since more than 80% of its raw materials are converted into waste and discharged into the water bodies

L. M. Del Gobbo
Facultad de Bioquímica, Química y Farmacia, Universidad Nacional de Tucumán, Tucumán, Argentina

V. L. Colin (✉)
Planta Piloto de Procesos Industriales Microbiológicos (PROIMI-CONICET), Tucumán, Argentina



R. Prasad, E. Aranda (eds.), *Approaches in Bioremediation*, Nanotechnology in the Life Sciences, https://doi.org/10.1007/978-3-030-02369-0_9

and soils, which induce undesirable physical, chemical, and biological changes to the environment (Colin et al. 2018).

Vinasse composition depends on the feedstock used, e.g., sugarcane in tropical areas (Argentina, Brazil, Colombia) and corn in areas such as the United States, the European Union, and China. The fermentation/distillation conditions also affect the final effluent composition (Vohra et al. 2014; Moran-Salazar et al. 2016). However, all vinasses share some similar properties such as a low pH (3.5–5.0), high concentrations of salts, and a high chemical oxygen demand (COD) and biochemical oxygen demand (BOD) (Rodrigues Reis and Hu 2017). According to available literature, lactic acid, acetic acid, glycerol, and ethanol are the major organic components of sugarcane vinasse (España-Gamboa et al. 2011). Potassium and recalcitrant colored organic compounds (phenolics from the feedstock and melanoidins) are also predominant constituents (España-Gamboa et al. 2011; Rajasundari and Murugesan 2011).

A traditional plant generates between 10 and 15 L of vinasse for each liter of alcohol produced (Cavalett et al. 2012). It has been projected a bioethanol production of 1.12 billion liters in Argentina for 2018 (Kenneth 2017), which is equivalent to an annual production of several billions of liters of vinasse. Although our province (Tucumán, Argentine) has achieved a substantial improvement with regard to the vinasse spills onto watercourses, several liters of effluent are accumulated in open-pit pools to the limit of their capacity. Hence, the accumulation of large volumes of vinasse remains a key bottleneck for production of bioethanol. Table 9.1 summarizes some potential effects of the distillery wastes on the environment, according to EPA Guidelines (2004).

A variety of physicochemical and microbiological technologies is continually evaluated to mitigate the environmental impact of vinasse. A schematic view of the conventional practices for vinasse management applied in the Northwest of Argentina is shown in Fig. 9.1. Agricultural use of vinasse is a frequent practice especially in Latin American countries since this residue contains all essential elements for crop growth (Fig. 9.1a). In order to avoid detrimental effects on the soil, this practice requires the previous dilution of the effluent, which has as the main disadvantage the consumption of water. However, the use of vinasse in a crude form has been documented as a promising practice for the recovery of unproductive saline soils (Mornadini and Quaia 2013). Alternatively, there exist different concentration methods (natural, thermal or by using membranes) to reduce the vinasse volume before effluent can be used as fertilizer (Fig. 9.1b).

Biological conditioning of vinasse has been reported as a most eco-friendly practice (Colin et al. 2016; Pires et al. 2016; Nair and Taherzadeh 2016). Certain studies have highlighted a possible recycling of the treated microbiologically effluent as irrigation water or fertilizer (Nitayavardhana et al. 2013; Colin et al. 2018) (Fig. 9.1c). Because of its high organic load, vinasse can also constitute a cheap feedstock to obtain a variety of bioproducts by microbial fermentation (Fig. 9.1c).

Among microbial technologies, fungal technology has made great contributions to the treatment of the distillery effluents since fungi possess an extraordinary ability to digest complex waste materials. In fact, it have been documented a diversity

**Table 9.1** Potential environmental impacts of distillery wastes

| Distillery waste constituent | Indicators | Effect |
|---|---|---|
| Organic matter | Total organic carbon, BOD, COD | Depletes oxygen when discharged into water causing hazardous conditions for aquatic life. Disagreeable odors generated by anaerobic decomposition |
| Alkaline or acidic | pH | Death of aquatic organisms at extreme pH values – affects the solubility of heavy metals in the soil and availability and/or toxicity in waters – affects crops growth |
| Nutrients | Nitrogen, phosphorous, potassium | Toxic to crops in large amounts – eutrophication when is discharged to water or stored in lagoons – presence of nitrate and nitrite in drinking water supply can be toxic to infants |
| Salinity | Electrical conductivity, total dissolved solids | Imparts undesirable taste to water – toxic to aquatic organisms – affects water uptake by crops |
| Heavy metals | Cadmium, chromium, cobalt, copper, nickel, lead, zinc, mercury | Toxic to plants and animals – neurotoxicity |
| Solids | Total suspended solids | Reduce soil porosity, reducing oxygen uptake – reduce light transmission in water, compromising ecosystem health |

Adapted from Kharayat (2012)
*BOD* biochemical oxygen demand, *COD* chemical oxygen demand

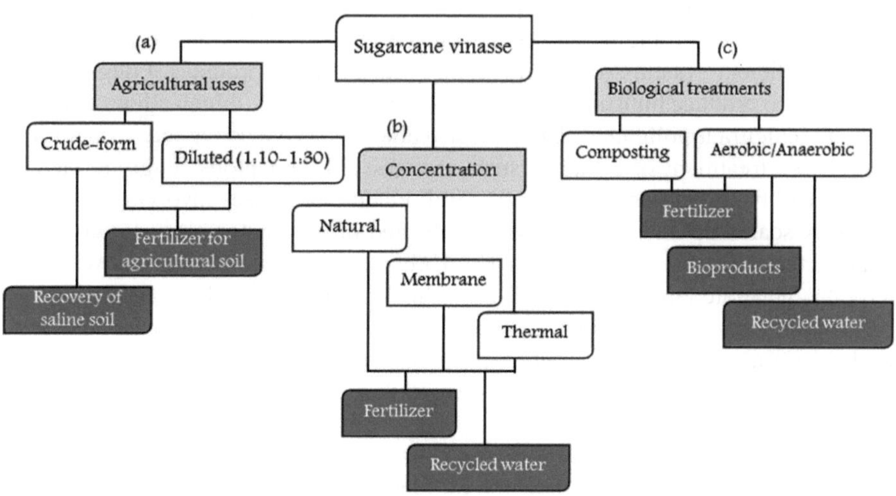


**Fig. 9.1** Conventional practices for management of sugarcane vinasse in the Northwest of Argentina: (**a**) Agricultural practices, (**b**) Concentration methods applied, and (**c**) Biological treatments available

of fungal strains with the ability to improve the vinasse quality (Romanholo Ferreira et al. 2011; Baldiris et al. 2012; Tapia-Tusell et al. 2015; España-Gamboa et al. 2015, 2017; Vilar et al. 2018).

The next sections provide an overview of the biological treatments of sugarcane vinasse carried out by filamentous fungi. In addition, the first advances of the potential of an indigenous fungus strain to degrade a local sugar cane vinasse sample are reported.

## 9.2 Fungal Technology Applied to the Treatment of Distillery Effluents: Basic Principles

Biological treatment of distillery waste often faces problems in meeting wastewater discharge standards. Hence, a variety of processes, both aerobic and anaerobic, has been examined in order to improve the quality of these effluents. Whether a treatment for the valorization of an effluent by microbial pathways is the most appropriate not only depends on the possibility to reduce its toxicity but also on the possibility to obtain valuable bioproducts with market acceptance (Colin et al. 2016). For example, anaerobic digestion of vinasse has been described as an effective and economical option as it eliminates the COD and it converts vinasse into methane (biogas), a green and renewable fuel readily usable in ethanol facilities. This process is carried out in four stages (hydrolysis, acidogenesis, acetogenesis, and methanogenesis), and each stage involves different bacterial genera (Kondusamy and Kalamdhad 2014). However, the presence of recalcitrant compounds in vinasse (phenols, melanoidins, sugar decomposition products, etc.) can be toxic to many anaerobic microorganisms and inhibit degradation processes.

Fungus-based aerobic processes have made great contributions to the development of efficient systems for treatment of complex waste products. Fungi possess the ability to produce high amounts of non-specific enzymes, which allows degradation of recalcitrant pollutants. They are also able to survive under acidic conditions that enable treatment of effluents such as vinasse. Filamentous fungi are easy to culture and handle at laboratory scale. Additionally, processes involving fungi can be easily scaled up and require less complicated processes for biomass separation than bacterial processes (Ferreira 2015). Unlike conventional treatments with bacteria that generate large amounts of low-value biomass, fungal biomass has a great biotechnological potential since it has a relatively high protein content (around 50% of dry weight). Therefore, the purpose of many fungal processes is to produce biomass for food/feed applications. In respect to this, Sartori et al. (2015) reported on four fungal species (*Pleurotus sajor-caju*, *Pleurotus albidus*, *Pleurotus ostreatus*, and *Pleurotus flabellatus*), whose dry mycelia obtained from vinasse are a possible dietary supplement in the fish industry. Similarly, Nair and Taherzadeh (2016) state that the use of fungal biomass cultured in vinasse can produce additional revenues to ethanol plants, because they provide feed and feed supplements for the production of livestock.

In addition to the biomass itself, certain high value-added fungal products could be recovered from vinasse-based supernatants. For example, Aguiar et al. (2010) reported the production of lignocellulolytic enzymes (endoglucanase, exoglucanase, laccase, manganese peroxidase, and peroxidase) by fungal strains such as *Pleurotus sajor-caju* CCB020, *Pleurotus ostreatus*, and *Trichoderma reesei*, grown on sugarcane bagasse combined with vinasse. These enzymes can be purified and used in diverse biotechnological applications, e.g., in the degradation of lignocellulosic waste materials, the most abundant biomass on the earth's surface. Oliveira et al. (2012) demonstrated the feasibility of *Aspergillus* CCT 4355 to produce organic acids using sugarcane bagasse impregnated with vinasse, as an alternative method to conventional submerged processes. More recently, Dorla et al. (2014) confirmed the production of carotenoid pigments by six different fungal strains (*Phycomyces blakesleeanus*, *Mucor circinelloides*, *Gibberella fujikuroi*, *Neurospora crassa*, *Aspergillus carbonarius*, and *Aspergillus giganteus*) cultured on vinasse agar. These natural pigments are found in fruits and vegetables, and they are known for their antioxidant and pro-vitamin A activity, properties that can be used to fight certain diseases and/or to delay their appearance.

Despite these advantages, the morphological complexity of filamentous fungi is a limiting factor in fungal fermentations. When fungi give rise to dispersed mycelial growth, common operational difficulties such as an increase in the medium viscosity, a decrease in oxygen supply, or wrapping of the dispersed filaments around agitators can occur. The morphological factor is a common bottleneck in industrial processes. Fungal development in pelleted form often eliminates these problems and facilitates downstream processing (Colin et al. 2013). Fungal pellets are spherical, ellipsoidal, or oval masses of intertwined hyphae whose size usually range from several hundreds of μm to several mm (Fig. 9.2). The characteristics of these pellets, including form, size, and fluffiness, depend on each particular strain as well as the culture conditions (Colin et al. 2013; Espinosa-Ortiz et al. 2016).

Certain pollutants have demonstrated higher removal efficiency with fungal pellets instead of other forms of fungal growth (Espinosa-Ortiz et al. 2016). However, one of the main issues in reactors operated with fungal pellets consists of maintaining the viability of these structures during long-term reactor processes. According to

**Fig. 9.2** Pellets of a fungal strain developed under different culture conditions after 96 h of incubation. Bar represents 1 mm

their size or fluffiness, large pellet structures can suffer diffusional limitation of oxygen and other nutrients and induce autolytic processes within big pellets. Therefore, small fluffy pellets are often assumed to be more suitable for high-performance fermentations (Colin et al. 2013). Despite these limitations, there exist a considerable number of studies in the literature concerning the treatment of sugarcane vinasse using filamentous fungi as biological agents. Some of the most valuable reports are considered in the following section.

## 9.3 Fungus-Treated Sugarcane Vinasse

Biodegradation studies conducted by Baldiris et al. (2012) evaluated the physicochemical changes in a vinasse sample after inoculation with *Trichoderma viride* or *Schizophyllum commune*, strains isolated at the Universidad Santiago de Cali (Colombia) from contaminated trees. Bioprocesses were conducted by inoculating 8 g of the mycelium of each strain in 100 mL of 5% (v/v) vinasse. Samples were incubated at 25 °C either under static conditions during 120 days or on an orbital shaker (150 rpm) for 20 days. In each case, vinasse samples without inoculation were used as abiotic controls. At the end of the incubation periods, physicochemical parameters of importance were determined in biologically treated supernatants and in abiotic controls. Microbial inoculations demonstrated neutralization of the vinasse, with an average pH of 7.2 ± 0.5. In addition, COD and BOD removal of the treated effluent ranged from 70–75% to 71–78%, respectively. Interestingly, the authors remarked that parameters such as phenol, potassium, and total nitrogen content showed higher removal percentages under static conditions than under shaking conditions, regardless of the strain assayed.

Tapia-Tussell et al. (2015) evaluated the expression of the laccase gene in the presence of phenolic compounds (guaiacol, ferulic acid, and vanillic acid) by *Trametes hirsuta* Bm-2, isolated in Yucatán, Mexico. The authors demonstrated that all compounds assayed had an inductive effect on the level of laccase activity compared with the control, and guaiacol showed the highest induction. Additionally, the effectiveness of *T. hirsuta* Bm-2 to remove phenolic compounds from different sugarcane vinasse concentrations (5%, 10%, 15%, and 20%) was examined using a final working volume of 100 mL. The total phenolic content and the percentage of discoloration were measured during 196 h of incubation on an orbital shaker (130 rpm) at 28 ± 2 °C. The highest discoloration (72.23%) was obtained in 10% (v/v) vinasse after 192 h of cultivation; higher vinasse concentrations showed less discoloration power for the strain. An increase in removal of phenol compounds coincided with a higher laccase activity, suggesting the potential of this strain to decolorize effluents.

España-Gamboa et al. (2015) assayed phenol and color removal of vinasse in an air-pulsed bioreactor using *Trametes versicolor*, a strain collection, provided by the Department of Chemical Engineering (Universidad Autónoma de Barcelona, Barcelona, Spain). This white-rot fungus is an excellent laccase producer, and it is, therefore, used in the degradation of a wide variety of recalcitrant pollutants, including textile dyes and phenols. *T. versicolor* pellets with a diameter of approximately 3 mm obtained in a 2% (w/v) malt extract medium were inoculated in vinasse diluted with distilled water at a 1:10 ratio (500 mL final working volume). The bioreactor was subsequently operated in the continuous mode for a period of 25 days, removing 80% of total phenols, 17% of the color, and until 60% of the COD. The highest laccase activity was recorded at the fifth day of the batch experiment (428 U/L), and a significant decrease was observed on day 7. Vinasse neutralization was detected because of the metabolism of this strain, with a pH value that ranged from 4.5 to 6.9. Recently, the authors provided new information regarding treatment of vinasse by *T. versicolor*, using a fluidized bed bioreactor (FBR) coupled to a UASB (upflow anaerobic sludge blanket) reactor under non-sterile conditions (España-Gamboa et al. 2017). Continuous operation of the FBR alone was carried out for 26 days, with a successful reduction in phenolic compounds (67%) and COD (38%). However, when they coupled the FBR to a UASB reactor, the authors observed a better effluent quality. They also reported methane content of 74% with a yield of 0.18 $m^3$ $CH_4$/kg COD removed. The authors concluded that coupling of an FBR to a UASB reactor could be a promising environmental technology for the treatment of vinasse. However, cost-benefit analysis is required in order to determine the feasibility of this process at full scale.

Diverse organisms can be used to predict the effectiveness of biological treatments applied to industrial waste materials. Romanholo Ferreira et al. (2011), for example, predicted the potential effects of sugarcane vinasse treated with a lignocellulolytic fungus, *Pleurotus sajor-caju* CCB020, using aquatic organisms as toxicological indicators. Firstly, Erlenmeyer flasks containing 100 mL of vinasse with an adjusted pH of 6.0 were inoculated with a standardized amount of fungal mycelium. Then, flasks were incubated for 15 days on an orbital shaker (180 rpm) at 28 ± 2 °C under no light conditions. A vinasse sample without inoculation was used as an abiotic control. Besides, a vinasse sample inoculated with the same amount of heat-killed biomass was used to rule out effluent discoloration produced by adsorption processes. At the end of the incubation period, physicochemical parameters were analyzed both in raw and microbiologically treated vinasse. Analysis of the effluent incubated with *P. sajor-caju* CCB020 revealed a significant reduction in parameters such as phenols (98.2%), total suspended solids (97.6%), phosphate (85.5%), calcium (69.5%), COD (82.7%), and BOD (75.3%). Other parameters analyzed showed less reduction compared to the crude effluent: reducing sugars (34.2%), magnesium (24.2%), sulfate (19.3%), conductivity (18.1%), total dissolved solids (9.0%), and potassium (7.9%). Secondly, populations of four aquatic organisms (*Pseudokirchneriella subcapitata*, *Daphnia magna*, *Daphnia similis*, and *Hydra attenuata*) were exposed to different concentrations of crude and treated vinasse in order to determine the following parameters: growth inhibition concentration ($IC_{50}$), lethal concentration ($LC_{50}$), and effective concentration ($EC_{50}$), all of which correspond to the concentration that causes effect in 50% of the organisms. As expected, the toxicity levels largely depended on the vinasse concentration assayed, and *D. magna* was the least sensitive organism. Only treated vinasse without dilution was extremely toxic to the organisms assayed. However, crude vinasse at concentrations of 12.5%, 25%, and 50% were also highly toxic to the indicator organisms. The authors concluded that *P. sajor-caju* CCB020 could be applied to bioprocessing of vinasse treatment, displaying an effective reduction in the effluent toxicity and improvement in the vinasse physicochemical properties.

More recently, Vilar et al. (2018) evaluated the effectiveness of a combined process for sugarcane vinasse degradation, consisting of a biological treatment with *P. sajor-caju* CCB020 followed by electrochemical oxidation (EO). After 15 days of fermentation, the biological treatment displayed an efficient decrease in color (97%), COD (50.6%), and the reduction in the total organic matter was 57.3%. However, the authors observed that biological treatment followed by EO, using a Ti/($RuO_2$) 0.7 ($IrO_2$)0.1($Sb_2O_3$) 0.2 anode, increased removal of color, and COD until 100%, and 61%, respectively. In order to evaluate the effectiveness of the biological and combined treatment for vinasse degradation, the authors carried out toxicity assays using lettuce seeds (*Lactuca sativa*) and *Raphidocelis subcapitata* microalgae as bioindicators. After exposure of the toxicological indicators to the different concentrations of treated and untreated vinasse, root relative growth (RRG), germination index (GI), and $EC_{50}$ were determined for *L. sativa*, and $IC_{50}$ was calculated for *R. subcapitata*. Based on this assay, they concluded that *R. subcapitata* was more susceptible than *L. sativa* to the toxic effect of vinasse. However, application of the combined treatment allowed a significant reduction in vinasse toxicity, with 0% lethality for either bioindicator.

## 9.4 Autochthonous Vinasse-Degrading Fungus

Previously, our work team conducted experiments using native actinobacteria from our collection of cultures for vinasse treatment (Colin et al. 2016; Aparicio et al. 2017). Four actinobacteria tested (*Streptomyces* sp. MC1, *Streptomyces* sp. A5, *Streptomyces* sp. M7, and *Amycolatopsis tucumanensis* DSM 45259$^T$) were effective in terms of BOD removal from sugarcane vinasse samples. However, bioprocesses required a prior vinasse neutralization to start the bacterial development.

Recently, our research group has isolated an autochthonous fungal strain from a soil from the Northwest of Argentina, named strain V1, which was able to grow under acid conditions (Del Gobbo et al. 2017). Determination of the vinasse-degrading potential of this strain could be an essential step to predict the efficiency of an eventual biological treatment of distillery effluents. Thereby, the current section presents the first advances in the treatment of a sugarcane vinasse sample, by using this autochthonous fungus.

### 9.4.1 Experimental Procedures

An economical solid medium consisting of vinasse only supplemented with 2% agar (agar-vinasse medium, AVM) was successfully used for the production of spores (Fig. 9.3). After 5–7 days of incubation at 30 °C, spores harvested from AVM were inoculated in Erlenmeyer flasks containing 30% (v/v) vinasse without the addition of exogenous nutrients.

To optimize the biodegradation process, two experimental factors, spore concentration (A) and incubation time (B), were evaluated using two levels for each factor (Table 9.2). Consequently, four different bioprocess experiments were conducted in orbital shakers (150 rpm at 30 °C) at a final working volume of 100 mL and without pH control. Non-inoculated vinasse samples were used as abiotic controls.

At the end of the incubation periods, biomass was harvested by centrifugation (10,000 g for 10 min at 4 °C) and washed with distilled water. Then, dry weight was determined at 80 °C in aluminum foil cups until constant weight (Colin et al. 2017). pH and BOD values of the supernatants inoculated with the strain V1 were determined according to the Standard Methods for the Examination of Water and Wastewater (A.P.H.A. AWWA et al. 2012) and compared with the respective abiotic controls. A two-factor full factorial experiment was designed in order to evaluate the main effects of the A and B factors and their interactions on biomass, pH, and BOD removal.

### 9.4.2 Data Interpretation and Statistical Approaches

Although the addition of carbon or nitrogen sources to vinasse is almost always necessary to promote microbial growth, the strain V1 was able to develop without the addition of exogenous nutrients. Fungal growth in the four bioprocesses was consistent with pelleted development, forming very small spherical masses that showed an approximate diameter of 100 μm (Fig. 9.4).

**Fig. 9.3** Fungus spores produced on agar-vinasse plates

**Table 9.2** Experimental factors and their levels

| Factor | Level (−) | Level (+) |
|---|---|---|
| Spore concentration (A) | $1 \times 10^6$ | $8 \times 10^6$ |
| Incubation time (B) | 72 h | 144 h |

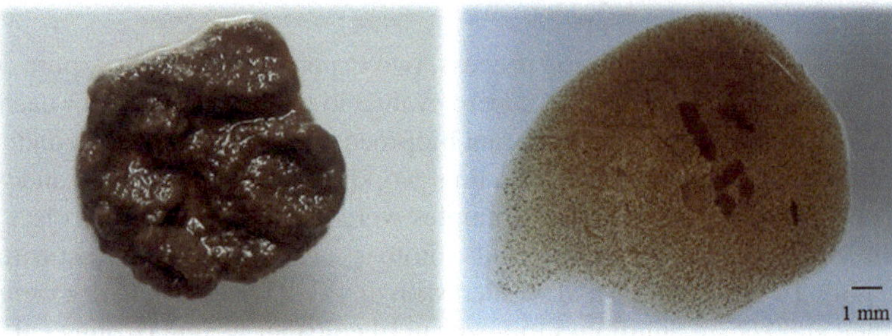


**Fig. 9.4** Pellets of the strain V1 observed at the end of each bioprocess

Experimental results are shown in Table 9.3. Biomass was higher than 1 g/L under all assay conditions. However, regardless of the initial inoculum concentration, processes conducted for 144 h practically doubled the biomass concentration compared to that for 72 h.

At the end of each experiment, the pH of the abiotic controls remained unchanged compared to the initial value (Table 9.3). Bioprocesses conducted for 72 h did not show any significant increase in the pH with respect to the abiotic controls. However, the pH significantly increased in bioprocesses conducted for 144 h, with values close to neutral (Table 9.3). Tiso and Schechte (2015) reported that aerobic neutralization of the pH is caused by the formation of ammonia in a process known as ammonification or mineralization. This process involves protein deamination and decomposition of other nitrogen compounds.

BOD removal mediated by strain V1 was also analyzed. Determinations after 72 h of incubation revealed a removal close to 50%, regardless of the initial spore concentration (Table 9.3). However, bioprocesses conducted for 144 h exhibited a much higher BOD removal, close to 80% (Table 9.3). It is key to remark that the application of Pearson's correlation coefficient confirmed a positive linear association between the biomass concentration and BOD removal ($r = 0.97$; $p = 0.0001$). From these results, it can be inferred that BOD removal would be associated with the growth of the strain V1.

The estimated effects analysis of the factor A and B for biomass, pH, and BOD removal is given in Table 9.4. From this analysis, it can be concluded that the biomass concentration was only affected by the incubation time (factor B). Consequently, spore concentration (factor A) and AB interaction had no significant effect on the growth of the strain V1 ($p > 0.05$). Similarly, the ability of the strain to neutralize vinasse and diminish the BOD was associated with factor B, while factor A and the AB interaction only had negligible effects on these parameters ($p > 0.05$).

**Table 9.3** Matrix for a two-factor full factorial design and experimental results

| | Experimental factors | | Experimental results | | |
|---|---|---|---|---|---|
| Run | A | B | Biomass (g/L) | pH values | BOD removal (%) |
| 1 | – | + | 2.61 ± 0.18[b] | 6.70 ± 0.52[b] | 78.0 ± 1.4[b] |
| 2 | + | – | 1.56 ± 0.07[a] | 4.52 ± 0.16[a] | 48.5 ± 3.5[a] |
| 3 | – | – | 1.44 ± 0.28[a] | 4.68 ± 0.08[a] | 45.0 ± 1.1[a] |
| 4 | + | + | 2.59 ± 0.13[b] | 6.99 ± 0.03[b] | 81.0 ± 1.4[b] |
| Abiotic controls | | | – | 4.19 ± 0.10[a] | – |

A: Spore concentration
B: Incubation time
*BOD* biochemical oxygen demand
Values that sharing the same letter were not significantly different (p > 0.05, Tukey post-test)

**Table 9.4** Estimated effects analysis of factors A and B for biomass, pH, and BOD removal

| | Biomass (g/L) | | pH-values | | BOD removal (%) | |
|---|---|---|---|---|---|---|
| Factors | Effects | *t*-values (*P*-values) | Effects | *t*-values (*P*-values) | Effects | *t*-values (*P*-values) |
| A | 0.050 | 0.39 ($p = 0.718$) | 0.068 | 0.35 ($p = 0.747$) | 3.250 | 2.18 ($p = 0.095$) |
| B | 1.100 | 8.53 ($p = 0.001$) | 2.248 | 11.51 ($p < 0.001$) | 32.750 | 21.97 ($p < 0.001$) |
| AB | −0.075 | −0.58 ($p = 0.592$) | 0.223 | 1.14 ($p = 0.318$) | −0.250 | −0.17 ($p = 0.875$) |
| $R\text{-}Sq_{(adj)}$ (%) | | 90.95 | | 94.93 | | 98.58 |

A: Spore concentration
B: Incubation time
*BOD* biochemical oxygen demand

## 9.5 Concluding Remarks

As mentioned throughout this chapter, fungus-based aerobic processes not only offer a possibility to reduce vinasse toxicity but also to obtain value-added bioproducts. Accordingly, distillery effluents could be considered valuable industrial by-products that can substitute expensive synthetic substrates conventionally used in microbial fermentations.

The present chapter also provides the first advances on the potential of an autochthonous fungus to degrade sugarcane vinasse, at laboratory scale. All conditions assayed presented a satisfactory biomass yield, which could have a valuable biotechnological potential for food purposes. In addition, it has been demonstrated the ability of this strain to neutralize vinasse and remove close to 80% of the BOD after only 144 h of treatment. Currently, complementary studies are being conducted in order to validate our findings at the pilot scale.

**Acknowledgments** The present study has been supported by Agencia Nacional de Promoción Científica y Tecnológica (ANPCyT) (PICT 2015 N° 0292) and CONICET.

## References

Aguiar MM, Romanholo Ferreira LF, Rosim Monteriro RT (2010) Use of vinasse and sugarcane bagasse for the production of enzymes by lignocellulolytic fungi. Braz Arch Biol Technol 53(5):1245–1254

Aparicio JD, Benimeli CS, Almeida CA, Polti MA, Colin VL (2017) Integral use of sugarcane vinasse for biomass production of Actinobacteria and their potential application in soil remediation technologies. Chemosphere 181:478–484

APHA, AWWA, WEF (2012) Standard methods for examination of water and wastewater, 22nd edn. American Public Health Association, 1360 pp, Washington, DC. isbn:978-087553-013-0

Baldiris LF, López E, Castillo J, Caicedo LD (2012) Biodegradation of sugarcane vinasse with strains of the fungus *Schyzophyllum commune* and *Trichoderma viride*. Ingenium 6(14):39–46

Cavalett O, Junqueira TL, Dias MOS, Jesus CDF, Mantelatto PE, Cunha MP (2012) Environmental and economic assessment of sugarcane first generation biorefineries in Brazil. Agric Ecosyst Environ 14:399–410

Colin VL, Baigorí MD, Pera LM (2013) Tailoring fungal morphology of *Aspergillus niger* MYA 135 by altering the hyphal morphology and the conidia adhesion capacity: biotechnological applications. AMB Express 3:1–13

Colin VL, Juárez Cortes AA, Aparicio JD, Amoroso MJ (2016) Potential application of a bioemulsifier-producing actinobacterium for treatment of vinasse. Chemosphere 144:842–847

Colin VL, Bourguignon N, Gómez JS, Gianni K, Ferrero MA, Amoroso MJ (2017) Production of surface-active compounds by a hydrocarbon-degrading actinobacterium: presumptive relationship with lipase activity. Water Air Soil Pollut 228:454

Colin VL, Rulli MM, Del Gobbo LM, Amoroso MJ (2018) Potential application of an indigenous actinobacterium to remove heavy metals from sugarcane vinasse. In: Donati RE (ed) Heavy metals in the environments: microorganisms and bioremediation. CRC Press Taylor & Francis Group, Boca Raton, pp 47–58

Del Gobbo LM, Rulli MM, Cuozzo SA, Colin VL (2017) Selection of vinasse-degrading microorganisms. Congreso Argentino de Microbiologia General, SAMIGE-2017

Dorla E, Caro Y, Fouillaud M, Dufossé L, Laurent P (2014) Valorization of vinasse, a rum distillery effluent, by the production of carotenoid pigments using filamentous fungi. Conference: 7th International Congress of Pigments in Food – New technologies towards health, through colors, Novara, Italy, June 18–21 2013, At Novara, Italy

EPA (2016) Resource Conservation and Recovery Act (RCRA) regulations. http://www.epa.gov/rcra/resource-conservation-and-recovery-act

EPA Guidelines for Wineries and Distilleries (2004) South Australian wine industry association environment committee. ISBN 1-876562-66-8

España-Gamboa E, Mijangos-Cortes J, Barahona-Perez L, Dominguez-Maldonado J, Hernandez-Zarate G, Alzate-Gaviria L (2011) Vinasse: characterization and treatments. Waste Manag Res 29(12):1235–1250

España-Gamboa E, Vicent T, Font X, Mijangos-Cortés J, Canto-Canché B, Alzate-Gaviria L (2015) Penhol and color removal in hydrous ethanol vinasse in an air-pulsed bioreactor using *Trametes versicolor*. J Biochem Technol 6(3):982–986

España-Gamboa E, Vicent T, Font X, Dominguez-Maldonado J, Canto-Canché B, Alzate-Gaviria L (2017) Pretreatment of vinasse from the sugar reninery industry under non-sterile conditions by *Trametes versicolor* in a fluidized bed bioreactor and its effect when coupled to an UASB reactor. J Biol Eng 11:6

Espinosa-Ortiz EJ, Rene ER, Pakshirajan K, van Hullebusch ED, Lens PNL (2016) Fungal pellets reactor in wastewater treatment: applications and perspectives. Chem Eng J 283:553–571

Ferreira JA (2015) Integration of filamentous fungi in ethanol dry-mill biorefinery. Ph.D. thesis, Swedish Centre for Resource Recovery, University of Borås, Borås, Sweden

Kenneth J (2017) Global agricultural information network. USDA United States Department of Agriculture Foreign Agricultural Service. pp 1–18

Kharayat Y (2012) Distillery wastewater: bioremediation approaches. J Integr Environ Sci 9(2):69–91
Kondusamy D, Kalamdhad A (2014) Pre-treatment and anaerobic digestion of food waste for high rate methane production – a review. J Environ Chem Eng 2:1821–1830
Moran-Salazar RG, Sanchez-Lizarraga AL, Rodriguez-Campos J, Davila-Vazquez G, Marino-Marmolejo EN, Dendooven L, Contreras-Ramos SM (2016) Utilization of vinasses as soil amendment: consequences and perspectives. SpringerPlus 5(1):1007
Mornadini M, Quaia E (2013) Alternativas para el aprovechamiento de la vinaza como subproducto de la actividad sucroalcoholera. Avanc Agroindustrial 34(2). EEAO-DOSSIER, pp 40–43
Nair RB, Taherzadeh MJ (2016) Valorization of sugar-to-ethanol process waste vinasse: a novel biorefinery approach using edible ascomycetes filamentous fungi. Bioresour Technol 221:469–476
Nitayavardhana S, Issarapayup K, Pavasant P, Khanal SK (2013) Production of protein-rich fungi biomass in an airlift bioreactor using vinasse as substrate. Bioresour Technol 113:301–306
Oliveira AF, Matos VC, Bastos RG (2012) Cultivation of *Aspergillus niger* on sugarcane bagasse with vinasse. Biosci J 28(6):889–894
Pires JF, Ferreira GMR, Reis KC, Schwan RF, Silva CF (2016) Mixed yeasts inocula for simultaneous production of SCP and treatment of vinasse to reduce soil and fresh water pollution. J Environ Manag 182:455–463
Rajasundari K, Murugesan R (2011) Decolourization of distillery waste water – role of microbes and their potential oxidative enzymes (review). J Appl Environ Biol Sci 1:54–68
Rodrigues Reis CE, Hu B (2017) Vinasse from sugarcane ethanol production: better treatment or better utilization? Front Energy Res 5:7
Romanholo Ferreira LF, Aguiar MM, Mesias TG, Pompeu GB, Queijeiro Lopez AM, Silva DP, Monteiro RT (2011) Evaluation of sugar-cane vinasse treated with *Pleurotus sajor-caju* utilizing aquatic organisms as toxicological indicators. Ecotoxicol Environ Saf 74:132–137
Sartori SB, Ferreira LFR, Messias TG, Souza G, Pompeu GB, Monteiro RTR (2015) Pleurotus biomass production on vinasse and its potential use for aquaculture feed. Mycology 6(1):28–34
Tapia-Tusell R, Pérez-Brito D, Torres-Calzada C, Cortéz-Velázquez A, Alzate-Gaviria L, Chablé-Villacís R, Solís-Pereira S (2015) Laccase gene expression and vinasse biodegradation by *Trametes hirsuta* strain Bm-2. Molecules 20:15147–15157
Tiso M, Schechter AN (2015) Nitrate reduction to nitrite, nitric oxide and amonia by gut bacteria under physiological conditions. PLoS One 10(3):e0119712
Vilar DS, Carvalho GO, Pupo MMS, Romanholo Ferreira LF (2018) Vinasse degradation using *Pleurotus sajor-caju* in a combined biological-electrochemical oxidation treatment. Sep Purif Technol 192:287–296
Vohra M, Manwar J, Manmode R, Padgilwar S, Patil S (2014) Bioethanol production: feedstock and current technologies. J Environ Chem Eng 2(1):573–584

# Chapter 10
# Constructed Wetlands to Treat Petroleum Wastewater

**Hassana Ibrahim Mustapha and Piet N. L. Lens**

## Contents



H. I. Mustapha (✉)
UNESCO-IHE, Institute for Water Education, Delft, The Netherlands

Department of Agricultural and Bioresources Engineering, Federal University of Technology Minna, Minna, Nigeria
e-mail: h.i.mustapha@futminna.edu.ng

P. N. L. Lens
UNESCO-IHE, Institute for Water Education, Delft, The Netherlands



R. Prasad, E. Aranda (eds.), *Approaches in Bioremediation*, Nanotechnology in the Life Sciences, https://doi.org/10.1007/978-3-030-02369-0_10

## 10.1 Petroleum Industry

Petroleum is made up of crude oil and natural gas. Crude oil is a complex mixture of hydrocarbons (Merkl et al. 2006; Abu and Dike 2008). Petroleum refining industry converts crude oil into finished products such as transportation fuels (gasoline, diesel fuel, jet fuel, compressed natural gas and propane) (Gousmi et al. 2016), heating fuels (propane, liquefied petroleum gas, kerosene, heating oil and natural gas), sources of electricity (natural gas and residual fuel oil) and petrochemicals (feedstock for plastics, clothing and building materials) (Hamza et al. 2012). Refined products from crude oil comprise a varied range of compounds such as hydrocarbons, heavy metals, dye additives, antioxidants and corrosion inhibitors (Adewuyi and Olowu 2012). Consequently, refined products can show higher toxicity compared to crude oil due to alteration of metal speciation and the metals added to the matrix during the refining processes (Adewuyi and Olowu 2012).

Based on their differential solubility in organic solvents (Abu and Dike 2008), there are four major petroleum compounds (Eke and Scholz 2008), namely, saturated hydrocarbons (which are the primary components), aromatic hydrocarbons, resins (pyridines, quinolones, carbazoles, sulfoxides and amides) and asphaltenes (phenols, fatty acids, ketones, esters and porphyrins) (Eke and Scholz 2008; Das and Chandran 2011). Hydrocarbons originate from natural and man-made sources (Achile and Yillian 2010). They are potentially toxic chemicals, highly soluble and neurotoxic (Ogunfowokan et al. 2003; Eke and Scholz 2008; Achile and Yillian 2010). Due to their toxicity, aromatic hydrocarbons are of serious concern (Ogunfowokan et al. 2003; Eke and Scholz 2008); they are highly soluble and have the ability to readily migrate into groundwater (Eke and Scholz 2008) and the environment (Grove and Stein 2005). Moreover, these contaminants are not easily degraded by conventional treatment (Saien and Shahrezaei 2012).

### *10.1.1 Petroleum Refining Wastewater Types*

Petroleum refining industries extract large volumes of freshwater in the process of refining crude oil and as cooling agent (Allen 2008; Shpiner et al. 2009a; Saien 2010; Nacheva 2011). Consequently, large volumes of wastewater are generated (Nwanyanwu and Abu 2010; Diya'uddeen et al. 2011, 2012; Hamza et al. 2012; Yu et al. 2017) as a result of production, storage, distribution and processing of petroleum or by accidents due to spills from water/fuel mixtures, oil well drilling, leaks from underground storage or water collected from secondary containment and sumps (Saien 2010; Diya'uddeen et al. 2011; Jeon et al. 2011; Ballesteros et al. 2016).

The characteristics of petroleum wastewater from different refineries vary from one region of the globe to another depending on the region in which the crude oil was drilled, type of crude oil, its chemical composition, the different processes and the employed treatment mechanisms (Valderrama et al. 2002; Wake 2005; Nwanyanwu and Abu 2010; Saien 2010; Nacheva 2011). This has significant impact on the character and quantity of contaminants entering a given refinery wastewater treatment system.

High rates of consumption of petroleum and its refined products will continue to generate effluents from petroleum refining processes, which, when discharged into water bodies, would result in environmental pollution (Diya'uddeen et al. 2011; Zhao and Li 2011). The effects of the discharge of effluents include eutrophication, accumulation of toxic compounds in biomass and sediments and loss of dissolved oxygen in water (Paul 2011) contamination of drinking water and groundwater resources, thus endangering aquatic resources and human health as well as destruction of the natural landscape (Yu et al. 2017).

The types of waste generated by the petroleum industry can be classified into oily wastewater, spent catalyst, spent chemicals, sour water and other residuals (Nacheva 2011; Adewuyi and Olowu 2012). These wastes pose major problems that are a challenge faced by the petroleum industry, imposing the need to recover oil as well as to prevent of the discharge of oily wastewater into the environment (Veil et al. 2004). Table 10.1 presents the types of wastewater generated in refining petroleum and their characteristics. These include refinery wastewater, brackish oilfield-produced water, heavy oil-produced water, petroleum hydrocarbon-contaminated sour water, produced water and diesel-contaminated wastewater.

Produced water is the largest volume of wastewater generated by the petroleum industry (Veil et al. 2004; Asatekin and Mayes 2009). It is described as water from an oil well after its separation from oil in American Petroleum Institute (API) separators (Murray-Gulde et al. 2003; Shpiner et al. 2009a). Refinery wastewater is the wastewater generated from refining crude oil and manufacturing fuels, lubricants and petroleum intermediates (Mustapha et al. 2015). Other types of petroleum-contaminated wastewaters may include ballast water from ships, storm water and runoff from roads (Veenstra et al. 1998).

## 10.2 Petroleum Contaminants

Petroleum-contaminated wastewaters contain different types of organic and inorganic pollutants with varying levels of contamination (Table 10.1). Petroleum wastewater is characterized by a range of pollutants (Mustapha et al. 2015) including organics, such as dispersed oil (Shpiner et al. 2009b; Nacheva 2011), oil and grease (Nwanyanwu and Abu 2010; Diya'uddeen et al. 2011) and heavy oil (viscosity >100 mPas) (Ji et al. 2007), polycyclic aromatic hydrocarbons (PAH) (Fountoulakis et al. 2009; Uzoekwe and Oghosanine 2011; Hamza et al. 2012; Tromp et al. 2012), phenols (Agarry et al. 2008; Otokunefor and Obiukwu 2010; Hamza et al. 2012) and inorganics such as ammonia ($NH_3$) (Wallace 2001; Mustapha et al. 2015) and heavy metals (Ali et al. 2005; Jadia and Fulekar 2009; Mustapha et al. 2011, 2018). Some crude oils contain small quantities of metals that may

**Table 10.1** Characteristics of produced petroleum refinery wastewaters (mg/L, all except pH)

| Types of wastewater | Influent concentrations | | | | | | | | | | | | | Reference |
|---|---|---|---|---|---|---|---|---|---|---|---|---|---|---|
| | COD | BOD | O&G | Benzene | Toluene | Other organics | BTEX | MTBE | P | N | TDS | Fe | pH | |
| Refinery wastewater | 232 (±121.2) | 106 (±58.9) | – | – | – | – | – | – | 4.0 (±2.0) | 1.81 (±1.6) | 255.5 (±70.3) | – | 7.3 (±1.6) | Mustapha et al. (2015) |
| | 79–130 | 34–93 | – | 13–32 | – | – | 1.2–5.9 | 1.2–5.9 | 0.1–2.7 | 37–63 | – | – | – | Seeger et al. (2011) |
| | 165–347 | 109–197 | 24–66 | – | – | – | – | – | – | – | 1167–2850 | 4–7.5 | 8.3–10.5 | Aslam et al. (2007) |
| | – | – | – | 13.96 (±20) | – | – | 2.97 (±0.82) | – | 1.20 (±0.75) | 51 (±9.34) | – | 6.73 (±2.36) | 7.45 (±0.35) | van Afferden et al. (2011) |
| | – | – | – | 0.17 | – | – | 0.47 | – | – | – | – | 33.3 | – | Wallace et al. (2011) |
| | 200–220 | – | – | – | – | – | – | – | – | – | – | – | – | Saien and Shahrezaei (2012) |
| | – | – | – | – | 169 | 40 | 1123 | – | – | – | – | – | – | Mazzeo et al. (2010) |
| Oilfield-produced water | 1050–1350 | – | – | – | – | – | – | – | – | – | – | – | – | Shpiner et al. (2009b) |
| Heavy oil-produced water | 390 (±124) | 32 (±8.5) | 20 (±4.6) | – | – | – | – | – | 0.07 (±0.03) | – | – | – | 7.88 (±0.40) | Ji et al. (2007) |
| Oil-refined wastewater | 132–196 | 26–78 | 36–61 | 8.97–9.30 | – | – | – | – | – | 22–29 | – | – | 7.10–7.21 | Xia et al. (2003) |
| Oilfield-produced water | – | 14,693 | 1213 | – | – | – | 12.393 | – | – | – | 43,048 | – | – | Davis et al. (2009) |
| Refinery diesel-range organics | – | 3–64 | – | 0.001–2.090 | <,0.001–1.519 | <0.001–0.522 | 0.127–6.680 | 1.12–1.38 | – | – | 2116–2818 | 4.3–22.0 | – | Bedessem et al. (2007) |
| Petroleum storage tank | – | 16,000 | – | – | – | – | 0.00047 | – | – | 230 | – | 40 | – | Wallace and Davis (2009) |

| | | | | | | | | | | | | | | |
|---|---|---|---|---|---|---|---|---|---|---|---|---|---|---|
| Petroleum-contaminated water | – | – | – | 10.2 ± 3.8 | 0.002 ± 0.001 | 0.019 ± 0.017[1] | 0.88 ± 0.32 | 0.009 ± 0.004 | 1.80 (±0.74) | 27.1 (±8.0) | – | 3.14 (±0.71) | – | Stefanakis et al. (2016) |
| Leaking underground petroleum wastewater | – | – | – | 66 | – | – | – | – | – | – | – | – | – | Ballesteros et al. (2016) |
| Refinery and chemical wastewater | – | – | – | 20 (±2) | – | – | 39 (±0.5) | – | – | – | – | – | – | De Biase et al. (2011) |
| Petroleum hydrocarbon-contaminated water | – | – | – | 0.3 | – | – | – | – | – | – | – | 40 | 8.3 | Wallace and Davis (2009) |
| Hydrocarbon-contaminated wastewater | – | – | – | 1300 | – | – | – | – | – | – | – | – | – | Eke and Scholz (2008) |
| Petroleum refinery wastewater | – | – | – | – | 169 | 40 | 1123 | – | – | – | – | – | – | Mazzeo et al. (2010) |
| Toxic levels (mg/L) | 40 | 10 | 10 | 0.05 | 0.5 | – | 0.5 | 0.01 | 0.30–1.30[a] | 45[b] | 2000 | 0.1 | 6.0–9.0 | WHO[c] |

[a]Phosphate in drinking water

[b]Nitrate in drinking water quality

[c]Recommended limits in drinking water quality

[1]The value “1” represent ethylbenzene. 0.019 + or - 0.017 represent avaerage and standard deviation for 12 samples collected during the preliminary phase of their study (Stefanakis et al. 2016)

require special equipment for refining the crude. In addition, oil and gas may contain sulphur and carbon dioxide that need to be removed before marketing (Uzoekwe and Oghosanine 2011).

Wastes containing petroleum compounds, nutrients and other toxic compounds should be properly treated prior to discharge into the receiving water bodies (Abdelwahab et al. 2009) because these substances may pose serious hazards to the environment (Diya'uddeen et al. 2011) as well as their immediate damages to the organisms (Brito et al. 2009). The toxicity of petroleum refinery effluent has been reported in many studies (Abdelwahab et al. 2009; Das and Chandran 2011; Diya'uddeen et al. 2011). However, the toxicity depends on a number of factors, including quantity, volume and variability of discharge (Nwanyanwu and Abu 2010).

### 10.2.1 *Organic Pollutants*

The discharge of wastewater with a high organic matter content into the aquatic environment results in the depletion of oxygen (Diya'uddeen et al. 2011). Organic pollutants produced by industrial activity such as BTEX (benzene, toluene, ethylbenzene and xylene) (Wallace 2001; Mazzeo et al. 2010), PAH and linear alkylbenzene sulfonates (Fountoulakis et al. 2009), chlorinated hydrocarbons (Haberl et al. 2003), benzene and methyl tert-butyl ether (MTBE) (De Biase et al. 2011) have been successfully removed or retained by CW systems (Table 10.1).

For example, Mustapha et al. (2015) analysed secondary refinery wastewater which contained 106 (±58.9) mg/L biological oxygen demand (BOD) and 232 (±121.2) mg/L chemical oxygen demand (COD) in the pretreated influent. Czudar et al. (2011) also analysed petrochemical wastewater with measured BOD and COD in the influent ranging between 42 and 131 mg/L in spring, 42 and 144 mg/L in summer and 32 and 101 mg/L in autumn. The refinery wastewater characterized by Aslam et al. (2007) had concentrations of 109–197 mg/L BOD, 200–258 mg/L COD and 6 mg/L phenol.

There are fewer publications on the treatment of organic pollutants in petroleum refining wastewater with CWs compared with other types of wastewater, i.e. domestic, tannery, textile, abattoir, food processing and agricultural. BOD and COD concentrations are indicators of the level of organic compounds in a wastewater (Nwanyanwu and Abu 2010). Excessive level of BOD/COD in wastewater released into water bodies will reduce the level of dissolved oxygen; low levels of dissolved oxygen can induce fishkills and reduce reproduction rates in aquatic life (Biswas 2013). De Biase et al. (2011) investigated volatile organic compounds in contaminated groundwater next to refineries and chemical plants using vertical-flow filters and vertical-flow CWs. Their influents contained 20 (±2) mg/L of benzene and 39.0 (±0.5) mg/L of MTBE.

Petroleum refinery wastewater contains aliphatic and aromatic petroleum hydrocarbons at different concentrations (Saien and Shahrezaei 2012). The analysis of pretreated refinery wastewater samples collected at the inlet of the biological treatment unit in the Kermanshah (Iran) refinery plant by Saien and Shahrezaei (2012) detected COD in the range of 200–220 mg/L, which consisted of methyl-tetrabutyl ether, phenol, 2,3,5,6-tetramethylphenol, naphthalene, xylene, tetradecane, 4-chloro-3-methylphenol and 3-*tert*-butylphenol.

Oil and grease (O&G) clog drain pipes and sewer liners, causing unpleasant odours, and also corrode sewer lines under anaerobic conditions (Diya'uddeen et al. 2011). In addition, O&G in wastewater can cause depletion of dissolved oxygen and loss of biodiversity in the receiving water bodies (Mohammed et al. 2013). PAHs are highly toxic, carcinogenic and mutagenic to microorganisms, organisms and humans (Fountoulakis et al. 2009; Zheng et al. 2013). Phenolic compounds are a serious problem due to their poor biodegradability and high toxicity (Abdelwahab et al. 2009). These compounds are harmful to organisms at concentrations as low as 0.05 mg/L (Nwanyanwu and Abu 2010) and carcinogenic to humans (Abdelwahab et al. 2009; El-Ashtoukhy et al. 2013; Zheng et al. 2013). In addition, phenol can reduce the growth and the reproductive capacity of aquatic organisms (Zheng et al. 2013).

### *10.2.2 Heavy Metals*

Metal contamination is a major environmental problem, especially in the aquatic environment (Chorom et al. 2012; Ho et al. 2012; Papaevangelou et al. 2017). Unlike organic pollutants, metals in wastewater are not degraded through biological processes (Yang et al. 2006). Many heavy metals are toxic both in elemental and soluble form (Jadia and Fulekar 2009). The most toxic metals are Cd, Pb, Hg, Ag and As (Hashim et al. 2011). Several authors have reported the presence of heavy metals in petroleum-contaminated wastewater (Gillespie et al. 2000; Moneke and Nwangwu 2011; Mustapha et al. 2018) as well as their hazardous effects (Calheiros et al. 2008; Hashim et al. 2011; Qasaimeh et al. 2015).

Cr, Cd and Pb are non-essential elements to plants and cause toxicity at multiple levels (Calheiros et al. 2008; Mustapha et al. 2018). Cu and Zn are essential elements for organisms (Song et al. 2011); however, these two metals become poisonous at excessive concentrations of 1–2 mg Cu/L and 3–5 mg Zn/L in water bodies (Korsah 2011; Mebrahtu and Zerabruk 2011; Kumar and Puri 2012). In addition, iron (Fe) is important for all forms of life (Jayaweera et al. 2008). However, excessive doses of Fe >1.6 mg/L can lead to haemorrhagic and sloughing of mucosa areas in the stomach of humans (Jayaweera et al. 2008). Ni is an essential trace element for plant growth and also a known human carcinogen at excessive levels (Chorom et al. 2012). Cr (VI) exhibits high toxicity, mobility and water solubility (Papaevangelou et al. 2017) and is carcinogenic (Rezaee et al. 2011). Wastewater that contains metals should be treated prior to discharge into the environment (Gillespie et al. 2000; Cheng et al. 2002; Rezaee et al. 2011).

Plants have the natural ability to uptake metals. The removal of heavy metals from CWs is by microbiota uptake (Khan et al. 2009), plant uptake as well as adsorption onto media and sediments in the system (Qasaimeh et al. 2015). Other processes that can contribute to heavy metal removal from wastewater in CWs are biosorption, bioaccumulation, redox transformation, dissimilatory sulphate reduction (Šíma et al. 2015) and precipitation as insoluble salts (Cheng et al. 2002).

### *10.2.3 Nutrients*

Several authors, such as Huddleston et al. (2000), Moreno et al. (2002), Nwanyanwu and Abu (2010), Moneke and Nwangwu (2011), Uzoekwe and Oghosanine (2011) and Mustapha et al. (2013, 2015), have reported the presence of nitrogen compounds in petroleum-contaminated wastewater. For example, Moreno et al. (2002) reported ammonia concentrations ranging from 3 to 20 mg N/L in oil refinery wastewater, while Huddleston et al. (2000) reported ammonia concentrations ranging from 2.14 to 8.6 mg/L in pretreated petroleum refinery effluent. In addition, nutrients are produced in wastewater due to the degradation of organic matter (Valderrama et al. 2002).

Nitrogen and phosphorus compounds in discharged wastewater may adversely contribute to eutrophication, depletion of oxygen and toxicity to humans, aquatic life and bacteria (Moreno et al. 2002; Abd-El-Haleem et al. 2003; Saeed and Guangzhi 2012) as well as acceleration of the corrosion of metals and construction materials (Taneva 2012). For instance, ammonia is one of the products resulting from purification of oil; 5.0 mg/L of ammonia is the maximum amount allowed in discharged oil refinery effluent according to the Brazilian Environmental Legislation limits (Coneglian et al. 2002). Ammonia creates a large oxygen demand in receiving waters (Abd-El-Haleem et al. 2003), and more than 45 mg/L of nitrate in drinking water may cause methemoglobinaemia, while over 8 mg/L $NO_2^-$ inhibits anoxic phosphate uptake (Abd-El-Haleem et al. 2003).

Phosphorus is essential for plant growth and a limiting factor for vegetative productivity (Rani et al. 2011). However, it is also important to remove phosphorus from wastewater since it is a major limiting nutrient for algae growth in freshwater ecosystems (Rani et al. 2011) and thus results in eutrophication of the receiving water bodies (Mulkerrins et al. 2004).

## 10.3 Constructed Wetlands for Treatment of Petroleum Refining Wastewater

### *10.3.1 Conventional Wastewater Treatment Technologies*

Petroleum-related industries have to comply with strict regulations by regulatory authorities. Besides, the concerns over discharge of produced water and reuse of wastewater are the major challenges facing these industries (Allen 2008; Shpiner et al. 2009a). These have led them to explore many treatment technologies as alternative methods for effluent management. Thus, it has expedited the technological advancements in the field of wastewater treatment in the past three decades.

The state-of-the-art technology for the treatment of petroleum-contaminated wastewater includes reverse osmosis (Murray-Gulde et al. 2003; Mant et al. 2005), membrane filtration techniques (Allen 2008; Ravanchi et al. 2009) such as microfiltration, ultrafiltration and nanofiltration (Saien 2010), natural sorbent zeolites (Mazeikiene et al. 2005), laccase and peroxidase enzymatic treatment and electrocoagulation (Abdelwahab et al. 2009), air floatation, chemical oxidation (Zhao and Li 2011), microbial degradation (Idise et al. 2010) and wastewater stabilization ponds (Shpiner et al. 2009a, b) and lagoons (Abdelwahab et al. 2009). Some of these technologies require high energy and a large capital investment (Baskar 2011). The technologies can lead to incomplete decomposition of contaminants (Das and Chandran 2011). Also, some of these methods are basically transfer of contaminants from one medium to another; and this process may require elimination of organic compounds by another treatment method (Diya'uddeen et al. 2011), or otherwise such systems are only suited for primary treatment of produced water (Shpiner et al. 2009a). Furthermore, the by-products of conventional method are often toxic to both humans and the environment (Ojumu et al. 2005; Wuyep et al. 2007; Baskar 2011). Toxic by-products can destroy the biological component of the soil and can change the chemical and physical soil characteristics (Tam and Wong 2008). Therefore, the by-products require further treatment as well (Prasad 2003).

Other disadvantages of conventional treatment systems include operational difficulties associated with wastewater flow rate and pollutant load (Ayaz 2008), complex procedures, poor performances and high management requirement without oil recovery (Zhao and Li 2011).

### *10.3.2 Constructed Wetland Design*

CWs are an example of phytoremediation (plant-assisted) systems that have been used for many years to treat wastewater (Yang et al. 2006; Dipu et al. 2010; Tromp et al. 2012). They are low cost, have a low energy consumption and low maintenance and are easily operated, and they provide effective treatment (Cheng et al. 2002; Xia et al. 2003; Wallace and Kadlec 2005; Eke and Scholz 2008; Mena et al. 2008; Fountoulakis et al. 2009; Mustapha et al. 2011; Spacil et al. 2011). CWs are chosen to treat many types of wastewater owing to their simplicity, low sludge production, high nutrient absorption capacity, process stability and its potential for creating biodiversity (Rani et al. 2011). In comparison, CWs have the advantages of providing a less intrusive approach than the conventional methods (Lin and Mendelssohn 2009); CWs are environmentally friendly processes where living plants can be considered as a solar-driven pump for the extraction of pollutants from wastewater (Wenzel 2009). CWs are economically viable options for wastewater management, especially for developing countries with limited water resources and

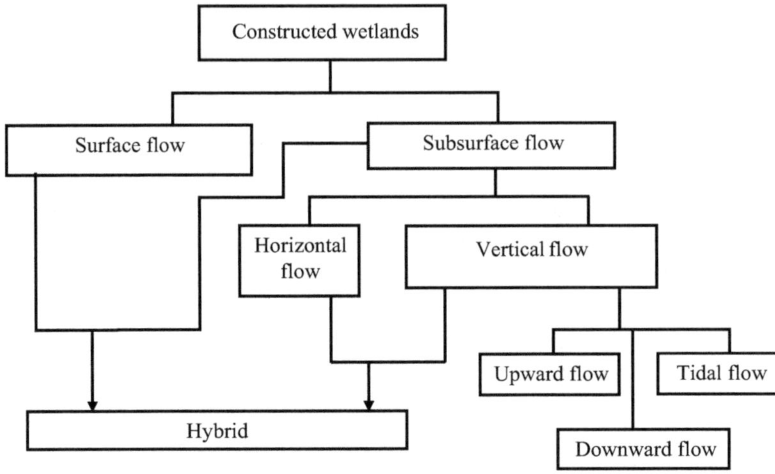


**Fig. 10.1** Classification of constructed wetlands

means (Mustapha et al. 2015). CWs are complex ecosystems (Maine et al. 2007; Fountoulakis et al. 2009) comprising wetland vegetation, hydric soils, microorganisms and prevailing flow patterns that assist in treating wastewater (Haberl et al. 2003; Fountoulakis et al. 2009; Vymazal 2010; Mustapha et al. 2015).

There are two main types of CWs (Fig. 10.1): (i) surface flow wetlands (those with water flowing above the substrate) that mimic natural wetlands (Fig. 10.2a) both in structures and mechanisms and (ii) subsurface flow wetlands (those with wastewater flowing through the gravel bed or porous media). The latter are further grouped into horizontal subsurface flow (HSSF) (Fig. 10.2c) and vertical subsurface flow wetlands (VSSF) (Fig. 10.2c). Based on the flow direction, VSSF CWs can be further categorized into two types: the upward flow type and the downward flow type (Stottmeister et al. 2003; Vymazal 2007) and tidal flow (Babatunde et al. 2008). Besides conventional (surface, subsurface – HSSF and VSSF), also nonconventional (hybrid) CWs have been used to successfully treat various types of contaminants from petroleum-contaminated wastewater (Wallace and Kadlec 2005; Ji et al. 2007; Eke and Scholz 2008; Davis et al. 2009; Aslam et al. 2010).

### *10.3.3 Constructed Wetlands for Petroleum Refinery Wastewater*

The petroleum industry has shown an increasing interest in applying CWs to manage wastewater at various sites, including refineries, oil and gas wells and pumping stations (Ji et al. 2007; Allen 2008). Further, refinery wastewater can be treated with CWs and reused for other purposes rather than the practice of some refineries pumping the water back into the aquifers at high pressure, which is energy intensive and

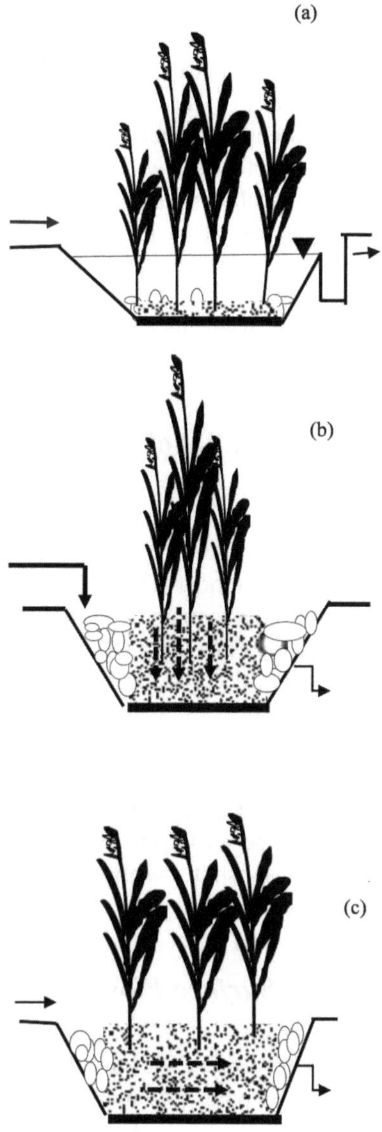


**Fig. 10.2** Diagram of a: (**a**) Surface flow constructed wetland; (**b**) vertical subsurface flow constructed wetland and (**c**) horizontal subsurface flow constructed wetland

expensive (Shpiner et al. 2009a). In addition, CW systems protect resources for future generations by preserving the ecosystem as well as protecting biodiversity (Schröder et al. 2007).

In spite of the potential to use CWs for the treatment of petroleum-contaminated wastewater, literature is scarce (Diya'uddeen et al. 2011) for studies in the tropics. Phytoremediation is favourable in the tropics because of suitable climatic conditions which support plant growth and microbial activity (Merkl et al. 2006). Moreover, countries located in the tropics (e.g. Venezuela, Nigeria and Indonesia) and subtropics (e.g. Saudi Arabia, Kuwait) consist of approximately 80% oil resource (Merkl et al. 2006). In contrast, most of the research on plant-assisted remediation of petroleum-contaminated water is rather conducted in the temperate regions (Merkl et al. 2006). Thus, there is a need for this research to be channelled towards the tropical and subtropical regions.

This chapter reviews literature specifically on the use of CW systems for treatment of petroleum-contaminated wastewater, addressing the efficiency of the treatment systems based on application, as well as organic and inorganic pollutant composition of the influent and effluent, and focusing on the three main components of CWs: macrophytes, microorganisms (found at the petroleum-contaminated sites and in the wastewater) as well as the substrate/filter media used.

## 10.4 Potential of CWs to Treat Petroleum Wastewater

The application of the different types of CWs for wastewater treatment is dependent on the objectives of the treatment; they can be used as either primary or as secondary or tertiary treatment (Haberl et al. 2003). For instance, subsurface flow (SSF) CWs are preferred as a main treatment system (Haberl et al. 2003), while for tertiary treatment, both surface flow (SF) and SSF CWs can be used. In addition, depending on the desired objectives, horizontal (HSSF) or vertical (VSSF) CWs are used. The limited oxygen in HSSF CWs provides good conditions for organic matter and denitrification processes (Jakubaszek and Saecka 2015), while COD and ammonia removal will be higher in the unsaturated VSSF CWs due to the better oxygen transfer capacity through specific design and operational conditions such as intermittent feeding and resting periods (Sgroi et al. 2018), although other mechanisms such as microbial processes, substrate sorption and plant uptake should be considered as well.

### *10.4.1 Surface Flow CWs*

Surface flow (SF) CWs have the ability to filter, absorb and retain particulate matter, nutrients or other pollutants from wastewater (Rani et al. 2011). One difference between SFCWs and subsurface (SSF) CWs is in the release of oxygen. Accordingly, Vymazal (2013) reported that the release of oxygen in SFCWs is insignificant, as most of the treatment processes occur in the water column and within the bottom layer. However, some researchers have used SF CWs to treat petroleum contaminants and have shown this to be a promising technological alternative for wastewater treatment. Horner et al. (2012) used a free water surface (FWS) flow CW planted with *Typha latifolia* operating at a 4-day hydraulic retention times (HRT) under the prevailing conditions at sub-Saharan Africa to treat produced oilfield water containing 0.08–0.40 mg/L Fe, 0.50–1.26 mg/L Mn, 0.37–1.44 mg/L Ni, 2.0–5.0 mg/L Zn and 704–1370 mg/L total dissolved solids and at varied O&G concentrations of 10,

25 and 50 mg/L. The FWS CWs have an ability to remove Fe (0–89.2%), Mn (88.3–98.0%), Ni (23.1–63.2%) and Zn (11.5–84%), while O&G removal was below the detection limit of 1.4 mg/L (Table 10.2). The effluent concentrations of O&G, Fe and Mn met the criteria for irrigation and livestock watering, while Ni concentrations met the livestock watering criteria. Further, Ji et al. (2007) used SF CWs to treat produced wastewater in the Liaohe oilfield (China) using two reed beds (reed bed # 1 and bed # 2) (Table 10.2). There were variations in the removal efficiency of the two reed beds; these differences may be due to the different HRT of 15 and 7.5 days, respectively.

Simi (2000) conducted a study on water quality assessment of a SF CW used for polishing BP Oil's Bulwer Island refinery wastewater in Australia, specifically for reducing the suspended solid loads associated with algal growth in the upstream wastewater treatment system (Table 10.2). The system had low removal rates, which were likely linked to the growth of algae in areas of open water with poor plant establishment. Simi (2000) therefore emphasized the importance of macrophyte selection for system optimization.

## *10.4.2 Subsurface Flow CWs*

Subsurface flow (SSF) CWs are reliable treatment systems with high treatment efficiencies for the removal of pollutants (Hoffmann et al. 2011). Davis et al. (2009) used SF and SSF engineered wetlands to treat petroleum hydrocarbons. They reported that SSF engineered wetlands can achieve more biological treatment due to the higher surface area present in a gravel bed. In operating a SSF CW, there is no contact between the water column and the atmosphere; this is safer from a public health perspective (Rani et al. 2011).

### 10.4.2.1 Vertical Subsurface Flow CWs

Petroleum wastes are recognized to naturally degrade in natural wetland environments (Wallace and Kadlec 2005; Eke and Scholz 2008; Davis et al. 2009; Wallace et al. 2011). Eke and Scholz (2008) studied 12 vertical-flow microcosm wetlands with different compositions to remove benzene (Table 10.2). Their results indicated that the systems had a mean removal efficiency of 85%, which was predominantly due to the volatilization with a 1-day retention time. They have, however, concluded that optimizing the wetlands by locating them in areas of high temperature would enhance the biodegradation rates. Eke and Scholz (2008) and Davis et al. (2009) confirm that aeration enhances volatilization and hydrocarbon degradation.

Mustapha et al. (2015) reported removal efficiencies of 43–85% for total dissolved solids (TDS), turbidity, BOD, COD, ammonium N, nitrate N and phosphate P, respectively, from *Cyperus alternifolius*-planted VSSF CWs and 42–82% for *Cynodon dactylon*-planted VSSF CWs (Table 10.2). *C. alternifolius*- and *C.*

**Table 10.2** Constructed wetlands for petroleum wastewater treatment

| Types of CW | Wastewater type | Range of removal (%) | Location | Reference |
|---|---|---|---|---|
| *Surface flow (SF)* | | | | |
| SF | Leaks underground petroleum storage | Benzene: 48 | Diliman, Quezon City, Philippines | Ballesteros et al. (2016) |
| SF | Oilfield-produced water | Fe, NR-89.2; Mn, 88.3–98; Ni, 23.1–63.2; Zn, 11.5–84; O&G:ND | Clemson, USA | Horner et al. (2012) |
| SF | Heavy oil-produced water | COD, 71–80; BOD, 92–93; TKN, 81–88; TP, 81–86 | Beijing, China | Ji et al. (2007) |
| SF | Refinery wastewater | BOD, 10.24; COD, 16.44; SS, 14.2; TKN, 14.36; $NH_4$-N, 1.32; TP, 13.44; SRP | Australia | Simi (2000) |
| *Vertical subsurface flow (VSSF)* | | | | |
| VSSF CWs | Refinery wastewater | BOD, 68–70; COD, 63–65 $NH_4$+-N, 49–68%; $NO_3$-N, 54–58; $PO_4^3$+P, 42–42 | Kaduna, Nigeria | Mustapha et al. (2015) |
| Vertical-flow soil filter systems-rough filter (RF) | Refinery wastewater | MTBE, 70, benzene, 98 | Leipzig, Germany | van Afferden et al. (2011) |
| Polishing filter (PF) | | MTBE, 99, benzene,100 | | |
| RF + PF (combined) | | MTBE, 100, benzene, 100 | | |
| VSSF CWs | Hydrocarbon-contaminated wastewater | 85–95 | Edinburgh, UK | Eke and Scholz (2008) |
| VSSF CWs | Oil refinery wastewater | Ammonia, 97.7; COD, 78.2; BOD, 91.4; oil, 95.35 | Guangzhou, China | Xia et al. (2003) |
| Compost-based and gravel-based vertical-flow wetlands | Refinery wastewater | TSS, 51–73 and 39–56; COD, 45–78 and 33–61; BOD, 35–83 and 35–69 | Rawalpindi, Pakistan | Aslam et al. (2007) |

| | | | | |
|---|---|---|---|---|
| *Horizontal subsurface flow (HSSF)* | | | | |
| HSSF CWs | Petroleum-contaminated wastewater | MTBE, 49.6–52.8%; benzene, 72.3–82.2%; ammonia, 40% | Leipzig, Germany | Stefanakis et al. (2016) |
| HSSF CWs | Diesel-contaminated wastewater | TPH, 72.5% | Selangor, Malaysia | Al-Baldawi et al. (2014) |
| HSSF CWs: | Groundwater contaminated with benzene and MTBE | (1) Benzene, 0–33%; MTBE, 0–33% | Leipzig, Germany | Chen et al. (2012) |
| (1) Unplanted | | (2) Benzene, 24–100%; MTBE, 16–93% | | |
| (2) Planted | | (3) Benzene, 22–100%; MTBE, 8–93% | | |
| (3) Plant root mat | | | | |
| HSSF CWs | Simulated refinery wastewater – diesel | Cd, 89.9–92.5%, Cr, 82.1–90.7%; Pb, 84.9–90.9% and Zn, 93.8–94.2% | Delft, Netherlands | Mustapha et al. (2011) |
| HSSF CWs: (1) planted gravel filter; (2) plant root mat | Groundwater contaminated | (1): Benzene, 81%, MTBE, 17%, $NH_4^+$-N, 54% (2): benzene, 99%; MTBE, 82%, $NH_4^+$-N, 41% | Leipzig, Germany | Seeger et al. (2011) |
| HSSF CWs | Produced water | To non-detectable concentration at 40 and 80 of the gravel bed length | USA | Davis et al. (2009) |
| | Refinery effluent | Total Zn recoverable, 38%; soluble Zn, 65% | Houston, USA | Gillespie et al. (2000) |
| *Hybrid* | | | | |
| 3 detention basins, oil/water separator, a pair of saltwater wetland cells in series, a reverse osmosis unit and 2 series of four freshwater wetland cells | Four categories of produced water: | *Fresh*: Cd, 25%; Cu, ND; Pb, ND; Zn, 96.3%; Cl, NR | Clemson, USA | Kanagy et al. (2008) |
| | (1) Fresh | *Brackish*: | | |
| | (2) Brackish | Cd, 39%; Cu, 89%; Pb, 93, Zn, 40%; Cl, 12% | | |
| | (3) Saline | *Saline*: | | |
| | (4) Hypersaline | Cd, 99.6%; Cu, 98.8; Pb, 97.7; Zn, 99%; Cl, 99% | | |
| | | *Hypersaline*: | | |
| | | Cd, 99.6%; Cu, >99.9%; Pb, 99.3%; Zn, 99.8%; Cl, 99.5% | | |

(continued)

**Table 10.2** (continued)

| Types of CW | Wastewater type | Range of removal (%) | Location | Reference |
|---|---|---|---|---|
| Hybrid reverse osmosis-constructed wetland treatment system | Brackish oilfield-produced water | Conductivity: 95; TDS 94 | Clemson, USA | Murray-Gulde et al. (2003) |
| Aerated systems | Hydrocarbon-contaminated | BTEX 100%, aniline 94%, nitrobenzene 93%, Fe 98% | Wellsville, New York, USA | Wallace et al. (2011) |
| Upward VF, HF, with aeration, without aeration – forced subsurface aeration | Petroleum refinery-contaminated groundwater | Fe, benzene, MTBE, TPH(DRO), 77%, BOD, TSS, TDS, alkalinity | Laramie, Wyoming, USA | Bedessem et al. (2007) |
| | | Aerated: Fe, total BTEX, 72.6–85.3%; MTBE, TPH(DRO), 94–96%; BOD; TSS; TDS; alkalinity | | |
| Upward vertical subsurface flow wetland system with forced subsurface aeration | Groundwater contaminated with petroleum hydrocarbons | Effluent with benzene concentrations <0.05 mg/L | Wyoming, USA | Ferro et al. (2002) |
| Subsurface flow constructed wetland (SSFCW) with forced bed aeration system built into the wetland bed | High-strength petroleum contact waste | 99% for $CBOD_5$, 98% for ammonia and BTEX was removed at the 40% of the bed length to non-detectable level | South Dakota, USA | Wallace (2001) |

*dactylon*-planted VSSF CWs were capable of treating secondary refinery effluent to discharge permit limits. According to Vymazal (2010), VSSF CWs provide a more effective removal of organics, suspended solids and ammonia, while HSSF CWs provide a higher removal of nitrate through the denitrification processes under anoxic/anaerobic conditions as the concentration of dissolved oxygen in the filtration beds is limited. On average, COD is reduced less effectively than $BOD_5$ in CWs (Baskar 2011), although more COD data are required to confirm this, especially for petrochemical wastewater treatment in wetlands (Knight et al. 1999). Moreno et al. (2002) obtained above 90% removal efficiencies for high ammonia inflow concentrations (>6 mg N/L) from *Phragmites australis*-planted vertical upflow CWs with an HRT of 5 h. VSSF CWs have successfully been employed to remove ammonia in refinery wastewater. For instance, Wallace and Davis (2009) reported 98% efficiency of ammonia reduction after treating wastewater that contained more than 230 mg/L ammonia, while a similar result has also been reported by Xia et al. (2003). Aslam et al. (2010) assessed the viability of treating refinery wastewater using VSSF CWs filled with coarse sand for the removal of heavy metals at a hydraulic loading rate (HLR) of 1.21 $m^3/m^2$ day. They achieved removal efficiencies of 49% for Fe, 53% for Cu and 59% for Zn.

CWs can be intensified with forced aeration to enhance aerobic biodegradation rates of petroleum hydrocarbons in contaminated wastewater to non-detectable levels (Davis et al. 2009). Wallace (2001) used a 1486 $m^2$ SSF CW with a forced bed aeration system built into the wetland bed to treat high-strength petroleum contact waste containing 10,000 mg/L carbonaceous biological oxygen demand ($CBOD_5$), 100 mg/L ammonia and 1000 μg/L total BTEX. The SSF CW treatment system achieved average removal rates of 99% for $CBOD_5$ and 98% for ammonia, and BTEX was removed to non-detectable levels. They reported that BTEX removal was largely due to enhanced volatilization as a result of the aeration system. Similarly, Ferro et al. (2002) built a pilot-scale SSF wetland system at the BP Amoco refinery site in Wyoming (USA) to treat recovered groundwater contaminated with petroleum hydrocarbons with forced subsurface aeration at influent benzene concentrations that ranged from 0.2 to 0.6 mg/L. Their system consisted of four treatment cells packed with sand and operated in an upward vertical-flow mode with a 1-day mean HRT. The aerated upward vertical SSFCW achieved an effluent with benzene concentrations less than 0.05 mg/L. Wallace and Kadlec (2005) constructed an engineered wetland system with aeration which enhanced the volatilization and aerobic biodegradation. These systems were built at pilot scale and later at full scale operating at 6000 $m^3$/day and achieved permit compliance within 1 week after start-up. These studies are significant because they demonstrate successful treatment of petroleum compounds with an inbuilt aeration system (Wallace 2001). Bedessem et al. (2007) used a pilot system consisting of four SSF CWs operated in parallel in different flow modes: upward vertical flow or horizontal flow with or without aeration for the treatment of refinery-affected groundwater. The treatment system effectively removed total petroleum hydrocarbon-diesel range organics (TPH-DRO) (77% without aeration and >95% with aeration), total benzene, toluene, ethylbenzene and *o*-, *m*-, and *p*-xylenes. The systems showed 77% removal for TPH-DRO without aeration and >95% with aeration, 80% for benzene and 88% for total benzene. However, the treatment system was not effective in reducing the MTBE concentrations. Results from this pilot study were instrumental in the development of an aesthetically pleasing full-scale wetland system for volatile organic removal with cascade aeration pretreatment.

#### 10.4.2.2 Horizontal Subsurface Flow CWs

A couple of studies have shown the success of the removal of organic and inorganic contaminants from petroleum wastewater using HSSF CWs (Table 10.2). However, most of the studies were focused on contaminated groundwater (Ferro et al. 2002; Wallace and Davis 2009; Jechalke et al. 2010; Seeger et al. 2011; Wallace et al. 2011; Chen et al. 2012; Stefanakis et al. 2016). Stefanakis et al. (2016) tested three pilot-scale HSSF CWs for the removal of phenolic compounds, MTBE, benzene and ammonia, from contaminated groundwater in a pump-and-treat remediation research facility in Germany. The results showed a complete removal of phenol and *m*-cresol and a removal efficiency that varied from 49.6% to 52.8% for MTBE, 72.3–82.2% for benzene and about 40% for ammonia. Stefanakis et al. (2016) pointed that microbial processes (nitrification, denitrification) dominated the transformation and removal of ammonium in HSSF CWs, while direct plant uptake is of secondary importance in the long term.

Davis et al. (2009) presented three field-scale applications of HSSF engineered wetlands with forced bed aeration in North America for pipeline terminal wastewater containing benzene, toluene, ethylbenzene and xylene (BTEX) and ammonia, along with two former refineries using HSSF engineered wetlands with a designed flow rate of 1.5 $m^3$/day. It treated 6000 $m^3$/day of BTEX and 1060 $m^3$/day Fe from BTEX-contaminated extracted groundwater, respectively. The systems reduced benzene concentrations from 300 to <10 μg/L (to non-detectable concentrations) at 40% and 80% of gravel bed length, respectively. The high rate of removal was due to enhanced volatilization as a result of the aeration system.

Al-Baldawi et al. (2014) investigated the optimum conditions for total petroleum hydrocarbon (TPH) removal from diesel-contaminated water with *Scirpus grossus* planted in HSSF CWs. Three operational variables were investigated, i.e. diesel concentration (0.1, 0.175 and 0.25% Vdiesel/Vwater), aeration rate (0, 1 and 2 L/min) and retention time (14, 43 and 72 days). They reported that the optimum conditions were found to be a diesel concentration of 0.25% (Vdiesel/Vwater), a retention time of 63 days and no aeration with an estimated maximum TPH removal from water of 76.3%. This showed that a longer retention time has a positive effect on the reduction of the TPH concentration in water, although the diesel concentration and aeration rate did not have much effect on the TPH removal efficiency. Chen et al. (2012) evaluated the performance of planted, unplanted and plant root mat pilot-scale HSSF CWs in the decontamination of groundwater polluted with benzene and MTBE. They reported that the plant root mat showed a similar treatment efficiency as the planted HSSF CW for benzene removal and a higher treatment efficiency for

MTBE removal was achieved in summer time. The main removal pathway in this study was oxidative microbial degradation. Ji et al. (2002) demonstrated the use of SSF CWs for heavy oil-produced water (Table 10.2). Thus, effective degradation of organic compounds is majorly by bacterial metabolism of both attached and free-living bacteria under anoxic and/or anaerobic conditions (Vymazal 2010).

Several authors have demonstrated heavy metal removal by CWs, though the application of HSSF CWs for metal removal as the main focus of treatment is rather limited (Kröpfelova et al. 2009). Using aerated and non-aerated HSSF CWs planted with *P. australis*, Mustapha et al. (2011) reported that Cd, Cr, Pb and Zn were removed from simulated refinery wastewater passing through the wetland systems with a 2-day HRT at a hydraulic loading rate of 11 L/day (Table 10.2). There were no large variations in the removal efficiency between the aerated and the non-aerated treatment systems (Mustapha et al. 2011). This may likely be due to the low influent diesel concentrations, and a 2-day retention time may be too short to bring about a high difference between the treatment systems. Gillespie et al. (2000) investigated the transfer and transformation of Zn in a refinery effluent. They used two pilot-scale CWs in parallel consisting of an alluvial flood plain sediment planted with *Scirpus californicus* operated at a 24-h nominal HRT. An average of 38% of the total recoverable and 65% of the soluble Zn was removed from the refinery effluent during the experiment.

### *10.4.3 Hybrid CWs*

Hybrid CWs are nonconventional CW in which either two or more CWs are combined in series. Thus, hybrid CWs provide a better effluent quality than that of single CW systems. In general, hybrid CWs combine either a VF CW at the first stage with a HF CW at the second stage or vice versa to treat effluents in an efficient manner. For instance, they combine their various advantages to compliment processes that produce an effluent lower in BOD and total nitrogen concentration (Vymazal 2005). The use of hybrid CWs (horizontal + vertical flow or vertical + horizontal flow) is an effective wastewater treatment method with reduced water loss potential (Melián et al. 2010).

Wallace et al. (2011) designed hybrid CWs to reduce metals and organic contaminants as well as buffer the pH of the recovered groundwater (Table 10.2). The Wellsville system was also very effective in iron removal, removing 98% of the iron despite relatively high influent concentrations (mean value of 33.3 mg/L). The high performance of the Wellsville wetland shows that CWs are a viable and cost-effective treatment alternative to mechanical treatment, even under cold climate conditions.

Kanagy et al. (2008) designed, built and used a modular pilot-scale CW (freshwater wetland and saltwater wetland) to treat four simulated waters (fresh, blackish, saline and hypersaline waters) representing the range of contaminant concentrations typical of actual produced waters. Freshwater wetland cells planted with

*Schoenoplectus californicus* and *T. latifolia* were used to treat the fresh and brackish waters, while saline and hypersaline waters were treated by saltwater wetland cells planted with *Spartina alterniflora* and by reverse osmosis (RO). Effective removal of cadmium, copper, lead and zinc was achieved by the pilot-scale system. The same authors reported that, although the metal concentrations met the targeted levels immediately following the treatment of saline and hypersaline waters by RO, pH levels were typically too low for discharge. Also, during the flow through the freshwater wetland cells, the pH increased to acceptable levels. For all the four types of gas storage produced waters, freshwater wetland cells improved the performance of the system by increasing the dissolved oxygen concentrations. This is because the freshwater wetland contained quartz sand as hydrosoil that promotes an oxidizing environment and was planted with *T. latifolia*, which has the ability to oxygenate its root zone, thus supporting an oxidizing environment.

Murray-Gulde et al. (2003) considered conductivity, total dissolved solids (TDS) and toxicity as parameters of concern in their study; *Ceriodaphnia dubia* and *Pimephales promelas* were used for the toxicity tests. No significant mortality was observed at 100% exposure to treated produced water when compared to the control organisms. Nonetheless, the system effectively decreased conductivity and TDS by 95% and 94%, respectively.

Plants in CWs can adsorb and accumulate metals. Cheng et al. (2002) used a twin-shaped CW comprising of a vertical-flow (inflow) chamber planted with *C. alternifolius*, followed by a reverse-vertical-flow (outflow) chamber planted with *Villarsia exaltata* to assess the decontamination of artificial wastewater polluted with Cd, Cu, Pb and Zn for over 150 days and with Al and Mn for 114 days. Heavy metals were undetected in the treated effluent with the exception of Mn. The inflow chamber was, therefore, seen as the predominant decontamination step of more toxic metal species with final concentrations far below the WHO drinking water standards. The lateral roots of *C. alternifolius* accumulated more than 4500 times higher amounts of Cu and Mn from the applied influent and 100–2200 times the amounts of the other metals studied.

## 10.5 Removal Pathways in Constructed Wetlands

CWs use natural geochemical, physical and biological processes (Fig. 10.3) in a wetland ecosystem to treat contaminants of concern. The bioremediation of contaminants takes place during the passage of raw or pretreated wastewater through the gravel layer and root zone of the CWs (Babatunde et al. 2008; Kadlec and Wallace 2009). The constituents of concern are removed by various mechanisms such as filtration and sedimentation of suspended particles (Shelef et al. 2013), adsorption to suspended matter, photolysis, volatilization, plant uptake (Zhang et al. 2011) and precipitation by biogeochemical processes (Barber et al. 2001; Stottmeister et al. 2003; Grove and Stein 2005; Farooqi et al. 2008; Shelef et al. 2013). The removal of contaminants occurs by microbial degradation, by physical

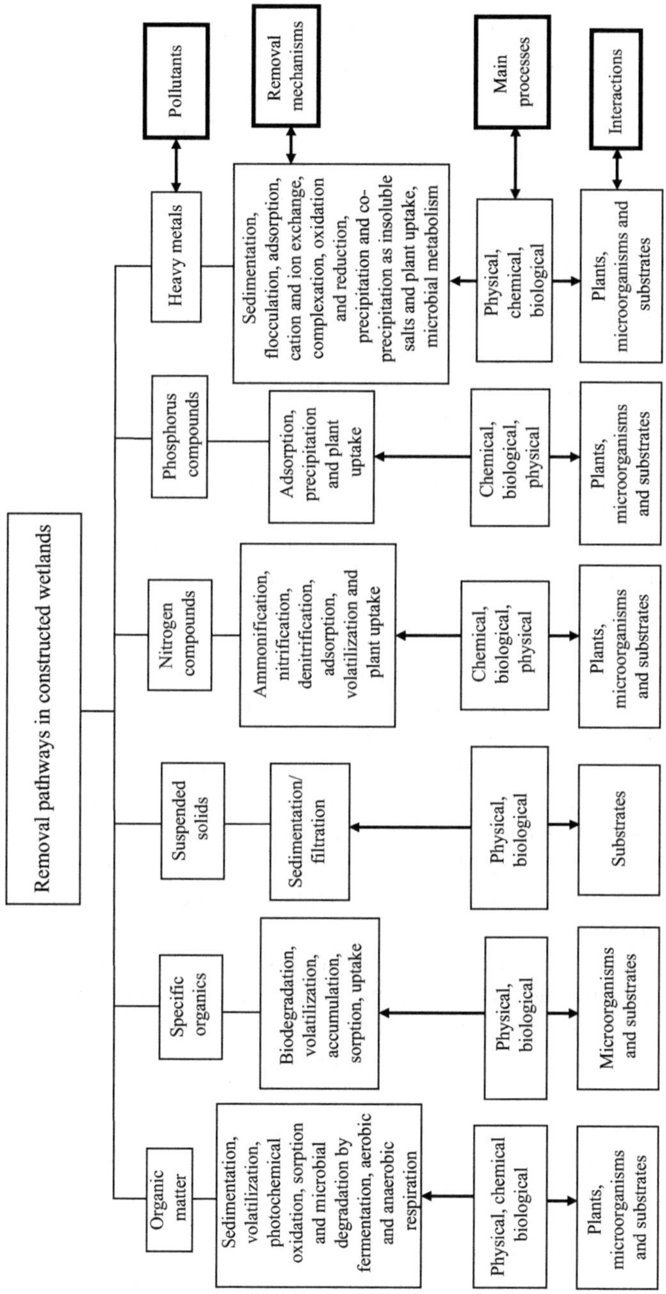


**Fig. 10.3** Removal mechanisms in constructed wetlands for petroleum wastewater treatment

and chemical processes in aerobic, anoxic as well as anaerobic zones (Vymazal and Brĕzinová 2016). The rhizosphere is the active zone where physicochemical and biological processes occur through interactions between plants, microorganisms and substrates to remove pollutants from wastewater (Khan et al. 2009; Saeed and Guangzhi 2012; Papaevangelou et al. 2017). The major removal mechanisms of organic matter are volatilization, photochemical oxidation, sedimentation, sorption and microbial degradation by fermentation, aerobic and anaerobic respiration (Haberl et al. 2003; Czudar et al. 2011). The mechanisms for nitrogen removal in CWs are nitrification, denitrification, plant uptake, volatilization and adsorption (Rani et al. 2011). The major processes responsible for phosphorus removal in SSF CWs are typically adsorption, precipitation and plant uptake (Rani et al. 2011).

Imfeld et al. (2009) reviewed the various mechanisms contributing to organic matter removal and the main degradation pathways for different groups of contaminants. The removal efficiency of the pollutants is influenced by a number of factors, including the substrate media, redox potential, loading rate, retention time, carbon source availability, electron acceptor concentrations, temperature and the plant species (Zhang et al. 2011).

Reddy and D'Angelo (1997) in their study on the biogeochemical processes of wetlands reported that CWs remediate pollutants because they sustain a number of aerobic and anaerobic biogeochemical processes that control the removal and/or retention of pollutants present in wastewater. These natural processes are also employed to treat petroleum refinery wastewaters (Knight et al. 1999; Xia et al. 2003; Ji et al. 2007; Spacil et al. 2011; Mustapha et al. 2015). They showed a good potential as a treatment method capable of removing organic and inorganic pollutants from petroleum refinery effluents (Campagna and da Motta Marques 2000; Cheng et al. 2002; Diya'uddeen et al. 2011; Mustapha et al. 2015, 2018). Yet, the efficiency of pollutant removal mechanisms depends on the hydraulic conductivity of the substrate, types and number of microorganisms, oxygen supply for microorganisms, chemical composition of substrate and hydraulic retention time (Barber et al. 2001; Haberl et al. 2003). For petroleum hydrocarbon removal, design parameters include the biodegradation rate coefficients, flow rate, hydraulic retention time (HRT), influent and required effluent concentrations (Davis et al. 2009).

The main route of heavy metal uptake in aquatic plants is through the roots in the case of emergent and surface floating plants, whereas roots as well as leaves take part in removing heavy metals and nutrients in submerged plants (Dhote and Dixit 2009). Besides, the removal mechanisms in HSSF CWs include binding to sediments and soils through sedimentation, flocculation, adsorption, cation and ion exchange, complexation, oxidation and reduction, precipitation and co-precipitation as insoluble salts and plant uptake and, to a lesser extent, microbial metabolism (Galletti et al. 2010). Numerous factors can affect the remediation processes of contaminated sites, including pH level of water and sediment, mobilization and uptake from the soil, compartmentalization and sequestration within the root, efficiency of xylem loading and transport (transfer factors), distribution between metal sinks in the aerial parts, sequestration and storage in leaf cells and plant growth and transpiration rates (Hadad et al. 2006; Khan et al. 2009).

## 10.6 Components of Constructed Wetland Treatment

CWs are made up of four main components: plants, substrate media, microbial biomass (Dordio and Carvalho 2013) and the aqueous phase (Shelef et al. 2013). The sediment and gravel provide nutrients and support to the root zone and plants (Stottmeister et al. 2003). The root zone is the active reaction zone of CWs, where physicochemical and biological processes are induced by the interaction of the pollutants with the plants, microorganisms and soil particles (Stottmeister et al. 2003). CWs degrade or remove the various pollutants, as a result of the synergetic actions of the system components (Stefanakis et al. 2016). However, the ability of CW to purify wastewater depends on naturally occurring physical, chemical and biological processes that take place within the system.

### *10.6.1 The Macrophyte Component*

Typically, macrophytes are conspicuous components of a wetland (Vymazal 2011). The macrophytes mainly used in SSF CWs are emergent. These are anchored to the substrate with shoots emerging from water to more than 1 m in height. Examples are *P. australis*, *C. alternifolius*, *C. papyrus* and *T. latifolia*. *P. australis* is widely used in CWs all over the world due to their productivity, wide distribution and variable wetland plant species in the world (Lee and Scholz 2007; Vymazal and Brězinová 2016). Surface-floating plants are another type of macrophytes; examples include *Lemna minor* and *Eichhornia crassipes*. The third group is called submerged macrophytes. Examples of this group include water lilies (*Nymphaea* sp.), *Potamogeton* sp., *Naja peclinata* and *Ceratophyllum* (Haberl et al. 2003; Allen 2008; Liu et al. 2008).

#### 10.6.1.1 Role and Macrophytes Used in Treatment Wetlands

Wetland systems support a dense growth of vascular plants adapted to saturated conditions (Campagna and da Motta Morques 2001). These plants are known to degrade, extract, contain or immobilize contaminants in soil and water (Chorom et al. 2012). Vegetation is an essential component of the design of a wetland system (Haberl et al. 2003; Lee and Scholz 2007; Vymazal and Brězinová 2016). The plant roots slow the movement of water, create microenvironments within the water column and provide attachment sites for the microbial community (Haberl et al. 2003; Lee and Scholz 2007).

Macrophytes are assumed to be the main biological component of wetlands (Hadad et al. 2006; Maine et al. 2007). They are important in their role of pollutant removal, nutrient uptake, accumulation of metals (Cheng et al. 2002; Weis and Weis 2004; LeDuc and Terry 2005), transfer of oxygen to the rhizosphere for growth of microorganisms and decomposition of organic matter (Zhang et al. 2010) and renewing the carbon supplies of metabolizing bacteria (O'Sullivan et al. 2004). Other vital roles played by macrophytes in the efficiency of a CW include flow velocity reduction (which aids in settling of particulates and adsorption of solutes), transportation of gases and solutes between above-ground and below-ground biomasses and uptake of inorganic compounds and organic pollutants and influence the microbial diversity and activity (Taylor et al. 2011). Some plant species can also be used for phytoextraction of heavy metals from contaminated water (Cheng et al. 2002). According to Lee and Scholz (2007), macrophytes have a negative impact on wetland management when they lose their leaves in fall; this will increase the BOD concentration due to the release of carbon, nutrients and other pollutants as well as heavy metals in the litter zone.

Macrophytes have the ability to improve the bioremediation process through diffusion of oxygen from the shoots to the roots as well as to the soil, for soil microbes to utilize it for aerobic respiration (Dowty et al. 2001). This, at times, corresponds to as much as 90% of the total oxygen entering a wetland substrate (Allen et al. 2002). Wetland macrophytes have unique characteristics of adaption to anaerobic soil conditions, such as developing internal air spaces (aerenchyma) for supporting supply of oxygen into the root zone (Reddy and D'Angelo 1997). Further, wetland plants have an intrinsic capacity to aerate the rhizosphere, i.e. these plants can transport approximately 90% of the oxygen available in the rhizosphere (Lee and Scholz 2007), thus potentially increasing both aerobic decomposition of organic matter and the growth of nitrifying bacteria (Lee and Scholz 2007; Lin and Mendelssohn 2009).

Wetland plants can incorporate pollutants directly into their tissues and act as catalysts for purification reactions by escalating the environmental diversity in the rhizosphere, thus promoting different types of chemical and biochemical reactions that can improve treatment processes (Maine et al. 2007). Additionally, plants can uptake pollutants (including polyaromatic hydrocarbons) into their rhizosphere to varying extents through the transpiration stream (Weyens et al. 2009). Additionally, all plants have the ability to accumulate essential metals from the soil solution for growth and development. This potential allows plants to also take up other non-essential metals like Al, As, Au, Cd, Hg, Pb, Pt, Sb, Te, T and U, which have no biological function (Jadia and Fulekar 2009). For wetlands constructed to treat petroleum-contaminated wastewater, a number of macrophytes have proven effective in the degradation of contaminants of concern. This is because of their high biomass production, assimilation and long-term storage of organic and inorganic pollutants (Campagna and da Motta Marques 2001) as well as their natural ability to treat wastewater contaminated with oil and grease (Xia et al. 2003; Ji et al. 2007; Davis et al. 2009).

Table 10.3 presents the different types of wetland vegetation used for the treatment of organic and inorganic contaminants found in petroleum refining wastewater. The removal of the contaminants varied from for the different contaminants as well as for the plants (Table 10.3). For natural wastewater treatment systems, plant productivity and pollutant removal efficiency are important in selecting a suitable plant for a given application (Haberl et al. 2003). Accordingly, Madera-Parra et al. (2015)

**Table 10.3** Examples of macrophytes used for petroleum-contaminated wastewater treatment

| Macrophytes | Removal efficiency | Reference |
|---|---|---|
| *Cyperus alternifolius* and *Cynodon dactylon* (L.) Pers. | TDS, 50–54%; BOD, 68–70%; COD, 63–65%; $NH_4^+$-N, 49–68%; $NO_3$-N, 54–58%; and $PO_4^{3-}$-P, 42–43% | Mustapha et al. (2015) |
| *Scirpus grossus* | 72.5% | Al-Baldawi et al. (2014) |
| *Juncus roemerianus* | PAH, 84–100%; *n*-alkanes, 85–99.8% | Lin and Mendelssohn (2009) |
| *Typha latifolia* | Fe, 49%; Cu, 53%; Zn, 59% | Aslam et al. (2010) |
| *Vetiveria zizanioides*, *Phragmites australis*, *Typha latifolia*, *Lepironia articulata* | Ammonia N, 97.7%; COD, 78.2%; BOD, 91.4%; oil, 95.35% | Xia et al. (2003) |
| *Typha latifolia* | BOD, 80%; $NH_3$-N, 95% | Huddleston et al. (2000) |
| *Phragmites australis* | 90% | Moreno et al. (2002) |
| *Phragmites* spp. | COD, 71–80%; BOD, 92–93%; TKN, 81–88%; mineral oil, 81–86% | Ji et al. (2007) |
| *Phragmites karka* | TSS, 51–73% and 39–56%; COD, 45–78% and 33–61%; BOD, 35–83% and 35–69% | Aslam et al. (2007) |
| *Typha latifolia*, *Scirpus californicus* | 95%; 94% | Murray-Gulde et al. (2003) |
| *Phragmites australis* | 10.24% (COD); 16.44% (SS); 14.2% (TKN); 14.36% ($NH_4^+$-N); 1.32% (TP); 13.44% (SRP) | Simi (2000) |
| *Scirpus californicus* | 66% of soluble Zn | Gillespie et al. (2000) |
| *Schoenoplectus californicus* and *Typha latifolia* | Cd, 25–99.6%; Cu, 89% -ND; Pb, 93% – ND; Zn, 40–99.8%; Cl, NR – 99.5% | Kanagy et al. (2008) |
| *Sparganium erectum*, *Scirpus* L, *Typha latifolia*, *Cornus sanguinea* | To non-detectable concentration at 40% and 80% of the gravel bed length | Davis et al. (2009) |

stated that appropriate plant selection is crucial to improve the heavy metal removal efficiency of CWs. Aside for heavy metal removal, Brisson and Chazarenc (2009) are of the opinion that plant species selection is fundamental to the overall pollutant removal efficiency of a CW.

Many studies have reported significantly higher removal efficiencies of pollutants and enhanced transformation of contaminants in planted CWs, compared to unplanted CWs (Tanner 2001; Merkl et al. 2006; Davis et al. 2009; Seeger et al. 2011; Taylor et al. 2011; Vymazal 2011; Noori et al. 2015; Papaevangelou et al. 2017; Cheng et al. 2017; McIntosh et al. 2017). This is attributable to the activity of the microbes in the rhizosphere (Haberl et al. 2003; Seeger et al. 2011; Cheng et al. 2017). Stefanakis et al. (2016) reported the performance for three HSSF CWs,

A and C (planted) and B (unplanted), operating under the same conditions. The planted beds achieved better removal efficiencies than the unplanted bed B with removal efficiencies of 52.8% and 82.2% for bed A, 49.6% and 72.3% for bed C and 41.2% and 66.1% for bed B, for MTBE and benzene, respectively. Similarly, in a study by Mustapha et al. (2015), VSSF CWs planted with *C. alternifolius* and *C. dactylon* were significantly more effective than the unplanted VSSF CWs. The planted VSSF CWs were able to reduce the concentrations of contaminants in the Kaduna refinery effluent to the compliance limits set by the World Health Organization (WHO) and the Federal Environmental Protection Agency (FEPA) of Nigeria. Also, from the results of an investigation conducted by Taylor et al. (2011) with monocultures of 19 plant species, the average COD removal in unplanted wetlands was 70%, while the same parameter of individual species was range from 70% to 97%. Also, differences in transformation of organic compounds were found in the rhizosphere of plants from the sedge and rush families (Cyperaceae and Juncaceae) compared to the grass (Poaceae) family (Taylor et al. 2011).

Merkl et al. (2006) reported that macrophytes can improve microbial degradation by supplying oxygen to the root area along loosened soil aggregates and that remediation of petroleum hydrocarbons is based on the stimulation of microbial degradation in the rhizosphere. Chapelle (1999) also stressed the significance of oxygen as well as nitrogen and phosphorus for the biodegradation of petroleum hydrocarbons. For example, Eke and Scholz (2008) used *P. australis* in their experiment for benzene removal. They reported that *P. australis* does not play an important role in removing benzene, in spite of being given supplementary oxygen through its rhizomes, except nutrients (including fertilizer) are provided. Dowty et al. (2001) further noted that if there is insufficient oxygen, microbial degradation of oil may produce compounds that are toxic to plants (such as hydrogen sulphide) or decrease limiting nutrients (nitrogen) to the extent of inhibiting plant growth.

## 10.7 Microorganisms

### *10.7.1 Microbial Ecology of Petroleum-Degrading Constructed Wetlands*

CW systems support an ideal environment for the growth of microorganisms (Saeed and Guangzhi 2012) which gives the microorganisms' tremendous potential to uptake and degrade pollutants (Wuyep et al. 2007). Microorganisms play the most crucial role in the transformation and mineralization of nutrients and organic pollutants (Dordio and Carvalho 2013). Indigenous microorganisms utilize petroleum contaminants of crude oil as carbon and energy source, thereby breaking down the hydrocarbons into simple non-toxic compounds such as $CO_2$ and $H_2O$ (De-qing et al. 2007).

Bioremediation is to a great extent enhanced by higher temperatures, humidity and soil radiation (Merkl et al. 2006). A diverse microbial community of bacteria, fungi, algae and protozoa present in the aerobic and anaerobic zones of a wetland

(Scholz 2003; McIntosh et al. 2017) is able to degrade volatile organics such as benzene, toluene, ethylbenzene and *p*-xylene (Eke and Scholz 2008; Sepahi et al. 2008; Davis et al. 2009). Dordio and Carvalho (2013) stated that biodegradation of petroleum organic compounds is increased under aerobic condition, while PCBs degrade faster under moderately reduced conditions.

It is assumed that the actual degradation process is performed by microorganisms in the rhizosphere (Merkl et al. 2006). Therefore, their activities are very important to the accomplishment of any treatment process in wetlands (Scholz 2003). Hydrocarbon-utilizing bacteria are present in almost all soil types, and their population will grow if the right feedstock is available (Ayotamuno et al. 2006; McIntosh et al. 2017). In the rhizosphere, microbial diversity, density and activity are more abundant, and this can promote increased phytoremediation activity (Hietala and Roane 2009). The enhanced remediation of contaminants in the rhizosphere is due to high microbial densities and metabolic activities in the rhizosphere, which can be attributable to microbial growth on root exudates and cell debris originating from the plant roots (Weyens et al. 2009).

To improve the biodegradation rate of hydrocarbons, some activities can be considered: (i) addition of nutrients, (ii) watering, (iii) tilling and (iv) bioaugmentation, i.e. addition of suitable microbiota (De-qing et al. 2007; McIntosh et al. 2017). Some scientists have reported that the action of adapted microorganisms depends on the microbial composition, contaminant type, geology of the polluted site and chemical conditions of the contaminated site (Sepahi et al. 2008). Ayotamuno et al. (2006) investigated the remediation of a crude oil spill by combining biostimulation with agricultural fertilizers. They concluded that enhanced degradation of petroleum hydrocarbons can be achieved by the addition of nutrients and oxygen. This was also confirmed by De-qing et al. (2007) and Sepahi et al. (2008). In addition, Fernandez-Luqueno et al. (2011) concluded in their review that indigenous microorganisms have the potential of remediating PAHs from a contaminated soil, although if the soil lacks nutrients, it could hinder microbial activity and consequently the mineralization of these contaminants.

### *10.7.2 Potential of Hydrocarbon-Degrading Microorganisms*

The complex mixture of hydrocarbons in crude oil hinders their complete degradation by a single strain of microorganisms. Therefore, degradation is mostly achieved by microbial consortia and their broad enzymatic capacity (Merkl et al. 2006; Scullion 2006). Examples of such diverse community members capable of utilizing crude oil as a source of carbon and energy include *Pseudomonas putida* and *Flavobacterium* spp. (Huang et al. 2005); *Pseudomonas fluorescens* (Ojumu et al. 2005); *Corynebacterium*, *Micrococcus*, *Acinetobacter* and *Aerococcus* (Ayotamuno et al. 2006); *Staphylococcus*, *Serratia*, *Chromobacterium* and *Alcaligenes* (Abu and Dike 2008); *Pseudomonas aeruginosa* and *Penicillium janthinellum* (Bako et al. 2008); *Mycobacterium parafortuitum* and *Sphingobium yanoikuyae* (Tam and Wong 2008); *Bacillus* spp. (Sepahi et al. 2008); and *Bacillus cereus* (Idise et al. 2010).

Abu and Dike (2008) monitored natural attenuation processes in a model microcosm wetland representing a typical Niger Delta (Nigeria) environment by comparing natural and enhanced processes. A total of 28 bacteria were isolated and identified to the genus level (Table 10.4). These bacteria have been reported by different researchers as crude oil degraders (Ojumu et al. 2005; Bako et al. 2008; Idise et al. 2010). Their results showed a higher prevalence of gram-negative rod forms. They noted that naturally occurring microorganisms in crude oil-impacted sediments can utilize hydrocarbons. However, oxygen was a limiting factor for biodegradation of oil in polluted wetlands.

BTEX compounds are highly soluble and are extremely mobile in groundwater (Jechalke et al. 2010). Benzene is the most toxic and water-soluble BTEX compound, while it can be degraded by many microorganisms under oxic and hypoxic conditions (Jechalke et al. 2010). Table 10.4 summarizes the potential of hydrocarbon-degrading bacteria from the literature.

Ojumu et al. (2005) studied *Pseudomonas aeruginosa* and *Pseudomonas fluorescens* for their bioremediation potential of phenol biodegradation in refinery effluent. Phenol was degraded completely by *P. aeruginosa* and *P. fluorescens* within 60 and 84 h, respectively. Bako et al. (2008) evaluated the potential of *P. aeruginosa* and *P. janthinellum*, and their mutants had effective degradation of crude oil in the water samples taken from River Kaduna (Nigeria) after 2 weeks of incubation at 30 °C. The study of Agarry et al. (2008) revealed a high potential of *P. aeruginosa* NCIB 950 and *P. fluorescens* NCIB 3756, with *P. aeruginosa* being more effective in degrading phenol in refinery effluent. Idise et al. (2010) emphasized the need to stimulate the strains with organic fertilizer to achieve a better outcome. *Bacillus* spp. (*Bacillus* S6 and *Bacillus* S35) isolated from crude oil were able to utilize crude oil as carbon and energy source with increased optical densities and total viable count concomitant with a pH decrease on the fifth day of the experiment (Sepahi et al. 2008). *Pseudomonas* species are also accountable for about 87% of gasoline degradation in contaminated aquifers (Eke and Scholz 2008).

Weyens et al. (2009) stressed the significance of plant-microbe partnerships for a successful removal of organic contaminants. They also demonstrated how plants depend on their associated microorganisms to efficiently remove organic compounds. These associated microorganisms enhance the capability for a stepwise transformation of organic (petroleum) contaminants by consortia and provide an environment that is favourable for genetic exchange and gene rearrangements (Weyens et al. 2009).

## 10.8 Role of the Substrate Media of the Constructed Wetland

The substrate medium is an important component of a CW (Dordio and Carvalho 2013; Papaevangelou et al. 2017). It provides surface area for plant and microbial film growth (Papaevangelou et al. 2017). Furthermore, plant roots and

**Table 10.4** Effectiveness of strains of microorganisms involved in bioremediation of crude oil-polluted medium

| Type of microorganism | Techniques | Comment | Reference |
|---|---|---|---|
| *Pseudomonas aeruginosa* and *Pseudomonas fluorescence* | Batch fermentation and continuous culture | *Pseudomonas aeruginosa* was able to completely remove phenol from the refinery effluent within 60 h of cultivation and while *Pseudomonas fluorescence* could only remove 73.1% of phenol within the same period | Ojumu et al. (2005) |
| *Pseudomonas aeruginosa* and *Penicillium janthinellum* | Incubation for 2 weeks | All consortia were observed to have significant decreases in contents of phenol, oil and grease, phosphates, ammonia, nitrates and sulphates after 2 weeks of incubation at 30 °C | Bako et al. (2008) |
| *Pseudomonas*, *Micrococcus*, *Flavobacterium*, *Staphylococcus*, *Serratia*, *Corynebacterium*, *Bacillus*, *Chromobacterium* and *Alcaligenes* | Natural attenuation | Microorganisms occurring naturally in crude oil-impacted sediment utilize hydrocarbons, and therefore, remediation of oil-polluted environment can be achieved | Abu and Dike (2008) |
| *Pseudomonas aeruginosa* NCIB 950 and *Pseudomonas fluorescence* NCIB 3756 | Batch fermentation process | Phenol was successfully degraded by both species; with *P. aeruginosa* more effective. They can both be used for bioremediation of petroleum refinery wastewater | Agarry et al. (2008) |
| *Bacillus cereus* and *Pseudomonas aeruginosa* | Organic fertilizer and modified strains | The strains achieved better results (98.25% for oil and grease and 87.34% for total petroleum hydrocarbons) when modified and treated with organic fertilizer (NPK 15-15-15) | Idise et al. (2010) |
| *Bacillus* spp. | Growth of isolated *Bacillus* on crude oil | The results of the test revealed that *Bacillus* spp. can utilize crude oil as a carbon and energy source | Sepahi et al. (2008) |
| *Bacillus* sp. and *Pseudomonas* sp. | Bioaugmentation with bacteria and biostimulation with poultry manure | Bioaugmentation was more effective than biostimulation | Ijah and Antai (2003) |
| *Bacillus*, *Corynebacterium*, *Pseudomonas*, *Flavobacterium*, *Micrococcus*, *Acinetobacter* and *Aerococcus* | Fertilizer application and oxygen exposure | With nutrient supplementation, bioremediation can achieve high rates of degradation of petroleum hydrocarbons in agricultural soils | Ayotamuno et al. (2006) |

microorganisms can have their supplies of water, air and nutrients as well as some moderation of environmental conditions that can influence their development, such as temperature or pH in the substrate media (Dordio and Carvalho 2013). The substrate medium is the primary sink for heavy metals present in the aquatic environment (Tirkey et al. 2012; Yadav et al. 2012; Papaevangelou et al. 2017) and oil contaminants if they are not degraded (Abu and Dike 2008). Furthermore, both the plant root zone and the substrate absorb ionic heavy metals (Chen et al. 2009), and substrate media can have a higher absorptive capacity than the plant roots (Galletti et al. 2010; Rani et al. 2011).

Fine-textured sediments can accumulate metals if they contain high amounts of organic matter; in contrast, coarse-textured materials have low affinity for metals (Galletti et al. 2010). Papaevangelou et al. (2017) reported that Cr removal was accomplished mainly through the substrate and attached organic matter rather than the plant itself through the phytoremediation processes.

Gravel and sand are the most common types of substrate media used in CWs. These substrate media have a limited adsorptive capacity, and the capability of a wetland to remove inorganic pollutants can greatly reduce overtime (Hua et al. 2015). Therefore, the capacity of substrate media can be greatly improved by using active filter materials, e.g. with reactive Fe/Al hydrous oxide adsorption surfaces (Hua et al. 2015).

## 10.9 Capital, Operation and Maintenance Costs

The investment costs to consider for the construction of wetlands are basically land acquisition, site survey, system design, site preparation, plastic liners for prevention of ground water contamination, filtration, rooting media, vegetation, hydraulic control structures and miscellaneous costs which may include fencing and access roads (Kivaisi 2001; Rousseau et al. 2008; Vymazal 2010). In addition, Haberl et al. (2003) included specific demands and circumstances of the site such as topography, distance to the receiving water, existing devices and availability of necessary area into the costs. The costs of construction greatly differ from one site-specific factor to the other, for example, flow control structures may vary from US \$2000 to \$80,000/10,000 m$^2$ for SF wetlands and up to US \$150,000/10,000 m$^2$ for SSF wetlands (Allen 2008) and approximately €2200/PE in Upper Austria (Haberl et al. 2003). Excavation costs vary between 7% and 27.4% of the total capital cost, while other cost goes to gravel (27–53%), lining (13–33%), plants (2–12%), plumbing (6–12%), control structures (3.1–5.7%) and miscellaneous (1.8–12%) (Vymazal 2010). The total investment costs vary even more globally, and the cost could be as low as US \$29/m$^2$ in India or US \$33/m$^2$ in Costa Rica or as high as €257/m$^2$ (Vymazal 2010) and €392/PE for SF and €1258/PE for SSF wetlands in Belgium (Rousseau et al. 2008).

The capital cost for CWs may be influenced by the choice of substrate, plant species, basin compartmentation, lining, flow structure and other CW components

(Calheiros et al. 2009). In general, the capital costs for FWS CWs are generally lower than for SSF CWs; this is primarily due to lower quantity of media required for rooting soil on the bottom of the beds (Vymazal 2010). The average capital costs for SF wetland systems are US \$200,000 per hectare, while the FWS systems cost approximately US \$50,000 per 10,000 $m^2$. The main cost difference between the two systems is in the expenses of acquiring the gravel media and transporting it to the site. The construction cost per hectare is higher for SF wetlands. The unit cost is US \$163/$m^2$ (US \$0.21/L) of wastewater treated for the SF type and US \$206/$m^2$ (\$0.21/L) for the FWS type (EPA 1993).

The cost of construction of CWs and that of conventional wastewater treatment plants are similar (Haberl et al. 2003). Thus, well-designed wetland systems have low operating costs, making them economically competitive with conventional treatment systems (Allen 2008). Operation and maintenance (O&M) costs of CW include pumping energy if necessary, compliance monitoring, maintenance of access roads and berms, pretreatment, weed control, vegetation harvesting, effluent sampling and control, cleaning of distribution systems and pumps, equipment replacement and repairs (Haberl et al. 2003; Rousseau et al. 2008; Vymazal 2010). Rousseau et al. (2008) reported an estimated range of US \$2500 and US \$5000/ha/year for O&M costs for SSF CW and about US \$1000/ha/year for median cost for SF CW. The basic costs are much lower, by a factor of 2–19, than those for competing technologies using concrete and steel constructions (Vymazal 2010).

## References

Abd-El-Haleem D, Beshay U, Abdelhamid AO, Moawad H, Zaki S (2003) Effects of mixed nitrogen sources on biodegradation of phenol by immobilized Acinetobacter sp. strain W-17. Afr J Biotechnol 2(1):8–12

Abdelwahab O, Amin NK, El-Ashtoukhy E-SZ (2009) Electrochemical removal of phenol from oil refinery wastewater. J Hazard Mater 163:711–716

Abu GO, Dike PO (2008) A study of natural attenuation processes involved in a microcosm model of crude oil-impacted wetland sediment in the Niger Delta. Bioresour Technol 99(11):4761–4767. https://doi.org/10.1016/j.biortech.2007.09.063

Achile GN, Yillian L (2010) Mineralization of organic compounds in wastewater contaminated with petroleum hydrocarbon using Fenton's reagent: a kinetic study. Am J Sci 6(4):58–66

Adewuyi GO, Olowu RA (2012) Assessment of oil and grease, total petroleum hydrocarbons and some heavy metals in surface and groundwater within the vicinity of NNPC oil depot in Aata, Ibadan metropolis, Nigeria. IJRRAS 13(1):166–174

Agarry SE, Durojaiye AO, Yusuf RO, Solomon BO, Majeed O (2008) Biodegradation of phenol in refinery wastewater by pure cultures of *Pseudomonas aeruginosa* NCIB 950 and *Pseudomonas fluorescence* NCIB 3756. Int J Environ Pollut 32(1):3–11

Al-Baldawi IA, Abdullah SR, Abu Hasan H, Suja F, Anuar N, Mushrifah I (2014) Optimized conditions for phytoremediation of diesel by *Scirpus grossus* in horizontal subsurface flow constructed wetlands (HSFCWs) using response surface methodology. J Environ Manag 140:152–159

Ali N, Oniye S, Balarabe M, Auta J (2005) Concentration of Fe, Cu, Cr, Zn and Pb in Makera-Drain, Kaduna, Nigeria. Chem Class J 2:69–73

Allen WE (2008) Process water treatment in Canada's oil sands industry: II. A review of emerging technologies. J Environ Eng Sci 7:499–524

Allen WC, Hook PB, Biederman JA, Stein OR (2002) Temperature and wetland plant species effects on wastewater treatment on wastewater treatment and root zone oxidation. J Environ Qual 31(3):1010–1016

Asatekin A, Mayes A (2009) Oil industry wastewater treatment with fouling resistant membranes containing amphiphilic comb copolymers. Environ Sci Technol 43(12):4487–4492

Aslam MM, Malik M, Baig M, Qazi I, Iqbal J (2007) Treatment performances of compost-based and gravel-based vertical flow wetlands operated identically for refinery wastewater treatment in Pakistan. Ecol Eng 30:34–42

Aslam MM, Malik M, Braig MA (2010) Removal of metals from refinery wastewater through vertical flow constructed wetlands. Int J Agric Biol 12:796–798

Ayaz S (2008) Post-treatment and reuse of tertiary treated wastewater by constructed wetlands. Desalination 226(1–3):249–255

Ayotamuno MJ, Kogbara RB, Ogaji SO, Robert SD (2006) Bioremediation of a crude-oil polluted agricultural soil at Port Harcourt, Nigeria. Appl Energy 83:1249–1257

Babatunde AO, Zhao Y, O'Neill M, O'Sullivan B (2008) Constructed wetlands for environmental pollution control: a review of developments, research and practice in Ireland. Environ Int 34:116–126

Bako SP, Chukwunonso D, Adamu AK (2008) Bioremediation of refinery effluents by strains of *Pseudomonas aeruginosa* and *Penicillium janthinellum*. Appl Ecol Environ Res 6(3):49–60

Ballesteros F Jr, Vuong TH, Secondes MF, Tuan PD (2016) Removal efficiencies of constructed wetland and efficacy of plant on treating benzene. Sustain Environ Res 26:93–96

Barber LB, Leenheer JA, Noyes TI, Stiles EA (2001) Nature and transformation of dissolved organic matter in treatment wetlands. Environ Sci Technol 35(24):4805–4816

Baskar G (2011) Studies on application of subsurface flow constructed wetland for wastewater treatment. SRM University, Kattankulathur –Doctoral Thesis<: 603 203, Department of Civil Engineering. Chennai, Tamil Nadu, India

Bedessem ME, Ferro AM, Hiegel T (2007) Pilot-scale constructed wetlands for petroleum-contaminated groundwater. Water Environ Res 79:581–586

Biswas J (2013) Evaluation of various method and efficiencies for treatment of effluent from Iron and Steel industry – a review. Int J Mech Eng Robot Res 2(3):67–73

Brisson J, Chazarenc F (2009) Maximizing pollutant removal in constructed wetlands: should we pay more attention to macrophyte species selection? Sci Total Environ 407(13):3923–3930

Brito EM, Duran R, Guyoneaud R, Goñi-Urriza M, Oteyza D, García de Oteyza T, Crapez MAC, Aleluia I, Wasserman JC (2009) A case study of in situ oil contamination in a mangrove swamp (Rio De Janeiro, Brazil). Mar Pollut Bull 58:418–423

Calheiros CS, Rangel AO, Castro PM (2008) The effects of tannery wastewater on the development of different plant species and chromium accumulation in *Phragmites australis*. Ach Environ Contam Toxicol 55(3):404–414

Calheiros CS, Rangel A, Castro PM (2009) Treatment of industrial wastewater with two-stage constructed wetlands planted with *Typha latifolia* and *Phragmites australis*. Bioresour Technol 100(12):3205–3215

Campagna A, da Motta Marques D (2000) The effect of habitat heterogeneity in experimental wetlands receiving petroleum refinery effluent on the size and clonal modules location of aquatic macrophytes *Scirpus californicus*, *Typha subulata* and *Zizaniopsis bonariensis*. 7th International Conference on Wetlands Systems for Water Pollution Control, Florida. pp 485–493

Campagna A, da Motta Morques D (2001) The effect of refinery effluent on the aquatic macrophytes *Scirpus californicus*, *Typha subulata* and *Zizaniopsis bonariensis*. Water Sci Technol 44(11–12):493–498

Chapelle F (1999) Bioremediation of petroleum hydrocarbon-contaminated ground water: the perspectives of history and hydrology. Ground Water 37(1):122–132

Chen M, Tang Y, Li X, Yu Z (2009) Study on the heavy metals removal efficiencies of constructed wetlands with different substrates. J Water Res Protect 1:1–7

Chen Z, Kuschk P, Reiche N, Borsdorf H, Kästner M, Köser H (2012) Comparative evaluation of pilot scale horizontal subsurface-flow constructed wetlands and plant root mats for treating groundwater contaminated with benzene and MTBE. J Hazard Mater 209–210:510–515

Cheng S, Grosse W, Karrenbrock F, Thoennessen M (2002) Efficiency of constructed wetlands in decontamination of water polluted by heavy metals. Ecol Eng 18(3):317–325

Cheng L, Wang Y, Cai Z, Liu J, Yu B, Zhou Q (2017) Phytoremediation of petroleum hydrocarbon-contaminated saline-alkali soil by wild ornamental *Iridaceae* species. Int J Phytoremediation 19(3):300–308

Chorom M, Parnian A, Jaafarzadeh N (2012) Nickel removal by the aquatic plant (*Ceratophyllum demersum* L.). Int J Environ Sci Dev 3(4):372–375

Coneglian CM, Conceição DM, de Angelis D, Marconato JC, Bidóia E (2002) Concentration reduction of ammonia in industrial oil residue of an oil refinery. Salusvita Bauru 21(1):43–50

Czudar A, Gyulai I, Kereszturi P, Csatari I, Serra-Paka S, Lakatos G (2011) Removal of organic materials and plant nutrients in a constructed wetland for petrochemical wastewater treatment. Stud Univ "Vasile Goldiş" Seria Ştiinţele Vieţii 21(1):109–114

Das N, Chandran P (2011) Microbial degradation of petroleum hydrocarbon contaminants: an overview. Biotechnol Res Int 2011:1–13

Davis BM, Wallace S, Willison R (2009) Pilot-scale engineered wetlands for produced water treatment. Soc Pet Eng 4(3):75–79 Retrieved December 22, 2016, from http://doi.org/10.2118/120257-PA

De Biase C, Reger D, Schmidt A, Jechalke S, Reiche N, Martínez-Lavanchy PM, Rosell M, Van Afferden M, Maier U, Oswald, Sascha E, Thullner M (2011) Treatment of volatile organic contaminants in a vertical flow filter: relevance of different removal processes. Ecol Eng 37:292–1303

De-qing S, Jian Z, Zhao-long G, Jian D, Tian-Li W, Kalyuzhnyi MS (2007) Bioremediation of oil sludge in shengli oilfield. Bioremediation. Water Air Soil Pollut 185:177–184

Dhote S, Dixit S (2009) Water quality improvement through macrophytes – a review. Environ Monit Assess 152(1):149–153

Dipu S, Anju A, Kumar V, Thanga SG (2010) Phytoremediation of dairy effluent by constructed wetland technology using wetland macrophytes. Global J Environ Res 4(2):90–100

Diya'uddeen BH, Wan Daud MA, Abdul Aziz AR (2011) Treatment technologies for petroleum refinery effluents: a review. Process Saf Environ Prot 89:95–105

Diya'uddeen BH, Abdul Aziz AR, Wan Daud WM (2012) Oxidative mineralisation of petroleum refinery effluent using Fenton-like process. Chem Eng Res Des 90:298–307

Dordio A, Carvalho A (2013) Organic xenobiotics removal in constructed wetlands, with emphasis on the importance of the support matrix. J Hazard Mater 252–253:272–292

Dowty RA, Shaffer GP, Hester MW, Childers GW, Campo FM, Greene MC (2001) Phytoremediation of small-scale oil spills in fresh marsh environments: a mesocosm simulation. Mar Environ Res 52(3):195–211

Eke PE, Scholz M (2008) Benzene removal with vertical – flow constructed treatment wetlands. J Chem Technol Biotechnol 83(1):55–63

El-Ashtoukhy E-S, El-Taweel Y, Abdelwahab O, Nassef E (2013) Treatment of petrochemical wastewater containing phenolic compounds by electrocoagulation using a fixed bed electrochemical reactor. Int J Electrochem Sci 8:1534–1550

Farooqi H, Basheer F, Chaudhari RJ (2008) Constructed wetland system (CWS) for wastewater treatment. In: Sengupta MA (ed) The 12th world lake conference, International Lake Environment Committee, Jaipur, Rajasthan, Indian, pp 1004–1009

Fernandez-Luqueno F, Valenzuela-Encinas C, Marsch R, Martinez-Suarez C, Vazquez-Nunez E, Dendoven L (2011) Microbial communities to mitigate contamination of PAHs in soil-possibilities and challenges: a review. Environ Sci Pollut Res 18:12–30

Ferro AM, Utah NL, Kadlec RH, Deschamp J, Wyoming C (2002) Constructed wetland system to treat wastewater at the BP Amoco former Casper refinery: pilot scale project, The 9th International Petroleum Environmental Conference. IPEC Consortium, Albuquerque Retrieved from http://ipec.utulsa.edu/Ipec/Conf2002/tech_sessions.html

Fountoulakis MS, Terzakis S, Kalogerakis N, Manios T (2009) Removal of polycyclic aromatic hydrocarbons and linear alkylbenzene sulfonates from domestic wastewater in pilot constructed wetlands and gravel filter. Ecol Eng 35(12):1702–1709

Galletti A, Verllicchi P, Rainieria E (2010) Removal and accumulation of Cu, Ni and Zn in horizontal subsurface flow constructed wetlands: contribution of vegetation and filling medium. Sci Total Environ 408:5097–5105

Gillespie WB Jr, Hawkins WB, Rodgers JH Jr, Cano ML, Dorn PB (2000) Transfers and transformations of zinc in constructed wetlands: mitigation of a refinery effluent. Ecol Eng 14:279–292

Gousmi N, Sahmi A, Li H, Poncin S, Djebbar R, Bensadok K (2016) Purification and detoxification of petroleum refinery wastewater by electrocoagulation process. Environ Technol 37(18):1–10

Grove JK, Stein OR (2005) Polar organic solvent removal in microcosm constructed wetlands. Water Res 39(16):4040–4050

Haberl R, Grego S, Langergraber G, Kadlec RH, Cicalini A-R, Dias SM, Novais JM, Aubert S, Gerth A, Thomas H, Hebner A (2003) Constructed wetlands for the treatment of organic pollutants. J Soils Sediments 3(2):109–124

Hadad HR, Maine M, Bonetto CA (2006) Macrophytes growth in a pilot constructed wetland for industrial wastewater treatment. Chemosphere 63:1744–1753

Hamza UD, Mohammed IA, Sale A (2012) Potentials of bacterial isolates in bioremediation of petroleum refinery wastewater. J Appl Phytotechnol Environ Sanita 1(3):131–138

Hashim M, Mukhopadhyay S, Sahu JN, Sengupta B (2011) Remediation technologies for heavy metal contaminated groundwater. J Environ Manag 92:2355–2388

Hietala KA, Roane TM (2009) Microbial remediation of metals in soils. Adv Appl Biorem Soil Biol 17:201–220

Ho Y, Show K, Guo X, Norli I, Abbas FA, Morad N (2012) Industrial discharge and their effect to the environment. In: Show PK-Y (ed) Industrial waste. InTECH, pp 1–32

Hoffmann H, Platzer C, Winker M, von Muench E (2011) Technology review of constructed wetlands Subsurface flow constructed wetlands for greywater and domestic wastewater treatment, Federal Ministryfor Economic Cooperation and Development, Division Water, Energy and Transport. Deutsche Gesellschaft für Internationale Zusammenarbeit (GIZ) GmbH Sustainable sanitation – ecosan program, Eschbom

Horner JE, Castle JW, Rodgers JH Jr, Murray Gulde C, Myers JE (2012) Design and performance of pilot-scale constructed wetland treatment systems for treating oilfield produced water from sub-saharan Africa. Water Air Soil Pollut 223:1945–1957

Hua T, Haynes R, Zhou Y-F, Boullemant A (2015) Potential for use of industrial waste materials as filter media for removal of Al, Mo, As, V and Ga from alkaline drainage in constructed wetlands – adsorption studies. Water Res 71:32–41

Huang X-D, El-Alawi Y, Gurska J, Glick BR, Greenberg BM (2005) A multi-process Phytoremediation system for decontamination of persistent total petroleum hydrocarbons (TPHs) from soils. Microchemical 81:139–147

Huddleston GM, Gillespie WB, Rodgers JH (2000) Using constructed wetlands to treat biochemical oxygen demand and ammonia associated with a refinery effluent. Ecotoxicol Environ Saf 45:188–193

Idise OE, Ameh JB, Yakubu SE, Okuofu CA (2010) Biodegradation of a refinery effluent treated with organic fertilizer by modified strains of *Bacillus cereus* and *Pseudomonas aeruginosa*. Afr J Biotechnol 9(22):3298–3302

Ijah UJ, Antai SP (2003) The potential use of chicken drop microorganisms for oil spillage remediation. Environmentalist 23(1):89–95

Imfeld G, Braeckevelt M, Kuschk P, Richnow HH (2009) Monitoring and assessing processes of organic chemicals removal in constructed wetlands. Chemosphere 74(3):349–362

Jadia C, Fulekar M (2009) Phytoremediation of heavy metals: recent techniques. Afr J Biotechnol 8(6):921–928

Jakubaszek A, Saecka Z (2015) The effectiveness of organic pollutants removal in constructed wetland with horizontal sub-surface flow. Civil Environ Eng Rep 16(1):69–82

Jayaweera MW, Kasturiarachchi JC, Kularatne RK, Wijeyekoon SL (2008) Contribution of water hyacinth (*Eichhornia crassipes* (*Mart.*) So*l*ms) grown under different nutrient conditions to Fe-removal mechanisms in constructed wetlands. J Environ Manag 87:450–460
Jechalke S, Vogt C, Reiche N, Franchini AG, Borsdorf H, Neu TR, Richnow HH (2010) Aerated treatment pond technology with biofilm promoting mats for the bioremediation of benzene, MTBE and ammonium contaminated groundwater. Water Res 44:1785–1796
Jeon BY, Jung L, Park DH (2011) Mineralization of petroleum contaminated wastewater by co-culture of petroleum-degrading bacterial community and biosurfactant-producing bacterium. Environ Prot 2(7):895–902
Ji GD, Sun TH, Oixing Z, Xin S, Shijun C, Peijun L (2002) Subsurface flow wetland for treating heavy oil-produced water of the Liohe Oilfield in China. Ecol Eng 18:459–465
Ji G, Sun T, Ni J (2007) Surface flow constructed wetland for heavy oil-produced water treatment. Bioresour Technol 98(2):436–441
Kadlec RH, Wallace SD (2009) Treatment wetlands, 2nd edn. CRC Press Taylor & Francis Group, New York
Kanagy LE, Johnson BM, Castle JW, Rodgers JH Jr (2008) Design and performance of a pilot-scale constructed wetland treatment system for natural gas storage produced water. Bioresour Technol 99:1877–1885
Khan S, Ahmad I, Shah MT, Rehman S, Khaliq A (2009) Use of constructed wetland for the removal of heavy metals from industrial wastewater. J Environ Manag 50(11):3451–3457
Kivaisi AK (2001) The potential for constructed wetlands for wastewater treatment and reuse in developing countries; a review. Ecol Eng 16:545–560
Knight R, Kadlec R, Ohlendorf HM (1999) The use of treatment wetlands for petroleum industry effluents. Environ Sci Technol 33(7):973–980
Korsah PE (2011) Phytoremediation of irrigation water using *Limnocharis flava*, *T. latifolia* and *Thalia geniculata* in a constructed wetland. Kwame Nkrumah University of Science and Technology, Department of Theoretical and Applied Biology. Department of Theoretical and Applied Biology Kumasi, Ghana
Kröpfelova L, Vymazal J, SŭehIa J, Štíchová J (2009) Removal of trace elements in three horizontal sub-surface flow constructed wetlands in the Czech Republic. Environ Pollut 157:1186–1194
Kumar M, Puri A (2012) A review of permissible limits of drinking water. Indian J Occup Environ Med 16(1):40–44
LeDuc DL, Terry N (2005) Phytoremediation of toxic trace elements in soil and water. J Ind Microbiol Biotechnol 32:514–520
Lee B-H, Scholz M (2007) What is the role of *Phragmites australis* in experimental constructed wetland filters treating urban runoff? Ecol Eng 29:87–95
Lin Q, Mendelssohn IA (2009) Potential of restoration and phytoremediation with *Juncus roemerianus* for diesel-contaminated coastal wetlands. Ecol Eng 35(1):85–91. http://doi.org/10.1016/j.ecoleng.2008.09.010
Liu D, Ge Y, Chang J, Peng C, Gu B, Chan GY, Wu X (2008) Constructed wetlands in China: recent developments and future challenges. Front Ecol Environ 7(5):261–268
Madera-Parra CA, Pena-Salamanca EJ, Pena M, Rousseau DP, Lens PN (2015) Phytoremediation of Landfill Leachate with *Colocasia esculenta*, *Gynerum sagittatum* and *Heliconia psittacorum* in constructed wetlands. Int J Phytoremediation 17:16–24
Maine MA, Suñe N, Haadad H, Sánchez G, Bonetto C (2007) Removal efficiency of a constructed wetland for wastewater treatment according to vegetation dominance. Chemosphere 68(6):1105–1113
Mant C, Coasta S, Williams J, Tambourgi E (2005) Studies of removal of chromium by model constructed wetlands. Braz J Chem Eng 22(3):381–387
Mazeikiene A, Rimeika M, Osikinis VV, Paskauskaite N, Brannvall E (2005) Removal of petroleum products from water using natural sorbent zeolite. Environ Eng Landsc Manag 13(4):187–191
Mazzeo DE, Levy CE, de Angelis DD, Marin-Morales MA (2010) BTEX biodegradation by bacteria from effluents of petroleum refinery. Sci Total Environ 408:4334–4340

McIntosh P, Schulthess CP, Kuzovkina YA, Guillard K (2017) Bioremediation and phytoremediation of total petroleum hydrocarbons (TPH) under various conditions. Int J Phytoremediation 19(8):755–764

Mebrahtu G, Zerabruk S (2011) Concentration of heavy metals in drinking water from urban areas of the Tigray region, Northern Ethiopia. CNCS. Mckelle Univ 3:105–121

Melián JH, Rodríguez AM, Arãna J, Díaz OG, Henríquez JG (2010) Hybrid constructed wetlands for wastewater treatment and reuse in the Canary Islands. Ecol Eng 36:891–899

Mena J, Rodriguez L, Numez J, Fermandez FJ, Villasenor J (2008) Design of horizontal and vertical subsurface flow constructed wetlands treating industrial wastewater. Water Pollut 111:555–557

Merkl N, Schultze-Kraft R, Arias M (2006) Effect of the tropical grass *Brachiaria brizantha* (*Hochst. ex A. Rich.*) *Stapf* on microbial population and activity in petroleum-contaminated soil. Microbiol Res 161:80–91. http://doi.org/10.1016/j.micres.2005.06.005

Mohammed Y, Aliyu A, Audu P (2013) Concentration of Oxirane and Naphthalene in selected petroleum based industrial effluents in Kaduna metropolis, Nigeria. Direct Res J Chem 1:1–5

Moneke A, Nwangwu C (2011) Studies on the bioutilization of some petroleum hydrocarbons by single and mixed cultures of some bacterial species. Afr J Microbiol Res 5(12):1457–1466

Moreno C, Farahbakhazad N, Morrison GM (2002) Ammonia removal from oil refinery effluent in vertical upflow macrophytes column systems. Water Air Soil Pollut 135:237–247

Mulkerrins D, Dobson A, Colleran E (2004) Parameters affecting biological phosphate removal from wastewaters. Environ Int 30:249–259

Murray-Gulde C, Heatley JE, Karanfil T, Rodgers JR Jr, Myers JE (2003) Performance of a hybrid reverse osmosis-constructed wetland treatment system for brackish oil field produced water. Water Res 37(3):705–713

Mustapha HI, Rousseau D, van Bruggen JJA, Lens PNL (2011) Treatment performance of horizontal subsurface flow constructed wetlands treating inorganic pollutants in simulated refinery effluent, 2nd Biennial Engineering Conference. School of Engineering and Engineering Technology, Federal University of Technology, Minna, Minna

Mustapha HI, van Bruggen JJA, Lens PNL (2013) Preliminary studies on the application of constructed wetlands for a treatment of refinery effluent in Nigeria: a mesocosm scale study, 2013 International Engineering Conference, Exhibition and Annual General Meeting. Nigerian Society of Engineers, Abuja

Mustapha HI, van Bruggen JJA, Lens PNL (2015) Vertical subsurface flow constructed wetlands for polishing secondary Kaduna refinery wastewater in Nigeria. Ecol Eng 85(20):588–595

Mustapha HI, van Bruggen JJA, Lens PNL (2018) Fate of heavy metals in vertical subsurface flow constructed wetlands treating secondary treated petroleum refinery wastewater in Kaduna, Nigeria. Int J Phytorem 20(1):44–53. http://doi.org/10.1080/15226514.2017.1337062

Nacheva PM (2011) Water management in the petroleum refining industry. In: Jha DM (ed) Water conservation. InTech, pp 105–128

Noori AS, Maivan HZ, Alaie E, Newman L (2015) *Leucanthemum vulgare Lam.* crude oil phytoremediation. Int J Phytorem https://doi.org/10.1080/15226514.2015.1045122

Nwanyanwu CE, Abu GO (2010) In vitro effects of petroleum refinery wastewater on dehydrogenase activity in marine bacterial strains. Ambi Agua Taubate 5(2):21–29

O'Sullivan AD, Moran BM, Otte ML (2004) Accumulation and fate of contaminants (Zn, Pb, Fe and S) in substrates of wetlands constructed for treating mine wastewater. Water Air Soil Pollut 30:1–20

Ogunfowokan AO, Asubiojo OI, Fatoki O (2003) Isolation and determination of polycyclic aromatic hydrocarbons in surface runoff and sediments. Water Air Soil Pollut 147(1):245–261

Ojumu TV, Beelo OO, Solomon BO (2005) Evaluation of microbial systems for bioremediation of petroleum refinery effluents in Nigeria. Afr J Biotechnol 4(1):31–35. http://doi.org/10.4314%2Fajb.v4i1.15048

Otokunefor T, Obiukwu C (2010) Efficacy of inorganic nutrients in bioremediation of a refinery effluent. Sci Afr 9(1):111–125

Papaevangelou VA, Gikas GD, Tsihrintzis VA (2017) Chromium removal from wastewater using HSF and VF pilot-scale constructed wetlands: overall performance, and fate and distribution of this element within the removal environment. Chemosphere 168:716–730

Paul W (2011) Impact of industrial effluents on water quality of receiving streams in Nakawa-Ntinda, Uganda. Environment and Natural Resources, Makerere University

Prasad MN (2003) Phytoremediation of metal-polluted ecosystems: hype for commercialization. Russ J Plant Physiol 50(5):686–701

Qasaimeh A, AlSharie H, Masoud T (2015) A review on constructed wetlands components and heavy metal removal from wastewater. J Environ Prot 6:710–718

Rani SH, Din MF, Yusof BM, Chelliapan S (2011) Overview of subsurface constructed wetlands application in tropical climates. Univ J Environ Res Technol 1(2):103–114 Retrieved from www.environmentaljournal.org

Ravanchi MT, Kaghazchi T, Kargari A (2009) Application of membrane separation processes in petrochemical industry: a review. Desalination 235(1–3):199–244

Reddy KR, D'Angelo EM (1997) Biogeochemical indicators to evaluate pollutant removal efficiency in constructed wetlands. Water Sci Technol 35(5):1–10

Rezaee A, Hossini H, Masoumbeigi H, Soltani RD (2011) Simultaneous removal of hexavalent chromium and nitrate from wastewater using electrocoagulation method. Int J Environ Sci Dev 2(4):294–298

Rousseau DP, Lesage E, Story A, Vanrolleghen PA, De Pauw N (2008) Constructed wetlands for water reclamation. Desalination 128:181–189

Saeed T, Guangzhi S (2012) A review on nitrogen and organics removal mechanisms in subsurface flow constructed wetlands: dependency on environmental parameters, operating conditions and supporting media. J Environ Manag 112:429–448

Saien J (2010) Treatment of the refinery wastewater by nano particles of TiO2. USA Patent No. Pub. No.: US 20100200515 A1

Saien J, Shahrezaei F (2012) Organic pollutants removal from petroleum refinery wastewater with nanotitania photocatalyst and UV light emission. Int J Photoenergy 2012:1–5

Scholz M (2003) Performance predictions of mature experimental constructed wetlands which treat urban water receiving high loads of lead and copper. Water Res 37(6):1270–1277

Schröder P, Navarro-Aviñó J, Azaizeh H, Azaizeh H, Goldhirsh AG, DiGregorio S, Komives T, Langergraber G, Lenz A, Maestri E, Memon AR, Ranalli A, Sebastiani L, Smrcek S, Vanek T, Vuilleumier S, Wissing F (2007) Using phytoremediation technologies to upgrade waste water treatment in Europe. Environ Sci Pollut Res Int 14(7):490–497

Scullion J (2006) Remediating polluted soils. Naturwissenschaften 93(2):51–65

Seeger EM, Kuschk P, Fazekas H, Grathwohl P, Kaestner M (2011) Bioremediation of benzene-, MTBE- and ammonia-contaminated groundwater with pilot-scale constructed wetlands. Environ Pollut 199:3769–3776

Sepahi AA, Golpasha ID, Emami A, Nakhoda MA (2008) Isolation and characterization of crude oil degrading *Bacillus spp*. Iran J Environ Health Sci Eng 5(3):150–152

Sgroi M, Pelissari C, Roccaro P, Sezerino PH, García J, Vagliasindi FG, Ávila C (2018) Removal of organic carbon, nitrogen, emerging contaminants and fluorescing organic matter in different constructed wetland configurations. Chem Eng J 332:619–627

Shelef O, Gross A, Rachmilevitch S (2013) Role of plants in a constructed wetland: current and new perspectives. Water 5:405–419

Shpiner R, Vathi S, Stuckey DC (2009a) Treatment of "produced water" by waste stabilization ponds: removal of heavy metals. Water Res 43:4258–4268

Shpiner R, Liu G, Stuckey DC (2009b) Treatment of oilfield produced water by waste stabilization ponds: biodegradation of petroleum-derived materials. Bioresour Technol 100:6229–6235

Šíma J, Krejsa J, Svoboda L (2015) Removal of mercury from wastewater using a constructed wetland. Croat Chem Acta 88(2):165–169

Simi A (2000) Water quality assessment of a surface flow constructed wetland treating oil refinery wastewater. In: Reddy KR (ed) 7th International conference on wetlands system for water pollution control, vol 3. IWA, Lake Buena Vista, pp 1295–1304

Song X, Yan D, Liu Z, Chen Y, Lu S, Wang D (2011) Performance of laboratory-scale constructed wetlands coupled with micro-electric field for heavy metal-contaminating wastewater treatment. Ecol Eng 37(12):2061–2065

Spacil M, Rodgers J, Castle J, Murray-Gulde C, Myers JE (2011) Treatment of Selenium in simulated refinery effluent using a pilot-scale constructed wetland treatment system. Water Air Soil Pollut 221(1):301–312

Stefanakis AI, Seeger E, Dorer C, Sinke A, Thullner M (2016) Performance of pilot-scale horizontal subsurface flow constructed wetlands treating groundwater contaminated with phenols and petroleum derivatives. Ecol Eng 95:514–526

Stottmeister U, Wießner A, Kuschk P, Kappelmeyer U, Kästner M, Bederski O, Müller RA, Moormann H (2003) Effects of plants and microorganisms in constructed wetlands for wastewater treatment. Biotechnol Adv 22:93–117

Tam NF, Wong YS (2008) Effectiveness of bacterial inoculum and mangrove plants on remediation of sediment contaminated with polycyclic aromatic hydrocarbons. Mater Pollut Bull 57:716–726

Taneva N (2012) Removal of ammonium and phosphates from aqueous solutions by activated and modified Bulgarian clinoptilolite. J Chem Eng Mater Sci 3(5):79–85

Tanner CC (2001) Plants as ecosystem engineers in subsurface-flow treatment wetlands. Water Sci Tehnol 44(11–12):9–17

Taylor CR, Hook PB, Stein OR, Zabinski CA (2011) Seasonal effects of 19 plant species on COD removal in subsurface wetland microcosms. Ecol Eng 37(5):703–710. http://doi.org/10.1016/j.ecoleng.2010.05.007

Tirkey A, Shrivastava P, Saxena A (2012) Bioaccumulation of heavy metals in different components of two lakes ecosystem. Curr World Environ 7(2):293–297

Tromp K, Lima AT, Barendregt VT (2012) Retention of heavy metals and poly-aromatic hydrocarbons from road water in a constructed wetland and the effect of de-icing. Hazard Mater 203–204:290–298

Uzoekwe SA, Oghosanine FA (2011) The effect of refinery and petrochemical effluent on water quality of Ubeji Creek Waar, Southern Nigeria. Ethopian J Environ Stud Manag 4(2):107–116

Valderrama LT, Del Campo CM, Rodriguez CM, de- Bashan LE, Bashan Y (2002) Treatment of recalcitrant wastewater from ethanol and citric acid production using microalga *Chlorella vulgaris* and the macrophyte *Lemna minuscula*. Water Res 36(17):4185–4192

van Afferden M, Rahman KZ, Mosig P, De Biase C, Thullner M, Oswald SE, Müller RA (2011) Remediation of groundwater contaminated with MTBE and benzene: the potential of vertical-flow soil filter systems. Water Res 45:5063–5074

Veenstra J, Mohr K, Sanders DA (1998) Refinery wastewater management using multiple-angle water separators. The International Petroleum Environment Conference. Albuquerque

Veil JA, Puder MG, Elcock D, Redweil Jr RJ (2004) A white paper describing produced water from production of crude oil, natural gas and coal bed methane. Department of Energy, National Energy Technology. US Department of Energy, National Energy Technology under contract W-31-109-Eng-38

Vymazal J (2005) Horizontal sub-surface flow and hybrid constructed wetlands systems for wastewater treatment. Ecol Eng 25:478–490

Vymazal J (2007) Removal of nutrients in various types of constructed wetlands. Sci Total Environ 380:48–65

Vymazal J (2010) Constructed wetlands for wastewater treatment. Water 2(3):530–549

Vymazal J (2011) Plants used in constructed wetlands with horizontal subsurface flow: a review. Hydrobiologia 674:133–156

Vymazal J (2013) Emergent plants used in free water surface constructed wetlands: a review. Ecol Eng 61P:582–592

Vymazal J, Březinová T (2016) Accumulation of heavy metals in aboveground biomass of *Phragmites australis* in horizontal flow constructed wetlands for wastewater treatment: a review. Chem Eng J 290:232–242

Wake H (2005) Oil refineries: a review of their ecological impacts on the aquatic environment. Estuar Coast Shelf Sci 62:131–140

Wallace SD (2001) On-site remediation of petroleum contact wastes using surface flow wetlands. 2nd International Conference on Wetlands and Remediation. Burlington

Wallace S, Davis BM (2009) Engineered wetland design and applications for on-site bioremediation of PHC groundwater and wastewater. Soc Pet Eng 4(1):1–8

Wallace S, Kadlec R (2005) BTEX degradation in a cold-climate wetland system. Water Sci Technol 51(9):165–171

Wallace S, Schmidt M, Larson E (2011) Long term hydrocarbon removal using treatment wetlands, SPE Annual Technical Conference and Exhibition. Society of Petroleum Engineers (SPE) International, Denver, pp 1–10

Weis JS, Weis P (2004) Review. Metal uptake, transport and release by wetland plants: implications for phytoremediation and restoration. Environ Int 30:685–700

Wenzel WW (2009) Rhizosphere processes and management in plant-assisted bioremediation (phytoremediation) of soils. Plant Soil 321(1):385–408

Weyens N, van der Lelie D, Taghavi S, Vangronsveld J (2009) Phytoremediation: plant-endophyte partnerships take the challenge. Curr Opin Biotechnol 20(2):248–254

Wuyep PA, Chuma AG, Awodi S, Nok AJ (2007) Biosorption of Cr, Mn, Fe, Ni, Cu and Pb metals from petroleum refinery effluent by calcium alginate immobilized mycelia of Polyporus squamosus. Sci Res Essay 2(7):217–221

Xia P, Ke H, Deng Z, Tang P (2003) Effectiveness of constructed wetlands for oil-refined wastewater purification. In: Truong PN, Hanping X (eds) The third international conference on vetiver and exhibition: vetiver and water: an eco-technology for water quality improvement, land stabilization, and environmental enhancement, vol 22. Guangdong Academy of Agricultural Sciences China Agriculture Press, Guangzhou, pp 649–654

Yadav AK, Abbassi R, Kumar N, Satya S, Sreekrishnan T (2012) The removal of heavy metals in wetland microcosms: effects of bed, plant species, and metal mobility. Chem Eng J 211–212:501–507

Yang B, Lan C, Yang C, Liao W, Chang H, Shu W (2006) Long-term efficiency and stability of wetlands for treating wastewater of a lead/zinc mine and the concurrent ecosystem development. Environ Pollut 143:466–512

Yu L, Han M, He F (2017) A review of treating oily wastewater. Arab J Chem 10:S1913–S1922

Zhang J-E, Liu J-L, Liu Ounyang Y, Liao B-W, Zhao B-L (2010) Removal of nutrients and heavy metals from wastewater with mangrove *Sonneratia apetala Buch-Ham.* Ecol Eng 36(10):807–812

Zhang C-B, Liu W-L, Wang J, Ge Y, Ge Y, Chang SX, Chang J (2011) Effects of monocot and dicot types and species richness in mesocosm constructed wetlands on removal of pollutants from wastewater. Bioresour Technol 102:10260–10265

Zhao H, Li G (2011) Application of fibrous coalescer in the treatment of oily wastewater. Procedia Environ Sci 10:158–162

Zheng C, Zhao L, Zhou X, Fu Z, Li A (2013) Treatment technologies for organic wastewater. In: Water treatment. InTECH, pp 249–286

# Chapter 11
# Strategies for Biodegradation of Fluorinated Compounds

**Irina S. Moreira, Catarina L. Amorim, Cormac D. Murphy, and Paula M. L. Castro**

## Contents



I. S. Moreira (✉) · P. M. L. Castro
Universidade Católica Portuguesa, CBQF-Centro de Biotecnologia e Química Fina, Laboratório Associado, Escola Superior de Biotecnologia, Rua Arquiteto Lobão Vital 172, 4200-374, Porto, Portugal
e-mail: ismoreira@porto.ucp.pt

C. L. Amorim
Universidade Católica Portuguesa, CBQF-Centro de Biotecnologia e Química Fina, Laboratório Associado, Escola Superior de Biotecnologia, Rua Arquiteto Lobão Vital 172, 4200-374, Porto, Portugal

Biology Department and CESAM, University of Aveiro, Campus Universitário de Santiago, 3810-193, Aveiro, Portugal

C. D. Murphy
UCD School of Biomolecular and Biomedical Science, University College Dublin, Belfield, Dublin 4, Ireland



R. Prasad, E. Aranda (eds.), *Approaches in Bioremediation*, Nanotechnology in the Life Sciences, https://doi.org/10.1007/978-3-030-02369-0_11

## 11.1 Introduction: Fluorinated Organic Compounds

Fluorinated organic compounds possess a unique chemistry. The incorporation of fluorine into organic compounds dramatically alters their properties such as stability, lipophilicity and biological activity. These biological and/or physical effects are due to the unique properties of the fluorine atom, such as small van der Waals radius (1.47 Å), strong electronegativity (4.0) and high strength of the carbon-fluorine bond (485 KJ $mol^{-1}$), which strongly contribute to the high stability of the fluorinated substrates. The fluorine atom is isosterically similar to an hydroxyl group, thus allowing its use as a substitute for hydroxyl groups in organic compounds to introduce greater structural stability without affecting the steric characteristics of the molecule (O'Hagan 2008). The most frequently fluorinated compounds do not occur naturally in the environment, being mainly introduced through their production by humans (Dolbier 2005; Sandford 2000). Fluorine's outstanding properties have led to the development and application of fluorinated organic compounds in vast number of industrial, agricultural, manufacturing and medical purposes. For the synthesis of compounds with biological activity, such as pharmaceuticals and agrochemicals, the incorporation of a fluorine atom is very attractive to improve metabolic stability, bioactivity, bioavailability and protein-ligand interactions (Kirk 2006; Purser et al. 2008). Currently, fluorinated organic molecules account for up to 40% of all agrochemicals and 20% of all pharmaceuticals on the market (Grushin 2010). Moreover, these compounds are also widely used in other industries for the production of propellants, surfactants, adhesives, refrigerants and fire retardants (Chaojie et al. 2007).

## 11.2 Environmental Concerns

The production and extensive use of fluorinated organic compounds have resulted in the release of thousands of tons of these pollutants into environmental matrices (Prevedouros et al. 2006). Recently, in Zhejiang Province (China), an annual production of approximately 10–20 million tons of fluorine industrial wastewater has been reported (Wang et al. 2013). Fluoroquinolones and fluoxetine are examples of synthetic fluorinated pharmaceuticals which are among the most widely prescribed drugs worldwide (Sukul and Spiteller 2007). Recently, several studies have indicated the occurrence of these fluorinated pharmaceuticals in various environmental matrices, such as effluents of wastewater treatment plants, rivers, soils and even in tap water (Benotti et al. 2009; Kolpin et al. 2002; Lajeunesse et al. 2008; Adachi et al. 2013; Yang et al. 2010; Yiruhan et al. 2010; Zuccato et al. 2010). Although usually present at low concentrations, ranging from ng $L^{-1}$ up to μg $L^{-1}$, their presence may disturb the microbial ecology in those matrices (Adachi et al. 2013; Amorim et al. 2018) or even pose threats to living organisms (Dorne et al. 2007; Robinson et al. 2005). Polyfluoroalkyl compounds have been detected in environmental matrices, such as air (Liu et al. 2015a; Shoeib et al. 2010), precipitation and surface water (Mahmoud et al. 2009; Nguyen et al. 2017), seawater (Brumovský et al. 2016), sediments (Benskin et al. 2012), soil (Sepulvado et al. 2011), wastewater (Weinberg et al. 2010) and landfill leachates (Fuertes et al. 2017) and also drinking water (Ericson et al. 2009). Moreover, the use of aqueous film-forming formulations containing polyfluoroalkyl chemicals for firefighting resulted in soil and groundwater contamination (Backe et al. 2013; Moody and Field 1999).

The widespread use combined with the high stability of organofluorine compounds turn them in ubiquitous environmental contaminants, with high recalcitrance to biotic and/or abiotic degradation (Xiao et al. 2017). The robustness of the carbon-fluorine bond makes it cleavage especially difficult (Young and Grushin 1999), major for perfluoralkyl substituents and trifluoromethyl groups (Vargas et al. 2000). However, despite the stability of these compounds, they are not completely inert in the environment; they are potentially biological active, e.g. pharmaceuticals and pesticides, which are designed to have a biological function. Moreover, these compounds can also be transformed into other fluorinated metabolites that may pose an hazard (Murphy 2010).

The widely distributed presence of these compounds in the environment is a problem of serious concern worldwide due to their bioaccumulation (Martin et al. 2003; Xu et al. 2014) and toxic potential. Many of these compounds are potentially carcinogenic, neurotoxic, hepatotoxic and endocrine or immune systems disruptors which could cause harmful effects in living organisms (Buser and Scinicariello 2016). In addition, some organofluorine compounds have been detected in the human body (Calafat et al. 2007) and wildlife (Letcher et al. 2015).

## 11.3 Naturally Produced Fluorinated Compounds

Fluorine is present in the Earth's crust, being the 13th most abundant element and the most abundant halogen (Murphy et al. 2009). Despite its abundance, fluorine plays a minor role in biological systems, mostly because it exists mainly in an insoluble form (e.g. as calcium fluoride) in nature, which makes it unavailable (Eisenhauer and Harnisch 1998). Consequently, only a few fluorinated natural products (30) have been described among the more than 4000 organohalogens (Gribble 2002). Biogenic organofluorine compounds (Fig. 11.1) are mostly fluorinated carboxylic acids. Fluoroacetic acid (fluoroacetate) is the most common natural organofluorine compound. It was first discovered in the South African shrub *Dichapetalum cymosum* (Marais 1944) and has since been isolated from several plants growing in tropical and subtropical regions (O'Hagan and Harper 1999). Fluoroacetate is also produced by the bacterium *Streptomyces cattleya*, in which it is co-biosynthesized with the fluorinated amino acid, 4-fluorothreonine (Sanada et al. 1986). The fluorinase enzyme that catalyses the carbon-fluorine bond formation was first isolated from *Streptomyces cattleya* (O'Hagan et al. 2002); it catalyses the nucleophilic substitution of methionine in *S*-adenosyl methionine by fluoride, generating

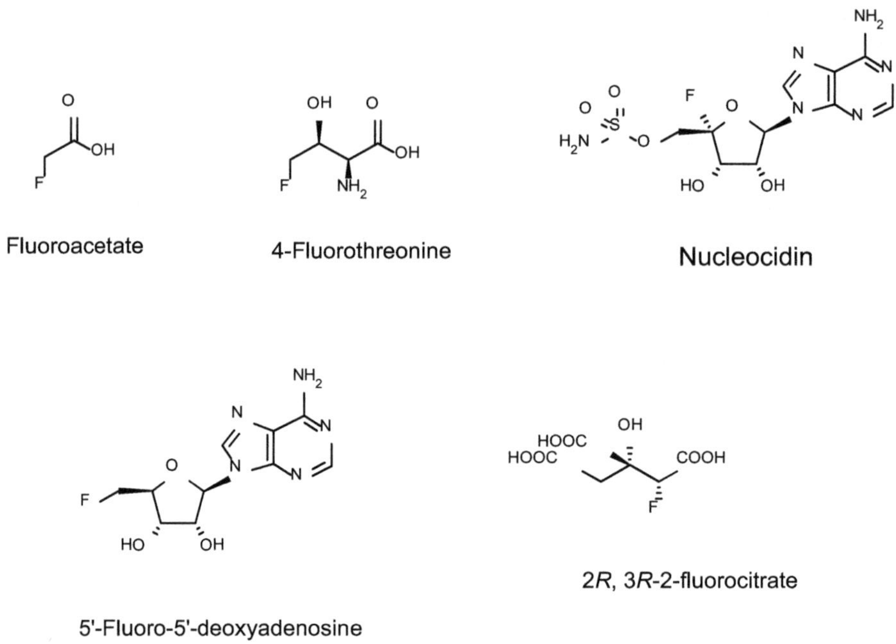


**Fig. 11.1** Examples of biogenic fluorinated compounds

5′-fluoro-5′-deoxyadenosine, which is the first intermediate of the fluorometabolite-producing pathway in that bacterium. Other fluorinase enzymes have since been identified in bacteria (Deng et al. 2014), but the key enzyme in *Streptomyces calvus*, which produces the fluorinated anti-trypanosomal agent nucleocidin (4′-fluoro-5′-O-sulphamoyladenosine), has yet to be uncovered (Bartholomé et al. 2017).

Many plants that accumulate fluoroacetate also produce fluorocitrate at lower concentrations by the citrate synthetase-catalysed condensation of fluoroacetyl-CoA with oxaloacetate (O'Hagan and Harper 1999). ω-Fluorinated fatty acids, which probably are also originated from fluoroacetyl CoA, were isolated from the seed oil of *Dichapetalum toxicarium* (Peters et al. 1960), the most abundant of which was identified as ω-fluorooleic acid (C18:1F), but small amounts of ω-fluoropalmitic acid (C16:0F) and minor additional ω-substituted fluoro acids were also identified (Christie et al. 1998; Harper et al. 1990). Fluoroacetate is highly toxic to animals owing to the in vivo synthesis of (2*R*, 3*R*) 2-fluorocitric acid as described above. This isomer of fluorocitrate is an inhibitor of the TCA-cycle enzyme aconitase (Lauble et al. 1996) and of citrate transport across the mitochondrial membrane (Kirsten et al. 1978). Despite its toxicity, some microorganisms can use fluoroacetate as a sole source of carbon and energy (Camboim et al. 2012; Walker and Lien 1981). A specific dehalogenase enzyme, which hydrolyses the carbon-fluorine bond yielding glycolate, has been isolated from different bacteria (Donnelly and Murphy 2009). Cattle deaths occur in regions where fluoroacetate-producing plants are indigenous; thus efforts have been made to circumvent the toxicity to animals by introducing genetically engineered rumen bacteria. For example, Gregg et al. (1998) cloned a gene encoding fluoroacetate dehalogenase into *Butyrivibrio fibrisolvens*, which was then introduced into sheep. Upon consuming a contaminated meal, these animals did not exhibit symptoms of fluoroacetate poisoning, suggesting that the modified bacterium expressed the dehalogenase, thus affording protection to the animals. Other researchers have investigated the adaptation of the rumen microbiome to non-lethal doses of fluoroacetate and demonstrated that transfaunation of the adapted rumen fluid to other animals resulted in the transfer of fluoroacetate resistance (Becker et al. 2016).

## 11.4 Biodegradation of Fluorinated Organic Compounds

### *11.4.1 Mechanisms of Biodegradation*

Biodegradation is the ability of microorganisms to transform or mineralize organic contaminants into less harmful, nonhazardous substances, which are then integrated into natural biogeochemical cycles (Margesin and Schinner 2001). Fungi, yeasts and, primarily, bacteria have a key role in the removal of those toxics from the environment (Neilson and Allard 2002). Biodegradation plays an important role in environmental clean-up, leading often to complete, cost-effective and eco-friendly elimination of pollutants, thus constituting an attractive way for the removal of recalcitrant compounds. Processes for the removal of pollutants can benefit from the capabilities of microbial enzymatic systems to reduce environmental pollution under gentle treatment conditions (Prasad 2017, 2018). Fluorinated organic compounds pose major challenges to microorganisms, the key recyclers of natural compounds, due to their novel and often complex structures. As discussed above, there are only a few known natural fluorinated organic compounds, in contrast with the myriad man-made organofluorines. Additionally, while natural fluorinated organics contain one fluorine atom, the synthetic ones often contain many fluorine substituents and sometimes are even fully fluorinated. Considering the high stability conferred by the strong carbon-fluorine bond, it is not surprising that little is known about the metabolism of these compounds. However, their stability, bioactivity, and potential for accumulation in the environment stress the importance of understanding their environmental fate and biodegradation mechanisms.

In relation to the degradation of fluorinated aliphatic compounds, microbial defluorination can be performed either by a dehalogenase or by a dehydrogenase. The first studies about the defluorination reactions performed by fluoroacetate dehalogenase started in the 1960s (Goldman 1965; Goldman and Milne 1966; Tonomura et al. 1965). Liu et al. (1998) proposed that the catalytic mechanism for fluoroacetate dehalogenase is similar to the mechanism of haloalkane dehalogenase; basically, the halogen atom is substituted by an aspartate that acted as nucleophile, followed by the hydrolysis of the ester intermediate formed by a water molecule activated by histidine. The conversion of 3-fluoropyruvate to acetate and fluoride ions was reported to be performed by a pyruvate dehydrogenase, a reaction in which fluoride is removed by β-elimination, the classical mechanism for dehydrogenases (Leung and Frey 1978). Defluorination can also occur due to the action of a reductase as is the case of maleylacetate reductase that catalyses the defluorination of 2-fluoromaleylacetate as well as other halomaleylacetates (Kaschabek and Reineke 1995). Fluorine removal in fluorinated cycloalkyl N-phenylcarbamates occurs via hydroxylation reactions (Haufe et al. 2003).

In the case of fluorinated aromatic compounds, the degradation pathway is similar to that for the degradation of aromatic compounds in general. In both aerobic and anaerobic catabolism, well-defined channels have evolved for the most commonly encountered aromatic compounds. Structurally diverse pollutants are first transformed into a few intermediates through peripheral pathways, which are then further channelled, via a few central pathways, to the central metabolism. The anaerobic catabolism is based on reductive reactions to attack the aromatic ring, in which the different aromatic compounds are converted into a few key intermediates, which are substrates for the corresponding dearomatizing reductases and then channelled by the cognate central pathways to the central metabolism (Carmona et al. 2009). It is important to notice that the anaerobic degradation may depend on the availability of electron acceptors (Vargas et al. 2000). In aerobic degradation, most peripheral pathways carry out oxygenation reactions catalysed by mono- or dioxygenases that convert the aromatic pollutants to dihydroxy aromatic intermediates, which are then cleaved by intradiol (ortho-cleavage) or extradiol (meta-cleavage) dioxygenases. The generated metabolites are then channelled to subsequent central pathways leading to the formation of Krebs cycle intermediates. The ring-cleavage enzymes from various bacteria display significant functional similarities. These peripheral enzymes, which recognize and convert different aromatic pollutants into several central metabolites, play a significant role in several pollutants degradation (Cao et al. 2009).

Since halogen substituents are to a large extent responsible for the properties of halogenated aromatic compounds, removal of these substituents constitutes a key step in the biodegradation pathway of those compounds. In most cases, dehalogenation occurs after the cleavage of the aromatic ring. However, direct dehalogenation without loss of aromaticity has been also demonstrated (Commandeur and Parsons 1990). Eight dehalogenation mechanisms were described for the metabolism of halogenated compounds: hydrolytic dehalogenation, thiolytic dehalogenation, intramolecular substitution, dehydrohalogenation, dehalogenation by hydration, dehalogenation by methyl transfer, oxidative dehalogenation and reductive dehalogenation (van Pée and Unversucht 2003). As a general rule, the resistance to enzymatic cleavage of the carbon-halogen bond increases with the increasing of the substituent's electronegativity, being also dependent on the specificity of the enzymes catalysing the cleavage (Fetzner and Lingens 1994).

Fungal metabolism of xenobiotic compounds have been widely studied as a model of mammals metabolism, since these organisms possess cytochrome P450 enzymes, as well as enzymes related with phase II metabolism, such as sulfotransferases and glycosyl transferases (Asha and Vidyavathi 2009). In this way, fungal species degrade fluorinated compounds by hydroxylation, *N*-oxidation and *N*-demethylation reactions, performed by cytochrome P450 enzymes, followed by conjugation, originating sulphated, glucoside and glucuronic acid conjugates (Amadio and Murphy 2010; Cerniglia et al. 1984). However, in these cases, there is only biotransformation of the fluorinated substrates, in a detoxification mechanism, without cleavage of the carbon-fluorine bond.

In relation to poly- and perfluoroalkyl substances (PFASs), the complete mineralization is not expected under natural conditions (Liu and Mejia Avendaño 2013). In aerobic biodegradation, defluorination occurs in multiple reaction steps and with removal of one or two perfluorocarbons, which suggests the possibility of partial mineralization of perfluoroalkyl carbon chain (Wang et al. 2009). Usually higher biodegradability is observed for the compounds with shorter fluorinated carbon chain, which results in higher water solubility and higher bioavailability (Kim et al. 2012).

### *11.4.2 Selection/Attainment of Degrading Organisms*

Human activities have added a plethora of new fluorinated chemicals to the environment, which novel structures pose major challenges to the microorganisms. Catabolic genes are often associated with integrase and invertase genes or insertion elements and are usually located on plasmids (Janssen et al. 2005). When microorganisms face a new organic chemical in their environment, the catabolic genes needed for its degradation can be obtained through conjugational or transformational events from other microorganisms, or microorganisms can undergo an adaptive process through selection and mutation events leading to the modification of existing genes (Chaudhry and Chapalamadugu 1991). Molecular engineering can be used to increase the expression of a specific protein, enzyme or metabolic pathway to increase degradation (Kumamaru et al. 1998). However, the application of genetically engineered microorganisms in natural environments is strictly restricted in many countries (Sivasubramaniam and Franks 2016). For this reason, isolation of degrading microorganisms from environmental matrices is still a major tool for remediation of contaminated compartments.

Isolation of degrading bacteria can be performed by traditional cultivation-dependent approaches through enrichment in selective media containing the target xenobiotic compound followed by agar plating methods. Selective media that contain the target compound as the sole carbon source enrich the active subset of degrading microorganisms. Additionally, the exposure to the pollutant induces slow changes of the metabolic pathway of persistent bacteria allowing them to survive and proliferate with the stress. The origin of the samples used for the enrichment, the composition of the cultivation media and the incubation conditions are important factors for the successful isolation of degrading microorganisms (Chen et al. 2017; Rathod and Archana 2013). Varying the selective pressures in enrichment cultures can have an effect on the genetic potential for degrading metabolism (Guerra et al. 2018). The most frequent approach is the use of synthetic mineral salts media, but there are a few studies that make use of other media, such as wastewater (Bahobail et al. 2016). Samples collected from contaminated environments are used, as microorganisms are already adapted to the presence of similar pollutants (Carvalho et al. 2002; Duque et al. 2012). The importance of using acclimated inoculum to enhance the potential biodegradation of the compounds of interest was highlighted in the study of Wang et al. (2011), in which 6:2 fluorotelomer sulfonic acid was found non-biodegradable with activated sludge from a WWTP but biodegradable by the sludge from a different WWTP. An important observation is that longer enrichment times are required with the increase of fluorine substitution and that the increase in fluorine substitution results in enriched consortia with lower microbial diversity (Zhao et al. 2015). Several microorganisms capable of degrading fluorinated organics have been obtained by this methodology (Table 11.1), helping to elucidate the mechanisms and the metabolic pathways of degradation of several organofluorine compounds.

Usually the enrichments are performed in batch flasks experiments, with several transfers, but there are some variations to this approach, such as the use of sequencing batch reactors (SBRs) (Zhao et al. 2015). Another option is the soil-charcoal perfusion system, in which a soil sample is mixed with autoclaved charcoal, as a microhabitat, and the medium is circulated using an air pump through the soil-charcoal layer in the perfusion apparatus. The enriched charcoal is then used for bacterial isolation. This method has been used for the isolation of bacteria with the ability to degrade halogenated pollutants (Sakakibara et al. 2011; Takagi et al. 2009; Wang et al. 2017). Chemostat-enrichment systems have also been used for the isolation of degrading microorganisms (Rajoo et al. 2013). The system contained a medium with the target compound to be degraded as the sole carbon source, and the culture broth is replaced with fresh culture medium at a fixed dilution rate, resulting in the washout of the non-degrading microorganisms, whose growth rate is lower than the dilution rate. Seralathan et al. (2015) reported the isolation of xenobiotic-degrading bacteria based on chemotaxis of microbes towards pollutants in a study conducted with a fabricated prototype made as a vertical soil matrix column.

Most of the time, when performing enrichments for the isolation of degrading strains, the target compound for which degradation is aimed, is used as the carbon source in the cultivation media. However, sometimes structural analogues are used to overcome problems related with bioavailability of the target compound (Garrido-Sanz et al. 2018). Sakakibara et al. (2011) isolated dieldrin-degrading bacteria using aldrin *trans*-diol, which has greater water solubility, as an enrichment substrate. In another study, hexachlorobenzene-degrading bacteria were isolated using pentachloronitrobenzene as the substrate (Takagi et al. 2009).

There are a number of biochemical tests that can be used for the identification of the degradation potential of bacterial strains, which can be employed to select the degrading candidates after enrichment or to the isolates obtained directly from the environmental sample. For example, the use of MST - MS medium supplemented with triphenyltetrazolium chloride, an electron acceptor, which is reduced to the red

**Table 11.1** Examples of microorganisms with the ability to degrade organofluorine compounds

| Bacteria | Compound | Reference |
|---|---|---|
| *Labrys portucalensis* F11 | Fluorobenzene | Carvalho et al. (2005, 2006) |
| *Labrys portucalensis* F11 | 1,2- and 1,3-Difluorobenzene | Moreira et al. (2012a) |
| *Labrys portucalensis* F11 | 2-, 3- and 4-Fluoroaniline | Amorim et al. (2013a) |
| *Labrys portucalensis* F11 | Fluoxetine | Moreira et al. (2014) |
| *Labrys portucalensis* F11 | Ofloxacin, norfloxacin and ciprofloxacin | Amorim et al. (2014b) |
| *Labrys portucalensis* F11 | Ofloxacin and levofloxacin | Maia et al. (2018) |
| *Labrys portucalensis* F11 | Moxifloxacin | Carvalho et al. (2016) |
| *Rhodococcus* sp. FP1 | 2-Fluorophenol | Duque et al. (2012) |
| *Rhodococcus* sp. S2 | 4-Fluorocinnamic acid | Amorim et al. (2014a) |
| *Rhodococcus* sp. 89 | Fluorophenols | Bondar et al. (1998) |
| *Rhodococcus corallinus* 135 | Fluorophenols | Bondar et al. (1998) |
| *Rhodococcus erythropolis* 1CP | Fluorophenols | Bondar et al. (1998) |
| *Rhodococcus opacus* 1G | Fluorophenols | Bondar et al. (1998) |
| *Rhodococcus opacus* 1cp | 2-, 3- and 4-Fluorophenol | Finkelstein et al. (2000) |
| *Sphingomonas* sp. HB-1 | 3-Fluorobenzoate | Boersma et al. (2004) |
| *Sphingomonas* sp. SS3 | 4-Fluorodiphenyl ether | Schmidt et al. (1992) |
| Pseudomonad | 2-Fluorobenzoate | Milne et al. (1968) |
| *Pseudomonas* (spp.) | 2-Fluorobenzoate | Vora et al. (1988) |
| *Pseudomonas* spp. | Pentafluorosulfanyl-substituted aminophenol | Saccomanno et al. (2018) |
| *Pseudomonas* sp. D2 | Difluoromethane sulfonate | Key et al. (1998) |
| *Pseudomonas* sp. OCY4 and OCW4 | 8:2 Fluorotelomer alcohol | Liu et al. (2007) |
| *Pseudomonas* sp. B13 | 4-Fluorobenzoate | Schreiber et al. (1980) |
| *Pseudomonas* sp. B13 | 2-Fluorobenzoate | Engesser et al. (1980) |
| *Pseudomonas* sp. RHO23, RHO24 and RHO26 | 4-Fluorobenzoate | Oltmanns et al. (1989) |
| *Pseudomonas* sp. MFA9 and MFA32 | Monofluoroacetate | Alexandrino et al. (2018) |

(continued)

**Table 11.1** (continued)

| | | |
|---|---|---|
| *Pseudomonas knackmussii* B13 | 4-Fluorobenzoate | Misiak et al. (2011) |
| *Pseudomonas stutzeri* | 2- and 4-Fluorobenzoate | Vargas et al. (2000) |
| *Pseudomonas stutzeri* | 2-, 3- and 4-Fluorobenzoate | Song et al. (2000) |
| *Pseudomonas oleovorans* | 2-Fluorobenzoate | Song et al. (2000) |
| *Pseudomonas oleovorans* | 4:2, 6:2 and 8:2 Fluorotelomer alcohols | Kim et al. (2012) |
| *Pseudomonas butanovora* | 4:2, 6:2 and 8:2 Fluorotelomer alcohols | Kim et al. (2012) |
| *Pseudomonas fluorescens* 26-K | 3,4-Difluoroaniline | Travkin et al. (2003) |
| *Pseudomonas fluorescens* DSM 8341 | 6:2 Fluorotelomer alcohol | Kim et al. (2014) |
| *Pseudomonas pseudoalcaligenes* KF707 | 2- and 4-Fluorobiphenyl | Murphy et al. (2008) |
| *Pseudomonas pseudoalcaligenes* KF707 | 2,3,4,5,6-Pentafluorobiphenyl and 4,4′-difluorobiphenyl | Hughes et al. (2011) |
| *Pseudomonas cepacia* | 4-Fluorobenzoate | Schlömann et al. (1990) |
| *Pseudomonas putida* MFA15 | Monofluoroacetate | Alexandrino et al. (2018) |
| *Alcaligenes* sp. RHO21 and RHO22 | 2-, 3- and 4-Fluorobenzoate | Oltmanns et al. (1989) |
| *Alcaligenes eutrophus* B9 | 2-Fluorobenzoate | Engesser et al. (1980) |
| *Alcaligenes eutrophus* 335 | 4-Fluorobenzoate | Schlömann et al. (1990) |
| *Burkholderia xenovorans* LB400 | 2,3,4,5,6-Pentafluorobiphenyl and 4,4′-difluorobiphenyl | Hughes et al. (2011) |
| *Burkholderia fungorum* FLU100 | Fluorobenzene | Strunk and Engesser (2013) |
| *Arthrobacter humicola* MFA12 | Monofluoroacetate | Alexandrino et al. (2018) |
| *Arthrobacter* sp. IF1 | 4-Fluorophenol | Ferreira et al. (2008) |
| *Ralstonia* sp. FD-1 | 4-Fluoroaniline | Song et al. (2014) |
| *Arthrobacter* sp. G1 + *Ralstonia* sp. H1 | 4-Fluorocinnamic acid | Hasan et al. (2011) |
| *Acinetobacter calcoaceticus* N.C.I.B 8250 | 2-, 3- and 4-Fluorobenzoate | Clarke et al. (1975) |
| *Acinetobacter* sp. TW | 4-Fluoroaniline | Wang et al. (2013) |
| *Thauera aromatica* | 2-, 3- and 4-Fluorobenzoate | Song et al. (2000) |
| *Thauera aromatica* | 4-Fluorobenzoate and 4-fluorotoluene | Tiedt et al. (2016) |
| *Bacillus thermoleovorans* A2 | 2-Trifluoromethylphenol | Reinscheid et al. (1998) |
| *Aureobacterium* sp. RHO25 | 4-Fluorobenzoate | Oltmanns et al. (1989) |
| *Azoarcus tolulyticus* | 2- and 4-Fluorobenzoate | Song et al. (2000) |

(continued)

**Table 11.1** (continued)

| | | |
|---|---|---|
| *Bradyrhizobium elkanii* | 2- and 3-Fluorobenzoate | Song et al. (2000) |
| *Mesorhizobium loti* | 2- and 4-Fluorobenzoate | Song et al. (2000) |
| *Acidovorax avenae* | 2-, 3- and 4-Fluorobenzoate | Song et al. (2000) |
| *Ensifer adhaerens* | 2-, 3- and 4-Fluorobenzoate | Song et al. (2000) |
| *Comamonas testosteroni* MFA1 and MFA35 | Monofluoroacetate | Alexandrino et al. (2018) |
| *Stenotrophomonas maltophilia* MFA2 | Monofluoroacetate | Alexandrino et al. (2018) |
| *Herbaspirillum frisingense* MFA4 | Monofluoroacetate | Alexandrino et al. (2018) |
| *Delftia acidovorans* MFA5 | Monofluoroacetate | Alexandrino et al. (2018) |
| *Achromobacter anxifer* MFA16 and MFA31 | Monofluoroacetate | Alexandrino et al. (2018) |
| *Variovorax paradoxus* MFA10 | Monofluoroacetate | Alexandrino et al. (2018) |
| *Chryseobacterium taeanense* MFA 25 | Monofluoroacetate | Alexandrino et al. (2018) |
| **Fungi** | **Compound** | **Reference** |
| *Cunninghamella elegans* | 4-Fluorobiphenyl | Amadio and Murphy (2010) |
| *Cunninghamella elegans* ATCC 36112 | 1-Fluoronaphthalene | Cerniglia et al. (1984) |
| *Exophiala jeanselmei* | Fluorophenols | Boersma et al. (1998) |
| *Penicillium frequentans* Bi 7/2 | 2,3-, 2,4-, 2,5- and 3,4-Difluorophenol | Wunderwald et al. (1998) |
| *Gloeophyllum striatum* | 2-Fluorophenol | Kramer et al. (2004) |
| *Pisolithus tinctorius* | 2- and 3-Fluorophenol | Franco et al. (2014) |

triphenylformazan by bacterial dehydrogenases, allowed for the identification of these enzymes, which are important for the degradation of halogenated organic compounds (Olga et al. 2008). A 2,6-dichlorophenol indophenol assay, oxidation-reduction indicator, has also been demonstrated to be suitable for the rapid selection of degrading microorganisms (Bučková et al. 2013; Hanson et al. 1993; Kubota et al. 2008). Moreover, the existence of biochemical tests based on colorimetric reactions, such as 4-aminoantipyrine assay for phenolic compounds, also allowed its application for high-throughput screening in 96-well microplates (Lu et al. 2015).

Enrichment through cultivation has limitations regarding the high proportion of unculturable microorganism in natural microbial communities. The emergence of cultivation-independent methodologies has overcome this drawback and complements the enrichment procedure. Genetic screenings for the presence of genes expressing degrading enzymes, such as oxygenases, have been applied (Guisado et al. 2015). Next-generation sequencing techniques, such as Illumina sequencing technology, have been recently applied as promising methods for the characterization of the diversity of genes expressed by complex microbial communities in different environments, allowing the identification of the catabolic genes expressed during degradation (Auffret et al. 2015; Omrani et al. 2018). In addition, proteomics-based methodologies have been useful in determining changes in the composition and abundance of proteins and in the identification of key proteins involved in the degradation of xenobiotics (Festa et al. 2017).

PCR-DGGE and high-throughput sequencing technologies have been used to follow and characterize the microbial community during and after enrichments (Zhao et al. 2014, 2015). It is common to observe that the most abundant bacterial OTUs obtained by next-generation sequencing of the enrichment cultures are not recovered via classical culturing techniques (Wang et al. 2017); however, this knowledge has the potential to guide the isolation of those degrading strains by traditional culture-dependent methods and thus improve outcomes. Another important advance was the use of DNA-based stable isotope probing (SIP). This method employs stable isotopes (13C or 15N) to identify functionally active members within microbial communities by analysing their isotope-enriched intracellular components (DNA, RNA or proteins), allowing for the identification of the microorganisms responsible for the degradation of contaminants (Leewis et al. 2016; Li et al. 2017; Yamasaki et al. 2012). Perruchon et al. (2017) described the utilization of a combination of culture-independent methods including antibiotics-driven selection with DNA/RNA-DGGE, q-PCR and stable isotope probing (SIP)-DGGE to identify the key degrading members of an enriched consortium. In a study developed by Garrido-Sanz et al. (2018), metagenomic analysis, combining sequencing of 16S rRNA gene and metagenome shotgun sequencing, of an enriched bacterial consortium led to identifying not only the microorganisms involved in the degradation process but also to assign specific reactions and metabolic pathways to the bacterial population. Functions of the bacterial communities can also be predicted using bioinformatics tools, such as PICRUSt (Phylogenetic Investigation of Communities by Reconstruction of Unobserved States), which allows the prediction of metagenome functional and metabolic profiles from 16S rRNA gene sequences (Langille et al. 2013; Mukherjee et al. 2017; Sun et al. 2016). Bargiela et al. (2015) discovered the biodegradation potential of bacterial communities from polluted sites based on taxonomic barcoding and metagenomic prediction platforms for inferring biodegradation activities; predicted activities were further validated by experimental degradation assays. The use of these cultivation-independent methods has the potential to improve strain isolation in the future, guiding the use of specific targeted isolation medium, for example.

In addition, most bacteria in nature are in a "viable but non-culturable" (VBNC) state (Oliver 2010), which can be studied by molecular analyses. Nevertheless, these methods do not allow the recovery and growth of those bacterial strains. The discovery of the resuscitation-promoting factor (Rpf), a bacterial cytokine secreted by *Micrococcus luteus* (Mukamolova et al. 1998), allowed the culturing of difficult-to-culture bacteria. This approach of promoting the growth of previously uncultured microorganisms can be applied for screening pollutant-degrading microorganisms, to obtain rare species and provide an opportunity to investigate the relevant biodegrading reactions and genotypes in pure cultures. Su et al. (2013) followed this approach to explore the biphenyl degradative potential of VBNC or uncultured indigenous bacteria through enrichment using PCB-contaminated soil samples. Addition of Rpf to the enrichment culture, to promote resuscitation and growth of bacteria in VBNC, was also used to obtain degrading fastidious bacteria for dye degradation (Jin et al. 2017).

Metagenomic strategies based on the construction and screening of constructed libraries can also be applied for accessing biodegradative genes from polluted sites. The metagenomic DNA can be extracted directly from the contaminated environment or after enrichment for the desired activity. Biodegradative genes can then be screened by function- or sequence-based approaches. The sequence-based approach is dependent on sequence information, in which the genes related with the biodegradation are screened based on conserved nucleotide sequences. In the function-based approach, the library is screened for clones exhibiting the desired phenotype; this approach has the potential to discover new classes of genes with known or new functions (Madhavan et al. 2017). Metagenomic technologies provide insights into the metabolic potential of the microbial communities. Additionally, the genes for the degradation pathway can be cloned and expressed in an appropriate host for the biodegradation of xenobiotics in confined environments, such as bioreactors.

With the advancement of genome sequencing technology and bioinformatics tools, it is possible to predict the structure and function of all genes in an organism, allowing the estimation of its potential metabolic activity. Comparison of sequences, gene finding and prediction of gene expression are the main techniques used by computational genomics to elucidate the specific function of a sequence. There are several databases available that provide information regarding metabolic pathways of microorganisms. The sequence of the putative enzymes can then be used for homology modelling, in order to get the predicted three-dimensional structure of the protein. Homology modelling enables a molecular model of the protein to be constructed from its gene sequence data using as a "template" the structure of a homologous protein (Schwede et al. 2003). Then, protein-ligand docking can be used to predict the potential of a compound to be degraded by the enzyme (Liu et al. 2018). Generally, to perform the docking, a library of appropriate molecules is created, and then docking is performed between the molecules in the library and the target enzyme (Dong et al. 2017). The results containing high binding scores are potential substrates for the selected enzyme. Concerning the application of this tool to biodegradation, up to date, it was used to predict the pollutants which can be potential targets for laccases, and some "in silico" results were supported by experimental data (Suresh et al. 2008). Molecular docking was used to screen various petroleum hydrocarbons with catechol 2,3-dioxygenase in order to predict the potential substrates (Ajao et al. 2012). Molecular docking has also been used to understand the interaction of azo dyes with azoreductase (Haghshenas et al. 2016) and between pyrene and the active site of naphthalene dioxygenase from *Pseudomonas* sp. JPN2 (Jin et al. 2016). These studies also allow to identify the candidate residues for substitution to improve the interactions between protein and ligand, which could be used to increase the biological activity of bacteria. A review on the application of molecular docking between organic pollutants and enzymes to understand the reaction mechanisms and its application on environmental remediation was recently published (Liu et al. 2018).

### 11.4.3 Effects of Co-contamination in Biodegradation

Contaminated ecosystems typically contain heterogeneous mixtures of organic compounds, frequently also combined with inorganic species such as heavy metals. Many studies indicate that contamination with metal ions may inhibit microorganisms by decreasing bacterial growth and biodegradation rates (Lin et al. 2006; Riis et al. 2002; Sokhn et al. 2001; Utgikar et al. 2003). The level of inhibition may depend on the nature, concentration and availability of the heavy metals (Amor et al. 2001; Hoffman et al. 2005). Heavy metals inhibit microorganisms by blocking essential functional groups or by interfering with incorporation of essential metal ions into biological molecules (Doelman et al. 1994; Gadd and Griffiths 1978). Heavy metals cations, especially those with high atomic numbers, as $Hg^{2+}$, $Cd^{2+}$ and $Ag^{+}$, tend to bind to -SH groups, inhibiting the activity of sensitive enzymes (Nies 1999). Inhibition of fluorobenzene biodegradation by *Labrys portucalensis* F11 due to the metals $Ag^{+}$ and $Cu^{2+}$ through inhibition of the enzyme catechol 1,2-dioxygenase, as indicated by the accumulation of catechol and 4-fluorocatechol on the culture media, has been observed (Moreira et al. 2013).

In the presence of mixtures of organic compounds, the removal or degradation of one substrate can be impacted by other substrates in the mixture. This happens not only for mixtures of toxic chemicals but also for mixtures of pollutants with easily degradable compounds, such as sugars. The positive effect of the presence of a second compound on target pollutant biodegradation has been demonstrated to occur due to the induction of suitable degrading enzymes (Alvarez and Vogel 1991) or due to an increase of the growth at low substrates levels (Delgadillo-Mirquez et al. 2011). Nevertheless, negative interactions may also occur and are attributed to competitive and/or non-competitive inhibition (Tsai and Juang 2006), toxicity (Reardon et al. 2000), formation of toxic intermediates (Bartels et al. 1984) and synergistic inhibition of bacterial growth (Smith et al. 1991). An example of a positive effect of the presence of a second compound was observed in a study for the degradation of fluorobenzenes, in which the presence of fluorobenzene accelerated the biodegradation of 1,3-difluorobenzene and stimulated the growth of bacterial cells. Moreover, degradation of 1,4-difluorobenzene was only achieved in the presence of fluorobenzene, which suggests that the addition of fluorobenzene is necessary to induce the appropriate catabolic enzymes. On the other hand, the presence of 1,2-difluorobenzene inhibited the degradation of fluorobenzene (Moreira et al. 2012a).

Among interactions between substrates during biodegradation, special attention has been paid to the mechanism of cometabolism. In cometabolism, a nongrowth substrate is transformed in the presence of a growth substrate or another transformable compound, which is capable of inducing enzymes or producing cofactors and/or metabolites, which in turn are required for the transformation and/or to support cell growth by generating energy and carbon polymers. The growth substrate can be an analogue of the toxic compound (structure-driven cometabolism), often leading to a competitive inhibition between growth and nongrowth substrates, when they share the same enzymatic pathway. Alternatively, the growth substrate can be an easily degradable carbon source as yeast extract, sugars or acids (energy-driven cometabolism) (Ziagova and Liakopoulou-Kyriakides 2007). Several publications have shown cometabolism as a successful mechanism to potentiate the biodegradation, especially of micropollutants. Supplementation with a growth substrate, such as acetate, resulted in an increased biodegradation of fluorinated pharmaceuticals like fluoxetine (Moreira et al. 2014) and fluoroquinolone antibiotics (Amorim et al. 2014b; Carvalho et al. 2016; Maia et al. 2018). Degradation of 2- and 3-fluorophenol by the ectomycorrhizal fungi *Pisolithus tinctorius* was also improved by cometabolism with glucose (Franco et al. 2014).

### 11.4.4 Enantioselectivity in Biodegradation

A significant number of the organic chemicals regulated by the US EPA are chiral, possessing at least one element of asymmetry and leading to the existence of two or more stereoisomers called enantiomers. In general, these chiral compounds are used and thus released to the environment as racemates (mixtures of equal amounts of enantiomers), but some are used in its enantiomerically pure forms. Enantiomers of chiral compounds have similar physical-chemical properties, but each selectively interacts with biological systems, which are usually enantioselective. This may result in enantioselective toxicity, mutagenicity, carcinogenicity and endocrine disruptive activity (Lewis et al. 1999; Liu et al. 2005; Wong 2006). In the environment, abiotic transformations of chiral compounds are mostly non-enantioselective, whereas biological degradation usually proceeds with high enantioselectivity (Müller and Kohler 2004; Wong 2006). Understanding the impact of stereoselectivity on the degradation and transformation of chiral organic compounds is very important to assess the environmental fate and susceptibility to biodegradation (Maia et al. 2017).

Enantioselective biodegradation implies that the enzymes involved in the conversion of such substrates are able to differentiate between the enantiomers. The biodegradation of enantiomers of chiral pollutants may proceed through different pathways, by two enantioselective enzymes or by one enzyme that degrades both enantiomers simultaneously, but at different rates, or degrades the enantiomers sequentially (Müller and Kohler 2004). In some enzymes, with overall conserved folding, a few amino acids residues determine the stereoselectivity. Other enzymes express opposite stereospecificity and have completely different folds and surprisingly similar active sites that are mirror-images. Theoretically, two enzymes with identical sequences but built from enantiomeric amino acids, one built with D-amino acids and other built with L-amino acids, should have opposite stereospecificity for chiral substances (Müller and Kohler 2004).

Enantioselectivity in the biodegradation of fluorinated organics was observed on the preferential degradation of the (*R*)-enantiomer of fluoxetine over the corresponding (*S*)-enantiomer, both at racemic and single enantiomeric supplementation, by *Labrys portucalensis* F11 (Moreira et al. 2014). On the other hand, non-enantioselective degradation was observed in biodegradation assays using real wastewater samples spiked with fluoxetine (Ribeiro et al. 2014a). These results highlight that the enantioselectivity in the degradation of chiral compounds depend on the microorganisms. Recently, Maia et al. (2018) reported the enrichment of the (*R*)-enantiomer of racemic ofloxacin and racemization of the pure (*S*)-ofloxacin (levofloxacin) by *Labrys portucalensis* F11. A similar pattern for the degradation of the same compounds by an activated sludge consortium was also observed (Maia et al. 2016). A study performed with an aerobic granular sludge-sequencing batch reactor (AGS-SBR) operated with simulated wastewater containing a mixture of chiral pharmaceuticals, including fluoxetine and norfluoxetine, reported preferential removal of the (*R*)-enantiomer for norfluoxetine, indicating the occurrence of enantioselective biologically mediated processes, while fluoxetine and other non-fluorinated pharmaceuticals were abiotically removed in an non-enantioselective manner, through adsorption (Amorim et al. 2016).

### *11.4.5 Mineralization Versus Biotransformation*

In an ideal scenario, biodegradation of fluorinated organic compounds would lead to complete mineralization, with $CO_2$, $H_2O$ and inorganic fluoride ion ($F^-$) as final products. The stoichiometric release of the fluoride ion into the medium is considered a strong indication of complete degradation, since dehalogenation is usually the limiting step in the biodegradation of halogenated compounds (Kiel and Engesser 2015). However, in some cases, the microorganisms might not be able to perform complete mineralization, resulting in the formation of dead-end metabolites that cannot be further metabolized, a mechanism which is more appropriately defined as biotransformation. This is commonly the case of compounds with more than one fluorine atom, such as compounds containing trifluoromethyl or pentafluorosulfanyl groups and also the case of PFAs.

Trifluoroacetic acid (TFA), which occurs naturally in thermal vents and in the troposphere via hydrofluorocarbon (HFC) reactions with radicals, is a pollutant that is potentially phytotoxic (Smit et al. 2009). Visscher et al. (1994) and Kim et al. (2000) showed that TFA is anaerobically biodegradable; however, the compound is metabolically stable under aerobic conditions (Alexandrino et al. 2018). The trifluoromethyl group is popular in drugs and agrochemicals (Fig. 11.2); however, this

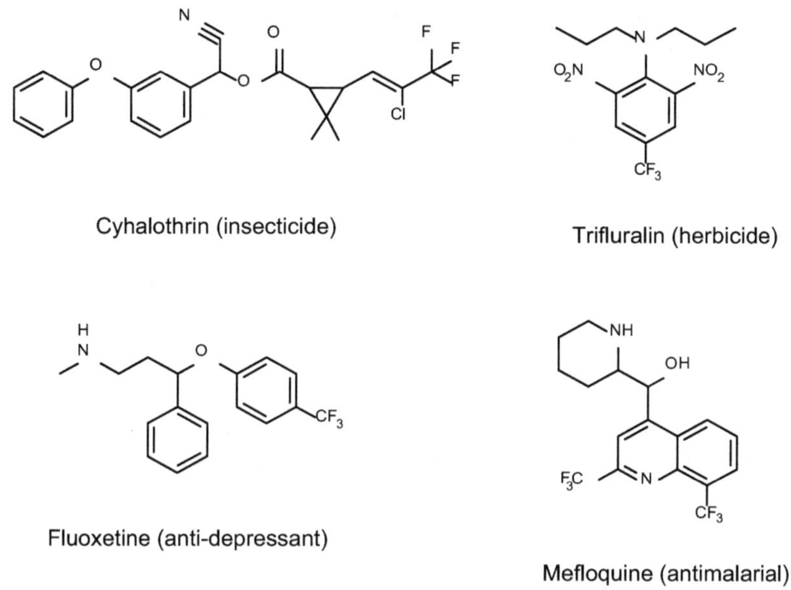


**Fig. 11.2** Examples of drugs and agrochemicals bearing the trifluoromethyl group

substituent is highly resistant to microbial attack. Earlier work by Engesser et al. (1988) demonstrated that 3- and 4-trifluoromethyl benzoates were biotransformed by *Pseudomonas putida* to the dead-end metabolite 2-hydroxy-6-oxo-7,7,7-trifluorohepta-2,4-dienoate via meta-cleavage. Recently, Palmer-Brown et al. (unpublished) demonstrated by 19F NMR that the trifluoromethyl group of the pesticide λ-cyhalothrin remained intact when this compound was incubated with the fungus *Cunninghamella elegans*, although the starting compound was biodegraded. Nevertheless, there are reports of defluorination of trifluoromethyl-containing substrates, for example, stoichiometric amounts of fluoride ion were detected upon incubation of the bacterium *Labrys portucalensis* F11 with low concentrations of fluoxetine (Moreira et al. 2014). Furthermore, Yano et al. (2015) isolated a bacterium, *Rhodococcus* sp. 065240, which defluorinates benzotrifluoride. These studies point out that whether a given fluorinated substrate is completely degraded or merely biotransformed depends not only on the structure of the compound but also on the enzymatic machinery of the microorganism.

The pentafluorosulfanyl group (-$SF_5$) is viewed as a potential replacement of the trifluoromethyl group in pharmaceuticals and agrochemicals owing to its enhanced stability, lipophilicity and electronegativity (Sowaileh et al. 2017). Improvements in synthetic strategies are enabling $SF_5$-derivatives of bioactive compounds to be prepared. For example, Moraski et al. (2017) synthesized eight $SF_5$-containing anti-tuberculosis compounds, which displayed similar activity to the corresponding $CF_3$-compounds but were more stable in microsomal preparations. Given its stability, it was a surprise that some *Pseudomonas* spp. were able to grow on

5-(pentafluorosulfanyl)2-aminophenol as a sole carbon and energy source (Saccomanno et al. 2018). Biotransformation of the parent compound was corroborated by the non-stoichiometric fluoride ion release. Furthermore, a $SF_5$-substituted catechol was identified as a metabolite and was shown to have inhibitory properties. However, other catabolic intermediates were not detected; thus the pathway of biodegradation requires further investigation. Unpublished microcosm studies showed a similar biotransformation pattern with the disappearance of the starting compound and formation of the $SF_5$-catechol. Interestingly, the corresponding $CF_3$-aminophenol was not a growth substrate for the bacteria.

Perfluorinated compounds, which are used in the manufacture of stain-repellent materials, are widespread pollutants, and numerous investigations focus on their biodegradation. The metabolites that have been detected in these studies indicate that enzymes from different catabolic pathways are involved in the biotransformation reactions (Butt et al. 2014). Recently, attention has focussed on the biodegradation of shorter fluorotelomer alcohols, reflecting the fluorochemical industry's move away from longer perfluoroalkyl chains. Kim et al. (2014) investigated the biodegradation of 6:2 fluorotelomer alcohol (6:2 FTOH) with three alkane-degrading bacteria and one fluoroacetate-degrading bacterium. Determination of the metabolic intermediates formed by each of selected strains indicated that the substrate was biotransformed into short chain per- and poly-fluoro acids (Fig. 11.3). The alcohol is initially oxidized, via aldehyde, to 6:2 fluorotelomer carboxylic acid (6:2 FTCA), from which HF is lost yielding 6:2 fluorotelomer unsaturated carboxylic acid (6:2 FTUCA). Further biodegradation proceeds via two distinct pathways: pathway I is via the 5:2 ketone forming perfluorinated hexanoic (PFHxA), pentanoic (PFPeA) and butyric acids (PFBA); pathway II progresses through 5:3 FTUCA to (PFPeA) and a 4:3 FTCA.

Depending on the mechanism of catabolism of each microorganism, it is also possible that a mixture of biotransformation and mineralization occurs. The biodegradation of 2-fluorobenzoic acid by *Sphingomonas* sp. HB-1 preferentially results in the mineralization of most of the initial substrate via 4-fluorocatechol, evidenced by the formation of fluoride as major product. Nevertheless, the formation of 2-fluoromuconic acid as a dead-end metabolite in a minor pathway through 3-fluorocatechol was concomitantly observed (Boersma et al. 2004). During the aerobic degradation of fluorobenzoic acids via catechol, the initial step, catalysed by benzoate dioxygenase, can result in hydroxylation at either C1/C2, resulting in spontaneous fluoride release, or at C1/C6, yielding a fluorinated dihydrodiol that is subsequently biotransformed to the dead-end metabolite 2-fluoromuconic acid (Fig. 11.4). The halobenzene-degrading bacterium, *Burkholderia fungorum* FLU100, possesses a novel and unique trait of productively metabolize fluorobenzene via 3-fluorocatechol and 2-fluoromuconic acid, which is later completely metabolized by the cells with the concomitant release of fluoride (Strunk and Engesser 2013). However, fluorobenzene metabolism does proceed nearly exclusively via 3-fluorocatechol as the principal catecholic metabolite, an alternative pathway for fluorobenzene degradation via 4-fluorocatechol also occurs, albeit at a minor degree. In fact, the path via 4-fluorocatechol is substantially slower and

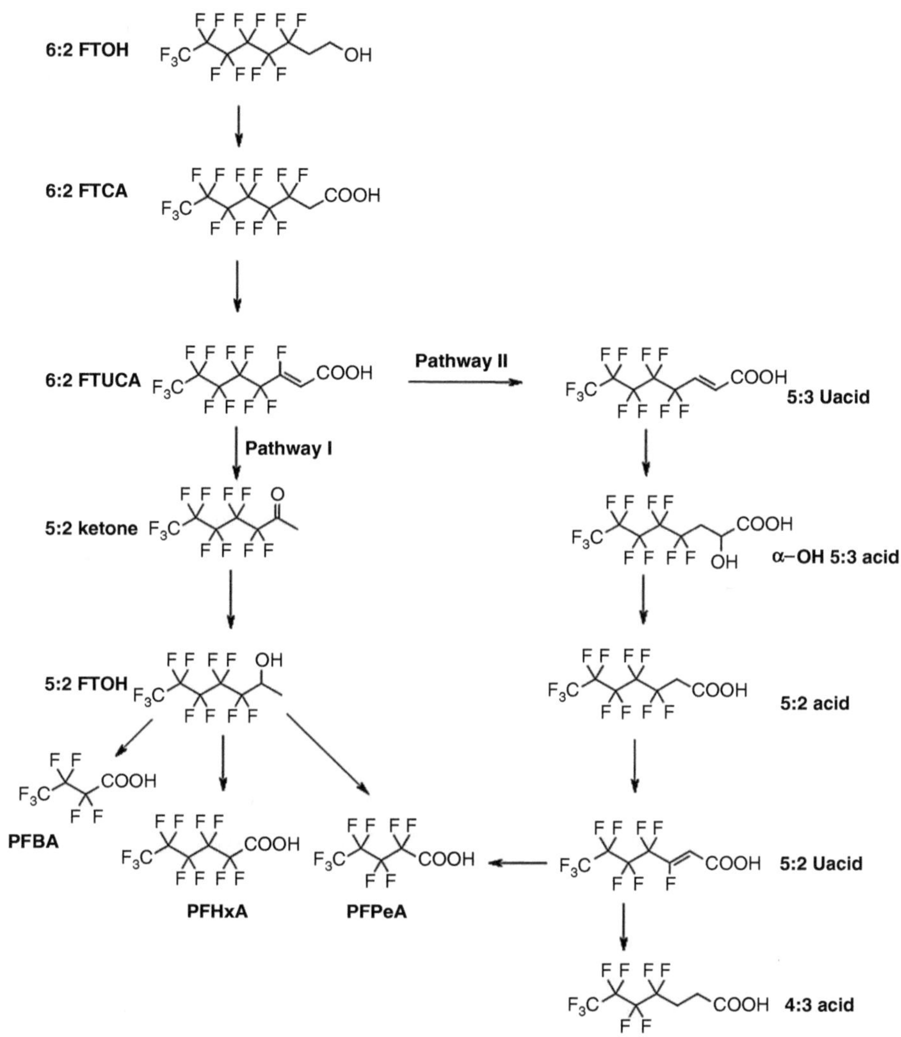


**Fig. 11.3** Proposed pathways for the biodegradation of 6:2 FTOH in alkane- and fluoroacetate-degrading bacteria (Kim et al. 2014). The compound abbreviations are defined in the text

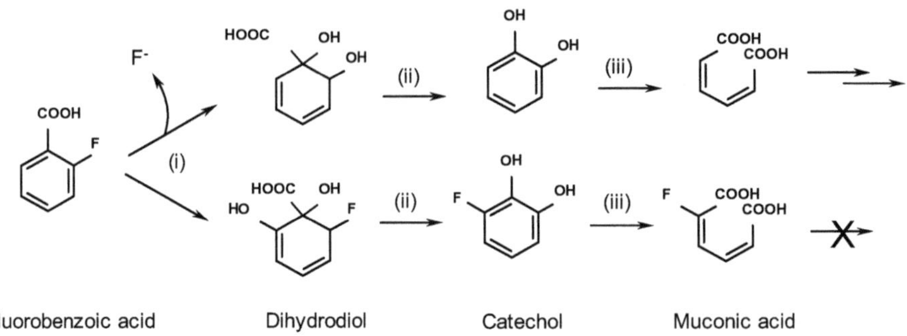


**Fig. 11.4** Biotransformation vs biodegradation. Bacterial catabolism of 2-fluorobenzoic acid via benzoate 1,2-dioxygenase (i), reductase (ii) and catechol dioxygenase (iii) yielding muconic acid. In some species 2-fluoromuconic acid is a dead-end metabolite

incomplete, leading to the accumulation of uncharacterized derivatives of muconic acid and muconolactone, thus being categorized as nonproductive. This finding contradicts the former assumption that the pathway via 3-fluorocatechol and 2-fluoromuconate is always nonproductive. Actually, it is opposite to what happens during the complete degradation of fluorobenzene by *Labrys portucalensis* F11 that proceeds via the ortho-cleavage of 4-fluorocatechol and catechol (Carvalho et al. 2006).

### *11.4.6 Analytical Methods Used for Monitoring the Degradation of Fluorinated Compounds*

There are several analytical methods that have been used to monitor the degradation of fluorinated organic compounds and to identify the catabolic intermediates, aiming at the elucidation of the degradative pathways.

Fluorine-19 nuclear magnetic resonance spectroscopy ($^{19}$F NMR) methods are very useful due to the properties of the $^{19}$F isotope, such as elevated sensitivity, absence of $^{19}$F NMR-relevant endogenous compounds in biological systems and broad chemical shift range of $^{19}$F nucleus (500 ppm). These properties allow the determination of fluorinated compounds at low (mM) concentrations and with an excellent signal-to-noise ratio. Moreover, modifications on its molecular surroundings cause severe and drastic changes in its chemical shift patterns simultaneously reducing chance of peak overlap (Bondar et al. 1998). In this way, $^{19}$F NMR analysis permits the characterization and quantification of the fluorometabolite(s) profile and the extent of defluorination, without extensive purification steps. If reference compounds are available, $^{19}$F NMR can be used for the elucidation of the metabolic pathway of fluorinated organics biodegradation. This technique was used to elucidate the biodegradation of 2- and 4-fluorobiphenyl by the bacterium *Pseudomonas pseudoalcaligenes* KF707 via analysis of culture supernatants (Murphy et al. 2008).

Notably, relatively large concentrations of fluorobenzoic acids were detected by $^{19}F$ NMR, which was consistent with the catabolism of fluorobiphenyl via the "upper" biphenyl pathway. In a study performed by Bondar et al. (1998), this method also allowed the elucidation of the regioselectivity of the hydroxylases that catalyse the first step of fluorophenols biodegradation by *Rhodococcus* strains. In addition, the $^{19}F$ NMR also possesses the advantage of allowing the determination of the exact substituent pattern of the fluorinated metabolites formed (Bondar et al. 1998; Finkelstein et al. 2000).

Other researchers have shown that continuous source molecular absorption spectrometry (CS-MAS) is an extremely sensitive method for the detection of organofluorine compounds. Qin et al. (2013) showed that this technique was able to act as a fluorine-specific detector when coupled to high performance liquid chromatography (HPLC) for the detection of fluorometabolites in complex growth medium. This technique has the potential to detect low concentration of metabolites arising from the biodegradation of fluorinated substrates, without the need for a pre-concentration step, which is often part of the workflow when using $^{19}F$ NMR as the analytical tool.

Beyond the methods mentioned above, which are specific for fluorinated compounds, gas chromatography (GC) and high performance liquid chromatography (HPLC) are the most common techniques of routine analysis used to quantify the degradation of pollutants by following the removal of the parent compound over time (Moreira et al. 2012a, b; 2014). Mass spectrometry (MS), especially preceded by liquid chromatography (LC) separation, have been an essential tool mainly due to its versatility, sensitivity and selectivity. It has been used either to assess degradation when the initial parent compound was supplied at very low concentrations or to identify the degradation metabolites (Amorim et al. 2014a; Franco et al. 2014). Elucidation of the structure of intermediates has benefited from the development of high-resolution mass spectrometry, such as time-of-flight detector analyser (Maia et al. 2014; Amorim et al. 2014b). Enantioselective HPLC methods have been developed to follow the degradation of enantiomers of chiral pollutants, such as the fluorinated pharmaceuticals fluoxetine (Ribeiro et al. 2014a, b; Moreira et al. 2014; Moreira et al. 2015) and fluoroquinolones (Maia et al. 2016). These methods rely on the use of a stationary phase able to discriminate enantiomers.

In addition, flow injection low-pressure chromatography methods have been developed to take the advantage from the combination of the potentialities of flow injection analysis, such as automation, speed and low cost, with analyte separation of chromatography. In a recent work, a simple low-pressure chromatography method with a high throughput was developed for monitoring the biodegradation of fluoroquinolones. The developed method was not as sensitive as, for example, HPLC-MS, but it displayed good resolution and good selectivity as no significant interferences were observed (Santos et al. 2016).

Monitoring fluoride ion generation using ion-selective electrodes is a quick method to assess the extent of defluorination, a critical step of organofluorine biodegradation (Amorim et al. 2014b; Moreira et al. 2014). The potentiometric determination of fluoride ion when coupled with flow injection systems allowed the on-line monitoring of the biodegradation of 2-fluorophenol on biological reactors (Mesquita et al. 2011). On-line determination of fluoride could be achieved by using a previously developed laboratory made tubular fluoride selective electrode (Santos et al. 2007).

## 11.5 Bioaugmentation as an Approach for the Degradation of Fluorinated Compounds

### *11.5.1 Principles and Strategies*

According to the successful results presented in previous sections, fluoroorganic-degrading strains could be of interest to accelerate the remediation of polluted zones. Apart from their catabolic potential, some microorganisms possess other adaptation mechanisms, such as the capacity to modify the cellular membrane to maintain their biological functions and/or the use of efflux pumps to decrease the concentration of toxics inside the cells. This will confer cells protection in contaminated environments and could make them even more interesting for use in bioremediation processes (Tyagi et al. 2011).

Bioaugmentation consists on the addition of selected degrading microorganisms/genes into a microbial community in an effort to promote a specific remediation process in that given environment (Fig. 11.1). Commonly, bioaugmentation processes involve the introduction of inocula consisting of pure microbial strains or microbial consortia; natural or genetically modified plasmid-bearing strains are also used (Fig. 11.5).

Bioaugmentation is a more eco-friendly and efficient environmental clean-up technique than chemical or physical treatments for the removal of hazardous compounds from contaminated soil or water. Nevertheless, although bioaugmentation has been regarded as a promising biotechnology approach, it is still not a well-

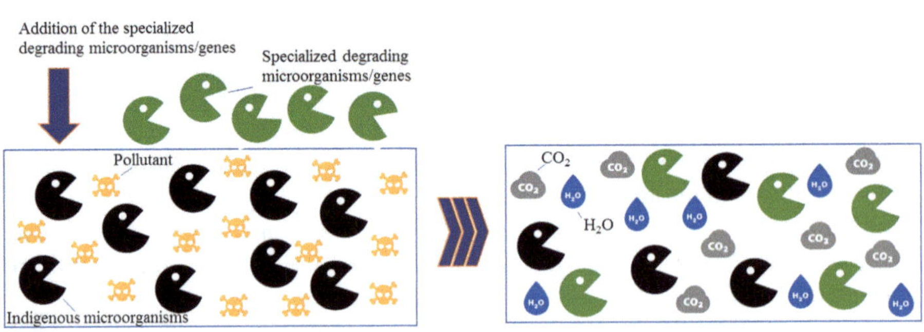


**Fig. 11.5** Schematic representation of the bioaugmentation process, in which the existing microbial indigenous community is supplemented with degrading microorganisms/genes to improve its pollutant removal ability

established strategy. In part, this is related with the capacity of the seed microorganism to survive and propagate in the foreign environment, competing with indigenous well-established microorganisms while maintaining its degrading capability. Thus, often, microorganisms that are efficient degraders under laboratory conditions are not necessarily effective when applied in the field. In fact, several studies reported that bioaugmentation using indigenous strains, isolated or enriched from the same ecological niche as the polluted area, is more likely to be efficient than allochthonous strains when reintroduced into that system (Hosokawa et al. 2009). To successfully obtain degrading microorganisms well-adapted to the contaminated environment, it is of major importance to conduct the enrichment process under similar conditions to those of the contaminated site where bioaugmentation will be performed. The evolution of the indigenous strains to achieve the capacity to degrade the pollutants involves mutations and horizontal gene transfer (HGT) and is often a relatively slow process.

In bioaugmentation processes, inocula consisting of single microbial strains or microbial consortia can be used. A priori, the application of microbial consortia consisting of several degrading microorganisms could be more effective than single strains application as a wide and versatile metabolic network is present. In this way, a wide spectrum of substrates can be potentially degraded; inclusive intermediates produced by one strain may be further degraded by other strains possessing suitable catabolic pathways (Mrozik and Piotrowska-Seget 2010). Furthermore, the use of microbial consortia instead of a pure culture could be particularly useful in field applications, where co-contamination by different pollutants often occurs. From an applied perspective, the addition of microbial consortia could provide a wider metabolic diversity and robustness needed to deal with the complexity of the natural habitats (Tyagi et al. 2011). Nevertheless, there are pure degrading microorganisms able to concomitantly degrade the target pollutant and other toxics (Amorim et al. 2013a; Moreira et al. 2012b) and non-toxic organic substrates (Amorim et al. 2014b; Moreira et al. 2014), feature that makes those strains potential candidates for bioaugmentation processes.

Plasmid-mediated bioaugmentation is an alternative process of augmentation. Natural or genetically modified strains carrying catabolic plasmid are imported to a contaminated system to enhance its capability to degrade a pollutant through the spread of the catabolic genes of interest (Garbisu et al. 2017). Especially when applied in biofilm systems, the genetic exchange by HGT can promote the acquisition of new catabolic genes by the native population which thereby may achieve the capacity to degrade the target contaminants. In this way, the phenomena of plasmid transfer from the donor strain to the indigenous bacteria could benefit the establishment of an effective degradative community. The use of genetically engineered microorganisms allowed conferring degradative abilities to existing microorganisms. Thus, a remediation process can be tailored through the selection of suitable degradative genes that, after being cloned, are inserted into a vector which is further introduced into a host cell. Nevertheless, an appropriate selection of the donor bacteria harbouring self-transmissible catabolic plasmids is required. In addition, the success of plasmid-mediated bioaugmentation does not rely only on an efficient plasmid transfer; the recipient cells should also be able to properly acquire and express the plasmid-harboured catabolic genes so that biodegradation can be attained (Garbisu et al. 2017). Nonetheless, irrespective of the potential of genetic engineering to confer new catabolic functions to microorganisms, the application of genetically modified microorganisms in the field raises some concerns due to their long-term environmental effects (Nzila et al. 2016).

The addition of specific pollutant-removal microorganisms/genes is in fact a good alternative for cleaning up contaminated systems, as a treatment can be tailored to a specific pollutant. However, failure of such process has been reported, especially during scale-up. A successful bioaugmentation process relies on the capacity of the added microorganism to persist within the contaminated system and simultaneously be functionally active under those environmental conditions (Tyagi et al. 2011). Typically, the treatment is only effective during the early stages of the bioaugmentation process as the cell numbers of the added microorganism usually decrease shortly after addition. Therefore, for an effective pollutant removal, treatment must sometimes be adjusted intermittently with the addition of more degrading microorganisms. The persistence of the inoculated degrading microorganism can be affected by different biotic factors, including competition between indigenous and inoculated microorganisms for energy and resources (e.g. nutrients and electron acceptors) and predation by protozoa and/or bacteriophages. Abiotic factors such as the pollutant concentration and availability and the physico-chemical characteristics of the contaminated environment (e.g. temperature, pH and others) should also be taken into consideration in the bioaugmentation process as they can be responsible for the lack of acclimation to the environmental conditions, preventing the pollutant removal (El Fantroussi and Agathos 2005). In addition, the environmental factors can also influence the transfer efficiency of catabolic plasmids from the donor to the indigenous microorganisms.

### *11.5.2 Delivery Approaches for Introduction of the Specific Degraders*

The delivery of the specialized degrading bacterial strain to the contaminated system can be performed using different approaches. Commonly, bioaugmentation is performed by inoculating a bacterial biomass suspension, the simplest approach. As the capacity of a strain to be retained within a system is crucial for an effective bioaugmentation, its addition as a biomass suspension is challenging due to bacteria washout. In general, the application of high initial dosages of suspended specialized degrading strain resulted in longer survival periods within the bioaugmented system (Martín-Hernández et al. 2012; Zhang et al. 2017). Nevertheless, several studies showed that this bioaugmentation strategy is not permanent, implying the reinoculation of the degrading strain on a regular basis to assure the continuous treatment efficacy in the system (Amorim et al. 2013b; Boon et al. 2000; Park et al. 2008).

From a practical point of view, the bioaugmentation in real scenarios with such high dosages of specialized strains and the need of regular re-supplementation would require a large-scale facility to continuously grow the specialized bacteria, thus increasing the economic costs of the process. A successful bioaugmentation with a suspended biomass relies on the capacity of the added inoculum to colonize the existent communities, which was modulated by quorum sensing. In bacteria, cell to cell communication is mediated by autoinducers (chemical signalling molecules). When those signals concentration reached a minimum threshold, it triggered changes in gene expression, thereby inducing the secretion of extracellular polymeric substances (EPS) and biofilm formation. Recently, Wang et al. (2013) determined the optimal conditions for 4-fluoroaniline degradation and autoinducer release in *Acinetobacter* sp. TW with the aim to develop a bioaugmentation system for treatment of fluorine-containing industrial wastewater. It was observed that the optimal conditions for pollutant degradation did not match those for autoinducer release. Therefore, at an initial stage after bioaugmentation, optimal conditions for autoinducer release should be applied to improve colonization, while those for pollutant degradation should be applied later to maintain system performance post-colonization.

Currently, bioaugmentation with encapsulated or immobilized cells of the degrading microorganism has been adopted as an alternative approach to promote bacteria persistence within the contaminated system. The carrier materials not only provide a physical support for the biomass but also extent the survival rate of the microorganisms (Mrozik and Piotrowska-Seget 2010). The immobilization materials offer protection from predation and competition by indigenous microorganisms, thus benefiting the specialized degrading strain persistence within a system. Additionally, under stress conditions, the immobilization materials shield the specialized strains, reducing their exposure to toxics, and thus avoiding cell membrane damages. In general, immobilized cells proved to be more efficient and rapid in toxics biodegradation when compared with its free-living counterparts (Chris Felshia et al. 2017; Schultz-Jensen et al. 2016). However, although cell immobilization can protect microorganisms from the harsh environment, tolerating higher toxics concentrations, it did not always increase the rate of toxics degradation. Chen and co-workers (2007) compared the degradation of trichloroethene using suspended and chitosan beads with immobilized *Pseudomonas putida*. Slower degradation kinetics was observed with the immobilized cells probably due to the limited diffusion of substrate molecules into the carrier. Therefore, the properties of the matrix should be considered for an effective and expeditious degradation of the target pollutant.

Carrier materials suitable to be used as an immobilization matrix must fulfil several requirements: (1) high cell loading capacity, (2) long cell viability and survival, (3) long-term storage stability, (4) non-toxic to immobilized cells nor bioaugmented system microflora, (5) adequate diffusion properties, (6) easily accessible and (7) inexpensive (Bayat et al. 2015; Dzionek et al. 2016). Currently, both natural and synthetic materials are being used as matrix for the immobilization of the specialized microorganisms. The former class of matrix materials include dextran, agar, agarose, alginate, chitosan and k-carrageenan, while the last include acrylate copolymers, polyurethane, polyvinyl alcohol and resins. Due to some intrinsic features, the use of natural materials as immobilization matrix has some advantages when compared with the synthetic ones, including its biodegradability, non-toxicity and availability in the natural environments. The major drawback in using some natural carriers is related with its low resistance to toxics and stability in a narrow pH range (Dzionek et al. 2016). Nevertheless, the interest in finding good natural immobilization materials from wastes of agricultural and food industries has been growing as this can be an opportunity for its economical valorization (Deng et al. 2016; Ivshina et al. 2013; Piccirillo et al. 2013; Plangklang and Reungsang 2009).

Recently, with the research focus of bioaugmentation as the preferred in situ strategy to clean-up contaminated systems, this process has attracted commercial interest. Emerging microbial degraders formulations and products are gaining attention and application, claiming fast removal rates. However, the effectiveness of bioaugmentation with such formulations is variable possibly due to the lack of metabolic activity, adaptability and ecological competence (Herrero and Stuckey 2015). Therefore, continuous research is needed to develop and refine the available microbial cultures before widespread application on a commercial basis.

Another strategy that has been adopted to introduce specific degraders into contaminated systems is the development of a granular biomass using cultures of the degrading strain. Adav and Lee (2008) made use the natural ability of *Bacillus thuringiensis* to form biofilm to successfully cultivate stable phenol-degrading granules in a sequencing batch reactor. The strain aggregates were formed due to the presence of secreted EPS among cells that allow binding them together. More recently, Liu et al. (2015b) investigated the granulation process using *Rhizobium* sp. NJUST18 pure culture, a pyridine-degrading bacterium. After 18 weeks of the initial inoculation, mature pyridine aerobic degrading granules were obtained. The developed granules could efficiently degrade pyridine over concentrations up to 4000 mg $L^{-1}$, exhibiting a maximum volumetric degradation much higher than that of the pure *Rhizobium* sp. NJUST18 bacterial suspension. Nevertheless, authors inferred that the auto-aggregation activity of *Rhizobium* sp. NJUST18 cells was the key factor for the occurrence of granulation. The presence of aggregating bacterial strains allowed the formation of primary matrixes which were crucial to initiate the granulation process. Therefore, degrading bacteria that do not have auto-aggregation capabilities will probably fail in granules formation.

### *11.5.3 Application of Bioaugmentation Processes to Improve Fluoroorganics Removal*

Bioaugmentation has been widely applied to remediate polluted contaminated soil and water systems and proven to be a valuable method in cleaning them up although it still faces many environmental problems. Several papers highlight the effectiveness of the bioaugmentation processes in laboratory- and field-scale to remediate contaminated systems. To date, the feasibility of field-scale bioaugmentation processes to enhance the removal of fluorinated pollutants from contaminated systems was not found in the scientific literature. Nevertheless, recently, the applicability and effectiveness of lab-scale bioaugmentation for the treatment of aqueous streams containing fluoroorganic pollutants have been addressed in different types of biofilm reactors (Amorim et al. 2013b; Duque et al. 2011a, b; Ramos et al. 2017).

The biological treatment of wastewater is challenging due to the variability in composition of the aqueous streams to be treated, especially those resulting from industrial facilities which usually contain toxics. The transient or continuous presence of such pollutants often impairs the microbial activity, resulting in process instability. A successful bioaugmentation process in a rotating biological contactor subject to shock loadings of 2-fluorophenol (2-FP) was reported by Duque et al. (2011a). Only after the inoculation with the 2-FP specialized degrading strain *Rhodococcus* sp. FP1, the mineralization of the target pollutant was observed. Moreover, the system showed to be robust towards starvation periods; after ca. 1 month of non-supply of the target pollutant, the reactor could resume 2-FP degradation. Furthermore, by the end of operation, the inoculated strain was successfully recovered from the biofilm by culture-dependent and culture-independent methods. The operation in batch mode after the addition of the strain FP1 seemed to be a key factor for the success of the bioaugmentation, as it probably allowed the attachment of the bacterial strain to the biofilm, thus avoiding its loss. Moreover, apart from strain FP1, six bacterial isolates extracted from the biofilm showed degradation activity towards 2-FP (Duque et al. 2014), suggesting the occurrence of HGT.

The study conducted by Amorim et al. (2013b) reinforces the need of bioaugmentation when biodegradation of highly recalcitrant pollutants is targeted. A rotating biological contactor was used to treat intermittent shock loadings of 4-fluorocinnamic acid (4-FCA). Although no 4-FCA was detected at the effluent, before the bioaugmentation process, only limited degradation of the pollutant was occurring, with its biotransformation into an intermediate metabolite, 4-fluorobenzoate. From the addition of the 4-FCA degrading bacterium (*Rhodococcus* sp. S2) onwards, complete mineralization of the pollutant was observed, corroborated by the fluoride ion release. However, a gradual decline in 4-FCA degradation performance occurred over time. Although strain S2 was retrieved from the biofilm 2 months after its inoculation, intermittent feeding of the pollutant may have led to diminishing strain numbers. In this case, in the absence of a selective pressure, the inoculated strain was probably outcompeted by the indigenous community.

The feasibility of treating 2-FP-containing wastewater using the aerobic granular sludge (AGS) technology has been investigated (Duque et al. 2011b). Acetate-fed aerobic granular sludge failed to degrade 2-FP, and its degradation was only achieved after bioaugmentation with *Rhodococcus* sp. FP1, demonstrating that the inoculated strain had a key role on 2-FP degradation. The degrading strain was recovered from the granular sludge by plating, suggesting that the AGS biomass was capable to incorporate and retain the specialized degrader. Moreover, DGGE analysis revealed that the bacterial community was dynamic, and the small shifts observed in the bacterial population seemed to be mainly due to 2-FP load, rather than the bioaugmentation process. More recently, it was investigated the effect of the bioaugmentation of the AGS-SBR treating synthetic wastewater containing 2-FP and with high ammonium content (Ramos et al. 2017). Simultaneous nitritation and 2-FP biodegradation started occurring after bioaugmentation with *Rhodococcus* sp. FP1. In fact, the addition of the degraders allowed the pollutant removal from the bulk liquid, thus reducing the toxicity on the microbial community, which in turn led to an improvement of the ammonium oxidation. Through DGGE, it was not possible to infer the presence of the degrading strain used to augment the bioreactor; however, by high-throughput sequencing analysis, the presence of phylotypes of the genus *Rhodococcus* at the end of operation was observed.

Bioaugmentation is widely used and can be a valuable mechanism, especially in emergency response to stress scenarios (Herrero and Stuckey 2015). Even though a successful bioaugmentation process in the laboratory under controlled conditions does not imply similar success when applied in the field, it is a step forward to solve potential problems in the environment. Furthermore, the efficiency of bioinocula to be applied in the field should be first evaluated in laboratory simulating the contaminated system. Lab-scale studies provide knowledge on the microorganisms and their growth requirements, which could be useful to overcome traits that could limit or prevent the pollutant biodegradation during on-site application.

## 11.6 Conclusion

Environmental contamination with fluorinated organic compounds is an issue of concern. The increasing number of fluorinated chemicals released to the environment stresses the importance to understand their fate and degradation potential and to find solutions for its efficient removal. Fortunately, several degrading strains have been isolated for the effective biodegradation of these pollutants, and there is an increasing availability of tools to foster for new degraders. Both lab- and field-scale studies clearly indicate that bioaugmentation is promising for in situ remediation. However, the complexity of the natural environment could impair its application. Therefore, finding competent microorganisms and exploring their catabolic potential for each contaminated system is probably the best solution for a successful bioaugmentation process. Moreover, the development of different methodologies to bioaugment contaminated environments is needed to turn this approach feasible for field-scale application.

**Acknowledgements** I.S. Moreira and C.L. Amorim acknowledge research grants from Fundação para a Ciência e Tecnologia (FCT), Portugal (SFRH/BPD/87251/2012 and SFRH/BPD/96481/2013) and Fundo Social Europeu (Programa Operacional Potencial Humano (POPH), Quadro de Referência Estratégico Nacional (QREN)). This work was supported by National Funds from FCT – Fundação para a Ciência e a Tecnologia – through project AGeNT - PTDC/BTA-BTA/31264/2017, (POCI-01-0145-FEDER-031264). We would also like to thank the scientific collaboration of CBQF under the FCT project UID/Multi/50016/2013.

## References

Adachi F, Yamamoto A, Takakura K-I, Kawahara R (2013) Occurrence of fluoroquinolones and fluoroquinolone-resistance genes in the aquatic environment. Sci Total Environ 444:508–514. https://doi.org/10.1016/j.scitotenv.2012.11.077

Adav SS, Lee DJ (2008) Aerobic granulation of pure bacterial strain *Bacillus thuringiensis*. Front Environ Sci Eng China 2:461–467. https://doi.org/10.1007/s11783-008-0066-0

Ajao A, Kannan M, Krishna R, Yakubu S, Umoh V, Ameh J (2012) Homology modeling, simulation and molecular docking studies of catechol-2, 3-Dioxygenase from Burkholderia cepacia: involved in degradation of petroleum hydrocarbons. Bioinformation 8:848–854. https://doi.org/10.6026/97320630008848

Alexandrino DAM, Ribeiro I, Pinto LM, Cambra R, Oliveira RS, Pereira F, Carvalho MF (2018) Biodegradation of mono-, di- and trifluoroacetate by microbial cultures with different origins. New Biotechnol 43:23–29. https://doi.org/10.1016/j.nbt.2017.08.005

Alvarez PJJ, Vogel TM (1991) Substrate Interactions of benzene, toluene, and para-xylene during microbial degradation by pure cultures and mixed culture aquifer slurries. Appl Environ Microbiol 57:2981–2985

Amadio J, Murphy CD (2010) Biotransformation of fluorobiphenyl by *Cunninghamella elegans*. Appl Microbiol Biotechnol 86:345–351. https://doi.org/10.1007/s00253-009-2346-4

Amor L, Kennes C, Veiga MC (2001) Kinetics of inhibition in the biodegradation of monoaromatic hydrocarbons in presence of heavy metals. Bioresour Technol 78:181–185. https://doi.org/10.1016/S0960-8524(00)00182-6

Amorim CL, Carvalho MF, Afonso CMM, Castro PML (2013a) Biodegradation of fluoroanilines by the wild strain *Labrys portucalensis*. Int Biodeterior Biodegrad 80:10–15. https://doi.org/10.1016/j.ibiod.2013.02.001

Amorim CL, Duque AF, Afonso CMM, Castro PML (2013b) Bioaugmentation for treating transient 4-fluorocinnamic acid shock loads in a rotating biological contactor. Bioresour Technol 144:554–562. https://doi.org/10.1016/j.biortech.2013.07.010

Amorim CL, Ferreira ACS, Carvalho MF, Afonso CMM, Castro PML (2014a) Mineralization of 4-fluorocinnamic acid by a *Rhodococcus* strain. Appl Microbiol Biotechnol 98:1893–1905. https://doi.org/10.1007/s00253-013-5149-6

Amorim CL, Moreira IS, Maia AS, Tiritan ME, Castro PML (2014b) Biodegradation of ofloxacin, norfloxacin, and ciprofloxacin as single and mixed substrates by *Labrys portucalensis* F11. Appl Microbiol Biotechnol 98:3181–3190. https://doi.org/10.1007/s00253-013-5333-8

Amorim CL, Moreira IS, Ribeiro AR, Santos LHMLM, Delerue-Matos C, Tiritan ME, Castro PML (2016) Treatment of a simulated wastewater amended with a chiral pharmaceuticals mixture by an aerobic granular sludge sequencing batch reactor. Int Biodeterior Biodegrad 115:277–285. https://doi.org/10.1016/j.ibiod.2016.09.009

Amorim CL, Alves M, Castro PML, Henriques I (2018) Bacterial community dynamics within an aerobic granular sludge reactor treating wastewater loaded with pharmaceuticals. Ecotoxicol Environ Saf 147:905–912. https://doi.org/10.1016/j.ecoenv.2017.09.060

Asha S, Vidyavathi M (2009) *Cunninghamella* – a microbial model for drug metabolism studies – a review. Biotechnol Adv 27:16–29. https://doi.org/10.1016/j.biotechadv.2008.07.005

Auffret MD, Yergeau E, Labbé D, Fayolle-Guichard F, Greer CW (2015) Importance of *Rhodococcus* strains in a bacterial consortium degrading a mixture of hydrocarbons, gasoline, and diesel oil additives revealed by metatranscriptomic analysis. Appl Microbiol Biotechnol 99:2419–2430. https://doi.org/10.1007/s00253-014-6159-8

Backe WJ, Day TC, Field JA (2013) Zwitterionic, cationic, and anionic fluorinated chemicals in aqueous film forming foam formulations and groundwater from U.S. military bases by non-aqueous large-volume injection HPLC-MS/MS. Environ Sci Technol 47:5226–5234. https://doi.org/10.1021/es3034999

Bahobail A, Gad El-Rab SMF, Amin GA (2016) Locally isolated bacterial strains with multiple degradation potential capabilities on petroleum hydrocarbon pollutants. Adv Microbiol 6:852–866. https://doi.org/10.4236/aim.2016.611081

Bargiela R, Mapelli F, Rojo D, Chouaia B, Tornés J, Borin S, Richter M, Del Pozo MV, Cappello S, Gertler C, Genovese M, Denaro R, Martínez-Martínez M, Fodelianakis S, Amer RA, Bigazzi D, Han X, Chen J, Chernikova TN, Golyshina OV, Mahjoubi M, Jaouanil A, Benzha F, Magagnini M, Hussein E, Al-Horani F, Cherif A, Blaghen M, Abdel-Fattah YR, Kalogerakis N, Barbas C, Malkawi HI, Golyshin PN, Yakimov MM, Daffonchio D, Ferrer M (2015) Bacterial population and biodegradation potential in chronically crude oil-contaminated marine sediments are strongly linked to temperature. Sci Rep 5:1–15. https://doi.org/10.1038/srep11651

Bartels I, Knackmuss H, Reineket W (1984) Suicide inactivation of catechol 2, 3-dioxygenase from *Pseudomonas putida* mt-2 by 3-halocatechols. Appl Environ Microbiol 47:500–505

Bartholomé A, Janso JE, Reilly U, O'Hagan D (2017) Fluorometabolite biosynthesis: isotopically labelled glycerol incorporations into the antibiotic nucleocidin in *Streptomyces calvus*. Org Biomol Chem 15:61–64. https://doi.org/10.1039/c6ob02291j

Bayat Z, Hassanshahian M, Cappello S (2015) Immobilization of microbes for bioremediation of crude oil polluted environments: a mini review. Open Microbiol J 9:48–54. https://doi.org/10.2174/1874285801509010048

Becker M, Carneiro FM, de Oliveira LP, da Silva MIV, Riet-Correa F, Lee ST, Pescador CA, Nakazato L, Colodel EM (2016) Induction and transfer of resistance to poisoning by *Amorimia pubiflora* in sheep whith non-toxic dosis of the plant and ruminal content. Ciênc Rural 46:674–680. https://doi.org/10.1590/0103-8478cr20141484

Benotti MJ, Trenholm RA, Vanderford BJ, Holady JC, Stanford BD, Snyder SA (2009) Pharmaceuticals and endocrine disrupting compounds in U.S. drinking water. Environ Sci Technol 43:597–603. https://doi.org/10.1021/es801845a

Benskin JP, Ikonomou MG, Gobas FAPC, Woudneh MB, Cosgrove JR (2012) Observation of a novel PFOS-precursor, the perfluorooctane sulfonamido ethanol-based phosphate (SAmPAP) diester, in marine sediments. Environ Sci Technol 46:6505–6514. https://doi.org/10.1021/es300823m

Boersma MG, Dinarieva TY, Wouter J, Berkel WJH Van, Doran J, Rietjens IMCM, Middelhoven WJ (1998) 19 F nuclear magnetic resonance as a tool to investigate microbial degradation of fluorophenols to fluorocatechols and fluoromuconates. Appl Environ Microbiol. 64(4): 1256–1263.

Boersma FGH, McRoberts WC, Cobb SL, Murphy CD (2004) A 19F NMR study of fluorobenzoate biodegradation by *Sphingomonas* sp. HB-1. FEMS Microbiol Lett 237:355–361. https://doi.org/10.1016/j.femsle.2004.06.052

Bondar VS, Boersma MG, Golovlev EL, Vervoort J, Van Berkel WJ, Finkelstein ZI, Solyanikova IP, Golovleva LA, Rietjens IM (1998) 19F NMR study on the biodegradation of fluorophenols by various *Rhodococcus* species. Biodegradation 9:475–486. https://doi.org/10.1023/A:1008391906885

Boon N, Goris J, De Vos P, Verstraete W, Top EM (2000) Bioaugmentation of activated sludge by an indigenous 3-chloroaniline-degrading *Comamonas testosteroni* strain, I2gfp. Appl Environ Microbiol 66:2906–2913

Brumovský M, Karásková P, Borghini M, Nizzetto L (2016) Per- and polyfluoroalkyl substances in the Western Mediterranean Sea waters. Chemosphere 159:308–316. https://doi.org/10.1016/j.chemosphere.2016.06.015

Bučková M, Puškarová A, Chovanová K, Kraková L, Ferianc P, Pangallo D (2013) A simple strategy for investigating the diversity and hydrocarbon degradation abilities of cultivable bacteria from contaminated soil. World J Microbiol Biotechnol 29:1085–1098. https://doi.org/10.1007/s11274-013-1277-5

Buser MC, Scinicariello F (2016) Perfluoroalkyl substances and food allergies in adolescents. Environ Int 88:74–79. https://doi.org/10.1016/j.envint.2015.12.020

Butt CM, Muir DCG, Mabury SA (2014) Biotransformation pathways of fluorotelomer-based polyfluoroalkyl substances: A review. Environ Toxicol Chem. 33(2):243–267. https://doi.org/10.1002/etc.2407

Calafat AM, Wong LY, Kuklenyik Z, Reidy JA, Needham LL (2007) Polyfluoroalkyl chemicals in the U.S. population: data from the national health and nutrition examination survey (NHANES) 2003–2004 and comparisons with NHANES 1999–2000. Environ Health Perspect 115:1596–1602. https://doi.org/10.1289/ehp.10598

Camboim EKA, Tadra-Sfeir MZ, De Souza EM, Pedrosa FDO, Andrade PP, McSweeney CS, Riet-Correa F, Melo MA (2012) Defluorination of sodium fluoroacetate by bacteria from soil and plants in Brazil. Sci World J. 2012: 149893. https://doi.org/10.1100/2012/149893

Cao B, Nagarajan K, Loh K-C (2009) Biodegradation of aromatic compounds: current status and opportunities for biomolecular approaches. Appl Microbiol Biotechnol 85:207–228. https://doi.org/10.1007/s00253-009-2192-4

Carmona M, Zamarro MT, Blázquez B, Durante-Rodríguez G, Juárez JF, Valderrama JA, Barragán MJL, García JL, Díaz E (2009) Anaerobic catabolism of aromatic compounds: a genetic and genomic view. Microbiol Mol Biol Rev 73:71–133. https://doi.org/10.1128/MMBR.00021-08

Carvalho MF, Alves CCT, Ferreira MIM, De Marco P, Castro PML (2002) Isolation and initial characterization of a bacterial consortium able to mineralize fluorobenzene. Appl Environ Microbiol 68:102–105. https://doi.org/10.1128/AEM.68.1.102

Carvalho MF, Ferreira Jorge R, Pacheco CC, De Marco P, Castro PML (2005) Isolation and properties of a pure bacterial strain capable of fluorobenzene degradation as sole carbon and energy source. Environ Microbiol 7:294–298. https://doi.org/10.1111/j.1462-2920.2004.00714.x

Carvalho MF, Ferreira MIM, Moreira IS, Castro PML, Janssen DB (2006) Degradation of fluorobenzene by *Rhizobiales* strain F11 via ortho cleavage of 4-fluorocatechol and catechol. Appl Environ Microbiol 72:7413–7417. https://doi.org/10.1128/AEM.01162-06

Carvalho MF, Maia AS, Tiritan ME, Castro PML (2016) Bacterial degradation of moxifloxacin in the presence of acetate as a bulk substrate. J Environ Manag 168:219–228. https://doi.org/10.1016/j.jenvman.2015.12.010

Cerniglia CE, Miller DW, Yang SK, Freeman JP (1984) Effects of a fluoro substituent on the fungal metabolism of 1-fluoronaphthalene. Appl Environ Microbiol 48:294–300

Chaojie Z, Qi Z, Ling C, Yuan Y, Hui Y (2007) Degradation of mono-fluorophenols by an acclimated activated sludge. Biodegradation 18:51–61. https://doi.org/10.1007/s10532-005-9035-5

Chaudhry GR, Chapalamadugu S (1991) Biodegradation of halogenated organic compounds. Microbiol Rev 55:59–79

Chen Y, Lin T, Huang C, Lin J, Hsieh F (2007) Degradation of phenol and TCE using suspended and chitosan-bead immobilized *Pseudomonas putida*. J Hazard Mater 148:660–670. https://doi.org/10.1016/j.jhazmat.2007.03.030

Chen C, Liu Q, Liu C, Yu J (2017) Effect of different enrichment strategies on microbial community structure in petroleum-contaminated marine sediment in Dalian, China. Mar Pollut Bull 117:274–282. https://doi.org/10.1016/j.marpolbul.2017.02.004

Chris Felshia S, Aswin Karthick N, Thilagam R, Chandralekha A, Raghavarao KSMS, Gnanamani A (2017) Efficacy of free and encapsulated *Bacillus lichenformis* strain SL10 on degradation of phenol: a comparative study of degradation kinetics. J Environ Manag 197:373–383. https://doi.org/10.1016/j.jenvman.2017.04.005

Christie WW, Hamilton JTG, Harper DB (1998) Mass spectrometry of fluorinated fatty acids in the seed oil of *Dichapetalum toxicarium*. Chem Phys Lipids 97:41–47. https://doi.org/10.1016/S0009-3084(98)00090-5

Clarke KF, Callely AG, Livingstone A, Fewson CA (1975) Metabolism of monofluorobenzoates by *Acinetobacter calcoaceticus* NCIB 8250. Formation of monofluorocatechols. Biochim Biophys Acta 404:169–179

Commandeur LC, Parsons JR (1990) Degradation of halogenated aromatic compounds. Biodegradation 1:207–220. https://doi.org/10.1007/BF00058837

Delgadillo-Mirquez L, Lardon L, Steyer J-P, Patureau D (2011) A new dynamic model for bioavailability and cometabolism of micropollutants during anaerobic digestion. Water Res 45:4511–4521. https://doi.org/10.1016/j.watres.2011.05.047

Deng H, Ma L, Bandaranayaka N, Qin Z, Mann G (2014) Identification of fluorinases from *Streptomyces* sp MA37, *Norcardia brasiliensis*, and *Actinoplanes* sp N902-109 by genome mining. 15:364–368. https://doi.org/10.1002/cbic.201300732

Deng F, Liao C, Yang C, Guo C, Dang Z (2016) Enhanced biodegradation of pyrene by immobilized bacteria on modified biomass materials. Int Biodeterior Biodegrad 110:46–52. https://doi.org/10.1016/j.ibiod.2016.02.016

Doelman P, Jansen E, Michels M, Til M (1994) Effects of heavy metals in soil on microbial diversity and activity as shown by the sensitivity-resistance index, an ecologically relevant parameter. Biol Fertil Soils 17:177–184. https://doi.org/10.1007/BF00336319

Dolbier WR (2005) Fluorine chemistry at the millennium. J Fluor Chem 126:157–163. https://doi.org/10.1016/j.jfluchem.2004.09.033

Dong H, Liu J, Liu X, Yu Y, Cao S (2017) Molecular docking and QSAR analyses of aromatic heterocycle thiosemicarbazone analogues for finding novel tyrosinase inhibitors. Bioorg Chem 75:106–117. https://doi.org/10.1016/j.bioorg.2017.07.002

Donnelly C, Murphy CD (2009) Purification and properties of fluoroacetate dehalogenase from *Pseudomonas fluorescens* DSM 8341. Biotechnol Lett 31:245–250. https://doi.org/10.1007/s10529-008-9849-4

Dorne JLCM, Skinner L, Frampton GK, Spurgeon DJ, Ragas AMJ (2007) Human and environmental risk assessment of pharmaceuticals: differences, similarities, lessons from toxicology. Anal Bioanal Chem 387:1259–1268. https://doi.org/10.1007/s00216-006-0963-7

Duque AF, Bessa VS, Carvalho MF, Castro PML (2011a) Bioaugmentation of a rotating biological contactor for degradation of 2-fluorophenol. Bioresour Technol 102:9300–9303. https://doi.org/10.1016/j.biortech.2011.07.003

Duque AF, Bessa VS, Carvalho MF, de Kreuk MK, van Loosdrecht MCM, Castro PML (2011b) 2-Fluorophenol degradation by aerobic granular sludge in a sequencing batch reactor. Water Res 45:6745–6752. https://doi.org/10.1016/j.watres.2011.10.033

Duque AF, Hasan SA, Bessa VS, Carvalho MF, Samin G, Janssen DB, Castro PML (2012) Isolation and characterization of a *Rhodococcus* strain able to degrade 2-fluorophenol. Appl Microbiol Biotechnol 95:511–520. https://doi.org/10.1007/s00253-011-3696-2

Duque AF, Bessa VS, Castro PML (2014) Bacterial community dynamics in a rotating biological contactor treating 2-fluorophenol-containing wastewater. J Ind Microbiol Biotechnol 41:97–104. https://doi.org/10.1007/s10295-013-1381-4

Dzionek A, Wojcieszy D, Guzik U (2016) Natural carriers in bioremediation: a review. Electron J Biotechnol 23:28–36. https://doi.org/10.1016/j.ejbt.2016.07.003

Eisenhauer A, Harnisch J (1998) Natural CF4 and on earth. Geophys Res Lett 25:2401–2404. https://doi.org/10.1029/98GL01779

El Fantroussi S, Agathos SN (2005) Is bioaugmentation a feasible strategy for pollutant removal and site remediation? Curr Opin Microbiol 8:268–275. https://doi.org/10.1016/j.mib.2005.04.011

Engesser K-H, Schmidt E, Knackmuss H-J (1980) Adaptation of *Alcaligenes eutrophus* B9 and *Pseudomonas* sp. B13 to 2-Fluorobenzoate as growth substrate. Appl Environ Microbiol 39:68–73

Engesser KH, Rubio MA, Ribbons DW (1988) Bacterial metabolism of side chain fluorinated aromatics: cometabolism of 4-trifluoromethyl(TFM)-benzoate by 4-isopropylbenzoate grown *Pseudomonas putida* JT strains. Arch Microbiol 149:198–206

Ericson I, Domingo JL, Nadal M, Bigas E, Llebaria X, Van Bavel B, Lindström G (2009) Levels of perfluorinated chemicals in municipal drinking water from Catalonia, Spain: public health implications. Arch Environ Contam Toxicol 57:631–638. https://doi.org/10.1007/s00244-009-9375-y

Ferreira MIM, Marchesi JR, Janssen DB (2008) Degradation of 4-fluorophenol by *Arthrobacter* sp. strain IF1. Appl Microbiol Biotechnol 78:709–717. https://doi.org/10.1007/s00253-008-1343-3

Festa S, Coppotelli BM, Madueño L, Loviso CL, Macchi M, Neme Tauil RM, Valacco MP, Morelli IS (2017) Assigning ecological roles to the populations belonging to a phenanthrene-degrading bacterial consortium using omic approaches. PLoS One 12:1–21. https://doi.org/10.1371/journal.pone.0184505
Fetzner S, Lingens F (1994) Bacterial dehalogenases: biochemistry, genetics, and biotechnological applications. Microbiol Rev 58:641–685
Finkelstein ZI, Baskunov BP, Boersma MG, Vervoort J, Golovlev EL, van Berkel WJ, Golovleva LA, Rietjens IM (2000) Identification of fluoropyrogallols as new intermediates in biotransformation of monofluorophenols in *Rhodococcus opacus* 1cp. Appl Environ Microbiol 66:2148–2153
Franco AR, Ferreira AC, Castro PML (2014) Co-metabolic degradation of mono-fluorophenols by the ectomycorrhizal fungi *Pisolithus tinctorius*. Chemosphere 111:260–265. https://doi.org/10.1016/j.chemosphere.2014.03.094
Fuertes I, Gómez-Lavín S, Elizalde MP, Urtiaga A (2017) Perfluorinated alkyl substances (PFASs) in northern Spain municipal solid waste landfill leachates. Chemosphere 168:399–407. https://doi.org/10.1016/j.chemosphere.2016.10.072
Gadd GM, Griffiths AJ (1978) Microorganisms and heavy metal toxicity. Microb Ecol 4:303–317. https://doi.org/10.1007/BF02013274
Garbisu C, Garaiyurrebaso O, Epelde L, Grohmann E, Alkorta I (2017) Plasmid-mediated bioaugmentation for the bioremediation of contaminated soils. Front Microbiol 8:1–13. https://doi.org/10.3389/fmicb.2017.01966
Garrido-Sanz D, Manzano J, Martín M, Redondo-Nieto M, Rivilla R (2018) Metagenomic analysis of a biphenyl-degrading soil bacterial consortium reveals the metabolic roles of specific populations. Front Microbiol 9:1–13. https://doi.org/10.3389/fmicb.2018.00232
Goldman P (1965) The enzymatic cleavage of the carbon – fluorine bond in fluoroacetate. J Biol Chem 240:3434–3438
Goldman P, Milne GWA (1966) Carbon-fluorine bond cleavage. II. Studies on the mechanism of the defluorination of fluoroacetate. J Biol Chem 241:5557–5560
Gregg K, Hamdorf B, Henderson K, Kopecny J, Wong C (1998) Genetically modified ruminal bacteria protect sheep from fluoroacetate poisoning. Appl Environ Microbiol 64:3496–3498
Gribble GW (2002) Naturally occurring organofluorines. In: Handb Environ Chem Vol 3, Part N Organofluorines 3:121–136. https://doi.org/10.1007/10721878_5
Grushin VV (2010) The organometallic fluorine chemistry of palladium and rhodium: studies toward aromatic fluorination. Acc Chem Res 43:160–171. https://doi.org/10.1021/ar9001763
Guerra AB, Oliveira JS, Silva-Portela RCB, Araújo W, Carlos AC, Vasconcelos ATR, Freitas AT, Domingos YS, de Farias MF, Fernandes GJT, Agnez-Lima LF (2018) Metagenome enrichment approach used for selection of oil-degrading bacteria consortia for drill cutting residue bioremediation. Environ Pollut 235:869–880. https://doi.org/10.1016/j.envpol.2018.01.014
Guisado IM, Purswani J, Gonzalez-Lopez J, Pozo C (2015) Physiological and genetic screening methods for the isolation of methyl tert-butyl ether-degrading bacteria for bioremediation purposes. Int Biodeterior Biodegrad 97:67–74. https://doi.org/10.1016/j.ibiod.2014.11.008
Haghshenas H, Kay M, Dehghanian F, Tavakol H (2016) Molecular dynamics study of biodegradation of azo dyes via their interactions with AzrC azoreductase. J Biomol Struct Dyn 34:453–462. https://doi.org/10.1080/07391102.2015.1039585
Hanson KG, Desai JD, Desai AJ (1993) A rapid and simple screening technique for potential crude oil degrading microorganisms. Biotechnol Tech 7:745–748. https://doi.org/10.1007/BF00152624
Harper DB, Hamilton JTG, O'Hagan D (1990) Identification of threo-18-fluoro-9,10-dihydroxystearic acid: a novel fluorinated fatty acid from *Dichapetalum toxicarium* seeds. Tetrahedron Lett 31:7661–7662
Hasan SA, Ferreira MIM, Koetsier MJ, Arif MI, Janssen DB (2011) Complete biodegradation of 4-fluorocinnamic acid by a consortium comprising *Arthrobacter* sp. strain G1 and *Ralstonia* sp. strain H1. Appl Environ Microbiol 77:572–579. https://doi.org/10.1128/AEM.00393-10

Haufe G, Pietz S, Wölker D, Fröhlich R (2003) Synthesis of fluorinated cycloalkyl N-phenylcarbamates and their microbial defluorination/oxygenation by *Beauveria bassiana*. Eur J Org Chem:2166–2175. https://doi.org/10.1002/ejoc.200200648

Herrero M, Stuckey DC (2015) Bioaugmentation and its application in wastewater treatment: a review. Chemosphere 140:119–128. https://doi.org/10.1016/j.chemosphere.2014.10.033

Hoffman DR, Okon JL, Sandrin TR (2005) Medium composition affects the degree and pattern of cadmium inhibition of naphthalene biodegradation. Chemosphere 59:919–927. https://doi.org/10.1016/j.chemosphere.2004.11.057

Hosokawa R, Nagai M, Morikawa M, Okuyama H (2009) Autochthonous bioaugmentation and its possible application to oil spills. World J Microbiol Biotechnol 25:1519–1528. https://doi.org/10.1007/s11274-009-0044-0

Hughes D, Clark BR, Murphy CD (2011) Biodegradation of polyfluorinated biphenyl in bacteria. Biodegradation 22:741–749. https://doi.org/10.1007/s10532-010-9411-7

Ivshina IB, Kuyukina MS, Krivoruchko AV, Plekhov OA, Naimark OB, Podorozhko EA, Lozinsky VI (2013) Biosurfactant-enhanced immobilization of hydrocarbon-oxidizing *Rhodococcus ruber* on sawdust. Appl Microbiol Biotechnol 97:5315–5327. https://doi.org/10.1007/s00253-013-4869-y

Janssen DB, Dinkla IJT, Poelarends GJ, Terpstra P (2005) Bacterial degradation of xenobiotic compounds: evolution and distribution of novel enzyme activities. Environ Microbiol 7:1868–1882. https://doi.org/10.1111/j.1462-2920.2005.00966.x

Jin J, Yao J, Zhang Q, Liu J (2016) Biodegradation of pyrene by *Pseudomonas* sp. JPN2 and its initial degrading mechanism study by combining the catabolic nahAc gene and structure-based analyses. Chemosphere 164:379–386. https://doi.org/10.1016/j.chemosphere.2016.08.113

Jin Y, Gan G, Yu X, Wu D, Zhang L, Yang N, Hu J, Liu Z, Zhang L, Hong H, Yan X, Liang Y, Ding L, Pan Y (2017) Isolation of viable but non-culturable bacteria from printing and dyeing wastewater bioreactor based on resuscitation promoting factor. Curr Microbiol 74:787–797. https://doi.org/10.1007/s00284-017-1240-z

Kaschabek SR, Reineke W (1995) Maleylacetate reductase of *Pseudomonas* sp. strain B13: specificity of substrate conversion and halide elimination. J Bacteriol 177:320–325. https://doi.ors/0021-9193/95/$04.00+0

Key BD, Howell RD, Criddle CS (1998) Defluorination of organofluorine sulfur compounds by *Pseudomonas* sp. strain D2. Environ Sci Technol 32:2283–2287. https://doi.org/10.1021/es9800129

Kiel M, Engesser K (2015) The biodegradation vs. biotransformation of fluorosubstituted aromatics. Appl Microbiol Biotechnol 99:7433–7464. https://doi.org/10.1007/s00253-015-6817-5

Kim BR, Suidan MT, Wallington TJ, Du X (2000) Biodegradability of trifluoroacetic acid. Environ Eng Sci 17 (6):337–342. https://doi.org/10.1089/ees.2000.17.337

Kim MH, Wang N, Mcdonald T, Chu KH (2012) Biodefluorination and biotransformation of fluorotelomer alcohols by two alkane-degrading *Pseudomonas* strains. Biotechnol Bioeng 109:3041–3048. https://doi.org/10.1002/bit.24561

Kim MH, Wang N, Chu KH (2014) 6:2 Fluorotelomer alcohol (6:2 FTOH) biodegradation by multiple microbial species under different physiological conditions. Appl Microbiol Biotechnol 98:1831–1840. https://doi.org/10.1007/s00253-013-5131-3

Kirk KL (2006) Fluorine in medicinal chemistry: recent therapeutic applications of fluorinated small molecules. J Fluor Chem 127:1013–1029. https://doi.org/10.1016/j.jfluchem.2006.06.007

Kirsten E, Sharma ML, Kun E (1978) Molecular toxicology of (-)-erythro-Fluorocitrate: selective inhibition of citrate transport in mitochondria and the binding of Fluorocitrate to mitochondrial proteins. Mol Pharmacol 14:172 LP–172184

Kolpin DW, Furlong ET, Meyer MT, Thurman EM, Zaugg SD, Barber LB, Buxton HT (2002) Pharmaceuticals, hormones, and other organic wastewater contaminants in U.S. streams, 1999–2000: a national reconnaissance. Environ Sci Technol 36:1202–1211. https://doi.org/10.1021/es011055j

Kramer C, Kreisel G, Fahr K, Käßbohrer J, Schlosser D (2004) Degradation of 2-fluorophenol by the brown-rot fungus *Gloeophyllum striatum*: evidence for the involvement of extracellular Fenton chemistry. Appl Microbiol Biotechnol 64:387–395. https://doi.org/10.1007/s00253-003-1445-x

Kubota K, Koma D, Matsumiya Y, Chung SY, Kubo M (2008) Phylogenetic analysis of long-chain hydrocarbon-degrading bacteria and evaluation of their hydrocarbon-degradation by the 2,6-DCPIP assay. Biodegradation 19:749–757. https://doi.org/10.1007/s10532-008-9179-1

Kumamaru T, Suenaga H, Mitsuoka M, Watanabe T, Furukawa K (1998) Enhanced degradation of polychlorinated biphenyls by directed evolution of biphenyl dioxygenase. Nat Biotechnol 16:663–666. https://doi.org/10.1038/nbt0798-663

Lajeunesse A, Gagnon C, Sauvé S (2008) Determination of basic antidepressants and their N-desmethyl metabolites in raw sewage and wastewater using solid-phase extraction and liquid chromatography-tandem mass spectrometry. Anal Chem 80:5325–5333. https://doi.org/10.1021/ac800162q

Langille MGI, Zaneveld J, Caporaso JG, McDonald D, Knights D, Reyes JA, Clemente JC, Burkepile DE, Vega Thurber RL, Knight R, Beiko RG, Huttenhower C (2013) Predictive functional profiling of microbial communities using 16S rRNA marker gene sequences. Nat Biotechnol 31:814–821. https://doi.org/10.1038/nbt.2676

Lauble H, Kennedy MC, Emptage MH, Beinert H, Stout CD (1996) The reaction of fluorocitrate with aconitase and the crystal structure of the enzyme-inhibitor complex. Proc Natl Acad Sci U S A 93:13699–13703. https://doi.org/10.1073/pnas.93.24.13699

Leewis MC, Uhlik O, Leigh MB (2016) Synergistic processing of biphenyl and benzoate: carbon flow through the bacterial community in polychlorinated-biphenyl-contaminated soil. Sci Rep 6:1–12. https://doi.org/10.1038/srep22145

Letcher RJ, Su G, Moore JN, Williams LL, Martin PA, De Solla SR, Bowerman WW (2015) Perfluorinated sulfonate and carboxylate compounds and precursors in herring gull eggs from across the Laurentian Great Lakes of North America: temporal and recent spatial comparisons and exposure implications. Sci Total Environ 538:468–477. https://doi.org/10.1016/j.scitotenv.2015.08.083

Leung LS, Frey PA (1978) Fluoropyruvate: an unusual substrate for *Escherichia coli* pyruvate dehydrogenase. Biochem Biophys Res Commun 81:274–279. https://doi.org/0006-291X/78/0812-0274$01,00/0

Lewis DL, Garrison a W, Wommack KE, Whittemore A, Steudler P, Melillo J (1999) Influence of environmental changes on degradation of chiral pollutants in soils. Nature 401:898–901. https://doi.org/10.1038/44801

Li J, Zhang D, Song M, Jiang L, Wang Y, Luo C, Zhang G (2017) Novel bacteria capable of degrading phenanthrene in activated sludge revealed by stable-isotope probing coupled with high-throughput sequencing. Biodegradation 28:423–436. https://doi.org/10.1007/s10532-017-9806-9

Lin C-W, Chen S-Y, Cheng Y-W (2006) Effect of metals on biodegradation kinetics for methyl tert-butyl ether. Biochem Eng J 32:25–32. https://doi.org/10.1016/j.bej.2006.07.010

Liu J, Mejia Avendaño S (2013) Microbial degradation of polyfluoroalkyl chemicals in the environment: a review. Environ Int 61:98–114. https://doi.org/10.1016/j.envint.2013.08.022

Liu JQ, Kurihara T, Ichiyama S, Miyagi M, Tsunasawa S, Kawasaki H, Soda K, Esaki N (1998) Reaction mechanism of fluoroacetate dehalogenase from *Moraxella* sp. B. J Biol Chem 273:30897–30902. https://doi.org/10.1074/jbc.273.47.30897

Liu W, Gan J, Schlenk D, Jury WA (2005) Enantioselectivity in environmental safety of current chiral insecticides. Proc Natl Acad Sci U S A 102:701–706. https://doi.org/10.1073/pnas.0408847102

Liu J, Lee LS, Nies LF, Nakatsu CH, Turco RF (2007) Biotransformation of 8: 2 fluorotelomer alcohol in soil and by soil bacteria isolates. Environ Sci Technol 41:8024–8030. https://doi.org/10.1021/es0708722

Liu B, Zhang H, Yao D, Li J, Xie L, Wang X, Wang Y, Liu G, Yang B (2015a) Perfluorinated compounds (PFCs) in the atmosphere of Shenzhen, China: spatial distribution, sources and health risk assessment. Chemosphere 138:511–518. https://doi.org/10.1016/j.chemosphere.2015.07.012

Liu X, Chen Y, Zhang X, Jiang X, Wu S, Shen J, Sun X, Li J, Lu L, Wang L (2015b) Aerobic granulation strategy for bioaugmentation of a sequencing batch reactor (SBR) treating high strength pyridine wastewater. J Hazard Mater. https://doi.org/10.1016/j.jhazmat.2015.04.025

Liu Z, Liu Y, Zeng G, Shao B, Chen M, Li Z, Jiang Y, Liu Y, Zhang Y, Zhong H (2018) Application of molecular docking for the degradation of organic pollutants in the environmental remediation: a review. Chemosphere 203:139–150. https://doi.org/10.1016/j.chemosphere.2018.03.179

Lu ZY, Guo XJ, Li H, Huang ZZ, Lin KF, Di LY (2015) High-throughput screening for a moderately halophilic phenol-degrading strain and its salt tolerance response. Int J Mol Sci 16:11834–11848. https://doi.org/10.3390/ijms160611834

Madhavan A, Sindhu R, Parameswaran B, Sukumaran RK, Pandey A (2017) Metagenome analysis: a powerful tool for enzyme bioprospecting. Appl Biochem Biotechnol 183(2):636-651. https://doi.org/10.1007/s12010-017-2568-3

Mahmoud MAM, Kärrman A, Oono S, Harada KH, Koizumi A (2009) Polyfluorinated telomers in precipitation and surface water in an urban area of Japan. Chemosphere 74:467–472. https://doi.org/10.1016/j.chemosphere.2008.08.029

Maia AS, Ribeiro AR, Amorim CL, Barreiro JC, Cass QB, Castro PML, Tiritan ME (2014) Degradation of fluoroquinolone antibiotics and identification of metabolites/transformation products by liquid chromatography–tandem mass spectrometry. J Chromatogr A 1333:87–98. https://doi.org/10.1016/j.chroma.2014.01.069

Maia AS, Castro PML, Tiritan ME (2016) Integrated liquid chromatography method in enantioselective studies: biodegradation of ofloxacin by an activated sludge consortium. J Chromatogr B 1029:174–183. https://doi.org/10.1016/j.jchromb.2016.06.026

Maia AS, Ribeiro AR, Castro PML, Tiritan ME (2017) Chiral analysis of pesticides and drugs of environmental concern: biodegradation and enantiomeric fraction. Symmetry 9(9):196. https://doi.org/10.3390/sym9090196

Maia AS, Tiritan ME, Castro PML (2018) Enantioselective degradation of ofloxacin and levofloxacin by the bacterial strains *Labrys portucalensis* F11 and *Rhodococcus* sp. FP1. Ecotoxicol Environ Saf 155:144–151. https://doi.org/10.1016/j.ecoenv.2018.02.067

Marais JSC (1944) Monofluoroacetic acid, the toxic principle of "gifblaar" in *Dichapetalum cymosum* (Hook) Engl. Onderstepoort J Vet Sci Anim Ind 20:67–73

Margesin R, Schinner F (2001) Biodegradation and bioremediation of hydrocarbons in extreme environments. Appl Microbiol Biotechnol 56:650–663. https://doi.org/10.1007/s002530100701

Martin JW, Mabury SA, Solomon KR, Muir DCG (2003) Bioconcentration and tissue distribution of perfluorinated acids in rainbow trout (*Oncorhynchus mykiss*). Environ Toxicol Chem 22:196–204. https://doi.org/10.1002/etc.5620220126

Martín-Hernández M, Suárez-Ojeda ME, Carrera J (2012) Bioaugmentation for treating transient or continuous p-nitrophenol shock loads in an aerobic sequencing batch reactor. Bioresour Technol 123:150–156. https://doi.org/10.1016/j.biortech.2012.07.014

Mesquita RBR, Santos IC, Pedrosa MFF, Duque AF, Castro PML, Rangel AOSS (2011) Development of flow injection potentiometric methods for the off-line and on-line determination of fluoride to monitor the biodegradation of a monofluorophenol in two bioreactors. Talanta 84:1291–1297. https://doi.org/10.1016/j.talanta.2011.01.028

Milne GWA, Goldman P, Holtzman JL (1968) The metabolism of 2-Fluorobenzoic acid II. Studies with $^{18}O_2$. J Biol Chem 243:5374–5376.

Misiak K, Casey E, Murphy CD (2011) Factors influencing 4-fluorobenzoate degradation in biofilm cultures of *Pseudomonas knackmussii* B13. Water Res 45:3512–3520. https://doi.org/10.1016/j.watres.2011.04.020

Moody CA, Field JA (1999) Determination of perfluorocarboxylates in groundwater impacted by fire-fighting activity determination of perfluorocarboxylates in groundwater impacted by fire-fighting activity. Environ Sci Technol 33:2800–2806. https://doi.org/10.1021/es981355+

Moraski GC, Bristol R, Seeger N, Boshoff HI, Tsang PSY, Miller MJ (2017) Preparation and evaluation of potent Pentafluorosulfanyl-substituted anti-tuberculosis compounds. ChemMedChem 12:1108–1115. https://doi.org/10.1002/cmdc.201700170

Moreira IS, Amorim CL, Carvalho MF, Castro PML (2012a) Degradation of difluorobenzenes by the wild strain *Labrys portucalensis*. Biodegradation 23:653–662. https://doi.org/10.1007/s10532-012-9541-1

Moreira IS, Amorim CL, Carvalho MF, Castro PML (2012b) Co-metabolic degradation of chlorobenzene by the fluorobenzene degrading wild strain *Labrys portucalensis*. Int Biodeterior Biodegrad 72:76–81. https://doi.org/10.1016/j.ibiod.2012.05.013

Moreira IS, Amorim CL, Carvalho MF, Ferreira AC, Afonso CM, Castro PML (2013) Effect of the metals iron, copper and silver on fluorobenzene biodegradation by *Labrys portucalensis*. Biodegradation 24:245–255. https://doi.org/10.1007/s10532-012-9581-6

Moreira IS, Ribeiro AR, Afonso CM, Tiritan ME, Castro PML (2014) Enantioselective biodegradation of fluoxetine by the bacterial strain *Labrys portucalensis* F11. Chemosphere 111:103–111. https://doi.org/10.1016/j.chemosphere.2014.03.022

Moreira IS, Amorim CL, Ribeiro AR, Mesquita RBR, Rangel AOSS, Van Loosdrecht MCM, Tiritan ME, Castro PML (2015) Removal of fluoxetine and its effects in the performance of an aerobic granular sludge sequential batch reactor. J Hazard Mater 287:93–101. https://doi.org/10.1016/j.jhazmat.2015.01.020

Mrozik A, Piotrowska-Seget Z (2010) Bioaugmentation as a strategy for cleaning up of soils contaminated with aromatic compounds. Microbiol Res 165:363–375. https://doi.org/10.1016/j.micres.2009.08.001

Mukamolova GV, Kaprelyants AS, Young DI, Young M, Kell DB (1998) A bacterial cytokine. Proc Natl Acad Sci U S A 95:8916–8921. https://doi.org/10.1073/pnas.95.15.8916

Mukherjee A, Chettri B, Langpoklakpam JS, Basak P, Prasad A, Mukherjee AK, Bhattacharyya M, Singh AK, Chattopadhyay D (2017) Bioinformatic approaches including predictive metagenomic profiling reveal characteristics of bacterial response to petroleum hydrocarbon contamination in diverse environments. Sci Rep 7:1108. https://doi.org/10.1038/s41598-017-01126-3

Müller TA, Kohler H-PE (2004) Chirality of pollutants – effects on metabolism and fate. Appl Microbiol Biotechnol 64:300–316. https://doi.org/10.1007/s00253-003-1511-4

Murphy CD (2010) Biodegradation and biotransformation of organofluorine compounds. Biotechnol Lett 32:351–359. https://doi.org/10.1007/s10529-009-0174-3

Murphy CD, Quirke S, Balogun O (2008) Degradation of fluorobiphenyl by *Pseudomonas pseudoalcaligenes* KF707. FEMS Microbiol Lett 286:45–49. https://doi.org/10.1111/j.1574-6968.2008.01243.x

Murphy CD, Clark BR, Amadio J (2009) Metabolism of fluoroorganic compounds in microorganisms: impacts for the environment and the production of fine chemicals. Appl Microbiol Biotechnol 84:617–629. https://doi.org/10.1007/s00253-009-2127-0

Neilson AH, Allard A-S (2002) Degradation and transformation of organic fluorine compounds. Handb Environ Chem 3(3): 137–189. https://doi.org/10.1007/10721878_6

Nguyen MA, Wiberg K, Ribeli E, Josefsson S, Futter M, Gustavsson J, Ahrens L (2017) Spatial distribution and source tracing of per- and polyfluoroalkyl substances (PFASs) in surface water in Northern Europe. Environ Pollut 220:1438–1446. https://doi.org/10.1016/j.envpol.2016.10.089

Nies DH (1999) Microbial heavy-metal resistance. Appl Microbiol Biotechnol 51:730–750. https://doi.org/10.1007/s002530051457

Nzila A, Razzak SA, Zhu J (2016) Bioaugmentation: an emerging strategy of industrial wastewater treatment for reuse and discharge. Int J Environ Res Public Health 13(9): 846. https://doi.org/10.3390/ijerph13090846

O'Hagan D (2008) Understanding organofluorine chemistry. An introduction to the C–F bond. Chem Soc Rev 37:308–319. https://doi.org/10.1039/B711844A

O'Hagan D, Harper DB (1999) Fluorine-containing natural products. J Fluor Chem 100:127–133. https://doi.org/10.1016/S0022-1139(99)00201-8

O'Hagan D, Schaffrath C, Cobb SL, Hamilton JTG (2002) Biosynthesis of an organofluorine molecule. A fluorinase enzyme has been discovered that catalyses carbon–fluorine bond formation. Nature 416:279. https://doi.org/10.1038/416279a

Olga P, Petar K, Jelena M, Srdjan R (2008) Screening method for detection of hydrocarbon-oxidizing bacteria in oil-contaminated water and soil specimens. J Microbiol Methods 74:110–113. https://doi.org/10.1016/j.mimet.2008.03.012

Oliver JD (2010) Recent findings on the viable but nonculturable state in pathogenic bacteria. FEMS Microbiol Rev 34:415–425. https://doi.org/10.1111/j.1574-6976.2009.00200.x

Oltmanns RH, Müller R, Otto MK, Lingens F (1989) Evidence for a new pathway in the bacterial degradation of 4-fluorobenzoate. Appl Environ Microbiol 55:2499–2504

Omrani R, Spini G, Puglisi E, Saidane D (2018) Modulation of microbial consortia enriched from different polluted environments during petroleum biodegradation. Biodegradation 29:187–209. https://doi.org/10.1007/s10532-018-9823-3

Park D, Lee DS, Mo Y, Park JM (2008) Bioaugmentation of cyanide-degrading microorganisms in a full-scale cokes wastewater treatment facility. Bioresour Technol 99:2092–2096. https://doi.org/10.1016/j.biortech.2007.03.027

Perruchon C, Chatzinotas A, Omirou M, Vasileiadis S, Menkissoglou-Spiroudi U, Karpouzas DG (2017) Isolation of a bacterial consortium able to degrade the fungicide thiabendazole: the key role of a *Sphingomonas* phylotype. Appl Microbiol Biotechnol 101:3881–3893. https://doi.org/10.1007/s00253-017-8128-5

Peters RA, Hall RJ, Ward PFV, Sheppard N (1960) The chemical nature of the toxic compounds containing fluorine in the seeds of *Dichapetalum toxicarium*. Biochem J 77:17–22.

Piccirillo C, Pereira SIA, Marques APGC, Pullar RC, Tobaldi DM, Pintado ME, Castro PML (2013) Bacteria immobilisation on hydroxyapatite surface for heavy metals removal. J Environ Manag 121:87–95. https://doi.org/10.1016/j.jenvman.2013.02.036

Plangklang P, Reungsang A (2009) Bioaugmentation of carbofuran residues in soil using *Burkholderia cepacia* PCL3 adsorbed on agricultural residues. Int Biodeterior Biodegrad 63:515–522. https://doi.org/10.1016/j.ibiod.2009.02.003

Prasad R (2017) Mycoremediation and environmental sustainability. Volume 1. Springer International Publishing (ISBN 978–3–319-68957-9) https://doi.org/10.1007/978-3-319-68957-9

Prasad R (2018) Mycoremediation and environmental sustainability, Volume 2. Springer International Publishing (ISBN 978–3–319-77386-5) https://www.springer.com/us/book/9783319773858

Prevedouros K, Cousins IT, Buck RC, Korzeniowski SH (2006) Sources, fate and transport of perfluorocarboxylates. Environ Sci Technol 40:32–44. http://doi.org/10.1021/es0512475

Purser S, Moore PR, Swallow S, Gouverneur V (2008) Fluorine in medicinal chemistry. Chem Soc Rev 37:320–330. https://doi.org/10.1039/b610213c

Qin Z, Raab A, Krupp E, Deng H, Feldmann J (2013) Mining complex bacteria media for all fluorinated compounds made possible by using HPLC coupled parallel to fluorine-specific and molecular specific detection. J Anal At Spectrom 28:877–882. https://doi.org/10.1039/c3ja50086a

Rajoo S, Ahn JO, Lee HW, Jung JK (2013) Isolation and characterization of a novel ε-caprolactam-degrading microbe, *Acinetobacter calcoaceticus*, from industrial wastewater by chemostat-enrichment. Biotechnol Lett 35:2069–2072. https://doi.org/10.1007/s10529-013-1307-2

Ramos C, Amorim CL, Mesquita DP, Ferreira EC, Carrera J, Castro PML (2017) Simultaneous partial nitrification and 2-fluorophenol biodegradation with aerobic granular biomass: reactor performance and microbial communities. Bioresour Technol. 238:232-240. https://doi.org/10.1016/j.biortech.2017.03.173

Rathod J, Archana G (2013) Molecular fingerprinting of bacterial communities in enriched azo dye (reactive violet 5R) decolorising native acclimatised bacterial consortia. Bioresour Technol 142:436–444. https://doi.org/10.1016/j.biortech.2013.05.057

Reardon KF, Mosteller DC, Bull Rogers JD (2000) Biodegradation kinetics of benzene, toluene, and phenol as single and mixed substrates for *Pseudomonas putida* F1. Biotechnol Bioeng 69:385–400. https://doi.org/10.1002/1097-0290(20000820)69:4<385::AID-BIT5>3.0.CO;2-Q

Reinscheid UM, Zuilhof H, Vervoort J (1998) Biological, thermal and photochemical transformation of 2-trifluoromethylphenol. Biodegradation 9:487–499. https://doi.org/10.1023/A:1008394115008

Ribeiro AR, Maia AS, Moreira IS, Afonso CM, Castro PML, Tiritan ME (2014a) Enantioselective quantification of fluoxetine and norfluoxetine by HPLC in wastewater effluents. Chemosphere 95:589-596. https://doi.org/10.1016/j.chemosphere.2013.09.118

Ribeiro AR, Santos LHMLM, Maia AS, Delerue-Matos C, Castro PML, Tiritan ME (2014b) Enantiomeric fraction evaluation of pharmaceuticals in environmental matrices by liquid chromatography-tandem mass spectrometry. J Chromatogr A 1363:226–235. https://doi.org/10.1016/j.chroma.2014.06.099

Riis V, Babel W, Pucci OH (2002) Influence of heavy metals on the microbial degradation of diesel fuel. Chemosphere 49:559–568. https://doi.org/10.1016/S0045-6535(02)00386-7

Robinson AA, Belden JB, Lydy MJ (2005) Toxicity of fluoroquinolone antibiotics to aquatic organisms. Environ Toxicol Chem 24:423–430. https://doi.org/10.1897/04-210R.1

Saccomanno M, Hussain S, O'Connor NK, Beier P, Somlyay M, Konrat R, Murphy CD (2018) Biodegradation of pentafluorosulfanyl-substituted aminophenol in *Pseudomonas* spp. Biodegradation 29:1–12. https://doi.org/10.1007/s10532-018-9827-z

Sakakibara F, Takagi K, Kataoka R, Kiyota H, Sato Y, Okada S (2011) Isolation and identification of dieldrin-degrading *Pseudonocardia* sp. strain KSF27 using a soil-charcoal perfusion method with aldrin trans-diol as a structural analog of dieldrin. Biochem Biophys Res Commun 411:76–81. https://doi.org/10.1016/j.bbrc.2011.06.096

Sanada M, Miyano T, Iwadare S, Williamson JM, Arison BH, Smith JL, Douglas AW, Liesch JM, Inamine E (1986) Biosynthesis of fluorothreonine and fluoroacetic acid by the thienamycin producer, *Streptomyces cattleya*. J Antibiot (Tokyo) 39:259–265. https://doi.org/10.7164/antibiotics.39.259

Sandford G (2000) Organofluorine chemistry. Philos Trans R Soc A Math Phys Eng Sci 358:455–471. https://doi.org/10.1098/rsta.2000.0541

Santos JR, Lapa RAS, Lima JLFC (2007) Development of a tubular fluoride potentiometric detector for flow analysis. Evaluation and analytical applications. Anal Chim Acta 583:429–436. https://doi.org/10.1016/j.aca.2006.10.020

Santos IC, Mesquita RBR, Amorim CL, Castro PML, Rangel AOSS (2016) Development of a low pressure chromatographic flow system for monitoring the biodegradation of ofloxacin and ciprofloxacin. Anal Methods 8:5457–5465. https://doi.org/10.1039/c6ay00476h

Schlömann M, Fischer P, Schmidt E, Knackmuss HJ (1990) Enzymatic formation, stability, and spontaneous reactions of 4-fluoromuconolactone, a metabolite of the bacterial degradation of 4-fluorobenzoate. J Bacteriol 172:5119–5129. http://doi.org/0021-9193/90/095119-11$02.00/0

Schmidt S, Wittich RM, Erdmann D, Wilkes H, Francke W, Fortnagel P (1992) Biodegradation of diphenyl ether and its monohalogenated derivatives by *Sphingomonas* sp. strain SS3. Appl Environ Microbiol 58:2744–2750. http://doi.org/0099-2240/92/092744-07$02.00/0

Schreiber A, Hellwig M, Dorn E, Reineke W, Knackmuss H-J (1980) Critical reactions in fluorobenzoic acid degradation by *Pseudomonas* sp. B13. Appl Environ Microbiol 39:58–67. https://doi.org/0099-2240/80/0/0058/12$02.00/0

Schultz-Jensen N, Aamand J, Sørensen SR (2016) Bioaugmentation potential of free and formulated 2,6-dichlorobenzamide (BAM) degrading *Aminobacter* sp. MSH1 in soil, sand and water. AMB Express 6:33. https://doi.org/10.1186/s13568-016-0204-1

Schwede T, Kopp J, Guex N, Peitsch MC (2003) SWISS-MODEL: an automated protein homology-modeling server. Nucleic Acids Res 31:3381–3385. https://doi.org/10.1093/nar/gkg520

Sepulvado JG, Blaine AC, Hundal LS, Higgins CP (2011) Occurrence and fate of perfluorochemicals in soil following the land application of municipal biosolids. Environ Sci Technol 45:8106–8112. https://doi.org/10.1021/es103903d

Seralathan MV, Sivanesan S, Nargunanathan S, Bafana A, Kannan K, Chakrabarti T (2015) Chemotaxis-based endosulfan biotransformation: enrichment and isolation of endosulfan-degrading bacteria. Environ Technol (United Kingdom) 36:60–67. https://doi.org/10.1080/09593330.2014.937464

Shoeib M, Vlahos P, Harner T, Peters A, Graustein M, Narayan J (2010) Survey of polyfluorinated chemicals (PFCs) in the atmosphere over the Northeast Atlantic Ocean. Atmos Environ 44:2887–2893. https://doi.org/10.1016/j.atmosenv.2010.04.056

Sivasubramaniam D, Franks AE (2016) Bioengineering microbial communities: their potential to help, hinder and disgust. Bioengineered 7:137–144. https://doi.org/10.1080/21655979.2016.1187346

Smit MF, van Heerden PDR, Pienaar JJ, Weissflog L, Strasser RJ, Krüger GHJ (2009) Effect of trifluoroacetate, a persistent degradation product of fluorinated hydrocarbons, on *Phaseolus vulgaris* and *Zea mays*. Plant Physiol Biochem 47 (7):623-634. https://doi.org/10.1016/j.plaphy.2009.02.003

Smith MR, Ewing M, Ratledge C (1991) The interactions of various aromatic substrates degraded by *Pseudomonas* sp. NCIB 10643: synergistic inhibition of growth by two compounds that serve as growth substrates. Appl Microbiol Biotechnol 34:536–538. https://doi.org/0175759891000208

Sokhn J, De Leij FA, Hart TD, Lynch JM (2001) Effect of copper on the degradation of phenanthrene by soil micro-organisms. Lett Appl Microbiol 33:164–168. https://doi.org/10.1046/j.1472-765x.2001.00972.x

Song B, Palleroni NJ, Max M (2000) Isolation and characterization of diverse halobenzoate-degrading denitrifying bacteria from soils and sediments isolation and characterization of diverse halobenzoate-degrading denitrifying bacteria from soils and sediments. Appl Environ Microbiol. 66(8): 3446–3453. https://doi.org/10.1128/AEM.66.8.3446-3453.2000

Song E, Wang M, Shen D (2014) Isolation, identification and characterization of a novel *Ralstonia* sp. FD-1, capable of degrading 4-fluoroaniline. Biodegradation 25:85–94. https://doi.org/10.1007/s10532-013-9642-5

Sowaileh MF, Hazlitt RA, Colby DA (2017) Application of the pentafluorosulfanyl group as a bioisosteric replacement. ChemMedChem 12:1481–1490. https://doi.org/10.1002/cmdc.201700356

Strunk N, Engesser K-H (2013) Degradation of fluorobenzene and its central metabolites 3-fluorocatechol and 2-fluoromuconate by *Burkholderia fungorum* FLU100. Appl Microbiol Biotechnol 97:5605–5614. https://doi.org/10.1007/s00253-012-4388-2

Su X, Shen H, Yao X, Ding L, Yu C, Shen C (2013) A novel approach to stimulate the biphenyl-degrading potential of bacterial community from PCBs-contaminated soil of e-waste recycling sites. Bioresour Technol 146:27–34. https://doi.org/10.1016/j.biortech.2013.07.028

Sukul P, Spiteller M (2007) Fluoroquinolone antibiotics in the environment. In: Reviews of environmental contamination and toxicology. Springer, New York, pp 131–162. https://doi.org/10.1007/978-0-387-69163-3_5

Sun D, Meng J, Xu EG, Chen W (2016) Microbial community structure and predicted bacterial metabolic functions in biochar pellets aged in soil after 34 months. Appl Soil Ecol 100:135–143. https://doi.org/10.1016/j.apsoil.2015.12.012

Suresh PS, Kumar A, Kumar R, Singh VP (2008) An In silico approach to bioremediation: laccase as a case study. J Mol Graph Model 26:845–849. https://doi.org/10.1016/j.jmgm.2007.05.005

Takagi K, Iwasaki A, Kamei I, Satsuma K, Yoshioka Y, Harada N (2009) Aerobic mineralization of hexachlorobenzene by newly isolated pentachloronitrobenzene-degrading *Nocardioides* sp. strain PD653. Appl Environ Microbiol 75:4452–4458. https://doi.org/10.1128/AEM.02329-08

Tiedt O, Mergelsberg M, Boll K, Müller M, Adrian L, Jehmlich N, von Bergen M, Boll M (2016) ATP-dependent C–F bond cleavage allows the complete degradation of 4-fluoroaromatics without oxygen. MBio 7:1–9. https://doi.org/10.1128/mBio.00990-16

Tonomura K, Futai F, Tanabe O, Yamaoka T (1965) Defluorination of monofluoroacetate by bacteria. Agric Biol Chem 29:124–128. https://doi.org/10.1080/00021369.1965.10858362

Travkin VM, Solyanikova IP, Rietjens IMCM, Vervoort J, van Berkel WJH, Golovleva LA (2003) Degradation of 3,4-dichloro- and 3,4-difluoroaniline by *Pseudomonas fluorescens* 26-K. J Environ Sci Health B 38:121–132. https://doi.org/10.1081/PFC-120018443

Tsai S-Y, Juang R-S (2006) Biodegradation of phenol and sodium salicylate mixtures by suspended *Pseudomonas putida* CCRC 14365. J Hazard Mater 138:125–132. https://doi.org/10.1016/j.jhazmat.2006.05.044

Tyagi M, da Fonseca MMR, de Carvalho CCCR (2011) Bioaugmentation and biostimulation strategies to improve the effectiveness of bioremediation processes. Biodegradation 22:231–241. https://doi.org/10.1007/s10532-010-9394-4

Utgikar VP, Tabak HH, Haines JR, Govind R (2003) Quantification of toxic and inhibitory impact of copper and zinc on mixed cultures of sulfate-reducing bacteria. Biotechnol Bioeng 82:306–312. https://doi.org/10.1002/bit.10575

van Pée K-H, Unversucht S (2003) Biological dehalogenation and halogenation reactions. Chemosphere 52:299–312. https://doi.org/10.1016/S0045-6535(03)00204-2

Vargas C, Song B, Camps M, Häggblom MM (2000) Anaerobic degradation of fluorinated aromatic compounds. Appl Microbiol Biotechnol 53:342–347. https://doi.org/10.1007/s002530050032

Visscher PT, Culbertson CW, Oremland RS (1994) Degradation of trifluoroacetate in oxic and anoxic sediments. Nature 369(6483):729–731. https://doi.org/10.1038/369729a0

Vora KA, Singh C, Modi VV (1988) Degradation of 2-fluorobenzoate by a pseudomonad. Curr Microbiol 17:249–254. https://doi.org/10.1007/BF01571323

Walker JRL, Lien BC (1981) Metabolism of fluoroacetate by a soil *Pseudomonas* sp. and *Fusarium solani*. Soil Biol Biochem 13:231–235. https://doi.org/10.1016/0038-0717(81)90026-2

Wang N, Szostek B, Buck RC, Folsom PW, Sulecki LM, Gannon JT (2009) 8:2 Fluorotelomer alcohol aerobic soil biodegradation: pathways, metabolites, and metabolite yields. Chemosphere 75:1089–1096. https://doi.org/10.1016/j.chemosphere.2009.01.033

Wang N, Liu J, Buck RC, Korzeniowski SH, Wolstenholme BW, Folsom PW, Sulecki LM (2011) 6:2 Fluorotelomer sulfonate aerobic biotransformation in activated sludge of waste water treatment plants. Chemosphere 82:853–858. https://doi.org/10.1016/j.chemosphere.2010.11.003

Wang M, Xu J, Wang J, Wang S, Feng H, Shentu J, Shen D (2013) Differences between 4-fluoroaniline degradation and autoinducer release by *Acinetobacter* sp. TW: implications for operating conditions in bacterial bioaugmentation. Environ Sci Pollut Res 20:6201–6209. https://doi.org/10.1007/s11356-013-1660-7

Wang B, Liu W, Liu X, Franks AE, Teng Y, Luo Y (2017) Comparative analysis of microbial communities during enrichment and isolation of DDT-degrading bacteria by culture-dependent and -independent methods. Sci Total Environ 590–591:297–303. https://doi.org/10.1016/j.scitotenv.2017.03.004

Weinberg I, Dreyer A, Ebinghaus R (2010) Waste water treatment plants as sources of polyfluorinated compounds, polybrominated diphenyl ethers and musk fragrances to ambient air. Environ Pollut 159:125–132. https://doi.org/10.1016/j.envpol.2010.09.023

Wong CS (2006) Environmental fate processes and biochemical transformations of chiral emerging organic pollutants. Anal Bioanal Chem 386:544–558. https://doi.org/10.1007/s00216-006-0424-3

Wunderwald U, Hofrichter M, Kreisell G, Fritsche W (1998) Transformation of difluorinated phenols by *Penicillium frequentans* Bi 7/2. Biodegradation 8:379–385. https://doi.org/10.1023/A:1008230926973

Xiao F, Field JA, Seow J, Houtz EF, Sutton R, Park JS, Sedlak M, Murphy CD (2017) Emerging poly- and perfluoroalkyl substances in the aquatic environment: a review of current literature. Water Res 95:643–691. https://doi.org/10.1016/j.watres.2017.07.024

Xu J, Guo CS, Zhang Y, Meng W (2014) Bioaccumulation and trophic transfer of perfluorinated compounds in a eutrophic freshwater food web. Environ Pollut 184:254–261. https://doi.org/10.1016/j.envpol.2013.09.011

Yamasaki S, Nomura N, Nakajima T, Uchiyama H (2012) Cultivation-independent identification of candidate dehalorespiring bacteria in tetrachloroethylene degradation. Environ Sci Technol 46:7709–7716. https://doi.org/10.1021/es301288y

Yang J-F, Ying G-G, Zhao J-L, Tao R, Su H-C, Chen F (2010) Simultaneous determination of four classes of antibiotics in sediments of the Pearl Rivers using RRLC–MS/MS. Sci Total Environ 408:3424–3432. https://doi.org/10.1016/j.scitotenv.2010.03.049

Yano K, Wachi M, Tsuchida S, Kitazume T, Iwai N (2015) Degradation of benzotrifluoride via the dioxygenase pathway in *Rhodococcus* sp. 065240. Biosci Biotechnol Biochem 79:496–504. https://doi.org/10.1080/09168451.2014.982502

Yiruhan, Wang Q-J, Mo C-H, Li Y-W, Gao P, Tai Y-P, Zhang Y, Ruan Z-L, Xu J-W (2010) Determination of four fluoroquinolone antibiotics in tap water in Guangzhou and Macao. Environ Pollut 158:2350–2358. https://doi.org/10.1016/j.envpol.2010.03.019

Young RJ, Grushin VV (1999) Catalytic C-F bond activation of nonactivated monofluoroarenes. Organometallics 18:294–296. https://doi.org/10.1021/om980887w

Zhang QQ, Yang GF, Zhang L, Zhang ZZ, Tian GM, Jin RC (2017) Bioaugmentation as a useful strategy for performance enhancement in biological wastewater treatment undergoing different stresses: application and mechanisms. Crit Rev Environ Sci Technol 47:1877–1899. https://doi.org/10.1080/10643389.2017.1400851

Zhao Z, Feng Y, Feng H, Ghulam A, Su Y, Shen D (2014) Anaerobic biotransformation of fluoronitrobenzenes and microbial communities in methanogenic systems. J Environ Sci Health A Tox Hazard Subst Environ Eng 49:1187–1197. https://doi.org/10.1080/10934529.2014.897537

Zhao Z-Q, Tian B-H, Zhang X, Ghulam A, Zheng T-C, Shen D-S (2015) Aerobic degradation study of three fluoroanilines and microbial community analysis : the effects of increased fluorine substitution. Biodegradation 26:1–14. https://doi.org/10.1007/s10532-014-9704-3

Ziagova M, Liakopoulou-Kyriakides M (2007) Comparison of cometabolic degradation of 1,2-dichlorobenzene by *Pseudomonas* sp. and *Staphylococcus xylosus*. Enzym Microb Technol 40:1244–1250. https://doi.org/10.1016/j.enzmictec.2006.09.004

Zuccato E, Castiglioni S, Bagnati R, Melis M, Fanelli R (2010) Source, occurrence and fate of antibiotics in the Italian aquatic environment. J Hazard Mater 179:1042–1048. https://doi.org/10.1016/j.jhazmat.2010.03.110

# Chapter 12
# Marine-Derived Fungi: Promising Candidates for Enhanced Bioremediation

**Anjana K. Vala, Bhumi Sachaniya, and Bharti P. Dave**

## Contents



## 12.1 Introduction

Extensive use and need for the more novel materials for rapidly increasing industrialization and population explosion in the world has resulted in accelerated degradation of environmental components on a large-scale (Deshmukh et al. 2016). Humans have synthesized various novel components which are proving to be contaminants/pollutants of the natural environment and are becoming a major cause of concern today. Various environmental compartments are being loaded with a large quantum of contaminants and recalcitrant compounds including plastics, petroleum oil products, materials containing heavy metals, etc. These compounds are major pollutants

A. K. Vala (✉) · B. Sachaniya
Department of Life Sciences, Maharaja Krishnakumarsinhji Bhavnagar University, Bhavnagar, India
e-mail: akv@mkbhavuni.edu.in

B. P. Dave (✉)
Department of Life Sciences, Maharaja Krishnakumarsinhji Bhavnagar University, Bhavnagar, India

School of Sciences, Indrashil University, Rajpur, Chhattisgarh, India

 
R. Prasad, E. Aranda (eds.), *Approaches in Bioremediation*, Nanotechnology in the Life Sciences, https://doi.org/10.1007/978-3-030-02369-0_12

found in the marine environments. Marine environment also represents relatively less explored niche with context to fungi which have potential to degrade or transform these pollutants (Prasad 2017, 2018).

Fungi comprise a group of heterotrophic microorganisms which have evolved biologically more precisely biochemically in a diverse manner which has resulted in tremendous metabolic potential of these organisms towards an array of recalcitrant compounds. Fungi have some distinct characteristic features that enable them to be potential degraders (Bennet et al. 2002). Marine fungi have the ability to adapt to high saline conditions and pH extremes compared to terrestrial fungi. The mycelial growth pattern of filamentous fungi benefits them with respect to invasion and colonization of insoluble substrates. Fungi produce a battery of extracellular degradative enzymes that can digest diverse substances enabling easy penetration of the mycelia. The filamentous mycelial growth form provides mechanical adjunct. High surface to cell ratio of filaments of fungi provides more wide contact to the substrate. As most of the fungal degradative enzymes are extracellular, insoluble substrates which cannot cross the cell membrane also can be easily utilized by fungi compared to bacteria. Majority of the fungal degradative enzymes are non-specific or can act upon diverse substrate, thus enabling fungi to metabolize diverse substances. Besides, fungi also play an important role in the treatment of different pollutants by being an important component of biofilm-forming microbial communities.

In addition to their application in hydrocarbons and heavy metals bioremediation, the potential of marine fungi in production of secondary metabolites, biosurfactants, novel enzymes, polysaccharides and polyunsaturated fatty acids has also been well documented (Damare et al. 2012). The pivotal role of marine fungi in biodegradation of various contaminants has been well described in literature by numerous researchers (Bonugli-Santos et al. 2012; Gao et al. 2013; Naranjo-Briceno et al. 2013). The present chapter focuses on the potentials of marine-derived fungi relevant to environmental clean-up including biofilm formation and degradation/transformation of prime hazardous pollutants such as plastics, synthetic and textile dyes, petroleum oil and heavy metal(oid)s.

## 12.2 Diverse Potentials of Marine-Derived Fungi Relevant to Bioremediation

### *12.2.1 Marine-Derived Fungi in Biofilm Formation*

Fungi do occur ubiquitously in marine environment and even the harshest habitats can be inhabited by many fungal species (Sole et al. 2008; Salamone et al. 2016). Fungi are also an important component of biofilm-forming microbial communities, and they could be potential candidates as bioindicators of ecosystem disturbance (Sole et al. 2008; Salamone et al. 2016).

With adhesion and subsequent differential gene expression as intrinsic and fundamental processes, fungi are regular biofilm-forming biota (Gutiérrez-Correa and Villena 2003; Mitra et al. 2013). Higher transcription of exoglucanase (exo) and xylanase (*xynB*) genes in the biofilm cultures than that in the submerged cultures of *Aspergillus niger* ATCC 10864 (Villena et al. 2009; Mitra et al. 2013) suggests significant role of biofilm-forming organisms in bioremediation processes.

Mycobiota in the marine environment may function as secondary producers; they help sustain many of the primary consumers supporting even the artificial reef habitats. Looking to the role of mycobiota in processing of organic matter, it has been suggested that the current marine microbial loop model should consider their important role (Gutierrez et al. 2011; Salamone et al. 2016).

Recently, Salamone et al. (2016) characterized marine biofilm-forming fungal communities associated with artificial reef. From the artificial reef biofilm samples, 295 isolates, belonging to 36 fungal genera, could be cultured. *Aspergillus*, *Aureobasidium*, *Cladosporium*, *Penicillium* and *Trichoderma* were among the most commonly occurring genera in the biofilms. The authors claim the work to be the first ever report on fungal biofilm communities from in situ artificial reef substrates.

*Yarrowia lipolytica* has been reported to be biotechnologically relevant and to play an important role in the treatment of effluents and hydrocarbons, and also it metabolizes explosives. Biofilm formation by tropical marine *Yarrowia lipolytica* NCIM 3589 on various materials have been examined (Dusane et al. 2008). With different carbon sources, the test isolate exhibited variation in morphology. The authors suggested possible occurrence of highly structured biofilms in diverse ecological niches that are the source for the yeast.

According to Zinjarde et al. (2014), the biofilm-forming capability of *Y. lipolytica* could play an important role under field conditions and in the course of treatment of different pollutants. During bioremediation and in waste treatment processes, occurrence of microbes as biofilms has intrinsic benefits over their planktonic counterparts. The biofilm mode allows cells to be shielded within extracellular polymeric substance leading to better chance of adaptation and endurance during stress period (Singh et al. 2006). For application of yeast for bioremediation, understanding of their morphological traits in presence of hydrocarbons is very important. For field application, the choice of strains entirely growing in yeast form or those developing biofilms may be very useful (Zinjarde et al. 2014).

Mitra et al. (2013) investigated biotransformation of polycyclic aromatic hydrocarbon fluoranthene by intertidally derived fungus *Cunninghamella elegans* under biofilm-based and niche-mimicking conditions. Increased level of cytochrome P450 (CYP) monooxygenase mRNA in the biofilm under the niche-mimicking condition was observed compared to that under submerged conditions that were observed. The most suitable condition for biotransformation of fluoranthene was formation of biofilm on a hydrophobic surface with alternating immersion and emersion in the fluid medium.

### 12.2.2 Marine-Derived Fungi in Heavy Metal(oid) Removal

In natural as well as synthetic environments, microorganisms interact with metals and minerals. Such interactions are important to microbial biomineralization processes (Gadd 2010).

In aquatic environment, heavy metal(oid)s get absorbed by aquatic biota, and such metals are likely to be accumulated in humans via food chain. The consequences of metal(oid) exposure to human vary according to type and concentration of metal(oid)s. For example, health implications including minor headache, allergy, respiratory tract disorder, various cancers, etc. are among the common consequences of metal exposure (Kaufman 1970; Katz and Salem 1993; Costa 2003; Park et al. 2005; Kotas and Stasicka 2000; Vala and Dave 2017; Panseriya et al. 2018). Looking to the devastating effects that metal(oid) may exert, it is mandatory to treat the metal-contaminated waste water before its release into the environment. The treatment methods conventionally being used (e.g. chemical precipitation, electrodialysis, evaporation, ion exchange, liquid extraction, membrane process and reverse osmosis) suffer from one or the other limitations, viz. high-energy requirement, inefficient removal and generation of toxic sludge; besides they are not eco-friendly. Many of them despite being effective are not practicable when used for large-scale applications (Akcil et al. 2015; Deshmukh et al. 2016). Hence, bioremediation has been considered as a favourable alternative (Eccles 1999; Paul 2017; Vala and Dave 2017). As fungi possess unique characteristics including greater growth capability, capacity to propagate in diverse habitats, reach owing to mycelial network, capability to produce a battery of enzymes and accumulate metals, etc., they hold prime importance as promising candidates for bioremediation among microbes (Vala et al. 2004; Deshmukh et al. 2016; Vala and Dave 2017).

Whether having biological role or otherwise, metals in bioavailable form above a threshold concentrations impart toxicity in a number of ways (Gadd 1993, 2007). Most of the mycobiota, however, are tolerant towards metal owing to diverse mechanisms which support their survival and growth even in sites extremely contaminated with metal (Gadd 2007). In contaminated soils, shift in population from unicellular bacteria to streptomycetes to fungal biota has been observed (Chander et al. 2001a, b; Khan and Scullion 2000; Gadd 2007; Vala and Dave 2017). Figure 12.1 portrays various fungal mechanisms involved in tolerance to and detoxification of metals.

Studies pertaining to metal removal from terrestrial environment have been advanced; however, their counterparts from marine environment have comparatively been less explored (Taboski et al. 2005; Vala 2010; Vala and Dave 2017). Fungi, the key components of the inshore microorganisms, routinely encounter ions and complexes of metals in large harbours. Explorations on these microorganisms for removal of metal contaminants are a fascinating area of research (Millward et al. 2001; Hyde et al. 1998; Newell and Barlocher 1993; Vala et al. 2004; Vala 2010; Vala and Dave 2017).

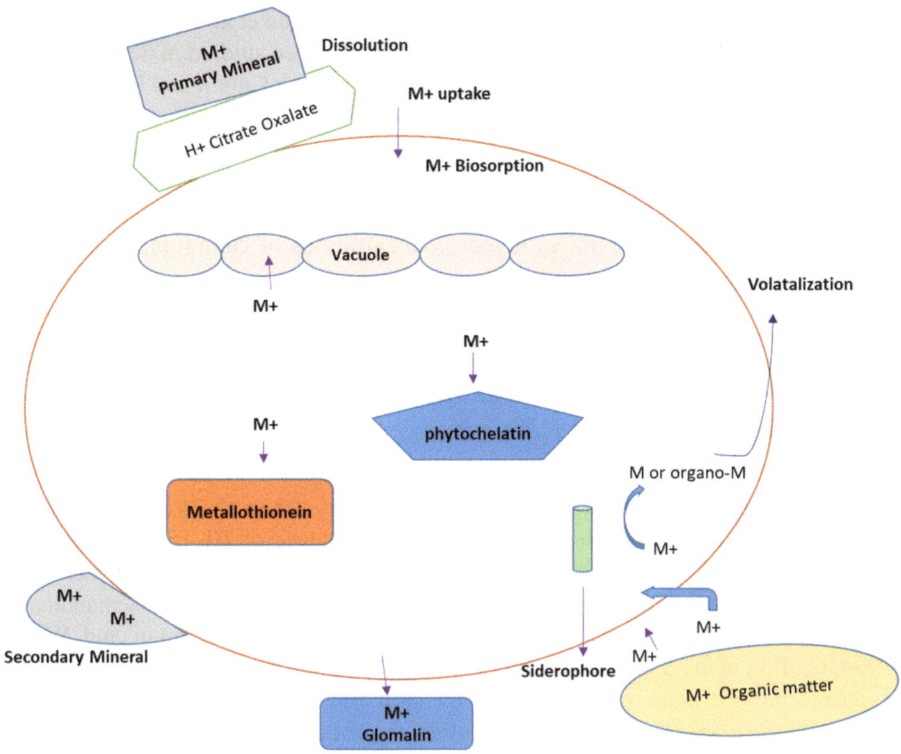


**Fig. 12.1** Diverse means of fungal-metal interactions. (Adapted from Harms et al. 2011, Nat Rev. Microbiol 9: 177–192)

Though comparatively less explored, the reports so far available have established the noteworthy role of marine-derived fungi in heavy metal(oid) removal and, hence, substantiate their role as potential bioremediation agents. Marine-derived fungi, viz. *Aspergillus flavus*, *A. niger*, *A. terreus*, *A. wentii* and *Trichoderma viride*, have been observed to tolerate and successfully remove hexavalent chromium, a carcinogen (Vala et al. 2004; El-Kassas and El-Taher 2009; Khambhaty et al. 2009).

Fungi in the marine environment play a noteworthy role in reducing the toxicity of arsenic, a Group A Category 1 human carcinogen (Bełdowski et al. 2013; US EPA 1997; International Association for Research on cancer IARC 2004; Vala 2010; Bahar et al. 2013). Marine fungi have been observed to be more tolerant to arsenic than their non-marine counterparts (Irvine and Jones 1975). Marine-derived aspergilli have been envisaged to be significant contributors in tackling the arsenic pollution (Vala 2010, 2018; Vala et al. 2010, 2011; Vala and Patel 2011; Vala and Sutariya 2012). Fungi have been reported playing role in As biosequestration even at seafloor hydrothermal fields (Dekov et al. 2013).

Yeasts from marine environment have also been observed to have bioremediation potential. Strains of marine yeast *Yarrowia lipolytica* have been reported as potential candidates for chromium removal (Bankar et al. 2009; Rao et al. 2013; Imandi et al. 2014). Among marine yeasts, *Rhodotorula rubra* has been studied in detail for arsenic metabolism (Button et al. 1973; Vidal and Vidal 1980; Cullen and Reimer 1989; Maher and Butler 1988). However, there are no recent reports on arsenic-marine yeast interactions (Vala and Dave 2017). Resting as well as growing cells of mercury resistant *Yarrowia* sp. have been reported to be applicable as a bioremediation agent (Oyetibo et al. 2015). Deep-sea psychrotolerant yeast *Cryptococcus* sp. was observed to tolerate and remove higher concentrations of cadmium, copper, lead and zinc (Singh et al. 2013).

### *12.2.3 Marine-Derived Fungi and Treatment of Synthetic Dyes and Textile-Dye Effluent*

Nearly 10,000 dyes are available commercially (Przystas et al. 2015). Dyeing and succeeding washing processes ultimately lead to release of 10–15% unused dyes into waste water. Such dye-laden waste water upon mixing with large water body weakens primary productivity, impairs diffusion of gases and influences man health besides imparting aesthetically unacceptable coloration (Baughman and Weber 1994; Ciullini et al. 2008; Rodriguez et al. 2015). Various health implications of dye exposure include nausea, ulceration of the skin and mucous membranes, haemorrhage and even severe damage to the kidney, reproductive system, brain, liver and central nervous system. Several synthetic dyes are carcinogenic, toxic and mutagenic (Vala and Dave 2017). Various physico-chemical processes like adsorption, coagulation, flocculation, flotation, precipitation, oxidation and reduction, membrane separation and ozonation are used for treatment of coloured effluents. However, they involve high cost and have drawbacks (Azmi et al. 1998; Robinson et al. 2001). Activated sludge process also is used widely; however, most of the dyes are poorly removed, and when mixed and treated together with sewage, it is inefficient in decolorizing textile effluents (Przystas et al. 2015).

Looking to various health risks due to dye exposure and limitations of most of the conventional treatment methods, it is very important to have a more-efficient and cost-effective alternative for treatment of dye waste water. Biodegradation (mineralization or biotransformation) and adsorption on biomass are the main means of dye removal. Among microbes mycobiota could be an excellent candidate for removal of dye (Przystas et al. 2015). Dye decolorization can be achieved by employing fungal metabolic potential; by oxidative reactions, fungi are able to generate non-toxic derivatives from toxic dyes (Ciullini et al. 2008). Among the battery of enzymes produced by fungi, lignin-degrading enzymes play a very important role in textile dye degradation due to their non-specific nature and efficiency to attack aromatic compounds sharing little similarity with lignin (Field et al. 1992; Arun et al. 2008; Vala and Dave 2017).

Marine-derived fungi due to their adaptability to extreme conditions are better suited for treatment of coloured effluents than their counterparts in the terrestrial environment (Raghukumar et al. 2004; Bonugli-Santos et al. 2015; Vala and Dave 2017). Efficient production of non-specific enzymes like lignin peroxidase, manganese peroxidase and laccase plays an important role in employing such fungi for dye decolorization.

Several marine-derived fungi have been reported to bring about dye degradation successfully, among them basidiomycetous fungi have been observed to be dominating the list. The use of marine-derived fungi for dye degradation involves the use of growing culture, immobilized cells and biosorbent prepared from cultures. *Cerrena unicolor* (D'Souza-Ticlo et al. 2009), *Penicillium janthinellum* (Wang et al. 2015), *Aspergillus niger* (Lu et al. 2016; Joshi et al. 2012), *Flavodon flavus* (Raghukumar et al. 1999; Mtui and Nakamura 2008), *Alternaria tenuissima* (El Aty et al. 2017), *Peniophora* sp. (Bonugli-Santos et al. 2016), *Tinctoporellus* sp. (CBMAI 1061), *Marasmiellus* sp. (CBMAI 1062), and *Peniophora* sp. (CBMAI 1063) (Bonugli-Santos et al. 2012) *Phialophora* sp. (MF 6), *Penicillium* sp. (MF 49), and *Cladosporium* sp. (Torres et al. 2011) are some of the examples of potential dye degraders.

Lalitha et al. (2011) examined bioremediation of synthetic, paper mill and colour photographic dyes using marine *Aspergillus flavus* isolated from Bay of Bengal and observed 80% and 90% removal of synthetic dyes and 100% removal in colour photographic effluents within 3–7 days and 8 days, respectively. During the study, the authors observed a correlation between biomass, sugar used and quality of protein produced. Dye decolorization is affected by availability of nutrients and physical parameters (Singh et al. 2013); dye structure also has been observed to govern the mechanism involving laccase (D'Souza et al. 2006; Bonugli-Santos et al. 2015).

Rapid decolorization and detoxification of an anthraquinone dye Reactive Blue 4 were achieved by the development of a two-step technique involving enzymatic oxidation followed by sorption of degraded products on marine-derived fungal biomass (Verma et al. 2012). Development of such processes would help in reducing time required for decolorization.

### *12.2.4 Marine-Derived Fungi and Plastic Degradation*

Plastics are basically synthetic organic polymers with high molecular mass, and generally these polymers include many types of aliphatic polyesters, including polyhydroxyalkanoates (PHAs), poly(e-caprolactone) (PCL), polyethylene (PE) and poly(l-lactide) (PLA). Some variants of plastics also contain hydrocarbons derived from petroleum feedstock (Law and Thompson 2014). The versatility, durability, low cost and ease of manufacturing of these materials have increased their use over the past three decades and have led to their involvement in almost all aspects of everyday life (Derraik 2002). Our dependence on plastics in our everyday life has also resulted in its ubiquitous presence as waste in various environmental compartments. Plastic can enter into the marine environment through various anthropogenic sources such as discharge of municipality waste, industrial raw materials, fragments of fishing nets and many more. Plastic pollution in aquatic environment has become a future research priority as it is extremely persistent in marine environment and it has been now recognized as an emerging global threat for social and environmental point of view.

Plastics in marine environment may pose hazardous impacts not only on the marine biota but on the human activities related to marine environment. Plastic debris may either float on the sea surface or may settle down to the seafloor posing deleterious effects on the marine environment. Figure 12.2 shows the potential hazardous impact of plastics on marine biota and humans.

Bacterial degradation of plastic polymers has been well documented; however, there are very few reports on the fungal degradation of plastic polymers and their relative hydrolytic enzymes.

### 12.2.4.1 Polyhydroxyalkanoates (PHAs)

Polyhydroxyalkanoates (PHAs) are natural polyesters produced intracellularly by numerous bacteria. PHAs are generally divided into two major groups based on the length of the carbon chain as short chain length (SCL)-PHAs having (R)-hydroxyalkanoates of $C_3$–$C_5$ and medium chain length (MCL)-PHAs having aliphatic and/or aromatic (R)-hydroxyalkanoates of $C_6$–$C_{14}$ (Kim and Rhee 2003). Many filamentous fungi and yeasts are reported to be the potential degraders of poly(3-hydroxybutyrate) PHB. A list of 95 different genera of PHAs degrading fungi has been listed by Neumeier (1994) which were isolated from soil and marine environments. Out of these 95 genera, majority (97%) was comprised of *Basidiomycotina*, *Deuteromycotina* and *Ascomycotina*. Species of *Aspergillus* and

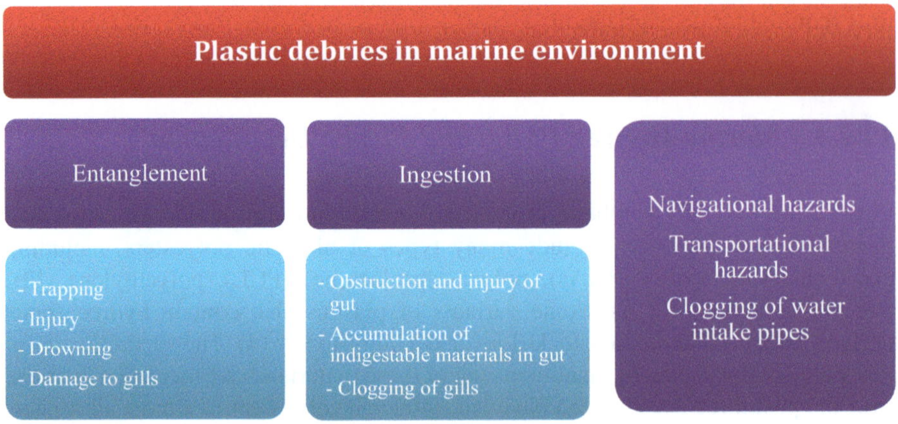


**Fig. 12.2** Impact of plastic pollution on marine biota and sea-based human activities

*Penicillium* have found to be contributing considerably in PHA degradation (Kim et al. 2000; Sang et al. 2002).

BIOPOL™ is one of the copolymers of PHAs and is produced on a large scale by Imperial Chemical Industries, England, today Zeneca, and is being used as packaging material. Gonda et al. (2000) have studied fungal degradation of BIOPOL™ by a filamentous fungus *Aspergillus ustus* and a yeast *Rhodosporidium sphaerocarpum* under simulated deep-sea environmental conditions. Matavulj and Molitoris (2009) have studied degradation of BIOPOL™ by 134 marine fungal strains. Out of 134 strains, *Debaryomyces hansenii* M-111 and *Fusarium merismoides* M-46 degraded BIOPOL™ when it is supplemented as sole source of carbon and energy. *Asteromyces cruciatus* strain M-1 demonstrated degradation of BIOPOL™ and PHB. However, *Candida guilliermondii* strain M-122 was able to depolymerise only PHB. *Nia vibrissa* strain M-167, fungus belonging to *Basidiomycota*, was able to depolymerize PHB. They have also concluded that the degrading activities are not evenly distributed among the different groups of marine fungi investigated, and their results indicated that the PHA-degrading activity in marine fungi is higher among the *Deuteromycotina* than in other groups.

PHA depolymerases are a class of serine hydrolases enzymes which are produced extracellularly by various microorganisms and are responsible for the degradation of PHAs. Based on the substrate specificity, these enzymes are divided into two major groups as SCL-PHA depolymerases which are able to degrade only PHB and its copolyesters and MCL-PHA depolymerases which act upon the aliphatic and aromatic PHAs consisting of 3-hydroxyalkanoates of C6–C14. PHA depolymerases from fungal origin have some distinct characteristics from bacterial PHA depolymerases. Fungal depolymerases have an acid to neutral pI values, and they are mostly glycoproteinic in nature. Fungal PHB depolymerases do not hydrolyse tributyrin, olive oil or p-nitrophenyl palmitate (Kim and Rhee 2003).

#### 12.2.4.2 Polyethylenes (PE)

Polyethylene (PE), comprising of long chains of the monomer ethylene, is a thermoplastic synthetic polymer. The major grades of PE are low-density polyethylene (LDPE) and high-density polyethylene (HDPE). The prime reason behind the resistance of PE to biodegradation is its high hydrophobicity and chemical inertness, and thus the biodegradation of PE is a very slow process. It also lacks functional groups in its structure to be recognizable by microbial enzymatic systems (Hamid 2000).

Kathiresan (2003) has isolated fungi *Aspergillus glaucus* and *Aspergillus niger* from the mangrove soil having potential to degrade polyethylene materials among which *Aspergillus glaucus* degraded 28.80% of polythene and 7.26% of plastics in 1-month period. Pramila and Ramesh (2011) have isolated *Aspergillus versicolor* and *Aspergillus* sp. from seawater and have studied degradation of LDPE in powdered form by these fungal isolates.

As plastic makes its entry into marine environment, various chemical, physical and biological processes result in the formation of microplastic fragments (Caruso

2015; Paço et al. 2017). These microplastics are more harmful than the original macroform because microplastics are mistaken as food by the smaller organisms such as zooplanktons, seabirds, fish, etc. and more readily ingested by them (Caruso 2015). Microplastics are hazardous, not only because they cause physical harm but also due to the adherence and absorbance of contaminants and pollutants on it (da Costa et al. 2016). Barata (2006) has isolated the fungus *Zalerion maritimum* from Mira river salt marsh in Portugal. This fungus was studied for its potential to degrade polyethylene microplastics by Paço et al. (2017). Their results indicated that naturally occurring fungus may actively contribute to the biodegradation of microplastics, requiring minimum nutrients.

Today, the worldwide utilization of plastic has increased greatly, and the production of plastics to satisfy the need of plastic also has increased. Thus, more intensive studies on the fungal degradation of plastics are expected. This will emphasize not only the ecological significance of fungi, but also on the development of more and more sustainable plastic biodegradation strategies.

### *12.2.5 Marine-Derived Fungi in Petroleum Oil Degradation*

Petroleum oil is a complex natural mixture of about 20,000 hydrocarbon and non-hydrocarbon compounds (Fig. 12.3) which, at appropriate concentration, possesses a measurable toxicity towards living systems (Marshall and Rodgers 2004; Al-Nasrawi 2012). These compounds are derived from crude oil during refining process and mainly contain hydrocarbons. Marine oil pollution is among the most portentous problem that the world is brawling today. Microbial degradation is the foremost route for the removal of oil products from contaminated marine environments in the nature especially fungi which have shown higher tolerance to the toxicity of hydrocarbons due to their physiology and adaptation to such variations in the marine environment. Majority of alkanes, asphaltanes and resins are degraded easily by microorganisms, while the hydrocarbon contents are highly recalcitrant and resistant to biodegradation. Compositional heterogeneity among different crude oils

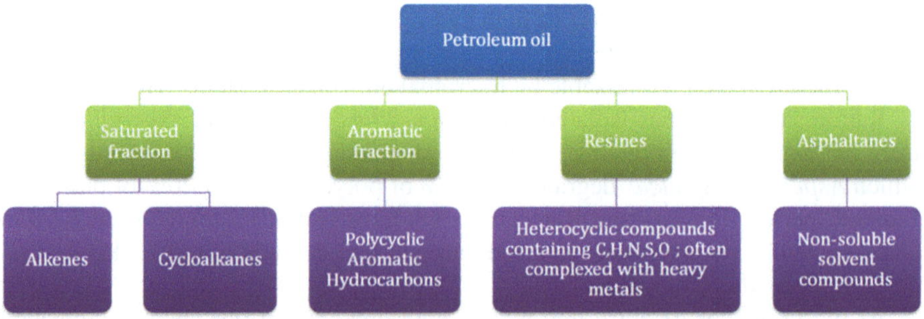


**Fig. 12.3** Major components of petroleum oil

and refined products influences the overall rate of biodegradation both of the oil and of its component fractions.

Different components of petroleum oil have different susceptibilities to biodegradation (Goodwin et al. 1983). During oil degradation, the early stages of oil biodegradation includes the loss of n-paraffins (n-alkanes or normal alkanes) followed by the amputation of acyclic isoprenoids. Other components like highly branched and cyclic saturated hydrocarbons as well as aromatic compounds are more resistant to biodegradation compared to other constituents of petroleum.

As most of the hydrocarbon constituents of petroleum oil are hydrophobic, they tend to get adsorb to the marine sediment particles. Thus, marine sediments are one of the desired ecological niches for fungi which can utilize oil hydrocarbons as carbon source. The pivotal role of fungi in degradation of petroleum products has been well reported by many researchers. Commonly studied fungi for petroleum biodegradation belonged to the genera *Alternaria*, *Aspergillus*, *Candida*, *Cephalosporium*, *Cladosporium*, *Fusarium*, *Geotrichum*, *Gliocladium*, *Mucor*, *Paecilomyces*, *Penicillium*, *Pleurotus*, *Polyporus*, *Rhizopus*, *Rhodotorula*, *Saccharomyces*, *Talaromyces* and *Torulopsis* (Saraswathy and Hallberg 2002; Atagana et al. 2006; Adekunle and Adebambo 2007; Gesinde et al. 2008; Husaini et al. 2008; Obire and Anyanwu 2009; Hadibarata and Tachibana 2009a, b; Romero et al. 2010).

Among the hydrocarbon components, polycyclic aromatic hydrocarbons (PAHs) are the major concerned component of petroleum oil which is presently becoming one of the major pollutants of marine environment. Thus, in this chapter, the major focus is given on the petroleum PAH-degrading fungi. PAHs are a group of numerous chemically related organic compounds consisting of two or more than two fused aromatic rings arranged in various structural configurations. These compounds have some unusual physico-chemical properties like high molecular weight, high water/octanol partitioning coefficient (log $K_{ow}$), high photo-sensitivity, etc., which make these compounds highly persistent in marine environment. Owing to these unusual characteristics, PAHs are proved to be potential carcinogens, mutagens and teratogens to the living organisms (Sachaniya et al. 2018).

*Aureobasidium*, *Candida*, *Rhodotorula* and *Sporobolomyces* are some of the most common marine isolates which can degrade petroleum oil (Leahy and Colwell 1990). Kirk and Gordon (1988) have studied true marine fungal genera *Corollospora*, *Dendryphiella*, *Lulworthia* and *Varicosporina* for the biodegradation of petroleum oil hydrocarbons. Walker and Colwell (1973) have studied petroleum oil degradation by the fungus *Cladosporium resinae* isolated from Chesapeake Bay and found that this fungus was responsible for 20–40% degradation of petroleum. Petroleum oil hydrocarbon-degrading fungi have also been reported to be important inhabitants of specialized niches like submerged wood (Kirk and Gordon 1988), the surface seawater, decomposing algae and the tarball surface (Ahearn and Crow 1986). Elshafie et al. (2007) have isolated marine fungi from tar balls collected from the beaches of Oman and have found *Aspergillus niger*, *A. terreus* and *Penicillium chrysogenum* to be efficient degraders of n-alkanes and crude oil.

Petroleum hydrocarbon-degrading marine fungi can be broadly divided into two categories as ligninolytic and non-ligninolytic fungi. Both of the categories utilize petroleum hydrocarbon specifically PAHs by their characteristic mechanisms (Fig. 12.4). Ligninolytic fungi are capable of producing enzymes collectively known as lignin-modifying enzymes (LMEs) which include lignin peroxidase (LP), manganese-dependent peroxidase (MnP), laccase, etc. Being non-specific, these enzymes can also be used for the degradation of PAHs. As LMEs are extracellular enzymes, they can degrade PAHs with low bioavailability. The aromatic benzene rings of PAHs are oxidized due to the generation of free hydroxyl radicals by LMEs. This, in turn, results in the formation of PAHs-quinones which by successive ring fission ultimately form $CO_2$ (Sutherland et al. 1995). PAHs degradation by ligninolytic fungi particularly white rot fungi has been studied intensively during past few years (Hofrichter et al. 1998; Cajthaml et al. 2008). Numerous experiments with LMEs have confirmed the ability of these enzymes to degrade PAHs (Hofrichter et al. 1998). The extent of PAHs degradation varies with variation in LMEs (Clemente et al. 2001). The degradation system of these fungi is inducible. Some of

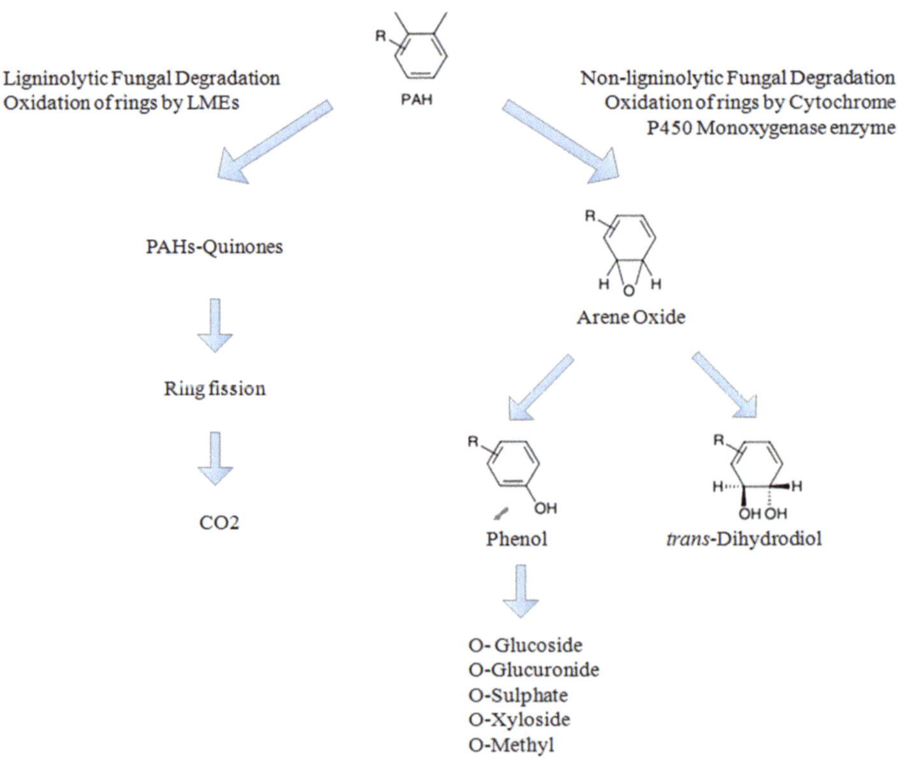


**Fig. 12.4** Mechanisms of fungal biodegradation of petroleum PAHs. (Adapted and modified from Bamforth and Singleton 2005)

the most studied ligninolytic fungi for PAHs degradation are *Aspergillus*, *Fusarium*, *Trichoderma*, *Phanerochaete*, *Pleurotus*, *Chlorella*, etc.

Most of the non-ligninolytic fungi cannot completely mineralize PAHs but the transformation products formed by these fungi are comparatively less toxic than parent PAHs. The enzyme, cytochrome $P_{450}$ monooxygenase, first oxidizes the aromatic benzene ring by incorporating only one oxygen atom in the ring to produce arene oxide. This mechanism is similar to the mammalian metabolism of PAHs. The subsequent reaction is then catalysed by epoxide hydrolase to form a *trans*-dihydrodiol (Jerina 1983). Phenol derivatives may also be produced through non-enzymatic rearrangements of the arene oxides. *Chrysosporium pannorum*, *Cunninghamella elegans*, *Aspergillus niger*, etc. are some examples of non-ligninolytic fungi that can degrade PAHs (Bamforth and Singleton 2005). Recently, from our laboratory, Bhatt et al. (2014) have isolated *Cochliobolus lunatus* strain CHR4D, a marine-derived ascomycete fungus from historically contaminated crude oil-polluted shoreline of Alang-Sosiya ship-breaking yard, at Bhavnagar coast, Gujarat. The fungus was capable of efficiently degrading chrysene, a four-ringed high molecular weight (HMW) PAH.

Marine yeasts have also proved to be efficient candidates for petroleum PAHs degradation. Ahearn et al. (1971) have isolated several yeasts from oil-polluted habitats and studied their ability to use hydrocarbons as sole source of carbon. *Trichosporon* species was found to emulsify the oil in their studies. Ahearn and Meyers (1972) have investigated the responses of yeast population to oil pollution in salt marsh in Louisiana. In their study, they have reported that in oil-polluted conditions, the normal predominated species of yeasts were replaced by hydrocarbonoclastic species of yeast mainly including *Pichia ohmeri* and *Trichosporon* sp. *Candida tropicalis* strains 7Y and 15Y are found to be efficient oil degraders by Palittapongarnpim et al. (1998). Zinjarde and Pant (2002) have studied petroleum oil degradation and found that yeasts isolated from marine mud and water around Mumbai were important degraders of the aliphatic fraction of crude oil and all the isolates belonged to the genus *Candida*.

Prior exposure of the marine microbial communities to the petroleum pollution is important in determining the rapidity of biodegradation of the subsequent petroleum component input. This is termed as the adaptation of that particular microbial community towards petroleum pollution (Spain et al. 1980). There are major three interrelated mechanisms for adaptation: (i) induction and/or depression of specific enzymes, (ii) genetic changes which result in new metabolic capabilities and (iii) selective enrichment of organisms able to transform the compound or compounds of interest (Leahy and Colwell 1990).

## 12.3 Conclusion

Marine-derived fungi exhibiting unique traits, enzymatic features and adaptations to extreme environments form an integral component of marine microbial diversity. Even in artificial reef, they have been observed to be important colonizing community. Marine-derived fungi have been observed to be able to interact with a variety of pollutants including highly recalcitrant ones. By various means including biofilm formation, production of degradative enzymes, etc., these organisms are playing important role in getting rid of hazardous and recalcitrant pollutants. Though several laboratories worldwide have been focusing on these groups of organisms and their involvement in enhanced bioremediation of pollutants, potentials of marine-derived fungi have been comparatively untapped, and they demand attention.

**Acknowledgement** Head, Department of Life Sciences, Maharaja Krishnakumarsinhji Bhavnagar University, Dr. B.M. Gohil is thankfully acknowledged for providing kind support.

## References

Ahearn DG, Meyers SP (1972) The role of fungi in the decomposition of hydrocarbons in the marine environment. Biodeterior Mater 2:12–18

Ahearn DG, Meyers SP, Standard PG (1971) The role of yeasts in the decomposition of oils in marine environments. Dev Ind Microbiol 12:126–134

Ahearn, D. G., & Crow, S. A. (1986). Fungi and hydrocarbons in the marine environment. The Biology of Marine Fungi, 11–18.

Akcil A, Erust C, Ozdemiroglu S, Fonti V, Beolchini F (2015) A review of approaches and techniques used in aquatic contaminated sediments: metal removal and stabilization by chemical and biotechnological processes. J Clean Prod 86:24–36

Al-Nasrawi H (2012) Biodegradation of crude oil by fungi isolated from Gulf of Mexico. J Bioremed Biodegr 3(4):147–152

Arun A, Raja PP, Arthi R, Ananthi M, Kumar KS, Eyini M (2008) Polycyclic aromatic hydrocarbons (PAHs) biodegradation by basidiomycetes fungi, *Pseudomonas* isolate, and their cocultures: comparative in vivo and in silico approach. Appl Biochem Biotechnol 151:132–142

Atagana HI, Haynes RJ, Wallis FW (2006) Fungal bioremediation of creosote- contaminated soil: a laboratory scale bioremediation study using indigenous soil fungi. Water Air Soil Pollut 172:201–219

Azmi W, Sani RK, Banerjee UC (1998) Biodegradation of triphenylmethane dyes. Enzym Microb Technol 22:185–191

Bahar MM, Megharaj M, Naidu R (2013) Bioremediation of arsenic-contaminated water: recent advances and future prospects. Water Air Soil Pollut 224:1722

Bamforth SM, Singleton I (2005) Bioremediation of polycyclic aromatic hydrocarbons: current knowledge and future directions. J Chem Technol Biotechnol 80(7):723–736

Bankar AV, Kumar AR, Zinjade SS (2009) Removal of chromium (VI) ions from aqueous solution by adsorption onto two marine isolates of *Yarrowia lipolytica*. J Hazard Mater 170(1):487–494

Barata M (2006) Marine fungi from Mira river salt marsh in Portugal. Rev Iberoam Micol 23(3):179

Baughman GL, Weber EJ (1994) Transformation of dyes and related compounds in anoxic sediment: kinetics and products. Environ Sci Technol 28:267–276

Bełdowski J, Szubska M, Emelyanov E (2013) Spatial variability of arsenic concentrations in Baltic sea surface sediments in relation to sea dumped chemical munitions. E3S Web of Conferences 1, 16002. https://doi.org/10.1051/e3sConf/20130116002

Bennet JW, Wunch KG, Faison BD (2002) Use of fungi in biodegradation. In: Environmental microbiology, 2nd edn. ASM Press, Washington, DC

Bhatt JK, Ghevariya CM, Dudhagara DR, Rajpara RK, Dave BP (2014) Application of response surface methodology for rapid chrysene biodegradation by newly isolated marine-derived fungus *Cochliobolus lunatus* strain CHR4D. J Microbiol 52(11):908–917

Bonugli-Santos RC, Durrant LR, Sette LD (2012) The production of ligninolytic enzymes by marine-derived basidiomycetes and their biotechnological potential in the biodegradation of recalcitrant pollutants and the treatment of textile effluents. Water Air Soil Pollut 223(5):2333–2345

Bonugli-Santos RC, dos Santos Vasconcelos MR, Passarini MRZ, Vieira GAL, Lopes VCP, Mainardi PH, dos Santos JA, de Azevedo Duarte L, Otero IVR, da Silva Yoshida AM, Feitosa VA, Pessoa A Jr, Sette LD (2015) Marine-derived fungi: diversity of enzymes and biotechnological applications. Front Microbiol 6:269. https://doi.org/10.3389/fmicb.2015.00269

Bonugli-Santos RC, Vieira GAL, Collins C, Fernandes TCC, Marin-Morales MA, Murray P, Sette LD (2016) Enhanced textile dye decolorization by marine-derived basidiomycete *Peniophora* sp. CBMAI 1063 using integrated statistical design. Environ Sci Pollut Res 23:8659–8668

Button DK, Dunker SS, Morse ML (1973) Continuous culture of *Rhodotorula rubra*: kinetics of phosphate-arsenate uptake, inhibition, and phosphate-limited growth. J Bacteriol 113(2):599–611

Cajthaml T, Erbanová P, Kollmann A, Novotný Č, Šašek V, Mougin C (2008) Degradation of PAHs by ligninolytic enzymes of *Irpex lacteus*. Folia Microbiol 53(4):289–294

Caruso G (2015) Plastic degrading microorganisms as a tool for bioremediation of plastic contamination in aquatic environments. J Pollut Eff Cont 3(3):1–2

Chander K, Dyckmans J, Joergensen R, Meyer B, Raubuch M (2001a) Different sources of heavy metals and their long-term effects on soil microbial properties. Biol Fertil Soils 34(4):241–247

Chander K, Dyckmans J, Höper H, Raubuch M (2001b) Long term effects on soil microbial properties of heavy metals from industrial exhaust deposition. J Pl Nutr Soil Sci 164(6):657–663

Ciullini I, Tilli S, Scozzafava A, Briganti F (2008) Fungal laccase, cellobiose dehydrogenase, and chemical mediators: combined actions for the decolorization of different classes of textile dyes. Bioresour Technol 99:7003–7010

Clemente AR, Anazawa TA, Durrant LR (2001) Biodegradation of polycyclic aromatic hydrocarbons by soil fungi. Braz J Microbiol 32(4):255–261

Costa M (2003) Potential hazards of hexavalent chromate in our drinking water. Toxicol Appl Pharmacol 188:1–5

Da Costa JP, Santos PS, Duarte AC, Rocha-Santos T (2016) (Nano) plastics in the environment-sources, fates and effects. Sci Total Environ 566:15–26

Cullen WR, Reimer KJ (1989) Arsenic speciation in the environment. Chem Rev 89:713–764

D'Souza DT, Tiwari R, Shah AK, Raghukumar C (2006) Enhanced production of laccase by a marine fungus during treatment of colored effluents and synthetic dyes. Enzym Microb Technol 38:504–511

D'Souza-Ticlo D, Sharma D, Raghukumar C (2009) A thermostable metal-tolerant laccase with bioremediation potential from a marine-derived fungus. Mar Biotechnol 11:725–737

Damare S, Singh P, Raghukumar S (2012) Biotechnology of marine fungi. In: Biology of marine fungi. Springer, Berlin, pp 277–297

Dekov VM, Bindi L, Burgaud G, Petersen S, Asael D, Rédou V et al (2013) Inorganic and biogenic As-sulfide precipitation at seafloor hydrothermal fields. Mar Geol 342:28–38

Derraik JG (2002) The pollution of the marine environment by plastic debris: a review. Marine Poll Bull 44(9):842–852

Deshmukh R, Khardenavis AA, Purohit HJ (2016) Diverse metabolic capacities of fungi for bioremediation. Indian J Microbiol 56(3):247–264

Dusane DH, Nancharaiah YV, Venugopalan VP, Kumar AR, Zinjarde SS (2008) Biofilm formation by a biotechnologically important tropical marine yeast isolate, *Yarrowia lipolytica* NCIM 3589. Water Sci Technol. https://doi.org/10.2166/wst.2008.479

Eccles H (1999) Treatment of metal-contaminated wastes: why select a biological process? Trends Biotechnol 17:462–465

El Aty AAA, Mostafa FA, Hassan ME, Hamed ER, Esawy MA (2017) Covalent immobilization of *Alternaria tenuissima* KM651985 laccase and some applied aspects. Biocatal Agricult Biotechnol 9:74–81

El-Kassas HY, El-Taher EM (2009) Optimization of batch process parameters by response surface methodology for mycoremediation of chrome-VI by a chromium resistant strain of marine *Trichoderma viride*. Am-Eurasian J Agric Environ Sci 5(5):676–681

Elshafie A, AlKindi AY, Al-Busaidi S, Bakheit C, Albahry SN (2007) Biodegradation of crude oil and n-alkanes by fungi isolated from Oman. Mar Pollut Bull 54(11):1692–1696

Field JA, de Jong ED, Costa GF, de Bond JAM (1992) Biodegradation of polycyclic aromatic hydrocarbons by new isolates of white rot fungi. Appl Environ Microbiol 58:2219–2226

Gadd GM (1993) Interactions of fungi with toxic metals. New Phytol 124:25–60

Gadd GM (2007) Geomycology: biogeochemical transformations of rocks, minerals, metals and radionuclides by fungi, bioweathering and bioremediation. Mycol Res 111(1):3–49

Gadd GM (2010) Metals, minerals and microbes: geomicrobiology and bioremediation. Microbiology 156:609–643

Gao GR, Yin YF, Yang DY, Yang DF (2013) Promoting behaviour of fungal degradation polychlorinated biphenyl by Maifanite. Adv Mater Res Trans Tech Publications 662:515–519

Gesinde AF, Agbo EB, Agho MO, Dike EFC (2008) Bioremediation of some Nigerian and Arabian crude oils by fungal isolates. Int J Pure Appl Sci 2(3):37–44

Gonda KE, Jendrossek D, Molitoris HP (2000) Fungal degradation of the thermoplastic polymer poly-b-hydroxybutyric acid (PHB) under simulated deep sea pressure. Hydrobiologia 426:173–183

Goodwin NS, Park PJD, Rawlinson AP (1983) Crude oil biodegradation under simulated and natural conditions. In: Bjorgy M et al (eds) Advances in organic geochemistry. Wiley, Chichester, pp 650–658

Gutierrez MH, Pantoja S, Tejos E, Quinones RA (2011) The role of fungi in processing marine organic matter in the upwelling ecosystem off Chile. Mar Biol 158:205–219

Gutiérrez-Correa M, Villena GK (2003) Surface adhesion fermentation: a new fermentation category. Rev Peru Biol 10:113–124

Hadibarata T, Tachibana S (2009a) Microbial degradation of crude oil by fungi pre-grown on wood meal. In: Interdisciplinary studies on environmental chemistry-environmental research in Asia, Terrapub, Tokyo, pp 317–322

Hadibarata T, Tachibana S (2009b) Identification of phenanthrene metabolites produced by Polyporus sp. S133. Interdisciplinary studies on environmental chemistry-environmental research in Asia. Terrapub, Tokyo, pp 293–299

Hamid SH (2000) Handbook of polymer degradation. Marcel Dekker, New York

Harms H, Schlosser D, Wick LY (2011) Untapped potential: exploiting fungi in bioremediation of hazardous chemicals. Nat Rev Microbiol 9:177–192

Hofrichter M, Scheibner K, Schneegaß I, Fritsche W (1998) Enzymatic combustion of aromatic and aliphatic compounds by manganese peroxidase from *Nematoloma frowardii*. Appl Environ Microbiol 64(2):399–404

Husaini A, Roslan HA, Hii KSY, Ang CH (2008) Biodegradation of aliphatic hydrocarbon by indigenous fungi isolated from used motor oil contaminated sites. World J Microbiol Biotechnol 24(12):2789–2797

Hyde KD, Gareth Jones EB, Leano E, Pointing SB, Poonyth AD, Vrijmoed LLP (1998) Role of fungi in marine ecosystems. Biodivers Conserv 7:1147–1161

Irvine J, Jones EBG (1975) The effect of a Copper-Chrome-Arsenate preservatives and its constituents on the growth of aquatic microorganisms. J Inst Wood Sci 7: 1–5

IARC (2004) IARC monographs on the evaluation of carcinogenic risks to humans. In: Some drinking-water disinfectants and contaminants, including Arsenic, vol 84. IARC Press, Lyon. 192 IARC Monographs, vol 86
Imandi SB, Chinthala R, Saka S, Vechalapu RR, Nalla KK (2014) Optimization of chromium biosorption in aqueous solution by marine yeast biomass of *Yarrowia lipolytica* using Doehlert experimental design. Afr J Biotechnol 13(12):1413–1422
Jerina DM (1983) The 1982 Bernard B. brodie award lecture. Metabolism of aromatic hydrocarbons by the cytochrome P-450 system and epoxide hydrolase. Drug Metab Dispos 11(1):1–4
Joshi HV, Vala AK, Bhatt PN, Anand N (2012) Treatment of CETP effluent by a seaweed associated *Aspergillus niger* isolate. In: Sahoo D, Kaushik BD (eds) Algal biotechnology and environment. I.K. International Publishing House, New Delhi, pp 193–197
Kathiresan K (2003) Polythene and plastic degrading microbes in an Indian mangrove soil. Rev Biol Trop 51:629–633
Katz SA, Salem H (1993) The toxicology of chromium with respect to its chemical speciation: a review. J Appl Toxicol 13:217–224
Kaufman BD (1970) Acute potassium dichromate poisoning in man. Am J Dis Child 119:374–379
Khambhaty Y, Mody K, Basha S, Jha B (2009) Biosorption of Cr(VI) onto marine *Aspergillus niger*: experimental studies and pseudo-second order kinetics. World J Microbiol Biotechnol 25:1413–1421
Khan M, Scullion J (2000) Effect of soil on microbial responses to metal contamination. Environ Pollut 110:115–125
Kim DY, Rhee YH (2003) Biodegradation of microbial and synthetic polyesters by fungi. Appl Microbiol Biotechnol 61(4):300–308
Kim M-N, Lee A-R, Yoon J-S, Chin I-J (2000) Biodegradation of poly(3-hydroxybutyrate), sky-green and mater-Bi by fungi isolated from soils. Eur Polym J 36:1677–1685
Kirk PW, Gordon AS (1988) Hydrocarbon degradation by filamentous marine higher fungi. Mycologia 80:776–782
Kotas J, Stasicka Z (2000) Chromium occurrence in the environment and methods of its speciation. Environ Pollut 107:263–283
Lalitha P, Nageswara Rao Reddy N, Arunalakshmi K (2011) Decolorization of synthetic dyes by *Aspergillus flavus*. Biorem J 15(2):121–132
Law KL, Thompson RC (2014) Microplastics in the seas. Science 345(6193):144–145
Leahy JG, Colwell RR (1990) Microbial degradation of hydrocarbons in the environment. Microbiol Rev 54(3):305–315
Lu T, Zhang Q, Yao S (2016) Efficient decolorization of dye-containing wastewater using mycelial pellets formed of marine-derived *Aspergillus niger*. Chin J Chem Eng A. https://doi.org/10.1016/j.cjche.2016.08.010
Maher W, Butler E (1988) Arsenic in the marine environment. Appl Organomet Chem 2:191–214
Marshall AG, Rodgers RP (2004) Petroleomics: the next grand challenge for chemical analysis. Acc Chem Res 37(1):53–59
Matavulj M, Molitoris HP (2009) Marine fungi: degraders of poly-3-hydroxyalkanoate based plastic materials. Zbornik Matice srpske za prirodne nauke 116:253–265
Millward RN, Carman KR, Fleeger JW, Gambrell RP, Powell RT, Rouse MA (2001) Linking ecological impact to metal concentrations and speciation: a microcosm experiment using a salt marsh meiofaunal community. Environ Toxicol Chem 20:2029–2037
Mitra S, Pramanik A, Banerjee S, Haldar S, Gachhui R, Mukherjee J (2013) Enhanced biotransformation of fluoranthene by intertidally derived *Cunninghamella elegans* under biofilm-based and niche-mimicking conditions. Appl Environ Microbiol 79(24):7922–7930
Mtui G, Nakamura Y (2008) Lignocellulosic enzymes from *Flavodon flavus*, a fungus isolated from Western Indian ocean off the coast of Dar es Salaam, Tanzania. Afr J Biotechnol 7(17):3066–3072

Naranjo-Briceño L, Pernía B, Guerra M, Demey JR, De Sisto Á, Inojosa Y, Yegres F (2013) Potential role of oxidative exoenzymes of the extremophilic fungus *Pestalotiopsis palmarum* BM-04 in biotransformation of extra-heavy crude oil. Microb Biotechnol 6(6):720–730

Neumeier S (1994) Abbau thermoplastischer Biopolymere auf Poly-b-Hydroxyalkanoat-Basis durch terrestrische und marine Pilze. PhD thesis, Universität Regensburg, Regensburg

Newell SY, Barlocher F (1993) Removal of fungal and total organic matter from decaying cordgrass leaves by shredder snails. J Exp Mar Biol Ecol 171:39–49

Obire O, Anyanwu EC (2009) Impact of various concentrations of crude oil on fungal populations of soil. Int J Environ Sci Technol 6(2):211–218

Oyetibo GO, Ishola ST, Ikeda-Ohtsubo W, Miyauchi K, Ilori MO, Endo G (2015) Mercury bioremoval by *Yarrowia* strains isolated from sediments of mercury-polluted estuarine water. Appl Microbiol Biotechnol 99(8):3651–3657

Paço A, Duarte K, da Costa JP, Santos PS, Pereira R, Pereira ME et al (2017) Biodegradation of polyethylene microplastics by the marine fungus *Zalerion maritimum*. Sci Total Environ 586:10–15

Palittapongarnpim M, Pokethitiyook P, Upatham ES, Tangbanluekal L (1998) Biodegradation of crude oil by soil microorganisms in the tropic. Biodegradation 9(2):83–90

Panseriya HZ, Sachaniya B, Gosai H, Vala AK, Dave BP (2018) Marine microbial mettle for heavy metal bioremediation: a perception. In: Pei D (ed) Marine pollution: current status, impacts and remedies. Bentham Publishers (Book under preparation), UAE

Park D, Yun YS, Park JM (2005) Studies on hexavalent chromium biosorption by chemically treated biomass of *Ecklonia* sp. Chemosphere 60:1356–1364

Paul D (2017) Research on heavy metal pollution of river Ganga: a review. Ann Agric Sci 15:278–286

Pramila R, Ramesh KV (2011) Biodegradation of low density polyethylene (LDPE) by fungi isolated from marine water a SEM analysis. Afr J Microbiol Res 5(28):5013–5018

Prasad R (2017) Mycoremediation and environmental sustainability, vol 1. Springer International Publishing, Cham, Switzerland https://doi.org/10.1007/978-3-319-68957-9

Prasad R (2018) Mycoremediation and environmental sustainability, vol 2. Springer International Publishing, Cham, Switzerland https://www.springer.com/us/book/9783319773858

Przystas W, Zablocka-Godlewska E, Grabinska-Sota E (2015) Efficacy of fungal decolorization of a mixture of dyes belonging to different classes. Braz J Microbiol 46(2):415–424

Raghukumar C, D'Souza TM, Thorn RG, Reddy CA (1999) Lignin-modifying enzymes of *Flavodon flavus*, a basidiomycete isolated from a coastal marine environment. Appl Environ Microbiol 65:2103–2111

Raghukumar C, Muraleedharan U, Gaud VR, Mishra R (2004) Simultaneous detoxification and decolorization of molasses spent wash by the immobilized white-rot fungus *Flavodon flavus* isolated from a marine habitat. Enzym Microb Technol 35:197–202

Rao A, Bankar A, Kumar AR, Gosavi S, Zinjarde S (2013) Removal of hexavalent chromium ions by *Yarrowia lipolytica* cells modified with phyto-inspired $Fe^0/Fe_3O_4$ nanoparticles. J Contam Hydrol 146:63–73

Robinson T, McMullan G, Marchant R et al (2001) Remediation of dyes in textile effluents: a critical review on current treatment technologies with a proposed alternative. Bioresour Technol 77:247–255

Rodriguez JPG, Williams DE, Sabater ID, Bonugli-Santos RC, Sette LD, Andersen RJ, Berlinck RGS (2015) The marine-derived fungus *Tinctoporellus* sp. CBMAI 1061 degrades the dye Remazol Brilliant Blue R producing anthraquinones and unique tremulane sesquiterpenes. RSC Adv 5(81):66360–66366

Romero MC, Urrutia MI, Reinoso HE, Kiernan MM (2010) Benzo[a]pyrene degradation by soil filamentous fungi. J Yeast Fungal Res 1:025–029

Sachaniya BK, Gosai HB, Panseriya HZ, Vala AK, Dave BP (2018) Polycyclic aromatic hydrocarbons (PAHs): occurrence and bioremediation in marine environment. In: Marine pollution: current status, impacts and remedies. Bentham Publishers, UAE (Accepted)

Salamone AL, Robicheau BM, Walker AK (2016) Fungal diversity of marine biofilms on artificial reefs in the North-Central Gulf of Mexico. Bot Mar 59(5):291–305
Sang B-I, Hori K, Tanji Y, Unno H (2002) Fungal contribution to in situ biodegradation of poly(3-hydroxybutyrate-co-3-hydroxyvalerate) film in soil. Appl Microbiol Biotechnol 58:241–247
Saraswathy A, Hallberg R (2002) Degradation of pyrene by indigenous fungi from a former gas-works site. FEMS Microbiol Lett 210(2):227–232
Singh R, Paul D, Jain RK (2006) Biofilms: implications in bioremediation. Trends Microbiol 14:389–397
Singh P, Raghukumar C, Parvatkar RR, Mascarenhas-Pereira MBL (2013) Heavy metal tolerance in the psychrotolerant *Cryptococcus* sp. isolated from deep-sea sediments of the Central Indian Basin. Yeast 30:93–101
Sole M, Fetzer I, Wennrich R, Sridhar KR, Harms H, Krauss G (2008) Aquatic hyphomycete communities as potential bioindicators for assessing anthropogenic stress. Sci Total Environ 389:557–565
Spain JC, Pritchard PH, Bourquin AW (1980) Effects of adaptation on biodegradation rates in sediment/water cores from estuarine and freshwater environments. Appl Environ Microbiol 40(4):726–734
Sutherland JB, Rafii F, Khan AA, Cerniglia CE (1995) Mechanisms of polycyclic aromatic hydrocarbon degradation. Microbial transformation and degradation of toxic organic chemicals (Eds L Y Young & C E Cerniglia) pp 269–306, Wiley-Liss, New York
Taboski MAS, Rand TG, Piorko A (2005) Lead and Cadmium uptake in the marine fungi *Corollosporalacera* and *Monodyctis pelagia*. FEMS Microbiol Ecol 53:445–453
Torres JMO, Cardenas CV, Moron LS, Guzman APA, dela Cruz TEE (2011) Dye decolorization activities of marine-derived fungi isolated from Manila Bay and Calatagan Bay. Philipp J Sci 140(2):133–143
US EPA (US Environmental Protection Agency) (1997) IRIS (Integrated Risk Information System) on-line database maintained in Toxicology Data Network (TOXNET) by the National Library of medicine. Bethesda
Vala AK (2010) Tolerance and removal of arsenic by a facultative marine fungus *Aspergillus candidus*. Bioresour Technol 101:2565–2567
Vala AK (2018) On the extreme tolerance and removal of arsenic by a facultative marine fungus *Aspergillus sydowii*. In: Gautam A, Pathak C (eds) Metallic contamination and its toxicity. Daya Publishing House, pp 37–44
Vala AK, Dave BP (2017) Marine-derived fungi: prospective candidates for bioremediation. In: Prasad R (ed) Mycoremediation and environmental sustainability. Springer, Cham, pp 17–37
Vala AK, Patel RJ (2011) Biosorption of trivalent arsenic by facultative marine *Aspergillus niger*. In: Mason A (ed) Bioremediation: biotechnology, engineering and environmental management. Nova Science Publishers, New York, pp 459–464
Vala AK, Sutariya V (2012) Trivalent arsenic tolerance and accumulation in two facultative marine fungi. Jundishapur J Microbiol 5:542–545
Vala AK, Anand N, Bhatt PN, Joshi HV (2004) Tolerance and accumulation of hexavalent chromium by two seaweed associated aspergilli. Mar Pollut Bull 48(9&10):983–985
Vala AK, Davariya V, Upadhyay R (2010) An investigation on tolerance and accumulation of a facultative marine fungus *Aspergillus flavus* to pentavalent arsenic. J Ocean Univ China 9:65–67
Vala AK, Sutariya V, Upadhyay RV (2011) Investigations on trivalent arsenic tolerance and removal potential of a facultative marine *Aspergillus niger*. Environ Prog Sustain Energy 30(4):586–588
Verma AK, Raghukumar C, Parvatkar RR, Naik CG (2012) A rapid two-step bioremediation of the anthraquinone dye, Reactive Blue 4 by a marine-derived fungus. Water Air Soil Pollut 223:3499–3509
Vidal FV, Vidal MV (1980) Arsenic metabolism in marine bacteria and yeast. Mar Biol 60(1):1–7
Villena GK, Fujikawa T, Tsuyumu S, Gutiérrez-Correa M (2009) Differential gene expression of some lignocellulolytic enzymes in *Aspergillus niger* biofilms. Rev Peru Biol 15:97–102

Walker JD, Colwell RR (1973) Microbial ecology of petroleum utilization in Chesapeake Bay. In: International Oil Spill Conference, 1, pp 685–690. American Petroleum Institute

Wang MX, Zhang QL, Yao SJ (2015) A novel biosorbent formed of marine-derived *Penicillium janthinellum* mycelial pellets for removing dyes from dye-containing wastewater. Chem Eng J 259:837–844

Zinjarde SS, Pant AA (2002) Hydrocarbon degraders from tropical marine environments. Marine Poll Bull 44(2):118–121

Zinjarde S, Apte M, Mohite P, Kumar AR (2014) *Yarrowia lipolytica* and pollutants: interactions and applications. Biotechnol Adv 32:920–933

# Chapter 13
# Environmental Nanotechnology: Applications of Nanoparticles for Bioremediation

**Geeta Bhandari**

## Contents



## 13.1 Introduction

Due to a widespread industrialization, urbanization, and modern agricultural practices, an extraordinary number of contaminants are released into the environment. These contaminants can pollute soil, air, and water, as well as cause deforestation, biodiversity losses, soil degradation and harm to human health. Examples of these pollutants are carbon monoxide (CO), chlorofluorocarbons (CFCs), heavy metals (arsenic, chromium, lead, cadmium, mercury, and zinc), hydrocarbons, nitrogen oxides, organic compounds (volatile organic compounds and dioxins), sulfur dioxide, and particulates. Many of these pollutants are known or suspected as carcinogens and mutagens and may alter ecosystem function. Thus, a variety of environmental cleanup techniques have been designed, using bioremediation,

G. Bhandari (✉)
Sardar Bhagwan Singh Post Graduate Institute of Biomedical Sciences and Research, Dehradun, Uttarakhand, India



R. Prasad, E. Aranda (eds.), *Approaches in Bioremediation*, Nanotechnology in the Life Sciences, https://doi.org/10.1007/978-3-030-02369-0_13

phytoremediation, and physical and chemical remediation. The recognition that "traditional" methods of treatment (e.g., disposal to landfill, isolation, pump-and-treat) are not sustainable lead to a massive increase in development of alternative treatment technologies for environmental remediation (Cundy et al. 2008). The development of new and more effective strategies for the remediation of contamination is of utmost importance and can help to preserve and restore the integrity of natural habitats (Prasad 2017, 2018).

## 13.2 Emergence of Nanotechnology

The emergence of nanotechnology has been the subject of extensive research in recent years, by intersecting with various other branches of science and involving all forms of life (Baker and Satish 2012). The concept of nanotechnology was first postulated by Richard Feynman in 1959 and is among the fastest growing areas of scientific research and technology development worldwide (Richard 1960). The area is often referred to as the "Next Industrial Revolution" (Roco 2005). Nanotechnology is an emerging field of applied science, focused on the design, synthesis, characterization, and application of materials and devices on the nanoscale. It plays a major role in the development of innovative methods to produce new products, to substitute existing production equipment, and to reformulate new materials and chemicals with improved performance resulting in less consumption of energy and materials and reduced harm to the environment as well as environmental remediation. The ability of nanotechnology to minimize pollution is in progress and could potentially catalyze the most revolutionary changes in the environmental field. Nanotechnology presents a number of potential environmental benefits such as treatment and remediation, sensing and detection, and pollution prevention (Watlington 2005).

Nanoparticles (nanoscale particles = NSPs) are atomic or molecular aggregates with dimension between 1 and 100 nm that can drastically modify their physico-chemical properties compared with the bulk material. These can be made from a variety of bulk materials and can act depending on chemical composition, size, or shape of the particles. They are more reactive and more mobile in nature. NMs show quantum effect; therefore less activation energy is required to make the chemical reactions feasible. Surface plasmon resonance is another phenomenon exhibited by NPs which can be used for the detection of toxic material.

Nanoparticles are broadly in two groups of organic and inorganic nanoparticles. Organic nanoparticles include carbon nanoparticles (fullerenes), while some of the inorganic nanoparticles include magnetic nanoparticles, noble metal nanoparticles (e.g., gold and silver), and semiconductor nanoparticles (e.g., titanium dioxide and zinc oxide). Ruffini-Castiglione and Cremonini (2009) identified three types of NSPs: natural (e.g., volcanic or lunar dust, mineral composites), incidental (resulting from anthropogenic activity, e.g., diesel exhaust, coal combustion, welding fumes), and engineered. The latter includes metal-based materials quantum dots, nanogold, nanozinc, nanoaluminium, $TiO_2$, ZnO, and $Al_2O_3$ (Lin and Xing 2007). Nanomaterials (NMs) have been characterized as efficient, cost-effective, and environment-friendly alternatives to existing treatment materials, in both resource conservation and environmental remediation (Friedrich et al. 1998; Dimitrov 2006). Although nanoparticles can be synthesized through an array of conventional methods, the biological route of synthesizing is advantageous because of ease of rapid synthesis, controlled toxicity, control of size characteristics, low cost, and eco-friendly approach (Prasad 2014; Prasad et al. 2016; Aziz et al. 2015, 2016). Nanoparticles are extensively used the in the areas of electronics, medicine, and agriculture (Kavitha et al. 2013; Ingale and Chaudhari 2013; Prasad et al. 2016).

## 13.3 Nanoremediation

Nanoremediation has evolved as a new and effective solution for environmental cleanup by playing a significant role in pollution prevention, detection, monitoring, and remediation (Rajan 2011) (Fig. 13.1). It has potential to address some of the challenges of site remediation and improve the overall efficiency of the remediation processes in a cost-effective manner. These nanoparticles have properties that enable both chemical reduction and catalysis to mitigate the pollutants of concern. Nanomaterials have highly desired properties for in situ applications. Due to their minute size and innovative surface coatings, nanoparticles may be able to pervade very small spaces in the subsurface and remain suspended in groundwater, allowing the particles to travel farther than larger, macro-sized particles and achieve wider distribution. The high surface area-to-mass ratios (i.e., specific surface area) of nanoparticles can benefit technologies that rely on reactions at solid-water and solid-gas interfaces. Such technologies include adsorption used for water and

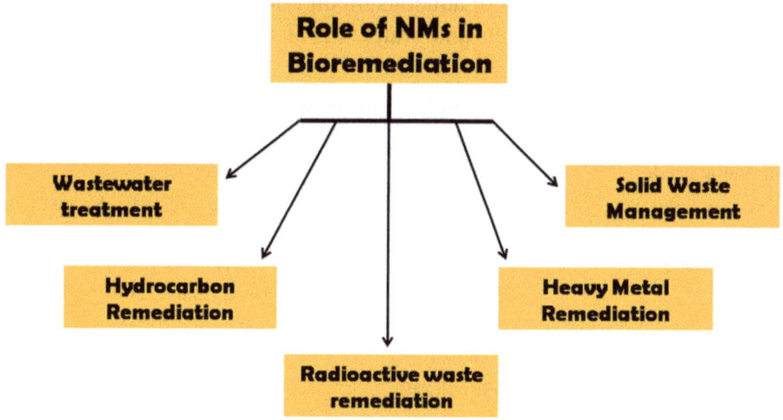


**Fig. 13.1** Applications of nanoparticles in bioremediation

exhaust-gas treatment as well as photocatalytic processes for contaminant degradation. Nanoscale sizes can also influence the chemical reactivity of materials by the predominance of near surface regions with compositions distinct from bulk regions, increasing the contribution of interfacial free energy to the free energy of dissolution-precipitation reactions (Kanatzidis and Poeppelmeier 2007). Nanoparticles can also provide very high flexibility for both in situ and ex situ remediations because they can be easily deployed in ex situ slurry reactors for the treatment of contaminated soils, sediments, and solid wastes. Alternatively, they can be anchored onto a solid matrix such as carbon, zeolite, or membrane for enhanced treatment of water, wastewater, or gaseous process streams. Nanoremediation methods involve the application of reactive NMs such as nanoscale zeolites, metal oxides, carbon nanotubes and fibers, and bimetallic nanoparticles for the transformation and detoxification of pollutants.

### *13.3.1 Nanoiron and Its Derivatives*

Among all, nanoscale zero-valent iron (nZVI) is currently the most widely used. Several authors have used nanoscale zero-valent iron (NZVI) for the removal various pollutants (Table 13.1). The high surface area associated with their high reactivity makes them an excellent agent, capable of transforming or degrading contaminants in soils and water. The other key factors favoring nZVIs include low standard reduction potential, favorable quantum size properties, and potential increase in transport efficiency through groundwater's underground matrix (Tosco et al. 2014). These NMs have properties that enable both chemical reduction and catalysis to mitigate a variety of pollutants of concern including chlorinated compounds, organochloride pesticides, polychlorinated biphenyls, heavy metal ions, and inorganic anions (Karn et al. 2009). Zero-valent iron removes aqueous contaminants by reductive dechlorination, in the case of chlorinated solvents, or by reducing to an insoluble from, in the case of aqueous metal ions. Iron also undergoes "redox" reactions with dissolved oxygen and water. nZVI has been used for removal of

**Table 13.1** Removal of pollutants using iron nanoparticles

| Type of nanoparticle | Contaminant removed | Reference |
|---|---|---|
| Zero-valent powder iron | Nitrate, arsenic (V) | Choe et al. (2000), Kanel et al. (2006) |
| Fe nanocomposite | Azo dye orange(II) | Feng et al. (2003) |
| Iron nanoparticle | Ni (II) | Li and Zhang (2007) |
| Colloidal zero-valent powder iron | Herbicide: molinate | Feitz et al. (2005) |
| Iron sulfide nanoparticle | Lindane | Paknikar et al. (2005) |
| CMC4-stabilized ZVI nanoparticle | Perchlorate | Xiong et al. (2007) |

heavy metals like arsenic and chromium, pesticides (lindane, DDT), and chlorinated solvents (PCE, TCE, DCE) and for transformation of organic compounds like nitrates (Karn et al. 2009). As (III) and As (V), which are highly toxic, mobile, and predominant arsenic species in anoxic groundwater, were removed using nanoscale zero-valent iron (Kanel et al. 2005, 2006). The supported zero-valent iron NPs "ferragels" rapidly separate and immobilize Cr (VI) and Pb (II) from aqueous solution, reducing the chromium to Cr (III) and the Pb to Pb (0) while oxidizing the Fe to goethite (α-FeOOH) (Ponder et al. 2000). Anionic, hydrophilic carbon (Fe/C), and poly (acrylic acid)-supported (Fe/PAA) zero-valent Fe-NPs are also reactive material for the dehalogenation of chlorinated hydrocarbons in groundwater as well as in soils and can be used to construct a reactive wall in the path of a contaminated groundwater plume to degrade halogenated organic compounds (Whang and Zhang 1997). High surface-area nickel-iron NPs have been used for dehalogenation of trichloroethylene (TCE) by (Schrick et al. 2002). The capability of powdered zerovalent iron to dechlorinate DDT and related compounds at room temperature has been investigated. Specifically, DDT, DDD [1,1-dichloro-2,2-bi's(p-chlorophenyl) ethane], and DDE [2,2-bi's(p-chlorophenyl)-1, 1-dichloroethylene] transformation by powdered zero-valent iron in buffered anaerobic aqueous solution was studied at 20 °C, with and without the presence of nonionic surfactant Triton X-114. The iron has been successful at dechlorinating DDT, DDD, and DDE (Sayles et al. 1997).

### *13.3.2 Nanocrystals and Carbon Nanotubes*

The exceptional and tunable properties of carbon-based nanomaterials such as nanocrystals and carbon nanotube(s) (CNT(s)) enable new technologies to identify and solve a broad range of environmental applications: sorbents, high-flux membranes, depth filters, antimicrobial agents, environmental sensors, renewable energy technologies, and pollution prevention strategies (Mauter and Elimelech 2008) (Table 13.2). Carbon nanotubes, in particular, hold tremendous potential for applications because of their unique properties, such as high thermal and electrical

**Table 13.2** Removal of pollutants using carbon nanotubes (CNTs)

| Type of nanoparticle | Contaminant removed | Reference |
|---|---|---|
| Carbon nanotubes | Pb(II) | Li et al. (2002, 2005) |
| Carbon nanotubes | Organic compounds (dyes, pesticides, pharmaceuticals/drugs) | Xu and Bhattacharyya (2005) |
| Carbon nanotubes | Trihalomethanes | Lu et al. (2005) |
| CNTs KMnO4 oxidized | Cd(II) | Li et al. (2003b) |
| MWCNTs | Organochlorines | Hua et al. (2017) |
| Carbon nanotubes | Trichloroethane | Kshitij et al. (2016) |

conductivities, high strength, high stiffness, and special adsorption properties. Carbon nanotubes have cylindrical pores, and adsorbent molecules interact with their carbon atoms on the surrounding walls. This interaction between molecules and solid surface depends on the pore size and geometry of pores. When a molecule is placed in between two flat surfaces, i.e., in a slit-shaped pore, it interacts with both surfaces, and the potentials on the two surfaces overlap. The extent of the overlap depends on the pore size. However, for cylindrical and spherical pores, the potentials are greater because more surface atoms interact with the adsorbed molecule. In addition, carbon nanotubes are highly graphitic (much more than the activated carbons). Hence, the carbon nanotubes can adsorb molecules much stronger than activated carbons, which have slit-shaped or wedge-shaped pores (Ralph and Yang 2003). Single-walled carbon nanotubes (SWCNTs), multiwalled carbon nanotubes (MWCNTs), and hybrid carbon nanotubes (HCNTs) have been evaluated as efficient and rapid adsorbents for ethylbenzene which possess good potential applications to maintain high-quality water. Thus these can be used for cleaning up environmental pollution to prevent ethylbenzene borne diseases (Bina et al. 2012). Single-walled or multiwalled CNTs were shown to transform organic compounds present in dyes, pesticides, pharmaceuticals, and drugs eluted in wastewater and activated sludge (Yu et al. 2014). In addition, CNTs have also been used for the removal of heavy metals like $Cr^{3+}$, $Pb^{2+}$, and $Zn^{2+}$ as well as metalloids such as arsenic compounds, organics biological impurities, volatile organic compounds, and dioxins (Li et al. 2003b; Rao et al. 2007). Calixarenes, thiacalixarenes, and CNT-based polymeric materials incorporating these molecules have been synthesized, characterized, and tested for removing both organic pollutants (p-nitrophenol) and inorganic pollutants ($Cd^{2+}$, $Pb^{2+}$) from water (Li et al. 2003a). An environment-friendly adsorbent, CNTs immobilized by calcium alginate (CNTs/CA) have been prepared, and their copper adsorption property has been investigated via equilibrium studies (Li et al. 2010).

### *13.3.3 Single-Enzyme Nanoparticles*

Enzymes function as biocatalysts in bioremediation. Due to lack of stability and short catalytic lifetimes of enzymes, their usefulness as cost-effective alternatives to synthetic catalysts is limited. An effective way to increase the stability, longevity, and reusability of the enzymes is to attach them to magnetic iron NPs. On attaching enzymes to the magnetic iron NPs, they can then be easily separated from reactants or products by applying a magnetic field. Qiang et al. (2007) have used two different catabolic enzymes, trypsin and peroxides, to form core shell magnetic nanoparticles (MNPs). MNP enzyme conjugates were found to be more stable, efficient, and economical.

### 13.3.4 Engineered Polymeric Nanoparticles

Sorption of hydrophobic organic contaminants, such as polycyclic aromatic hydrocarbons (PAHs), to the soil has been shown to limit their solubilization rate and mobility. In addition, sequestration of contaminants by sorption to soil and by partitioning in nonaqueous phase liquids (NAPLs) reduces their bioavailability. Polymer nano-network particles have been demonstrated to increase the "effective" solubility of a hydrophobic organic contaminant, PHEN, and to enhance the release of phenanthrene from the contaminated aquifer material (Tungittiplakorn et al. 2005).

### 13.3.5 Biogenic Uraninite Nanoparticles

Biogenic uraninite has great importance in bioremediation due to its small particle size and biological origin. Recent researchers have found the chemical/structural complexities of this important natural NM. The molecular-scale structure, energetics, and surface-area normalized dissolution rates of hydrated biogenic uraninite have been found to be similar to those of coarser-particle, abiotic, stoichiometric $UO_2$ (Bargar et al. 2008).

### 13.3.6 Dendrimers

The word "dendrimers" is derived from the Greek words where "dendri" means like a branch of tree and "meros" means part of tree. Dendrimers are highly branched and monodisperse macromolecules which have been recently recognized as members of the polymer field (Buhleier et al. 1978; Tomalia 1994; Newkome et al. 1985). In other words, dendrimers or cascade molecules have branching construction similar to a tree, in which one trunk forms several large branches, each forming smaller branches, and so on. The roots of the tree also have the same branching mode of growth. These nanostructures can be designed to encapsulate metal ions and zerovalent metals, enabling them to dissolve in suitable media or bind to appropriate surfaces. Dendrimers have numerous potential environmental applications (Table 13.3). Dendrimers are relatively monodispersed and highly branched macromolecules with controlled composition and architecture consisting of three components: central core, interior branch cells or radial symmetry, and terminal branch cell or peripheral group (Undre et al. 2013a). Dendrimers-NPs composite can be used to enhance catalytic activity due to more reactivity and more surface area and less toxicity (Undre et al. 2013b). Polyamidoamine (PAMAM) dendrimers are a new class of nanoscale materials that can be carried as water-soluble chelators. Usually, PAMAM macromolecules are synthesized by repeatedly attaching

**Table 13.3** Removal of pollutants using dendrimers

| | | |
|---|---|---|
| PAMAM(G3) | Cu(II) | Diallo et al. (1999) |
| PAMAM(G4) | $Ni^{2+}$, $Fe^{2+}$ and $Fe^{3+}$ | Castillo et al. (2014) |
| PAMAM(G5) | Cu(II) | Diallo et al. (1999) |
| PAMAM(G8) | Cu(II) | Diallo et al. (1999) |
| Modified benzyl ether dendrimer | Pyrene | Rebecca et al. (2015) |
| Octylurea DAB-32 | Pyrene | Arkas et al. (2003) |
| Octylurea DAB-32 | Pyrene | Arkas et al. (2003) |

**Table 13.4** Removal of pollutants using $TiO_2$ nanoparticles

| Type of nanoparticle | Contaminant removed | Reference |
|---|---|---|
| Arginine-modified $TiO_2$ | Nitrobenzene | Makarova et al. (2000) |
| $TiO_2$ nanoparticle | Rhodamine 6G | Chen et al. (2003) |
| Anatase $TiO_2$ | Phenol | Andersson et al. (2002) |
| Mesoporous titania nanohybrid (naohybrid-I)[4] | 4-Chlorophenol | Paek et al. (2006) |
| Mesoporous titania nanohybrid (naohybrid-I)[4] | Methyl orange | Paek et al. (2006) |
| Biocompatible $TiO_2$ nanoparticles | Rhodamine B, Congo red, and methylene blue | Bharati et al. (2017) |
| $TiO_2$ nanoparticle | Polyethylene | Thomas et al. (2013) |

amidoamine monomers in their radial branched layers, termed "generations," to a starting ammonia core (Diallo et al. 1999). Guo et al. (2012) have developed PAMAM dendrimers which can be used in water treatment. They have also developed simple filtration unit for the removal of organic pollutants by utilizing $TiO_2$ porous ceramic filters of which the pore was impregnated with an alkylated poly(propylene imine) dendrimer, poly(ethyleneimine) hyperbranched polymer, or $\beta$-cyclodextrin, thus resulting in hybrid organic/inorganic filter modules of high mechanical strength and high surface area.

### *13.3.7 Titanium Dioxide ($TiO_2$)-Based Nanoparticles*

Titanium dioxides possess semiconducting, photocatalytic, energy-converting, electronic, and gas-sensing properties. Three different polymorphs of titanium dioxide crystals present in nature are rutile, anatase, and brookite. These nanoparticles are readily available, inexpensive, and less toxic and hence are used to remove organic contaminants (Table 13.4). The semiconducting property of $TiO_2$ is necessary for the removal of different organic pollutants through excitation of $TiO_2$ semiconductor with a light energy greater than its band gap, which could generate electron hole pairs. This property can be exploited in different reduction processes at the semiconductor/solution interface (Mansoori et al. 2008). Stathatos et al. (1999) used reverse micelle technique to produce $TiO_2$ nanoparticles and deposited them as thin films. The films demonstrated high capacity in adsorption of several dyes from aqueous and alcoholic solutions. It also showed a rapid degradation of the adsorbed dyes on exposure of the colored films to the visible light. Phenol is one of the toxic materials found in municipal and waste waters. Andersson et al. (2002) have used synthesized titanium dioxide nanoparticles of both anatase and rutile forms for wet oxidation of phenols by hydrothermal treatment of microemulsions and their photocatalytic activity. Ilisz et al. (2004) used a combination of $TiO_2$-based photocatalysis and adsorption for the decomposition of 2-chlorophenol (2-CP). Asilturk et al. (2006) examined the behavior of anatase nano-$TiO_2$ in catalytic decomposition of rhodamine B (RB) dye. Rhodamine B was fully decomposed with the catalytic action of nano-$TiO_2$ in a short time of about 60 min. It was also found that the nano-$TiO_2$ could be repeatedly used with increasing the photocatalytic activity. The photocatalytic properties of both modified and unmodified $TiO_2$ has been investigated for the degradation of polyethylene (Thomas et al. 2013).

### *13.3.8 Bimetallic Nanoparticles*

Metal such as zinc and tin can transform halogenated organic compounds (HOCs) quicker than iron (Table 13.5). The deposition of small amounts of a second metal, such as Pd, Pt, Ag, Ni, and Cu, on iron has been shown to accelerate the reaction rate (Liou et al. 2005). Palladized iron with its superior catalytic ability has been found to completely dechlorinate many chlorinated aliphatic compounds to hydrocarbons. Synthesized nanoscale iron and palladized iron particles are also being used for degradation of chlorinated compounds (Wang and Zhang 1997; Lien and Zhang 2001). Ni (II) has been found to prevent formation of toxic by-products by dehalogenation of chlorinated compounds via hydrogen reduction rather than electron transfer (Feng and Lim 2005). Joo and Zhao (2007) prepared Fe-Pd bimetallic with 0.2% w/w of sodium carboxymethyl cellulose (CMC) as stabilizer and used them for the degradation of lindane and atrazine, the chlorinated herbicides.

**Table 13.5** Removal of pollutants using bimetallic nanoparticles

| Type of nanoparticle | Contaminant removed | Reference |
|---|---|---|
| Pd/Fe nanoparticle | Trichloroethene, tetrachloroethene | Wang and Zhang (1997), Lien and Zhang (2001) |
| Pd/Fe nanoparticle | Heavy metals (Pd, Cr, Cu) | Yan et al. (2010) |
| Ni/Fe nanoparticle | Carbon tetrachloride | Feng and Lim (2005) |
| Cu/Fe nanoparticle | Nitrate | Liou et al. (2005) |
| Fe/Pd nanoparticle | Lindane and atrazine | Joo and Zhao (2007) |

## 13.4 Potential Harmful Effects of Nanoparticles

Nanotechnology is emerging replacement of current practices for site remediation; however, nanoparticles present potential risks in terms of (i) dispersal, ability to disperse in the environment including potential long-range transport; (ii) ecotoxicity, ability to cause adverse effects to organisms in the environment; (iii) persistency, ability to remain in the environment; (iv) bioaccumulation, ability to bioaccumulate or bioconcentrate in higher-order organisms; and (v) reversibility, ability to remove or to reverse their original introduction from environment (Grieger et al. 2010). According to Lee et al. (2008), the reductive high power nZVI can denature lipopolysaccharides, as well as the ionic transport proteins, thus impairing permeability of the membrane and facilitating the $Fe^{2+}$ entry into the cell. Once inside the cell, the $Fe^{2+}$ can react with the hydrogen peroxide produced by mitochondria, forming reactive oxygen species and promoting oxidative stress and subsequent cell death. Human exposure to nanoparticles leads to number of effects including oxidative stress, lipid peroxidation, genotoxicity, lung diseases, inflammation, pulmonary pathological changes, etc. (Oberdörster et al. 2005; Sayes et al. 2005; Sharma et al. 2007).

In the last couple of years, a "greener" approach for the production of various nanoparticles has been employed to overcome the risk associated with chemically synthesized nanoparticles. Green synthesis of nanoparticles by plants is gaining importance nowadays because of single step biosynthesis process, absence of toxicants, and occurrence of natural capping agents (Bhuyan et al. 2015; Joshi et al. 2018). Machado et al. (2013) reported the successful application of nZVI synthesized from grape marc, black tea, and vine leaves for the remediation of soils contaminated with the common anti-inflammatory drug, ibuprofen. Kharissova et al. (2013) also reported ways of synthesizing nanoparticles like silver, gold, iron, metal alloys, oxides, and salts from various plants and plant parts such as tea, coffee, banana, table sugar and glucose as reductants and capping agents. Several microbe-derived nanoparticles (bio-nanoparticles) from bacteria, fungi, actinomycetes, yeasts, and algae have also been reported to degrade organic compounds effectively (Dhillon et al. 2012; Johnson et al. 2013; Shedbalkar et al. 2014). Study conducted by Johnson et al. (2013) has shown that bacteria from members of the genus *Clostridia* can reduce palladium, Pd (II), ions to form metallic Pd nanoparticles (bio-Pd) where cultures of *C. pasteurianum* BC1 were used to generate bio-Pd, which is primarily formed on the microbial cell wall. These cultures efficiently carried out the reduction of organic azo dyes, methyl orange, and Evans blue, with little dye reduction. Another approach is to synthesize "smarter" engineered NMs for environmental remediation is through the use of sophisticated NMs with new coatings or functional groups which could enhance stability, mobility, and the overall performance of nanoparticles that can become benign when the remediation process is over (Thatai et al. 2014; Von der Kammer et al. 2012).

## 13.5 Conclusion

Environmental damage due to increasing population and industrialization is a serious cause for concern. Nanotechnology has immense potential to revolutionize existing technologies used in various sectors. Nanotechnology can also provide eco-friendly alternatives and effective solutions for many pollution-related problems such as heavy metal contamination, adverse effects of chemical pollutants, oil pollution, and so on. The advent of nanoremediation, using smarter engineered NMs, can deliver cost-effective and timesaving in situ cleanup procedures for large-scale contaminated sites. NMs directly catalyze degradation of waste and toxic materials and also helps enhance the efficiency of microbial degradation of waste and toxic materials. Due to its powerful potential, it is expected that applications of nanoparticles will increase in the near future, and it will play a critical role in sustainable development. However various risks associated with NMs are an area of concern. Several ecological implications and health hazards may limit the widespread applications of NMs for environmental remediation. Hence, to make this technology more beneficial than harmful, monitoring and intervention measures need to be implemented sooner. Future research on both fundamental and practical aspects of nanoremediation is further recommended.

## References

Andersson M, Osterlund L, Ljungstrom S, Palmqvist A (2002) Preparation of nanosize anatase and rutile $TiO_2$ by hydrothermal treatment of microemulsions and their activity for photocatalytic wet oxidation of phenol. J Phys Chem B 106:10674–10679

Arkas M, Tsiourvas D, Paleos CM (2003) Functional dendrimeric "Nanosponges" for the removal of polycyclic aromatic hydrocarbons from water. Chem Mater 15:2844–2847

Asilturk M, Sayýlkan F, Erdemoglu S, Akarsu M, Sayýlkan H, Erdemoglu M, Arpac E (2006) Characterization of hydrothermally synthesized nano-$TiO_2$ crystallite and the photocatalytic degradation of Rhodamine B. J Hazard Mater B 129:164–170

Aziz N, Faraz M, Pandey R, Sakir M, Fatma T, Varma A, Barman I, Prasad R (2015) Facile algae-derived route to biogenic silver nanoparticles: synthesis, antibacterial and photocatalytic properties. Langmuir 31:11605–11612. https://doi.org/10.1021/acs.langmuir.5b03081

Aziz N, Pandey R, Barman I, Prasad R (2016) Leveraging the attributes of *Mucor hiemalis*-derived silver nanoparticles for a synergistic broad-spectrum antimicrobial platform. Front Microbiol 7:1984. https://doi.org/10.3389/fmicb.2016.01984

Baker S, Satish S (2012) Endophytes: toward a vision in synthesis of nanoparticles for future therapeutic agents. Int J Bio-Inorg Hybd Nanomater 1:1–11

Bargar JR, Bernier-Latmani R, Giammar DE, Tebo BM (2008) Biogenic uraninite nanoparticles and their importance for uranium remediation. Elements 4:407–412

Bharati B, Sonkar AK, Singh N, Dash D, Chandana R (2017) Enhanced photocatalytic degradation of dyes under sunlight using biocompatible $TiO_2$ nanoparticles. Mater Res Express 4(8):085503

Bhuyan T, Mishra K, Khanuja M, Prasad R, Varma A (2015) Biosynthesis of zinc oxide nanoparticles from *Azadirachta indica* for antibacterial and photocatalytic applications. Mater Sci Semicond Process 32:55–61

Bina B, Pourzamani H, Rashidi A, Amin MM (2012) Ethylbenzene removal by carbon nanotubes from aqueous solution. J Environ Sci Public Health 2012:817187. https://doi.org/10.1155/2012/817187
Buhleier E, Wehner W, Vogtle F (1978) Cascade and "nonskid chain- like" syntheses of molecular cavity topologies. Synthesis 2:155–158
Castillo VA, Barakat MA, Ramadan MH, Woodcock HL, Kuhn JN (2014) Metal ion remediation by polyamidoamine dendrimers: a comparison of metal ion, oxidation state, and titania immobilization. Int J Environ Sci Technol 11:1497–1502
Chen J, Liu M, Zhang L, Zhang J, Jin L (2003) Application of nano $TiO_2$ towards polluted water treatment combined with electro-photochemical method. Water Res 37:3815–3820
Choe S, Chang YY, Hwang KY, Khim J (2000) Kinetics of reductive denitrification by nanoscale zero-valent iron. Chemosphere 41:1307–1311
Cundy AB, Hopkinson L, Whitby RLD (2008) Use of iron-based technologies in contaminated land and groundwater remediation: a review. Sci Total Environ 400:42–51
Dhillon GS, Brar SK, Kaur S, Verma M (2012) Green approach for nanoparticle biosynthesis by fungi: current trends and applications. Crit Rev Biotechnol 32:49–73
Diallo MS, Balogh L, Shafagati A, Johnson JH, Goddard WA, Tomalia DA (1999) Poly(amidoamine) Dendrimers:Â A New Class of High Capacity Chelating Agents for Cu(II) Ions. Environmental Science & Technology 33 (5):820–824
Dimitrov D (2006) Interactions of antibody conjugated nanoparticles with biological surfaces. Colloids Surf A 8:282–283
Feitz AJ, Joo SH, Guana J, Suna Q, Sedlak DL, Waite TD (2005) Oxidative transformation of contaminants using colloidal zerovalent iron. Colloids Surf A 265:88–94
Feng J, Lim TT (2005) Pathways and kinetics of carbon tetrachloride and chloroform reductions by nano-scale Fe and Fe/Ni particles: comparison with commercial micro-scale Fe and Zn. Chemosphere 59:1267–1277
Feng J, Hu X, Yue PL, Zhu HY, Lu GQ (2003) Degradation of azo-dye orange II by a photoassisted Fenton reaction using a novel composite of iron oxide and silicate nanoparticles as a catalyst. Ind Eng Chem Res 42:2058–2066
Friedrich KA, Henglein F, Stimming U, Unkauf W (1998) Investigations of Pt particles on gold substrates by IR spectroscopy. Colloids Surf A 134:193–206
Grieger KD, Fjordøge A, Hartmann NB, Eriksson E, Bjerg PL, Baun A (2010) Environmental benefits and risks of zero-valent iron particles (nZVI) for in situ remediation: risk mitigation or trade-off? J Contam Hydrol 118:165–183
Guo R, Guo X, Yu D, Hu J (2012) Application research in water treatment of PAMAM dendrimer. Chem Ind Eng Prog 31:671–675
Hua S, Gong JL, Zeng GM, Yao FB, Guo M, Ou XM (2017) Remediation of organochlorine pesticides contaminated lake sediment using activated carbon and carbon nanotubes. Chemosphere 77:65–76
Ilisz I, Dombi A, Mogyorósi K, Dékány I (2004) Photocataltyic water treatment with different $TiO_2$ nanoparticles and hydrophilic/hydrophobic. Colloids Surf A Physicochem Eng Asp 230:89–97
Ingale AG, Chaudhari AN (2013) Biogenic synthesis of nanoparticles and potential applications: an eco- friendly approach. J Nanomed Nanotechnol 4:165–171
Johnson A, Merilis G, Hasting J, Palmer EM, Fitts JP, Chidambaram D (2013) Reductive degradation of organic compounds using microbial nanotechnology. J Electrochem Soc 160:G27–G31
Joo SH, Zhao D (2007) Destruction of lindane and atrazine using stabilized iron nanoparticles under aerobic and anaerobic conditions: effects of catalyst and stabilizer. Chemosphere 70:418–425
Joshi N, Jain N, Pathak A, Singh J, Prasad R, Upadhyaya CP (2018) Biosynthesis of silver nanoparticles using *Carissa carandas* berries and its potential antibacterial activities. J Sol-Gel Sci Technol 86:682–689. https://doi.org/10.1007/s10971-018-4666-2

Kanatzidis MG, Poeppelmeier KR (2007) Report from the third workshop on future directions of solid-state chemistry: the status of solid-state chemistry and its impact in the physical sciences. Prog Solid State Chem 36:1–133

Kanel SR, Manning B, Charlet L, Choi H (2005) Removal of arsenic (III) from groundwater by nanoscale zero-valent iron. Environ Sci Technol 39:1291–1298

Kanel SR, Greneche JM, Choi H (2006) Arsenic (V) removal from groundwater using nano scale zero-valent iron as a colloidal reactive barrier material. Environ Sci Technol 40:2045–2050

Karn B, Kuiken T, Otto M (2009) Nanotechnology and in situ remediation: a review of the benefits and potential risks. Environ Health Perspect 117:1823–1831

Kavitha KS, Syed B, Rakshith D, Kavitha HU, Yashwantha Rao HC, Harini BP, Satish S (2013) Plants as green source towards synthesis of nanoparticles. Int Res J Bio Sci 2:66–76

Kharissova OV, Dias Rasika HV, Kharisov BI, Pérez BO, Pérez Jiménez VM (2013) The greener synthesis of nanoparticles. Trends Biotechnol 31:240–248

Kshitij CJ, Zhuonan L, Hema V, Mallikarjuna N, Sharmila MK, Mesfin T (2016) Carbon nanotube based groundwater remediation: the case of trichloroethylene. Molecules 21:953–967

Lee C, Kim JY, Lee WI, Nelson KL, Yoon J, Sedlak DL (2008) Bactericidal effect of zero-valent iron nanoparticles on *Escherichia coli*. Environ Sci Technol 42:4927–4933

Li XQ, Zhang WX (2007) Iron nanoparticles: the core shell structure and unique properties for Ni(II) sequestration. Langmuir 22:4638–4642

Li YH, Wang S, Wei J, Zhang X, Xu C, Luan Z, Wu D, Wei B (2002) Lead adsorption on carbon nanotubes. Chem Phy Lett 357:263–266

Li YH, Dinga J, Luanb Z, Dia Z, Zhua Y, Xua C, Wu D, Wei B (2003a) Competitive adsorption of $Pb^{2+}$ $Cu^{2+}$ and $Cd^{2+}$ ions from aqueous solutions by multiwalled carbon nanotubes. Carbon 41:2787–2792

Li YH, Wang S, Luan Z, Ding J, Xu C, Wu D (2003b) Adsorption of cadmium(II) from aqueous solution by surface oxidized carbon nanotubes. Carbon 41:1057–1062

Li YH, Di Z, Ding J, Wu D, Luan Z, Zhu Y (2005) Adsorption thermodynamic, kinetic and desorption studies of $Pb^{2+}$ on carbon nanotubes. Water Res 39:605–609

Li Y, Liu F, Xia B (2010) Removal of copper from aqueous solution by carbon nanotube/calcium alginate composites. J Hazard Mater 177:876–880

Lien HL, Zhang WX (2001) Nanoscale iron particles for complete reduction of chlorinated ethenes. Colloids Surf A Physicochem Eng Asp 191:97–105

Lin DH, Xing BS (2007) Phytotoxicity of nanoparticles: inhibition of seed germination and root growth. Environ Poll 150:243–250

Liou YH, Lo SL, Lin CJ, Hu CY, Kuan WH, Weng SC (2005) Methods for accelerating nitrate reduction using zerovalent iron at near-neutral pH: effects of $H_2$-reducing pretreatment and copper deposition. Environ Sci Technol 39:9643–9648

Lu C, Chung YL, Chang KF (2005) Adsorption of trihalomethanes from water with carbon nanotubes. Water Res 39:1183–1189

Machado S, Pinto SL, Grosso JP, Nouws HPA, Albergaria JT, Delerue-Matos C (2013) Green production of zero-valent iron nanoparticles using tree leaf extracts. Sci Total Environ 445:1–8

Makarova OV, Rajh T, Thurnauer MC, Martin A, Kemme PA, Cropek D (2000) Surface modification of $TiO_2$ nanoparticles for photochemical reduction of nitrobenzene. Environ Sci Technol 34:4797–4803

Mansoori GA, Rohani-Bastami T, Ahmadpour A, Eshaghi Z (2008) Environmental application of nanotechnology. Annu Rev Nano Res 2:1–73

Mauter MS, Elimelech M (2008) Environmental applications of carbon-based nanomaterials. Environ Sci Technol 42:5843–5859

Newkome GR, Yao ZQ, Baker GR, Gupta VK (1985) Cascade molecules: a new approach to micelles. J Org Chem 50:2003–2004

Oberdörster G, Oberdörster E, Oberdörster J (2005) Nanotoxicology: An Emerging Discipline Evolving from Studies of Ultrafine Particles. Environmental Health Perspectives 113 (7):823–839

Paek SM, Jung H, Lee YJ, Park M, Hwang SJ, Choy JH (2006) Exfoliation and reassembling route to mesoporous titania nanohybrids. Chem Mater 18:1134–1140

Paknikar KM, Nagpal V, Pethkar AV, Rajwade JM (2005) Degradation of lindane from aqueous solutions using iron sulfide nanoparticles stabilized by biopolymers. Technol Adv Mater 6:370–374

Ponder SM, Darab JG, Mallouk TE (2000) Remediation of Cr(VI) and Pb(II) aqueous solutions using supported, nanoscale zero-valent iron. Environ Sci Technol 34:2564–2569

Prasad R (2014) Synthesis of silver nanoparticles in photosynthetic plants. J Nanopart 2014:963961. https://doi.org/10.1155/2014/963961

Prasad R (2017) Mycoremediation and environmental sustainability, vol 1. Springer International Publishing, Cham. (ISBN 978-3-319-68957-9) https://link.springer.com/book/10.1007/978-3-319-68957-9

Prasad R (2018) Mycoremediation and environmental sustainability, vol 2. Springer International Publishing, Cham. (ISBN 978-3-319-77386-5). https://www.springer.com/us/book/9783319773858

Prasad R, Pandey R, Barman I (2016) Engineering tailored nanoparticles with microbes: quo vadis. WIREs Nanomed Nanobiotechnol 8:316–330. https://doi.org/10.1002/wnan.1363

Qiang Y, Sharma A, Paszczynski A, Meyer D (2007) Conjugates of magnetic nanoparticle-enzyme for bioremediation. Proc 2007 NSTI Nanotechnol Conf Trade Show 4:656–659

Rajan S (2011) Nanotechnology in groundwater remediation. Int J Environ Sci Dev 2:182–187

Ralph T, Yang G (2003) Adsorbents: fundamentals and applications. Wiley, New Jersey

Rao GP, Lu C, Su F (2007) Sorption of divalent metal ions from aqueous solutions by carbon nanotubes: a review. Sep Purif Technol 58:224–231

Rebecca MT, Michele LP, Amy MB (2015) Effect of dendrimeric composition on the removal of pyrene from water. SpringerPlus 4:511–522

Richard F (1960) There's plenty of room at the bottom. California Institute of Technology. Caltech Eng Sci 23:22–36

Roco MC (2005) The emergence and policy implications of converging new technologies integrated from the nanoscale. J Nanopart Res 7:129–143

Ruffini-Castiglione M, Cremonini R (2009) Nanoparticles and higher plants. Caryologia 62:161–165

Sayes CM, Gobin AM, Ausman KD, Mendez J, West JL, Colvin VL (2005) Nano-C60 cytotoxicity is due to lipid peroxidation. Biomaterials 26:7587–7595

Sayles GD, You G, Wang M, Kupferle MJ (1997) DDT, DDD, and DDE dechlorination by zero-valent iron. Environ Sci Technol 31:3448–3454

Schrick B, Blough JL, Jones AD, Mallouk TE (2002) Hydrodechlorination of trichloroethylene to hydrocarbons using bimetallic nickel-iron nanoparticles. Chem Mater 14:5140–5147

Sharma CS, Sarkar S, Periyakaruppan A, Barr J, Wise K, Thomas R (2007) Single-walled carbon nanotubes induces oxidative stress in rat lungcepithelial cells. J Nanosci Nanotechnol 7:2466–2472

Shedbalkar U, Singh R, Wadhwani S, Gaidhani S, Chopade BA (2014) Microbial synthesis of gold nanoparticles: Current status and future prospects. Advances in Colloid and Interface Science 209:40–48

Stathatos E, Tsiourvas D, Lianos P (1999) Titanium dioxide films made from reverse micelles and their use for photo catalytic degradation of adsorbed dyes. Colloids Surf A Physicochem Eng Asp 149:49–56

Thatai S, Khurana P, Boken J (2014) Nanoparticles and core–shell nanocomposite based new generation water remediation materials and analytical techniques: a review. J Microchem 116:62–76

Thomas RT, Nair V, Sandhyarani N (2013) $TiO_2$ nanoparticles assisted solid phase photocatalytic degradation of polythene film: a mechanistic investigation. Colloids Surf A 422:1–9

Tomalia DA (1994) Starburst/Cascade Dendrimers: Fundamental building blocks for a new nanoscopic chemistry set. Advanced Materials 6 (7-8):529–539

Tosco T, Papini MP, Viggi CC, Sethi R (2014) Nanoscale zero valent iron particles for groundwater remediation: a review. J Clean Prod 77:10–21

Tungittiplakorn W, Cohen C, Lion LW (2005) Engineered Polymeric Nanoparticles for Bioremediation of Hydrophobic Contaminants. Environmental Science & Technology 39 (5):1354–1358

Undre SB, Singh M, Kale RK (2013a) Interaction behaviour of trimesoyl chloride derived 1st tier dendrimers determined with structural and physicochemical properties required for drug designing. J Mol Liq 182:106–120

Undre SB, Singh M, Kale RK, Rizwan M (2013b) Silibinin binding and release activities moderated by interstices of trimesoyl, tridimethyl, and tridiethyl malonate first-tier dendrimers. J Appl Polymer Sci 130:3537–3554

Von der Kammer F, Ferguson P, Holden P, Masion A, Rogers K, Klaine S, Koelmans A, Horne N, Unrine J (2012) Analysis of nanomaterials in complex matrices (environment and biota): general considerations and conceptual case studies. Environ Toxicol Chem 31:32–49

Wang CB, Zhang WX (1997) Synthesizing nanoscale iron particles for rapid and complete dechlorination of TCE and PCBs. Environ Sci Technol 31:2154–2156

Watlington K (2005) U.S. Environmental Protection Agency. www.epa.gov www.clu-in.org

Xiong Z, Zhao D, Pan G (2007) Rapid and complete destruction of perchlorate in water and ion-exchange brine using stabilized zero-valent iron nanoparticles. Water Res 41:3497–3505

Xu J, Bhattacharyya SD (2005) Membrane based bimetallic nanoparticles for environmental remediation: synthesis and reactive properties. Environ Prog 24:358–366

Yan W, Herzing AA, Li XQ, Kiely CJ, Zhang WX (2010) Structural evolution of Pd-doped nanoscale zero-valent iron (nZVI) in aqueous media and implications for particle ageing and reactivity. Environ Sci Technol 44:4288–4294

Yu JG, Zhao XH, Yang H, Chen XH, Yang Q, Yu LY, Jiang JH, Chen XQ (2014) Aqueous adsorption and removal of organic contaminants by carbon nanotubes. Sci Total Environ 483:241–251

# Chapter 14
# Fungal Nanoparticles Formed in Saline Environments Are Conducive to Soil Health and Remediation

Yi Wei, Li-Na Chen, Zi-Yu Zhang, Chi Zhu, and Shi-Hong Zhang

## Contents



## 14.1 Introduction

The main source of all salts in the soil is the primary minerals in the exposed layer of the earth's crust. The salt constituents are gradually released and made soluble during the soil-forming process. The released salts are transported away from their source of origin through surface or groundwater streams. The salts in the groundwater stream are gradually concentrated as the water with dissolved salts moves from the more humid to arid and semiarid areas, which is the primary cause of the soil salinization. Therefore, soil has the possibility to become salinized soil again. In addition, natural events and, more markedly, anthropogenic activities such as industrial activities, release of toxic compounds, application of excessive fertilizers, and uncontrolled utilization of natural resources may have a drastic impact on the environment and the quality of life and speed up the process of soil salinization.

Soil salinization leads to serious environmental problems on a global scale (Wang et al. 2003; Yadav et al. 2011; Liang et al. 2015). The problems of soil salinity are most widespread in the arid and semiarid regions, but salt-affected soils also occur extensively in subhumid and humid climates, particularly in the coastal regions where the ingress of seawater through estuaries and rivers and through groundwater causes large-scale salinization. Soil salinity is also a serious problem

Y. Wei · L.-N. Chen · Z.-Y. Zhang · C. Zhu · S.-H. Zhang (✉)
College of Plant Sciences, Jilin University, Changchun, People's Republic of China
e-mail: zhang_sh@jlu.edu.cn

 
R. Prasad, E. Aranda (eds.), *Approaches in Bioremediation*, Nanotechnology in the Life Sciences, https://doi.org/10.1007/978-3-030-02369-0_14

in areas where groundwater of high-salt content is used for irrigation. The most serious salinity problems are being faced in the irrigated arid and semiarid regions of the world, and it is in these very regions that irrigation is essential to increase agricultural production to satisfy food requirements.

Saline and saline-alkaline soils are generally called the two main groups of the salt-affected soils (Szabolcs 1994). As a matter of fact, the various sodium salts in nature do not occur absolutely separately, but in most cases either the neutral salts or the ones capable of alkaline hydrolysis exercise a dominant role on the soil-forming processes and therefore in determining soil properties. In most agricultural cases, the jeopardizing of soda-affected soil is more serious than that of other saline soils. It is the accumulation of solutes, primarily $Na_2CO_3$ and $NaHCO_3$, that induces primary soil alkalization: soda saline-alkaline soil leads to many negative effects on soil organic matter decomposition and uptake of available nutrients (Rietz and Haynes 2003; Karlen et al. 2008), which subsequently affect plant survival, health, and development (Rady 2011). Therefore, accumulation of excess salts in the root zone results in a partial or complete loss of soil productivity.

Soda saline-alkaline soils occur within the boundaries of at least 75 countries (Szabolcs 1994), and the severity of this issue has increased steadily in several major agricultural areas around the world (Ghassemi et al. 1995). The well-known typical saline soils are, respectively, located in Victoria in Australia, California in the United States, Mexico City in Mexico, and Baicheng City in China (Wang et al. 2009). In Victoria, sodic soils are estimated to occupy at least 13.4 Mha, representing at least 73% of Victoria's agricultural land, and the largest sodicity class is "alkaline sodic," dominated by a diverse range of soils (Ford et al. 1993). The soil of the former Lake Texcoco in Mexico is a unique extreme environment (called a soda desert) located near one of the biggest cities in the world, Mexico City. Large parts are saline-alkaline with pH more than 10 and electrolytic conductivity (EC) more than 150 dS $m^{-1}$ (Dendooven et al. 2010). Nowhere in China is the issue more serious than in the Songnen Plain of Northeast China (Fig. 14.1). Soil alkali is the major ecological gradient in the Songnen Plain, as well as the primary factor limiting its food security (Gao et al. 1996). Therefore, effective strategies to remediate soda saline-alkaline soil are urgently needed.

Physical tillage operations, leaching with water, chemical amendments, and plant-associated phytoremediation have been utilized to attempt to ameliorate soil salinity (Qadir et al. 2007). As a simple, convenient, reliable, and effective technology, leaching with municipal and industrial wastewaters has been widely recognized and applied for a long time. However, because of various kinds of metal contaminants contained in wastewater, land irrigation with such water has begun to be forbidden in agricultural areas (Harner et al. 1999). On account of the significant ecological, environmental, and economic effects of the former three techniques, phytoremediation is widely considered to be the best method for ameliorating soil salinity (Ilyas et al. 1993; Ghaly 2002; Nouri et al. 2017). Thus far, the primary factors influencing the success of phytoremediation have been the selection and application of appropriate plants, such as salt-resistant or salt-tolerant species, and

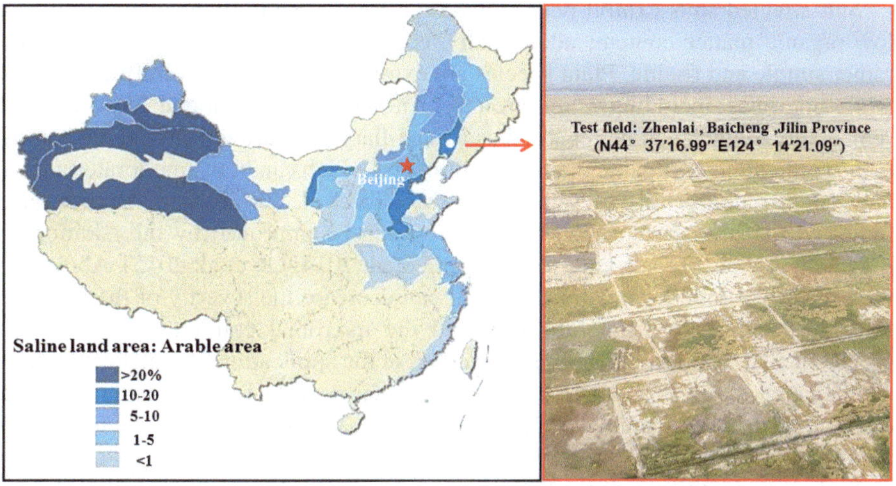


**Fig. 14.1** Distribution of saline and arable lands in the mainland of China. The saline and arable lands in the mainland of China are shown in the left picture; in the left picture, the white spot represents our experiment field that is located in Songyuan soda saline-alkaline land of Zhenlai County, Baicheng city, Jilin Province China; the right picture shows the several plots of the test field (the aerial photography was taken by an unmanned plane at 150 meters in the air): The white snow-like crusts on the land are soda salt that has returned to the soil surface (in a soil level extending from 0 to 20 cm depth: sodium salt = 15 g/kg, pH = 9.8; measured by Dr. Shi Yang)

their cropping sequence. At this point, the upper limit of plant resistance to salt restricts the application range of soil phytoremediation.

As far as healthy soil is concerned, clay, one of the mineralogy of the finest soil particles, is especially important. As well known, a very important property of clay minerals is their ability to adsorb and hold salt ions such as calcium, magnesium, sodium, and potassium and the acid element hydrogen. Clay minerals, because of their chemical composition and structure, possess a net negative charge. The bases are positively charged when in the soil solution and are attracted to and held on the surfaces of the negatively charged clay particles. The adsorbed cations are held tightly enough to retard their movement from the soil by leaching, yet loosely enough to be replaced by other cations, a process called cation exchange. The absolute quantity of cations that may be held in a soil by the clay fraction in exchangeable form is the cation-exchange capacity. One of the most significant features of the cation-exchange capacity of a soil is that it provides temporary storage of large quantities of plant nutrients such as calcium, magnesium, and potassium. When ammonium sulfate fertilizer is added to a soil, it dissolves in the soil solution forming the ammonium cation and the negatively charged sulfate ion. The ammonium then may replace some of the exchangeable cations and be held on the clay, silt, and sand surfaces until used by plants. This reaction occurs most readily on clay particles, to a lesser extent on silt, and to a much lesser extent on sand.

Salt-affected soils exhibit poor structural stability due to high sodium salts and low organic matter content; additionally, ecosystems in severely saline soils are rather simple and fragile. Plant species are extremely scarce in severely saline soil, while microbes, including fungi, are rare. An alternative technique for saline soil remediation, which can be regarded as an auxiliary measure for phytoremediation, is the application of organic matter conditioners, which can both ameliorate salinity and increase the fertility of saline soils (Melero et al. 2007). Some studies have indicated that the structural stability of soil can be improved by the addition of organic materials (Tejada et al. 2006; Wang et al. 2014; Oo et al. 2015). Above all, the addition of maize straw to saline soil can decrease the severity of the negative effects of salinity on mineralization and the microbial community in the soil (Wichern et al. 2006). Interestingly, the effect of the application of organic materials supplemented with fermentation microbes in saline field is quite better than the disposable decayed composts (Liu and Huang 2010), reflecting the additional actions of microbe beyond fermentation.

Soil microorganisms generally have the ability to adapt to or tolerate salinity; and examples of microbes thriving in ponds with very high-salt concentrations demonstrate the evolutionary potential of microorganisms (Casamayor et al. 2002). The biodiversity of microorganisms in soda environments has indicated that abundant bacterial communities, which also act as primary producers, are usually dominated by cyanobacteria species (Antony et al. 2013). In addition, the N-fixing cyanobacterium, *Anabaena torulosa*, has been applied in remediating soil salinity during crop growth (Apte and Thomas 1997). However, it has been reported that long-term salt stress reduces fungal diversity (Bruggen et al. 2000). In addition, fungi tend to be sensitive to salt stress, as indicated by the increasing salt content in soil (Sardinha et al. 2003). In general, the negative impact of elevated salinity on fungi is stronger than its effect on bacteria. The negative effects of salt stress on soil fungi reduce the microbial biomass and microbial activity of the soil and impair turnover of organic matter, which creates a vicious cycle that reduces soil fertility and eventually produces soil incapable of supporting crops. Obviously, in order to remediate the saline-alkaline soil, our primary task must be to increase the beneficial fungi that can survive in the saline-alkaline land.

Haloalkaliphilic fungi are excellent biological resources for soil mycoremediation. Soil-inhabitant fungi play individual or joint roles in improving salinity and alkalinity soil conditions through different mechanisms such as absorbing, immobilizing and/or deactivating sodium ions and heavy metal ions, secreting organic acids and/or macromolecular degradation enzymes, and other benefits of biomasses for soil healthy. *Fusarium oxysporum* f. sp. *lycopersici* has the ability to produce the nanoparticles of varying size (10–100 nm) and shape at both extracellular and intercellular levels when incubated with platinum chloride solutions. The particles precipitate out of the solution and bioaccumulate by nucleation either intercellularly, on the cell wall/membrane, or extracellularly in the surrounding medium (Riddin et al. 2006). In saline soil environment, our lab found that some halophilic and alkaliphilic fungi can also produce nanoparticles, which are mixed and linked with the enzymes or metabolites secreted by these microorganisms.

These nanoparticles, whether in vivo or in vitro, can directly or indirectly improve the structure and texture of soil aggregates (Figs. 14.2 and 14.3).

Microbial application for amelioration of saline soils is gaining popularity due to its better amelioration and reduction in economic and environmental costs. Fortunately, within the last few decades, a series of halophilic and alkaliphilic fungi capable of living in highly saline and alkaline environments (or both) have been identified. This chapter is focused on the isolation and characterizations of extreme haloalkaliphilic fungi and their roles in saline-alkaline soil mycoremediation mediated by nanoparticles synthesized by halotolerant or halophilic fungi. In addition, remediation mechanisms of nanoparticles are discussed.

## 14.2 Isolation and Characterizations of Halotolerant or Halophilic Fungi

The major lineage of fungi was believed to have first arisen about 1000 million or so years ago, which was followed by land plants in approximately 700 million years ago (Heckman et al. 2001). Immediately after the Cretaceous-Permian extinction that famously killed off most dinosaurs, there is a dramatic increase in evidence of fungi, apparently the death of most plant and animal species leading to a huge fungal bloom like "a massive compost heap" (Casadevall 2012). From the biological

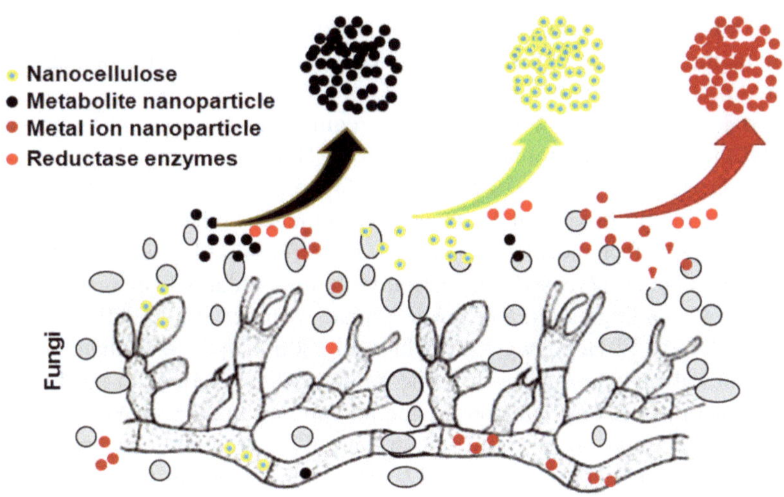


**Fig. 14.2** The proposed MIMI nanoparticles (microbial metabolites-integrated nanoparticles synthesis) for mycoremediation. In saline-alkaline soil, while haloalkaliphilic fungi and crop straw were mixed, a kind of integrated nanoparticles induced by soil structure and composition occurs, which involves not only salts or other soil components but also metabolites such as the proteins, enzymes, polypeptides, organic acids, or other metabolites secreted from soil-related microbial cells

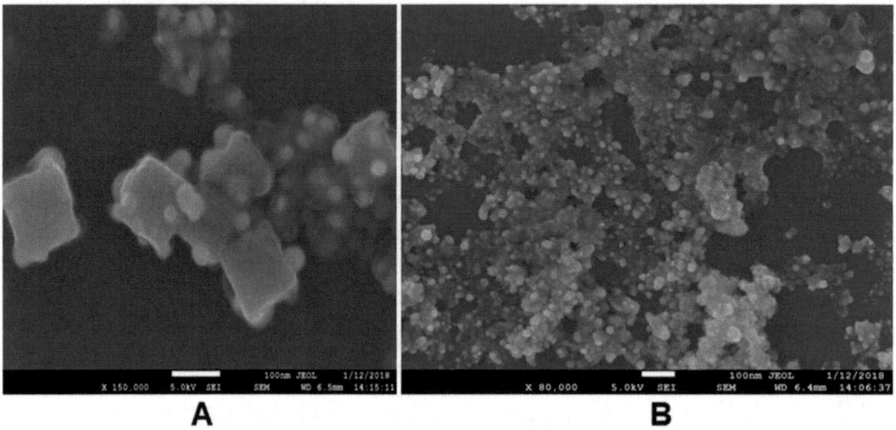


**Fig. 14.3** The morphologies of MIMI nanoparticles. (**a**) Fermentation filtrates of *A. glaucus* CCHA were mixed with physically grated corn stalks (particle size 0.5–2 mm); (**b**) *A. glaucus* CCHA and its fermentation filtrate were added in the saline soil mixed with physically grated corn stalks (particle size 0.5–2 mm). Multiple scales of tetrahedron-, sphere-, and wire-shaped nanoparticles were analyzed by SEM-EDS: the samples were first freeze-dried in a vacuum freeze dryer and then examined using a scanning electron microscope (Zeiss Super 55VP, Germany) (Samples prepared and SEM-EDS taken by the author Li-Na Chen unpublished)

and environmental evolution perspective, fungi are one of the earliest eukaryotes to colonize the ancient earth (Horodyski and Knauth 1994). Considering the harsh physical environments on the ancient earth, to ensure the chances of survival, fungi need to be more tolerant or resistant to adverse environmental factors than the latter appeared plants or animals.

Halotolerant fungi do not necessarily require certain concentrations of salt, although they were often found in saline areas. Halophilic fungi, however, require salt concentrations of at least 0.3 M (sodium salt, e.g., NaCl) to grow optimally, and they are capable of thriving in high-salt environments. To halotolerant fungi, salinity can directly affect sporulation and growth of fungi: at higher salinities (>5%), there tends to be an increased sporulation with more chlamydospores observed, an inhibition of conidiogenesis, and fewer hyphae (Mulder et al. 1989; Mahdy et al. 1996; Mulder and El-Hendawy 1999; Mandeel 2006). On the other hand, halophilic fungi do not always have to be in saline habitats; thus, there is no need to make a strict distinction between halotolerant and halophilic fungi (Arakaki et al. 2013).

Alkaliphilic fungi are a class of extremophilic microbes that are capable of survival in alkaline (pH roughly 8.5–11.0) environments and grow optimally even at a pH of approximately 10. Halophilic fungi growing in alkaline environments that are adapted to high pH and high concentrations of ions are described as haloalkaliphilic, rather than merely halophilic or alkaliphilic; thus, haloalkaliphilic fungi are alkaliphilic fungi that also require the same high-salt content to survive (Horikosh 1999).

Most halophilic fungi live in marine aquatic bodies, seashore, and inland terrestrial soils with high-salt concentrations, such as the Dead Sea, the Antarctic Ocean, and the Great Salt Plains, and a large number of studies on biodiversity and physiology have focused on the characterization of halophilic fungi present in the saline and hypersaline ecosystems, among which species of *Ascomycetes*, as well as some *Basidiomycetes*, have been described in detail (Gunde-Cimerman et al. 2000; Butinar et al. 2005a, b; Zalar et al. 2005; Evans et al. 2013; Gunde-Cimerman and Zalar 2014; Zajc et al. 2014a, b; Gonçalves et al. 2017). Hypersaline fungal communities are dominated by *Aspergillus* and *Penicillium* species, with melanized dematiaceous forms commonly observed in inland lands (Moubasher et al. 1990; Grum-Grzhimaylo et al. 2016), similar to the communities observed in marine environments (Buchalo et al. 1998, 2000; Gunde-Cimerman et al. 2000; Butinar et al. 2005a, b; Kis-Papo et al. 2003, 2014; Gunde-Cimerman and Zalar 2014).

The Dead Sea, a typical high-salt habitat for microorganisms, contains 340 g/L of dissolved salt; a variety of filamentous fungi have been isolated from the Dead Sea by the Nevo Group. *Gymnascella marismortui* is a remarkable salt-tolerant fungus that has been isolated from the surface water down to a depth of 300 m in the Dead Sea (Buchalo et al. 1998). *G. marismortui* grows optimally at NaCl concentrations between 0.5 and 2 M (Buchalo et al. 1998, 2000), suggesting that it is adapted to high-salt conditions and requires high-salt concentrations. Among 476 fungal isolates from the Dead Sea, *Aspergillus terreus*, *Aspergillus sydowii*, *Aspergillus versicolor*, *Eurotium herbariorum*, *Penicillium westlingii*, *Cladosporium cladosporioides*, and *Cladosporium sphaerospermum* were isolated consistently and probably form the stable core of the fungal community. In another study, approximately 43% of fungal isolates from the Dead Sea were found to belong to the genera *Eurotium* and *Aspergillus* (Yan et al. 2005).

The large diversity of the fungal species have been reported to inhabit high-salt environments; however, most of them can be regarded either as halotolerant or as extremely halotolerant. Halotolerant fungi can grow without NaCl added to the medium but tolerate up to saturated NaCl levels (30%) (Gunde-Cimerman et al. 2000). Up till today, only *Wallemia ichthyophaga*, *Wallemia muriae*, *Phialosimplex salinarum*, *Aspergillus baarnensis*, *Aspergillus salisburgensis*, and *Aspergillus atacamensis* are obligate halophilic fungi that strictly require NaCl from 5% to 10% (Piñar et al. 2016). Actually, *Gymnascella marismortui* (Buchalo et al. 1998), *Trichosporium* spp. (Elmeleigy et al. 2010), *Aspergillus unguis* (Nazareth et al. 2012), and *Aspergillus penicillioides* (Nazareth and Gonsalves 2014) have also been reported to be obligate halophiles according to their minimum saline requirement.

*Aspergillus penicillioides* are commonly found in saline habitats, suggesting that the species are extensively adaptable to varied environments. Among 39 tested isolates of *A. penicillioides*, most strains had a minimum salt requirement of 5% for growth; one strain grew only on media supplemented with at least 10% solar salt (Nazareth and Gonsalves 2014). Given that *A. penicillioides* species do not reproduce sexually (Tamura et al. 1999; Gostinčar et al. 2010; Gostinčar and Turk 2012), which consequently inhibits their gene flow, this species has significant promise in environmental remediation applications.

As mentioned above, some halophilic fungi, such as *A. niger* and *C. cladosporioides*, have been isolated from sand and mud on the shore of salty aquatic bodies or from inflowing fresh water from floods and springs (Kis-Papo et al. 2003, 2014; Grum-Grzhimaylo et al. 2016). We also isolated the halophilic fungus *Aspergillus glaucus* CCHA from air-dried wild vegetation from the surface periphery of a solar salt field (Liu et al. 2011); this species shows extreme salt tolerance, with a salinity range of 5–32% (NaCl) required for growth (Liu et al. 2011; Xie 2013). To our surprise, *A. glaucus* CCHA survives in solutions with a broad pH range of 2.0–11.5, indicating that it is a haloalkaliphilic fungus (Liu et al. 2014). Further investigation indicated that increasing the pH value (>8.0) can induce *A. glaucus* CCHA to produce a variety of organic acids, including citric acid, oxalic acid, and malic acid. In addition, *A. glaucus* CCHA shows resistance to aridity, heavy metal ions, and high temperature (Liu et al. 2015). The extremophilic nature of *A. glaucus* CCHA suggests that it has great promise in soil remediation applications.

Just like the proportion of the halophilic and halotolerant fungi isolated from saline environment, fewer alkaliphilic fungi have been identified in comparison with alkalitolerants. Hozzein and colleagues isolated 117 alkaliphilic and alkaline-resistant microorganisms from 30 soil samples collected from 6 localities around Wadi Araba, Egypt. By adjusting the pH to 10 after sterilization (using sterilized 10% $Na_2CO_3$ solution), they only identified 4 fungal isolates among 117 alkaliphilic and alkaline-resistant microorganisms (Hozzein et al. 2013); unfortunately, the authors did not determine the species of the isolates. Alkaliphilic fungi have also been isolated from industrial effluents. For example, *Aspergillus nidulans* KK–99 (isolated from the industrial effluents of Shreyans Paper Industry Limited, Ahmedgarh, Punjab, India) is adapted for growth in an alkalescent environment (pH 10.0) (Taneja et al. 2002). Another alkaliphilic fungus, *Myrothecium* sp. IMER1 also grows well under alkali conditions (pH 9.0) (Zhang et al. 2007).

Grum-Grzhimaylo and collaborators (2016) identified more than 100 strains of alkalitolerant and alkaliphilic fungi isolated from the alkaline soils with different degrees of salinity in Russia, Mongolia, Kazakhstan, Kenya, Tanzania, and Armenia. They found the alkaliphilic/strong alkali-tolerant phenotype in about 2/3 of our recovered strains from soda soils and uncovered that the alkaliphilic trait in filamentous fungi has evolved several times through phylogenetic analyses. Among the alkaliphilic/strong alkali-tolerant fungi, the *Sodiomyces* species (*Plectosphaerellaceae*), *Acrostalagmus luteoalbus* (*Plectosphaerellaceae*), *Emericellopsis* alkaline (*Hypocreales*), *Thielavia* sp. (*Chaetomiaceae*), and *Alternaria sect. Soda* (*Pleosporaceae*) grew best at high ambient pH, but the pH tolerance of *Chordomyces antarcticum*, *Acrostalagmus luteoalbus*, and some other species was largely affected by the presence of extra Na ion in the growth medium, further suggesting that the frequency of alkaliphilic fungi is low, while alkalitolerants seem to be far more widespread in soil (Grum-Grzhimaylo et al. 2016).

Research aimed at isolating and characterizing halophilic fungi has progressed rapidly in China. A series of promising halophilic fungi, including *A. glaucus* CCHA, have been reported. Three marine-derived isolates were collected in Wenchang, Hainan Province, China, and identified as extremely halotolerant fungi:

*Wallemia sebi* PXP-89 (Peng et al. 2011a), *Penicillium chrysogenum* PXP-55 (Peng et al. 2011b), and *Cladosporium cladosporioides* PXP-49 (Xu et al. 2011). The work focusing on isolating halotolerant/alkaliphilic/haloalkaliphilic fungi is being carried out in our laboratory, and actually several halotolerant species with alkaliphilic trait, such as *Aspergillus* sp., were recently identified based on a specimen collected from the saline-alkaline soils in Songnen Plain of Northeast China. China has remarkable biodiversity and many typical hypersaline environments, including Chaka Salt Lake and Qarhan Salt Lake in Qinghai, Barkol Salt Lake in Xinjiang, Yuncheng Salt Lake in Shanxi, and the Baicheng soda saline-alkaline area in Jilin. All of these environments are suitable for extremophilic fungi and other microorganisms; therefore, isolating and identifying extremophilic fungi in China could lead to the development of promising new methods of remediating saline-alkaline soil.

## 14.3 Nanoparticles Synthesized by Halotolerant or Halophilic Fungi

Fungi can actively or passively absorb, collect, or immobilize inorganic metallic ions from their close environment. Some inorganic metallic ions are beneficial nutrient elements and necessary for organisms, while others such as heavy metal ions are not, and sometimes these ions are even harmful to organisms, especially when they are excessively accumulated. On the other hand, fungi, when exposed to stress environments like saline soils, have the ability of producing extracellular metabolites that serve as protective factors for their own survival. Some researches clearly demonstrated the relevance of the toxic metal ion resistance and the formation of nanoparticles. During the synthesis of metal nanoparticles by a fungus, the fungal mycelium is exposed to the metal salt solution, which prompts the fungus to produce enzymes and metabolites for its own survival (Rajakumar et al. 2012). In this process, the toxic metal ions are reduced to the nontoxic metallic solid nanoparticles through the catalytic effect of the extracellular enzyme and metabolites of the fungus (Vahabi et al. 2011). It is because fungi play a crucial role in synthesis of metal nanoparticles that they are being considered as useful tools for producing nanoparticles today; in addition, more and more studies on synthesis of metal nanoparticles by fungi have been reported.

Nanoparticles are divided into intracellular and extracellular nanoparticles according to their accumulation locations. In the *Verticillium* fungus, the silver or gold nanoparticles were found to be deposited and bounded to the surface of the cytoplasmic membrane, suggesting the intracellular synthesis of gold or silver nanoparticles (Mukherjee et al. 2001; Sastry et al. 2003). And from then on, the various intracellular metal nanoparticles were successively reported in the fungus *Phoma* sp. 32883 (Chen et al. 2003), *Trichothecium* spp. (Ahmad et al. 2005), *Verticillium luteoalbum* (Gericke and Pinches 2006), *Penicillium chrysogenum*

(Sheikhloo and Salouti 2011), *Fusarium oxysporum* f. sp. *lycopersici* (Riddin et al. 2006), and *Aspergillus flavus* (Vala et al. 2014). As a matter of fact, there are much more researches focusing on the synthesis of extracellular nanoparticles. Extracellular nanoparticles can be divided into several subtypes on account of the capping components such as the proteins (enzymes), polypeptides, organic acids, or other metabolites secreted from the cells (Prasad et al. 2016).

The extracellular nanoparticles have been extensively studied in various fungi. The silver nanoparticles were synthesized outside the cells of the salt-tolerant yeast strain (MKY3), when MKY3 were treated with 1 mM soluble silver (Kowshik et al. 2003). And then biosynthesis of crystalline silver or gold nanoparticles was also reported by extremophilic yeasts (Mourato et al. 2011; Namasivayam et al. 2011). However, much more extracellular nanoparticles were described in filamentous fungi, for example, spherical-, rod-, square-, pentagonal-, and hexagonal-shaped nanoparticles formed by the *Alternaria* species in 1 mM concentration of chloroaurate solution (Dhanasekar et al. 2015), the smaller spherical nanoparticles formed by *Aspergillus tamarii* PFL2, and the larger spherical nanoparticles formed by *Aspergillus niger* PFR6 and *Penicillium ochrochloron* PFR8 (Devi and Joshi 2015). The formation of these nanoparticles is mostly associated with heavy metal ions, and the fungal producer should have the characteristics of salt tolerance and resistance, which have been embodied in the filamentous fungi such as *F. oxysporum* (Durán et al. 2005, Kumar et al. 2007a; Namasivayam et al. 2011); *F. acuminatum* (Ingle et al. 2008); *F. solani* (Ingle et al. 2009); *F. semitectum* (Basavaraja et al. 2007); *Trichoderma asperellum* (Mukherjee et al. 2008); *A. flavus* (Jain et al. 2011); *A. niger* (Gade et al. 2008); *Phoma glomerata* (Birla et al. 2009); *A. clavatus* (Verma et al. 2010); *Aspergillus* sp. (Pavani et al. 2012); *Trichoderma viride* (Fayaz et al. 2009); *Pestalotia* sp. (Raheman et al. 2011); *A. terreus* (Li et al. 2012a, b); *Trichophyton rubrum*, *Trichophyton mentagrophytes*, *Microsporum canis*, etc. (Moazeni et al. 2011); *Helminthosporium tetramera* (Shelar and Chavan 2014); and *P. gardeniae* (Rai et al. 2015).

In general, multiple stressful factors exist in saline soils and saline-alkaline soils that mostly involve heavy metal salts. Even if there is no heavy metal salt, other stressors such as sodium salt toxicity, lack of organic nutrition, and drought will seriously affect the survival of organisms in saline environments, particularly in soda-affected soil, which is more serious than any other saline soils and leads to many negative effects on soil organic matter decomposition, acid-base imbalance, and uptake of available nutrients (Rietz and Haynes 2003; Karlen et al. 2008). Therefore, in order to survive, soil organisms must endure or resist much more stress factors beyond heavy metal ions. The synthesis of a kind of integrated nanoparticles induced by soil structure and composition occurs, which involves not only salts or other soil components but also metabolites such as the proteins, enzymes, polypeptides, organic acids, or other metabolites secreted from soil-related microbial cells. In order to make it easy to express and remember, here we term MIMI nanoparticles as the microbial metabolites-integrated nanoparticles (Fig. 14.2).

As mentioned above, microbes are able to secrete considerable amounts of proteins, peptides, organic acids, and any other organic metabolites, which might directly or indirectly affect the formation and/or substantial mass productivity of metal or nonmetal nanoparticles. Both nitrate reductase and sulfite reductase are the main enzymes that directly impact on the metal nanoparticles through reducing metal ions (Kumar et al. 2007b; Gholami-Shabani et al. 2014), which will be presented in the next part. However, in saline-alkaline soils, the hydrolytic enzymes and organic acids that are beneficial to the soil health are important for the synthesis of MIMI nanoparticles (Fig. 14.2).

Hydrolytic enzymes involve cellulose, hemicellulose, lignin, and pectin, which can degrade plant organic matter (e.g., maize, wheat, or rice straw) (Castillo and Demoulin 1997; Santos et al. 2004; Arakaki et al. 2013; Batista-García et al. 2014; Li et al. 2018; Wei and Zhang 2018). The enzymatic hydrolysis of cellulose, particularly hydrogen-bonded and ordered crystalline regions, is very complex. The hydrolytic system generally includes endoglucanases (EC 3.2.1.4), exoglucanases (EC 3.2.1.91), and β-glucosidases (EC 3.2.1.21). Endoglucanases randomly attack the internal chain of cellulose to produce cello-oligosaccharides. Exoglucanases catalyze the hydrolysis of crystalline cellulose from the ends of the cellulose chain to produce cellobiose, which is ultimately hydrolyzed to glucose by β-glucosidases (Béguin and Aubert 1994; Tomme et al. 1995). The enzymatic hydrolysis process of cellulose is very slow. Multiple scales of products of enzymatic hydrolysis including nanocellulose are going to be formed during the microbe- cellulose interactions (Fig. 14.2).

*Trichoderma reesei* and *Penicillium janthinellum* are known to be excellent cellulase producers, but their cellulases are not stable under alkali conditions (Mernitz et al. 1996; Wang et al. 2005; Qin et al. 2008). *Aspergillus niger*, one of the most efficient identified cellulose-degrading microorganisms, secretes large amounts of different cellulases during fermentation (Schuster et al. 2002). Endoglucanase B (EGLB), encoded by the endoglucanase gene (GenBank GQ292753) of *Aspergillus niger* BCRC31494, has been used in the fermentation industry because of its alkaline and thermal tolerance (Li et al. 2012a). EGLB is a member of glycosyl hydrolase family 5 of the cellulase superfamily. When the recombinant EGLB cDNA was expressed in *Pichia pastoris*, a purified protein of 51 kDa in size was obtained. The enzyme was specific for substrates with β-1,3 and β-1,4 linkages, and it exhibited optimal activity at 70 °C and pH 4 (Li et al. 2012a). Interestingly, the relative activity of recombinant EGLB at pH 9 was significantly better than that of wild-type EGLB. The advantages of endoglucanase *EGLB*, particularly its tolerance to a broad range of pH values, indicate that this enzyme has significant promise as a means of genetically improving fungi for haloalkaline soil remediation.

The soft-rot ascomycete fungus *T. reesei* is utilized for industrial production of secreted enzymes, especially lignocellulose-degrading enzymes. *T. reesei* uses several different enzymes for the degradation of plant cell wall-derived material, including 9 characterized cellulases, 15 characterized hemicellulases, and at least 42 genes predicted to encode cellulolytic or hemicellulolytic activities (Häkkinen et al. 2014). The family 7 cellobiohydrolase (Cel7A) of *T. reesei* is comprised of a

36 amino acid CBM, a linker domain with O-glycan, and a large catalytic domain with N-linked glycan and a 50 Å tunnel for processing cellulose chains. The possibility of controlled hydrolysis of microcrystalline cellulose by *T. reesei* has been analyzed. The penetration of fungus into the ordered regions of microcrystalline cellulose during incubation resulted in reduced crystallinity of nanocellulose prepared by microbial hydrolysis compared to that of acid hydrolysis (Satyamurthy et al. 2011). However, in comparison with the fungal hydrolysis system, the anaerobic bacteria consortium is much more efficient in hydrolyzing microcrystalline cellulose to produce nanocellulose in a span of 7 days with a maximum yield of 12.3%; and nanocellulose prepared by this process has a bimodal particle size distribution (43 ± 13 and 119 ± 9 nm) (Satyamurthy and Vigneshwaran 2013). Thus, more efficient fungal hydrolyzing system must be strengthened.

Based on an analysis of the genomic sequence of haloalkaliphilic fungus *A. glaucus* CCHA, we found that *A. glaucus* CCHA expresses only one gene belonging to the GH5 family, AgCel5A. The open reading frame of *Agcel5A* consists of 1509 base pairs that encode a polypeptide of 502 amino acids. AgCel5A has four potential N-glycosylation sites and three O-glycosylation sites, which indicate high similarity to the characterized GH5 β-glucosidases from *Aspergillus niger* (65%) and *Trichoderma reesei* (31%). AgCel5A was cloned and heterologously expressed in *Pichia pastoris* GS115. Recombinant AgCel5A exhibited maximal activity at pH 5.0. AgCel5A is much more stable than PdCel5C from *Penicillium decumbens* (Liu et al. 2013); it retains more than 70% of its maximum activity at pH 8.0–10.0. In addition, AgCel5A exhibited stable degradation activity under high-salt (NaCl) conditions. In the presence of 4 M NaCl, AgCel5A retained 90% activity even after 4 h of preincubation. Interestingly, the activity of AgCel5A increased as the NaCl concentration was increased. The high resistance of AgCel5A to saline and alkaline conditions suggests that the *AgCel5A* gene is an ideal candidate for genetic improvement of soil fungi and industrial applications (Zhang et al. 2016; Li et al. 2018). As this fungus is being exploited for saline-alkaline soil remediation, nanocellulose production using this fungus is being analyzed in our lab. In a preliminary simulation test, multi-scale and multi-type nanoparticles were detected when *A. glaucus* CCHA and its fermentation filtrate were added in the saline soil mixed with physically grated corn stalks (particle size 0.5–2 mm) (Fig. 14.3). The complex nanoparticles should be MIMI nanoparticles as described above.

Fungi, like *A. glaucus* CCHA, have the ability to secrete considerable amounts of organic acids. Furthermore, with the increase of salinity and pH value, the organic acids secreted from fungi gradually increase (Wei and Zhang 2018). Several organic acids, such as gallic acid, gluconic acid, citric acid, itaconic acid, kojic acid, and malic acid, have been detected in the fermentation filtrate of *A. glaucus* CCHA while treated with 5% $NaHCO_3$. Interestingly, molecular cross-linking can occur within gallic acids or gluconic acids under high-salt and pH conditions (Ohno et al. 2001; Sanae et al. 2003). Both gallic and gluconic acids are efficient metal ion-masking agents, because the cross-linked organic acids tend to be further aggregated into nanoparticles (Fig. 14.3). We convince that the cross-linked organic acid-assisted nanoparticles contribute to the synthesis of MIMI nanoparticles, despite the lack of direct evidence currently (Fig. 14.2).

## 14.4 Roles of Fungi and Nanoparticles in Soil Mycoremediation and Health

The biotic synthesized nanoparticles are metal related, and these metals at mostly involve silver, gold, cadmium sulfite, copper oxide, platinum, and zinc oxide. Some filamentous fungi secreted a series of enzymes under stress circumstances. Metal-associated enzymes, reducing cofactors, and organic materials have significant roles in reducing agents, which help in providing natural capping to synthesize nanoparticles, thereby preventing the aggregation of nanoparticles and helping them to remain stable for a long time, thus providing additional stability (Alani et al. 2012; Birla et al. 2009; Kumar et al. 2007c; Narayanan and Sakthivel 2010; Aziz et al. 2016). The mechanism of extracellular production of metal nanoparticles by fungi is mostly found to be involving the action of oxidoreductases. These enzymes could reduce the toxicity in the method of metal nanoparticle synthesis by reduction of the metal ions (Gholami-Shabani et al. 2013, 2014; Huang et al. 2015). Among the oxidoreductases, both nitrate reductase and sulfite reductase are intensively studied over the past decades. A number of researchers supported oxidoreductase for cell-free synthesis of metal nanoparticles (Kumar et al. 2007b; Gholami-Shabani et al. 2013, 2014; Prasad et al. 2014, 2016). The nitrate reductase secreted from the fungi is responsible for the reduction of metal ions and the following synthesis of metal nanoparticles. When nitrate reductase is used, the color of the mixture turned reddish from white when tested with fungal filtrate demonstrating the existence of nitrate reductase.

The beneficial effect of microbial application on saline-alkaline soil has been reported by Sahin et al. (2011). In the study, suspensions of three fungal isolates (*Aspergillus* spp. FS 9, 11, and *Alternaria* spp. FS 8) and two bacterial strains (*Bacillus subtilis* OSU 142 and *Bacillus megaterium* M3) at $10^4$ spore/ml and $10^9$ CFU/ml, respectively, were mixed with leaching water and applied to the soil columns in the Igdir Plain of northeastern Turkey (Sahin et al. 2011). Gypsum is an economical alternative for replacing sodium with calcium in remediating saline-alkaline soils (Gharaibeh et al. 2009; Oad et al. 2002). In the experimental process, gypsum was applied for the saline-alkaline soil pretreatment, and the microorganisms are not halotolerant or halophilic (Aslantas et al. 2007; Turan et al. 2006); thus, the final results they obtained should not just be out of the function of microbes. Anyway, this study gives us an enlighten example for mycoremediation of saline-alkaline soil by using haloalkaliphilic fungi.

Organisms at simultaneous high-salt concentration and high pH value require special adaptive mechanisms, which during the course of evolution would be both facilitative and essential for life-supporting processes. Few researches focus on how haloalkaliphilic fungi cope with extremes of salt and pH value. We assume that haloalkaliphilic fungi adopt comprehensive strategies to survive the extreme environment; in other words, under saline-alkali conditions, soil fungi must possess certain mechanisms to alleviate the influence or damage of both salt and alkali. In terms of soil effects, only reducing soil soluble salt and regulating the pH value of soil solution can achieve the purpose of restoring saline-alkaline soils.

1. Soil fungi have the ability to accumulate cation contents in cells. Saline soil will be improved with the accumulation of salt cation contents into fungal cells. *Hortaea werneckii*, the black yeast-like fungus isolated from hypersaline waters of salterns as their natural ecological niche, has been previously defined as halophilic fungus (Butinar et al. 2005a, b). *H. werneckii* cells were grown in liquid media at different salinities, ranging from 0% to 25% NaCl. The measurements of cation contents in cells grown at constant salt concentration have shown that the amounts of $K^+$ and $Na^+$ in *H. werneckii* were changing according to the NaCl concentration of the medium. When *H. werneckii* was grown in a medium without added NaCl, it accumulated a very low amount of $Na^+$. But with the increasing NaCl concentration of the medium, the amounts of the $Na^+$ content increased and in the end reached a higher value (Kogej et al. 2005).
2. Soil fungi produce different organic acid patterns (Scervino et al. 2010). The released organic acids allow the formation of organic-mineral complexes (Richardson et al. 2001); on the other hand, with the release of organic acids, protons are produced that contribute to the acidification of the alkaline soil solution.

The saline-alkaline soils and most cultivated soils are deficient in available forms of phosphorus. The release of these organic acids and other compounds in the rhizosphere by these microorganisms may be important in the solubilization of various inorganic phosphorus compounds (Scervino et al. 2010). In spite of this, based on the principle of acid-base neutralization, the organic acids also adjust the pH value of soil solution to a lower level.

The reactions of the citric acid cycle are carried out by eight enzymes that completely oxidize acetate, in the form of acetyl-CoA, into two molecules each of carbon dioxide and water. Organic acid citrate, isocitrate, succinate, fumarate, malate, and oxaloacetate are produced during each turn of the cycle. The high pH tolerance of *A. glaucus* has led to its utilization as an organic production strain (Barnes and Weitzman 1986). When *A. glaucus* CCHA, *A. terreus* S108, and *A. niger* S211were cultured in an alkaline medium, key enzymes (e.g., citrate synthase, isocitrate dehydrogenase, succinyl-CoA synthetase, malate dehydrogenase) of the citric acid cycle were significantly upregulated, suggesting that these genes contribute to the high pH tolerance of *A. glaucus* (Wei et al. 2013; Liu 2014; Zhou 2016; Wei and Zhang 2018). Accordingly, alkali resistance might be improved in all these saline-alkaline resistance fungi by overexpressing enzymes involved in the citric acid cycle. Actually, a series of the corresponding organic acids like citric acid, itaconic acid, and malic acid have been detected; in addition, both gallic acid and gluconic acid, which are associated with MIMI nanoparticles, have been investigated as well.

3. The roles played by most of soil fungi (non-mycorrhizal fungi) at the cellular level to tolerate soil salt ions are probably similar to some of the strategies employed by ectomycorrhizal fungus, namely, binding to extracellular materials (Tam 1995; Aggangan et al. 2010; Gomes et al. 2018) or sequestration in the vacuolar compartment (Blaudez et al. 2000). We clearly observed that *A. glaucus* CCHA mycelia covered around the roots of rice like a net; but the mycelia did not penetrate and extend to the rice cortex. We think this should also be an effective protection to the plants living under saline soil conditions.
4. Halophilic and alkaliphilic fungi are of biotechnological interest, as they produce extremozymes, which are useful in medical and environmental field because of their ability to remain active under the severe saline and alkaline conditions (Tiquia-Arashiro and Rodrigues 2016). The enzymes secreted by haloalkaliphilic fungi possess the bioreduction effect on salt ions of soil. This bioreduction of metal particles by certain biomasses is regarded as an organism's survival mechanism against toxic metal ions and occurs via an active or passive process or a combination of both (Ibrahim et al. 2001; Durán et al. 2005). Correlation between soil properties and soil enzymes from fungi or other microorganisms has been substantiated; and these enzyme activities are now widely used as important indicators of soil quality and soil biological activities (Rietz and Haynes 2003). As described above, high-salt and pH value induce the secretion of organic acids. Similar to this case, cellulases and other so-called soil enzymes are also induced with the increasing salt concentration and pH value. When hydrolytic enzymes are secreted into soil solution, soil properties will be improved accordingly. Take cellulases, for instance; on one hand, soil cellulases can enhance the organic matters by degrading cellulose; on the other hand, cellulases in salt soils or salt solutions have been detected to form biotical nanoparticles (Riddin et al. 2006; Tiquia-Arashiro and Rodrigues 2016; Mohite et al. 2017). Recently, nanoparticles of varying size (10–300 nm) and shape (hexagons, pentagons, circles, squares, rectangles) were produced at extracellular levels by *A. glaucus* CCHA in our lab (unpublished data), further indicating that the formation of nanoparticles by haloalkaliphilic fungus is associated with saline-alkaline soil remediation.

Overall, soil fungi buffer salinity and alkalinity by absorbing and/or constraining salt ions, secreting organic acids and/or macromolecule degradation enzymes, generating nanoparticles, and providing biomass; all of these effects of fungi reduce plant stress (Fig. 14.3). Therefore, haloalkaliphilic fungi are excellent biological resources for soil mycoremediation.

## 14.5 Remarks and Prospects

Saline soil remediation by using salt-tolerant or salt-resistant plants is one of the most effective methods because of its significant ecological, environmental, and economic effects (Ilyas et al. 1993; Ghaly 2002; Nouri et al. 2017), but the method requires persistent management to produce meaningful changes in soil characteristics. Application of organic matter conditioners, which can ameliorate and increase the fertility of saline soils, is an alternative soil remediation technique (Wang et al. 2014).

Fertile soil is a vital complex that involves numerous species and immense biomass; soil organisms, particularly soil fungi, have significant effects on the soil ecosystem. The biosynthesis of fungal nanoparticles is beneficial to saline soil's remediation and health; and microbe nanoparticles-integrated technology (MiNIT) is easier, low cost, nontoxic, green, and organic for saline soil remediation.

Soil-inhabitant fungi build a metabolic bridge between insoluble organic matter and soil nutrients by producing cellulose degradation enzymes such as cellulase, as well as performing other biological processes. However, saline-alkaline soils generally lack fungi, which ordinarily play important roles in degrading insoluble organic matter such as crop straw into soluble and easily absorbed nutrients; therefore, applying organic matter supplemented with fermentation fungi to saline-alkaline soil is a feasible strategy for soil remediation.

Haloalkaliphilic fungi are excellent biological candidates for soil mycoremediation, but to date very few species with both abilities to produce effective soil enzymes and to grow in saline-alkaline environments have been reported. Our lab currently isolated a series of halotolerant/halophilic fungi. Microbe isolation and identification are time-consuming and tedious tasks, and their applications require more laborious field experiments. Based on the relationship between the remediation effect of saline-alkaline soil and the production capacity of nanoparticles, and considering the fact that nitrate reductase and sulfite reductase are main enzymes for synthesizing nanoparticles (Asmathunisha and Kathiresan 2013), we optimized a simple and quick method for identifying saline remediation fungi. Generally, with the increase of NaCl concentration, the upregulation level and efficiency of the nitrate reductase and sulfite reductase genes as well increase, which are positively correlated with the yield of nanoparticles (Fig. 14.4). Through designing degenerate primers of the both genes and real-time PCR method, we can obtain the target fungal isolates.

On the other hand, to get better remediation effect, natural soil fungi require to be genetically modified at their degradation ability or saline-alkaline resistance. Generally, with the several enzymes involved in salt and/or alkali resistance, such as the alkaline-stable endoglucanases B from *Aspergillus niger* BCRC31494 (Li et al. 2012a, b), the alkaline xylanase from *Aspergillus nidulans* KK–99 (Taneja et al. 2002), and the bilirubin oxidase from *Myrothecium* sp. IMER1 (Zhang et al. 2007), are highly abundant in fungi found in saline-alkaline soil, but such fungi usually have a relatively low capacity for cellulose degradation, whereas fungi found in

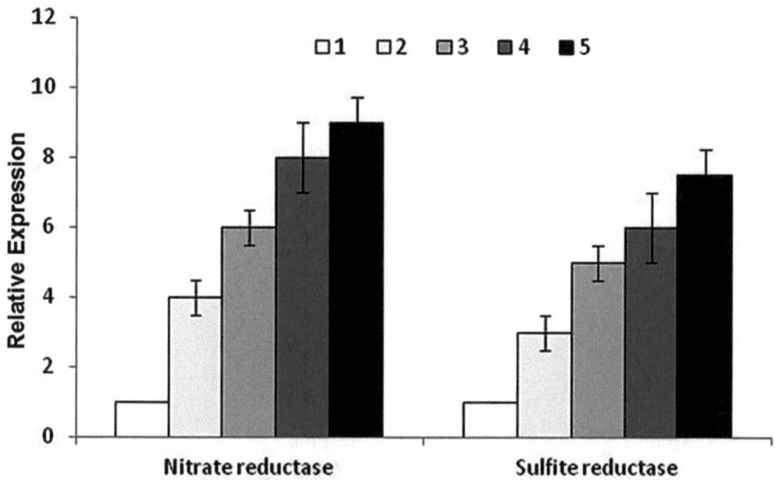


**Fig. 14.4** Gene expression of the nitrate reductase and sulfite reductase. The expression of nitrate reductase and sulfite reductase genes at different concentrations of NaCl in *A. glaucus* CCHA was analyzed using aRT-PCR. The CCHA strains were cultured in PD liquid medium (1~5% NaCl), and pH value was adjusted to 8, respectively; all cultures were performed at a temperature of 35 °C for 3 days, and then mRNA were extracted for qRT-PCR; the nitrate reductase and sulfite reductase genes were detected through qRT-PCR. Each gene was deduced according to NRRL3_11178 and NRRL3_09602 of *Aspergillus niger* NRRL3 (https://blast.ncbi.nlm.nih.gov/Blast) (The unpublished result was provided by the author Yi Wei)

fertile soil show opposite characteristics. Thus, genetically modified strategy to create novel haloalkaliphilic fungi with high cellulase activity is a good choice. Fungi endowed with high resistance to saline-alkaline environments or other beneficial genes would be promising candidates for saline-alkaline soil remediation.

A series of salt and/or alkali resistance (or tolerance) genes have been characterized to provide a list of candidate genes to be applied in efforts to genetically improve soil fungi (Fang et al. 2014; Zhang 2016). In order to enhance the cellulose degradation ability of haloalkaliphilic fungi, additional cellulases with salt and alkali stability must be identified. Using cellulases with salt and alkali tolerance, two strategies can be employed to obtain saline-/alkaline-resistant fungi with enhanced enzyme secretion. Indeed, natural strains remain the first choice for soil remediation; therefore, isolating and screening suitable strains from extreme natural environments are still an important long-term task.

Haloalkaliphilic fungus *Aspergillus glaucus* CCHA, a fungal species with extreme tolerance to saline and alkaline conditions, has significant potential value in industrial and agricultural applications. Our group has been assessing the potential of *Aspergillus glaucus* CCHA in the mycoremediation of saline-alkaline soil in the Songnen Plain of Northeast China (one of the three most famous saline and alkaline lands in the world) for 3 years (Shi and Zhu 2016). This study primarily indicates that the applied amendments mixed with haloalkaliphilic fungi significantly encourage steady growth and yield of rice in comparison with that achieved in the control plot.

**Acknowledgments** The related work in our lab was partially supported by grants from the National Natural Science Foundation of China (grant nos. 31671972 and 31670141) and a project of the Ministry of Science and Technology of China (grant no. 2016YFD0300703). The authors would like to thank the Zhang Lab members, who provided the photographic pictures taken at their spare time. The authors are also grateful to former labmates Dr. Zheng-Qun LI, Dr. Yang SHI, Mr. Sen-Lin ZHANG, and Zhi-Yuan GONG (visiting research fellow from the Northeast Forestry University, Harbin), who contributed to fungal isolation and field trials, as well as to collaborators Zhen-Dong CHEN and Run-Zhi TAO who provided encouragement and assistance in promoting our scientific and technological achievements regarding saline-alkaline soil mycoremediation using haloalkaliphilic fungi.

## References

Aggangan NS, Moon HK, Han SH (2010) Growth response of *Acacia mangium* wild seedlings to arbuscular mycorrhizal fungi and four isolates of the ectomycorrhizal fungus *Pisolithus tinctorius* (Pers.) Coker and Couch. New For 39:215–230

Ahmad A, Senapati S, Khan MI, Kumar R, Sastry M (2005) Extra−/intracellular biosynthesis of gold nanoparticles by an alkalitolerant fungus *Trichothecium* sp. J Biomed Nanotechnol 1:47–53

Alani F, Moo-Young M, Anderson W (2012) Biosynthesis of silver nanoparticles by a new strain of *Streptomyces sp.* compared with *Aspergillus fumigatus*. World J Microbiol Biotechnol 28:1081–1086

Antony CP, Kumaresan D, Hunger S, Drake HL, Murrell JC, Shouche YS (2013) Microbiology of Lonar Lake and other soda lakes. ISME J 7:468–476. https://doi.org/10.1038/ismej.2012.137

Apte SK, Thomas J (1997) Possible amelioration of coastal soil salinity using halotolerant nitrogen-fixing cyanobacteria. Plant Soil 189:205–211

Arakaki R, Monteiro D, Boscolo R, Gomes E (2013) Halotolerance, ligninase production and herbicide degradation ability of basidiomycetes strains. Braz J Microbiol 44:1207–1214

Aslantas R, Cakmakci R, Sahin F (2007) Effect of plant growth promoting rhizobacteria on young apple tree growth and fruit yield under orchard conditions. Sci Hortic 111:371–377

Asmathunisha N, Kathiresan K (2013) A review on biosynthesis of nanoparticles by marine organisms. Coll Surf B Biointerfaces 103:283–287. https://doi.org/10.1016/j.colsurfb.2012.10.030

Aziz N, Pandey R, Barman I, Prasad R (2016) Leveraging the attributes of *Mucor hiemalis*-derived silver nanoparticles for a synergistic broad-spectrum antimicrobial platform. Front Microbiol 7:1984. https://doi.org/10.3389/fmicb.2016.01984

Barnes SJ, Weitzman PD (1986) Organization of citric acid cycle enzymes into a multienzyme cluster. FEBS0 Lett 201:267–270

Basavaraja S, Balaji SD, Lagashetty A, Rajasab AH, Venkataraman A (2007) Extracellular biosynthesis of silver nanoparticles using the fungus Fusarium *semitectum*. Mater Res Bull 43:1164–1170

Batista-García RA, Balcázar-López E, Miranda-Miranda E, Sánchez-Reyes A, Cuervo-Soto L, Aceves-Zamudio D, Atriztán-Hernández K, Morales-Herrera C, Rodríguez-Hernández R, Folch-Mallol J (2014) Characterization of lignocellulolytic activities from a moderate halophile strain of *Aspergillus caesiellus* isolated from a sugarcane bagasse fermentation. PLoS One 9(8):e105893. https://doi.org/10.1371/journal.pone.0105893

Béguin P, Aubert JP (1994) The biological degradation of cellulose. FEMS Microbiol Rev 13:25–58

Birla SS, Tiwari VV, Gade AK, Ingle AP, Yadav AP, Rai MK (2009) Fabrication of silver nanoparticles by *Phoma glomerata* and its combined effect against *Escherichia coli, Pseudomonas aeruginosa* and *Staphylococcus aureus*. Lett Appl Microbiol 48:173–179

Blaudez D, Botton B, Chalot M (2000) Cadmium uptake and subcellular compartmentation in the ectomycorrhizal fungus *Paxillus involutus*. Microbiology 146:1109–1117

Bruggen AHCV, Semenov AM, Zeiss MR (2000) In search of biological indicators for soil health and disease suppression. Appl Soil Ecol 15:13–24

Buchalo AS, Nevo E, Wasser SP, Oren A, Molitoris HP (1998) Fungal life in the extremely hypersaline water of the Dead Sea: first records. Proc R Soc Lond B 265:1461–1465

Buchalo AS, Nevo E, Wasser SP, Oren A, Molitoris HP, Volz PA (2000) Fungi discovered in the Dead Sea. Mycol Res News 104:132–133

Butinar L, Sonjak S, Zalar P, Plemenitaš A, Gunde-Cimerman N (2005a) Melanized halophilic fungi are eukaryotic members of microbial communities in hypersaline waters of solar salterns. Bot Mar 48:73–79

Butinar L, Zalar P, Frisvad JC, Gunde-Cimerman N (2005b) The genus Eurotium–members of indigenous fungal community in hypersaline waters of salterns. FEMS Microbiol Ecol 51:155–166

Casadevall A (2012) Fungi and the rise of mammals. PLoS Pathog 8(8):e1002808. https://doi.org/10.1371/journal

Casamayor EO, Massana R, Benlloch S, Øvreas L, Diez B, Goddard VJ, Gasol JM, Joint I, Rodríguez-Valera F, Pedrós-Alió C (2002) Changes in archaeal, bacterial and eukaryal assemblages along a salinity gradient by comparison of genetic fingerprinting methods in a multipond solar saltern. Environ Microbiol 4:338–348

Castillo G, Demoulin V (1997) NaCl salinity and temperature effects on growth of three wood–rotting basidiomycetes from a Papua New Guinea coastal forest. Mycol Res 101:341–344

Chen JC, Lin ZH, Ma XX (2003) Evidence of the production of silver nanoparticles via pretreatment of Phoma sp. 32883 with silver nitrate. Lett Appl Microbiol 37:105–108

Dendooven L, Alcántara-Hernández RJ, Valenzuela-Encinas C, Luna-Guido ML, Perez-Guevara F, Marsch R (2010) Dynamics of carbon and nitrogen in an extreme alkaline saline soil: a review. Soil Biol Biochem 42:865–877

Devi LS, Joshi SR (2015) Ultrastructures of silver nanoparticles biosynthesized using endophytic fungi. J Microsc Ultrastruct 3:29–37

Dhanasekar NN, Rahul G, Narayanan KB, Raman G, Sakthivel N (2015) Green chemistry approach for the synthesis of gold nanoparticles using the fungus *Alternaria* sp. J Microbiol Biotechnol. https://doi.org/10.4014/jmb.1410.10036

Durán N, Marcato PD, Alves OL, Souza GI, Esposito E (2005) Mechanistic aspects of biosynthesis of silver nanoparticles by several *Fusarium oxysporum* strains. J Nanobiotechnol 3:1–8. https://doi.org/10.1186/1477-3155-3-8

Elmeleigy MA, Hoseiny EN, Ahmed SA, Alhoseiny AM (2010) Isolation, identification, morphogenesis and ultrastructure of obligate halophilic fungi. J Appl Sci Environ Sanit 5:201–202

Evans S, Hansen RW, Schneegurt MA (2013) Isolation and characterization of halotolerant soil fungi from the great salt plains of Oklahoma. Cryptogam Mycol 34:329–341. https://doi.org/10.7872/crym.v34.iss4.2013.329

Fang J, Han X, Xie L, Liu M, Qiao G, Jiang J, Zhuo R (2014) Isolation of salt stress–related genes from *Aspergillus glaucus* CCHA by random overexpression in *Escherichia coli*. Sci World J 39:620959

Fayaz AM, Balaji K, Girilal M, Yadav R, Kalaichelvan PT, Venketesan R (2009) Biogenic synthesis of silver nanoparticles and their synergistic effect with antibiotics: a study against gram-positive and gram-negative bacteria. Nanomedicine 6:103–109

Ford GW, Martin JJ, Rengasamy P, Boucher SC, Ellington A (1993) Soil sodicity in Victoria. Aust J Soil Res 31:869–909

Gade AK, Bonde P, Ingle AP, Marcato PD, Durán N, Rai MK (2008) Exploitation of *Aspergillus* niger for synthesis of silver nanoparticles. J Biobased Mater Bioenerg 2:243–247

Gao Q, Yang XS, Yun R, Li CP (1996) MAGE, a dynamic model of alkaline grassland ecosystems with variable soil characteristics. Ecol Model 93:19–32

Gericke M, Pinches A (2006) Microbial production of gold nanoparticles. Gold Bull 39:22–28
Ghaly FM (2002) Role of natural vegetation in improving salt affected soil in northern Egypt. Soil Tillage Res 64:173–178
Gharaibeh MA, Eltaif NI, Shunnar OF (2009) Leaching and reclamation of calcareous saline–sodic soil by moderately saline and moderate-SAR water using gypsum and calcium chloride. J Plant Nutr Soil Sci 172:713–719
Ghassemi F, Jakeman AJ, Nix HA (1995) Salinisation of land and water resources: human causes, extent, management and case studies. CABI Publishing, Wallingford
Gholami-Shabani M, Akbarzadeh A, Mortazavi M, Emzadeh MK (2013) Evaluation of the antibacterial properties of silver nanoparticles synthesized with *Fusarium oxysporum* and *Escherichia coli*. Int J Life Sci Biotechnol Pharma Res 2:333–348
Gholami-Shabani M, Akbarzadeh A, Norouzian D, Amini A, Gholami-Shabani Z, Imani A, Chiani M, Riazi G, Shams-Ghahfarokhi M, Razzaghi-Abyaneh M (2014) Antimicrobial activity and physical characterization of silver nanoparticles green synthesized using nitrate reductase from *Fusarium oxysporum*. Appl Biochem Biotechnol 172:4084–4098. https://doi.org/10.1007/s12010-014-0809-2
Gomes ECQ, Godinho VM, Silva DAS, de Paula MTR, Vitoreli GA, Zani CL, Alves TMA, Junior PAS, Murta SMF, Barbosa EC, Oliveira JG, Oliveira FS, Carvalho CR, Ferreira MC, Rosa CA, Rosa LH (2018) Cultivable fungi present in Antarctic soils: taxonomy, phylogeny, diversity, and bioprospecting of antiparasitic and herbicidal metabolites. Extremophiles. https://doi.org/10.1007/s00792-018-1003-1
Gonçalves VN, Vitoreli GA, de Menezes GCA, Mendes CRB, Secchi ER, Rosa CA, Rosa LH (2017) Taxonomy, phylogeny and ecology of cultivable fungi present in seawater gradients across the Northern Antarctica Peninsula. Extremophiles 21:1005. https://doi.org/10.1007/s00792-017-0959-6
Gostinčar C, Turk M (2012) Extremotolerant fungi as genetic resources for biotechnology. Bioengineered 3:293–297
Gostinčar C, Grube M, De Hoog S, Zalar P, Gunde-Cimerman N (2010) Extremotolerance in fungi: evolution on the edge. FEMS Microbiol Ecol 71:2–11
Grum-Grzhimaylo AA, Georgieva ML, Bondarenko SA, Debets AJM, Bilanenko EN (2016) On the diversity of fungi from soda soils. Fungal Divers 76:27–74
Gunde-Cimerman N, Zalar P (2014) Extremely halotolerant and halophilic fungi inhabit brine in solar salterns around the globe. Food Technol Biotechnol 52:170–179
Gunde-Cimerman N, Zalar P, de Hoog GS, Plemenitaš A (2000) Hypersaline waters in salterns: natural ecological niches for halophilic black yeasts. FEMS Microbiol Ecol 32:235–240
Häkkinen M, Valkonen MJ, Westerholm-Parvinen A, Aro N, Arvas M, Vitikainen M, Penttilä M, Saloheimo M, Pakula TM (2014) Screening of candidate regulators for cellulase and hemicellulose production in *Trichoderma reesei* and identification of a factor essential for cellulase production. Biotechnol Biofuels 7(14):1–21
Harner T, Wideman JL, Jantunen LM, Bidleman TF, Parkhurst WJ (1999) Residues of organochlorine pesticides in Alabama soils. Environ Pollut 106(3):323–332
Heckman DS, Geiser DM, Eidell BR, Stauffer RL, Kardos NL, Hedges SB (2001) Molecular evidence for the early colonization of land by fungi and plants. Science 293:1129–1133
Horikosh K (1999) Alkaliphiles: some applications of their products for biotechnology. Microbiol Mol Biol Rev 63:735–750
Horodyski RJ, Knauth PL (1994) Life on land in the Precambrian. Science 263(5146):494–498
Hozzein WN, Ali MIA, Ahmed MS (2013) Antimicrobial activities of some alkaliphilic and alkaline-resistant microorganisms isolated from Wadi Araba, the eastern desert of Egypt. Life Sci J 10:1823–1828
Huang J, Lin L, Sun D, Chen H, Yang D, Li Q (2015) Bio-inspired synthesis of metal nanoparticles and application. Chem Soc Rev 44:6330–6374
Ibrahim Z, Ahmad A, Baba B (2001) Bioaccumulation of silver and the isolation of metal-binding protein from Pseudomonas diminuta. Braz Arch Biol Tech 44:223–225

Ilyas M, Miller RW, Qureshi RH (1993) Hydraulic conductivity of saline-sodic soil after gypsum application and cropping. Soil Sci Soc Am J 57:1580–1585

Ingle A, Gade A, Pierrat S, Sonnichsen C, Rai M (2008) Mycosynthesis of silver nanoparticles using the fungus *Fusarium acuminatum* and its activity against some human pathogenic bacteria. Curr Nanosci 4:141–144

Ingle A, Rai MK, Gade A, Bawaskar M (2009) *Fusarium solani*: a novel biological agent for the extracellular synthesis of silver nanoparticles. J Nanopart Res 11:2079–2085

Jain N, Bhargava A, Majumdar S, Tarafdar JC, Panwar J (2011) Extracellular biosynthesis and characterization of silver nanoparticles using *Aspergillus flavus* NJP08: a mechanism perspective. Nanoscale 3:635–641

Karlen DL, Andrews SS, Wienhold BJ, Zobeck TM (2008) Soil quality assessment: past, present and future. J Integr Biosci 6:3–14

Kis-Papo T, Grishkan I, Oren A, Wasser SP, Nevo E (2003) Survival of filamentous fungi in hypersaline Dead Sea water. Microb Ecol 45:183–190

Kis-Papo T, Weig AR, Riley R, Peršoh D, Salamov A, Sun H, Lipzen A, Wasser SP, Rambold G, Grigoriev IV, Nevo E (2014) Genomic adaptations of the halophilic Dead Sea filamentous fungus *Eurotium rubrum*. Nat Commun 5:3745. https://doi.org/10.1038/ncomms4745

Kogej T, Ramos J, Plemenitas A, Gunde-Cimerman N (2005) The halophilic fungus *Hortaea werneckii* and the halotolerant fungus *Aureobasidium pullulans* maintain low intracellular cation concentrations in hypersaline environments. Appl Environ Microbiol 71:6600–6605

Kowshik M, Ashtaputre S, Kharrazi S, Vogel W, Urban J, Kulkarni SK, Paknikar M (2003) Extracellular synthesis of silver nanoparticles by a silver-tolerant yeast strain MKY3. Nanotechnology 14:95–100

Kumar AS, Ansari AA, Ahmad A, Khan MI (2007a) Extracellular biosynthesis of CdS quantum dots by the fungus *Fusarium oxysporum*. J Biomed Nanotechnol 3:190–194

Kumar SA, Abyaneh MK, Gosavi SW, Kulkarni SK, Ahmad A, Khan MI (2007b) Sulfite reductase mediated synthesis of gold nanoparticles capped with phytochelatin. Biotechnol Appl Biochem 47:191–195. https://doi.org/10.1042/BA20060205

Kumar SA, Abyaneh MK, Gosavi SW, Kulkarni SK, Pasricha R, Ahmad A, Khan MI (2007c) Nitrate reductase-mediated synthesis of silver nanoparticles from $AgNO_3$. Biotechnol Lett 29:439–445

Li CH, Wang HR, Yan TR (2012a) Cloning, purification, and characterization of a heat– and alkaline–stable endoglucanase B from *Aspergillus niger* BCRC31494. Molecules 17:9774–9789

Li G, He D, Qian Y, Guan B, Gao S, Cui Y, Yokoyama K, Wang L (2012b) Fungus mediated green synthesis of silver nanoparticles using *Aspergillus terreus*. Int J Mol Sci 13:466–476

Li ZQ, Pei X, Zhang ZY, Wei Y, Song YY, Chen LN, Liu SA, Zhang S-H (2018) The unique GH5 cellulase member in the extreme halotolerant fungus *Aspergillus glaucus* CCHA is an endoglucanase with multiple tolerances to salt, alkali and heat: prospects for straw degradation applications. Extremophiles. https://doi.org/10.1007/s00792-018-1028-5

Liang X, Liu Y, Xie L, Liu X, Wei Y, Zhou X, Zhang SH (2015) A ribosomal protein AgRPS3aE from halophilic *Aspergillus glaucus* confers salt tolerance in heterologous organisms. Int J Mol Sci 16:3058–3070

Liu XD (2014) Two stress tolerance genes in halophilic *Aspergillus*: functional analysis and their application. Dissertation, Jilin University (in Chinese). http://kns.cnki.net/KCMS/detail/detail.aspx?dbcode=CDFD&dbname=CDFDTEMP&filename=1015507655.nh

Liu CX, Huang WY (2010) Influence of saline-alkali -tolerant bacteria combined with organic matter on formation of saline-alkali soil aggregates. Soils 42(1):111–116 (in Chinese, abstract in English)

Liu XD, Liu JL, Wei Y, Tian YP, Fan FF, Pan HY, Zhang SH (2011) Isolation, identification and biologic characteristics of an extreme halotolerant *Aspergillus* sp. J Jilin Univ 49:548–552 (In Chinese; abstract in English)

Liu G, Qin Y, Hu Y, Gao M, Peng S, Qu Y (2013) An endo–1,4–β–glucanase PdCel5C from cellulolytic fungus *Penicillium decumbens* with distinctive domain composition and hydrolysis product profile. Enzym Microb Technol 52:190–195

Liu XD, Xie L, Wei Y, Zhou XY, Jia B, Liu J, Zhang SH (2014) Abiotic stress resistance, a novel moonlighting function of ribosomal protein RPL44 in the halophilic fungus *Aspergillus glaucus*. Appl Environ Microbiol 80:4294–4300

Liu XD, Wei Y, Zhou XY, Pei X, Zhang SH (2015) *Aspergillus glaucus* aquaglyceroporin gene glpF confers high osmosis tolerance in heterologous organisms. Appl Environ Microbiol 81:6926–6937

Mahdy HM, el-Sheikh HH, Ahmed MS, Refaat BM (1996) Physiological and biochemical changes induced by osmolarity in halotolerant aspergilli. Acta Microbiol Pol 45:55–65

Mandeel QA (2006) Biodiversity of the genus *Fusarium* in saline soil habitats. J Basic Microbiol 46:480–494

Melero S, Madejon E, Ruiz JC, Herencia JF (2007) Chemical and biochemical properties of a clay soil under dryland agriculture system as affected by organic fertilization. Eur J Agron 26:327–334

Mernitz G, Koch A, Henrissat B, Schulz G (1996) Endoglucanase II (EGII) of *Penicillium janthinellum*: cDNA sequence, heterologous expression and promoter analysis. Curr Genet 29:490–495

Moazeni M, Rashidi N, Shahverdi AR, Noorbakhsh F, Rezaie S (2011) Extracellular production of silver nanoparticles by using three common species of dermatophytes: *Trichophyton rubrum, Trichophyton mentagrophytes* and *Microsporum canis*. Iran Biomed J 16(1):332–333

Mohite P, Kumar AR, Zinjarde S (2017) Relationship between salt tolerance and nanoparticle synthesis by *Williopsis saturnus* NCIM 3298. World J Microbiol Biotechnol 33:163. https://doi.org/10.1007/s11274-017-2329-z

Moubasher A, Abdel-Hafez S, Bagy M, Abdel-Satar M (1990) Halophilic and halotolerant fungi in cultivated desert and salt marsh soils from Egypt. Acta Mycol 26:65–81

Mourato A, Gadanho M, Lino AR, Tenreiro R (2011) Biosynthesis of crystalline silver and gold nanoparticles by extremophilic yeasts. Bioinorg Chem Appl 2011:546074. https://doi.org/10.1155/2011/546074

Mukherjee P, Ahmad A, Mandal D, Senapati S, Sainkar SR, Khan MI, Ramani R, Parischa R, Ajayakumar PV, Alam M, Sastry M, Kumar R (2001) Bioreduction of $AuCl_4$ – ions by the fungus *Verticillium* sp. and surface trapping of the gold nanoparticles formed. Angew Chem Int Ed 40:3585–3588

Mukherjee P, Roy M, Mandal BP, Dey GK, Mukherjee PK, Ghatak J, Tyagi AK, Kale SP (2008) Green synthesis of highly stabilized nanocrystalline silver particles by a nonpathogenic and agriculturally important fungus *T. asperellum*. Nanotechnology 19:103–110

Mulder JL, El-Hendawy H (1999) Microfungi under stress in Kuwait's coastal saline depressions. Kuwait J Sci Eng 26:157–172

Mulder JL, Ghannoum MA, Khamis L, Elteen KA (1989) Growth and lipid composition of some dematiaceous hyphomycete fungi grown at different salinities. Microbiology 135:3393–3404

Namasivayam SKR, Ganesh S, Avimanyu S (2011) Evaluation of anti-bacterial activity of silver nanoparticles synthesized from *Candida glabrata* and *Fusarium oxysporum*. Int J Med Microbiol Res 1:130–136

Narayanan KB, Sakthivel N (2010) Biological synthesis of metal nanoparticles by microbes. Adv Colloid Interf Sci 156:1–13

Nazareth S, Gonsalves V (2014) *Aspergillus penicillioides* – a true halophile existing in hypersaline and polyhaline econiches. Ann Microbiol 64:397–402

Nazareth S, Gonsalves V, Nayak S (2012) A first record of obligate halophilic aspergilli from the Dead Sea. Indian J Microbiol 52:22–27

Nouri H, Borujeni CS, Nirola R, Hassanli A, Beecham S, Alaghmand S, Saint C, Mulcahy D (2017) Application of green remediation on soil salinity treatment: a review on halophytoremediation. Process Saf Environ 107:94–107

Oad FC, Samo MA, Soomro A, Oad DL, Oad NL, Siyal AG (2002) Amelioration of salt affected soils. Pak J Appl Sci 2:1–9

Ohno T, Inoue M, Ogihara Y (2001) Cytotoxic activity of gallic acid against liver metastasis of mastocytoma cells P-815. Anticancer Res 21(6A):3875–3880

Oo AN, Iwai CB, Saenjan P (2015) Soil properties and maize growth in saline and nonsaline soils using cassava–industrial waste compost and vermicompost with or without earthworms. Land Degrad Dev 26: 300-310

Pavani KV, Kumar NS, Sangameswaran BB (2012) Synthesis of lead nanoparticles by *Aspergillus* species. Pol J Microbiol 61:61–63

Peng XP, Wang Y, Liu PP, Hong K, Chen H, Yin X, Zhu WM (2011a) Aromatic compounds from the halotolerant fungal strain of *Wallemia sebi* PXP-89 in a hypersaline medium. Arch Pharm Res 34:907–912

Peng XP, Wang Y, Sun K, Liu PP, Yin X, Zhu WM (2011b) Cerebrosides and 2-pyridone alkaloids from the halotolerant fungus Penicillium *chrysogenum* grown in a hypersaline medium. J Nat Prod 74:1298–1302

Piñar G, Dalnodar D, Voitl C, Reschreiter H, Sterflinger K (2016) Biodeterioration risk threatens the 3100 year old staircase of hallstatt (Austria): possible involvement of halophilic microorganisms. PLoS One 11(2):e0148279. https://doi.org/10.1371/journal.pone.0148279

Prasad R, Kumar V, Prasad KS (2014) Nanotechnology in sustainable agriculture: present concerns and future aspects. Afr J Biotechnol 13(6):705–713

Prasad R, Pandey R, Barman I (2016) Engineering tailored nanoparticles with microbes: quo vadis. WIREs Nanomed Nanobiotechnol 8:316–330. https://doi.org/10.1002/wnan.1363

Qadir M, Oster JD, Schubert S, Noble AD, Sahrawat KL (2007) Phytoremediation of sodic and saline-sodic soils. Adv Agron 96:197–247

Qin Y, Wei X, Song X, Qu Y (2008) Engineering endoglucanase II from *Trichoderma reesei* to improve the catalytic efficiency at a higher pH optimum. J Biotechnol 135:190–195

Rady MM (2011) Effects on growth, yield, and fruit quality in tomato (*Lycopersicon esculentum* Mill.) using a mixture of potassium humate and farmyard manure as an alternative to mineral–N fertilizer. J Hortic Sci Biotechnol 86:249–254

Raheman F, Deshmukh S, Ingle A, Gade A, Rai M (2011) Silver nanoparticles: novel antimicrobial agent synthesized from an endophytic fungus Pestalotia sp. isolated from leaves of *Syzygium cumini* (L). Nano Biomed Eng 3:174–178

Rai M, Ribeiro C, Mattoso L, Duran N (2015) Nanotechnologies in food and agriculture. Springer, Germany. https://doi.org/10.1007/978-3-319-14024-7

Rajakumar G, Rahuman AA, Roopan SM, Khanna VG, Elango G, Kamaraj C, Velayutham K (2012) Fungus-mediated biosynthesis and characterization of $TiO_2$ nanoparticles and their activity against pathogenic bacteria. Spectrochim Acta A Mol Biomol Spectrosc 91:23–29. https://doi.org/10.1016/j.saa.2012.01.011

Richardson AE, Hadobas PA, Hayes JE, O'Hara CP, Simpson RJ (2001) Utilization of phosphorus by pasture plants supplied with myo-inositol hexaphosphate is enhanced by the presence of soil microorganisms. Plant Soil 229:47–56

Riddin TL, Gericke M, Whiteley CG (2006) Analysis of the inter- and extracellular formation of platinum nanoparticles by *Fusarium oxysporum* f. sp. lycopersici using response surface methodology. Nanotechnology 17:3482–3489

Rietz DN, Haynes RJ (2003) Effects of irrigation-induced salinity and sodicity on soil microbial activity. Soil Biol Biochem 35:845–854

Sahin U, Eroğlum S, Sahin F (2011) Microbial application with gypsum increases the saturated hydraulic conductivity of saline–sodic soils. Appl Soil Ecol 48:247–250

Sanae F, Miyaichi Y, Hayashi H (2003) Endothelium-dependent contraction of rat thoracic aorta induced by gallic acid. Phytother Res 17(2):187

Santos SX, Carvalho CC, Bonfa MR, Silva R, Gomes E (2004) Screening for pectinolytic activity of wood–rotting basidiomycetes and characterization of the enzymes. Folia Microbiol 49:46–52

Sardinha M, Müller T, Schmeisky H, Joergensen RG (2003) Microbial performance in soils along a salinity gradient under acidic conditions. Appl Soil Ecol 23:237–244

Sastry M, Ahmad A, Khan MI, Kumar R (2003) Biosynthesis of metal nanoparticles using fungi and actinomycetes. Curr Sci 85:162–170

Satyamurthy P, Vigneshwaran N (2013) A novel process for synthesis of spherical nanocellulose by controlled hydrolysis of microcrystalline cellulose using anaerobic microbial consortium. Enzym Microb Technol 52(1):20–25

Satyamurthy P, Jain P, Balasubramanya RH, Vigneshwaran N (2011) Preparation and characterization of cellulose nanowhiskers from cotton fibres by controlled microbial hydrolysis. Carbohydr Polym 83(1):122–129

Scervino JM, Mesa MP, Della Mónica I, Recchi M, Moreno NS, Godeas A (2010) Soil fungal isolates produce different organic acid patterns involved in phosphate salts solubilization. Biol Fertil Soils 46:755–763

Schuster E, Dunn-Coleman N, Frisvad JC, van Dijck PW (2002) On the safety of *Aspergillus niger*– a review. Appl Microbiol Biotechnol 59:426–435

Sheikhloo Z, Salouti M (2011) Intracellular biosynthesis of gold nanoparticles by the fungus Penicillium chrysogenum. Int J Nanosci Nanotechnol 7:102–105

Shelar GB, Chavan AM (2014) Fungus–mediated biosynthesis of silver nanoparticles and its antibacterial activity. Arch App Sci Res 6:111–114

Shi Y, Zhu J (2016) Saline-alkali soil improvement fertilizer and preparation method and use method thereof. Chinese Patent, CN 105237293 A (in Chinese). http://patentool.wanfangdata.com.cn/Patent/Details?id=CN201510597529.0

Szabolcs I (1994) Soils and salinization. In: Pessarakli M (ed) Handbook of plant and crop stress, 1st edn. Marcel Dekker, New York, pp 3–11

Tam PCF (1995) Heavy-metal tolerance by ectomycorrhizal fungi and metal amelioration by *Pisolithus tinctorius*. Mycorrhiza 5:181–187

Tamura M, Kawasaki H, Sugiyama J (1999) Identity of the xerophilic species *Aspergillus penicillioides*: integrated analysis of the genotypic and phenotypic. J Gen Appl Microbiol 45:29–37

Taneja K, Gupta S, Kuhad RC (2002) Properties and application of a partially purified alkaline xylanase from an alkalophilic fungus *Aspergillus nidulans* KK-99. Bioresour Technol 85:39–42

Tejada M, Garcia C, Gonzalez JL, Hernandez MT (2006) Use of organic amendment as a strategy for saline soil remediation: influence on the physical, chemical and biological properties of soil. Soil Biol Biochem 38:1413–1421

Tiquia-Arashiro SM, Rodrigues DF (2016) Extremophiles: applications in biotechnology. Springer briefs in microbiology: extremophilic microorganisms. Springer International Publishing. ISBN 978-3-319-45214-2, ISBN 978-3-319-45215-9 (eBook). https://doi.org/10.1007/978-3-319-45215-9

Tomme P, Warren RAJ, Gilkes NR (1995) Cellulose hydrolysis by bacteria and fungi. Adv Microb Physiol 37:1–81

Turan M, Ataoglu N, Sahin F (2006) Evaluation of the capacity of phosphate solubilizing bacteria and fungi on different forms of phosphorus in liquid culture. J Sustain Agric 28:99–108

Vahabi K, Mansoori GA, Karimi S (2011) Biosynthesis of silver nanoparticles by the fungus Trichoderma reesei. Insciences J 1:65–79

Vala AK, Shah S, Patel R (2014) Biogenesis of silver nanoparticles by marine derived fungus *Aspergillus flavus* from Bhavnagar coast, gulf of Khambat, India. J Mar Biol Oceanogr 3(1):1–3

Verma VC, Kharwar RN, Gange AC (2010) Biosynthesis of antimicrobial silver nanoparticles by endophytic fungus *Aspergillus clavatus*. Biomedicine 5:33–40

Wang W, Vinocur B, Altman A (2003) Plant responses to drought, salinity and extreme temperatures: towards genetic engineering for stress tolerance. Planta 218:1–14

Wang T, Liu X, Yu Q, Zhang X, Qu Y, Gao P, Wang T (2005) Directed evolution for engineering pH profile of endoglucanase III from *Trichoderma reesei*. Biomol Eng 22:89–94

Wang L, Seki K, Miyazaki T, Ishihama Y (2009) The causes of soil alkalinization in the Songnen Plain of Northeast China. Paddy Water Environ 7:259–270
Wang LL, Sun XY, Li SY, Zhang T, Zhang W, Zhai PH (2014) Application of organic amendments to a coastal saline soil in North China: effects on soil physical and chemical properties and tree growth. PLoS One 9(2):e89185. https://doi.org/10.1371/journal.pone.0089185
Wei Y, Zhang SH (2018) Abiostress resistance and cellulose degradation abilities of haloalkaliphilic fungi: applications for saline-alkaline remediation. Extremophiles 22:155–164. https://doi.org/10.1007/s00792-017-0986-3
Wei Y, Liu XD, Jia B, Zhang SH, Liu JL, Gao W (2013) Alkaline-tolerant and halophilic *Aspergillus* strain and application thereof in environmental management. Chinese Patent, CN 103436450 A (in Chinese). http://patentool.wanfangdata.com.cn/Patent/Details?id=CN201310162347.1
Wichern J, Wichern F, Joergensen RG (2006) Impact of salinity on soil microbial communities and the decomposition of maize in acidic soils. Geoderma 137:100–108
Xie LX (2013) Cloning and stress resistance analysis of ribosomal protein genes (SpRPS3ae and SpRPL44) in extreme halotolerant *Aspergillus* sp. Dissertation. Jilin University (in Chinese). http://kns.cnki.net/KCMS/detail/detail.aspx?dbcode=CMFD&dbname=CMFDTEMP&filename=1013196212.nh
Xu ZH, Peng XP, Wang Y, Zhu WM (2011) (22E, 24R)-3ß,5α,9α-Trihydroxyergosta-7,22-dien-6-one monohydrate. Acta Cryst E67:o1141–o1142
Yadav S, Irfan M, Ahmad A, Hayat S (2011) Causes of salinity and plant manifestations to salt stress: a review. J Environ Biol 32:667–685
Yan J, Song WN, Nevo E (2005) A MAPK gene from Dead Sea fungus confers stress tolerance to lithium salt and freezing–thawing: prospects for saline agriculture. Proc Natl Acad Sci U S A 102:18992–18997
Zajc J, Džeroski S, Kocev D, Oren A, Sonjak S, Tkavc R, Gunde-Cimerman N (2014a) Chaophilic or chaotolerant fungi: a new category of extremophiles? Front Microbiol 5:708-1–708-5. https://doi.org/10.3389/fmicb.2014.00708
Zajc J, Kogej T, Galinski EA, Ramos J, Gunde-Cimerman N (2014b) Osmoadaptation strategy of the most halophilic fungus, *Wallemia ichthyophaga*, growing optimally at salinities above 15% NaCl. Appl Environ Microbiol 80(1):247–256. https://doi.org/10.1128/AEM.02702-13
Zalar P, Kocuvan MA, Plemenitas A, Gunde-Cimerman N (2005) Halophilic black yeasts colonize wood immersed in hypersaline water. Bot Mar 48:323–326
Zhang SH (2016) The genetic basis of abiotic stress resistance in extremophilic fungi: the genes cloning and application. In: Purchase D (ed) Fungal applications in sustainable environmental biotechnology, 1st edn. Springer Press, pp 29–42. ISBN 978-3-319-42850-5. http://www.springer.com/gb/book/9783319428505#aboutBook
Zhang X, Liu Y, Yan K, Wu H (2007) Decolorization of an anthraquinone–type dye by a bilirubin oxidase–producing nonligninolytic fungus *Myrothecium* sp. IMER1. J Biosci Bioeng 104:104–114
Zhang SH, Li ZQ, Wei Y, Chen LN, Liu SS, Zhou XY, Song YY, Pei X (2016) Cellulase gene from extreme saline-alkali resistant *Aspergillus* and application. Chinese Patent, CN105420259A (in Chinese). http://patentool.wanfangdata.com.cn/Patent/Details?id=CN201610005024.5
Zhou XY (2016)Cloning and abiotic functional analysis of salt-tolerant genes in Halophilic *Aspergillus glaucus*. Dissertation. Jilin University (in Chinese). http://kns.cnki.net/KCMS/detail/detail.aspx?filename=1016091320.nh&dbname=CMFDTEMP

# Chapter 15
# Nanobioremediation: An Innovative Approach to Fluoride (F) Contamination

**Neha Singh and Suphiya Khan**

**Contents**



## 15.1 Introduction

Fluoride (F) is a major environmental concern as a water pollutant owing to its abundance and severe effects on human health. It was reported in 2015 that about 663 million people lack access to pure drinking water worldwide (Kumari and Khan 2017). Approximately 200 million people face the problem of F contamination of water globally (Kumari and Khan 2017). In India, 62 million people and 17 provinces are severely influenced by different fluorosis problems (Susheela 1999). Rajasthan is reported to have the most highly F-polluted groundwater (Chaudhary et al. 2009). According to World Health Organization (WHO) guidelines, the permissible limit for F is 1.5 mg/L and excessive intake causes dental and skeletal fluorosis, infertility, brain damage, and thyroid disorders (Meenakshi and Maheshwari 2006; Bhatnagar et al. 2011). Research has showed that people diagnosed with bladder cancers may have a history of exposure to high F concentrations (Islam and Patel 2011). The main sources of F pollution in drinking water are F-containing rocks such as fluorspar, cryolite, fluorapatite, and hydroxylapatite. Various conventional

N. Singh · S. Khan (✉)
Department of Bioscience and Biotechnology, Banasthali Vidyapith Tonk, Rajasthan, India

 
R. Prasad, E. Aranda (eds.), *Approaches in Bioremediation*, Nanotechnology in the Life Sciences, https://doi.org/10.1007/978-3-030-02369-0_15

technologies have been developed for F remediation of soil, such as phytoremediation, electrokinetic systems, excavation, and landfills (Zhu et al. 2009). Among these approaches, a phytoremediation process has proved to be more promising as it is affordable, user friendly, and environmentally friendly in nature. Phytoremediation is a method of using green plants for removal of different pollutants from soil (Johnson et al. 2015). The techniques available for removing F from groundwater and drinking water have various drawbacks such as limited effectiveness, expense and lack of proper disposal; thus, attempts should be made to manufacture cost-effective materials. Nowadays, nanomaterials (NMs)/nanoparticles (NPs) are recommended for both resource management and environmental remediation as being low cost, eco-friendly, and effective (Zare et al. 2013; Gupta et al. 2015; Prasad 2014; Prasad et al. 2014, 2017). NPs exhibit an extensive range of applications due to their unique optical, thermal, electrical, chemical, and physical properties (Panigrahi et al. 2004; Prasad et al. 2016, 2017). Elimination of ecological contaminants from polluted locations by means of NMs produced by plants, fungi, and bacteria, with the aid of nanotechnology, is called nanobioremediation. The advantages of bioremediation over traditional techniques are its low cost, less chemical and biological by-products, metal specificity, no need for additional nutrients, renewal of biosorbents, and the option of metal recovery (Kratochvil and Volesky 1998; Prasad 2017, 2018). Metal NPs can be synthesized by a sonochemical method, laser ablation method, radiolysis, and a green synthesis method (Latha and Gowri 2014). NPs can be synthesized through green routes from different parts of the plant, and this approach has various advantages such as low toxicity, organized size, accuracy, and eco-friendliness (Shameli et al. 2012; Mei et al. 2014; Prasad 2014).

Several conventional adsorbents such as activated carbon, zeolites, and bone char were evaluated, but NMs proved to be more capable for F remediation as they have a large surface area to volume ratio (Kumari and Khan 2017). Adsorption seems to be enhanced by many NM properties, such as the surface charge, hydrophobicity, and new functional group addition (Matlochová et al. 2013). Yet, researchers have faced problems with separation of NPs from suspension in aqueous solutions after adsorption. So, to overcome this problem, in the current research, absorbents such as graphene and hydrogels could be advantageous as they have large surface areas and volumes to retain water for removal of F. Graphene (GN) is a single carbon sheet of $sp^2$-attached carbon atoms densely packed in a honeycomb lattice (Geng et al. 2015). Moreover, the properties of GN can be changed according to the functional groups attached to it—such as carboxyl, hydroxyl, epoxy, or other organic groups—to facilitate its use in composite materials. In general, graphene oxide (GO) is synthesized from GN for utilization in F remediation. A hydrogel is a crosslinked polymeric network and can retain a significant amount of water within its three-dimensional structure. In a hydrogel, the hydrophilic functional group attached to its polymeric backbone confers its capacity to absorb water, and its absorption capacity is up to 10 g/g (Kabiri et al. 2011). Now, research needs to be focused on development of eco-friendly, selective, cost-effective, and sustainable approaches for green bioremediation. NMs developed with the help of nanotechnology are used in the remediation of in situ and ex situ pollutants at contaminant sites. They generally have unique properties that increase their efficiency in removing contaminants and are comparatively cost efficient as compared with traditional methods such as activated carbon. However, further work is essential to develop efficient, ex situ, and accurate processes for manufacturing NMs and checking their effectiveness for large-scale resources.

## 15.2 Sources of Fluoride

The constant increase in F compound emissions into the environment leads to F accumulation in living organisms, and so the risk of the entire ecosystem suffering from F pollution continues to increase. F pollution in the environment—occurring via water, soil, air, or amalgamation of all three—has severe consequences for fauna, flora, and humankind. Generally, F is released into the environment through natural processes such as weathering of rock and disintegration of minerals, volcanic emissions, coal burning, and industrial sludge (Ayoob and Gupta 2006). Fluorides and other hazardous chemical emissions are produced by the burning of coal and accumulate in the environment (Ando et al. 1998). The magnitude of F pollution can be examined through factors such as sources of F; concentrations in coal, minerals, soil, and groundwater; and the geographical distribution of F and its dispersion (Ando et al. 2001; Symonds et al. 1988). Obviously, F pollution can adversely impact wildlife, and these industrial atmospheric emissions are known for their ability to contaminate soil, water, and vegetation in both the immediate vicinity and distant areas.

Examples of industries that emit F pollution (Swarup and Dwivedi 2002) are aluminum smelters; blast furnace manufacturing; cement and phosphate fertilizer units; coal and fuel combustion; enamel, pottery, glass, and brick works; oil refineries; plastic, pharmaceutical, toothpaste, chemical, and automobile industries; steel and iron foundries; welding, refrigeration, and rust removal activities; and zinc production (Fig. 15.1).

## 15.3 Effects of Fluoride on Life-Forms

Chronic excessive intake of F by humans and animals leads to endemic disease ("fluorosis") in a region with high F content in the environment. F is known to cause several infectious diseases such as osteoporosis, crippling skeletal fluorosis, thyroid imbalance, growth obstruction, impaired kidney function, and pineal gland dysfunction, and in some cases it causes mortality (Ozsvath 2006). In developed countries the health of the human population is affected by uncontrolled consumption of fluoridated water and F-containing food supplements and dental products (Pizzo et al. 2007; Vestergaard et al. 2008). In rural communities, fluorosis is prominently

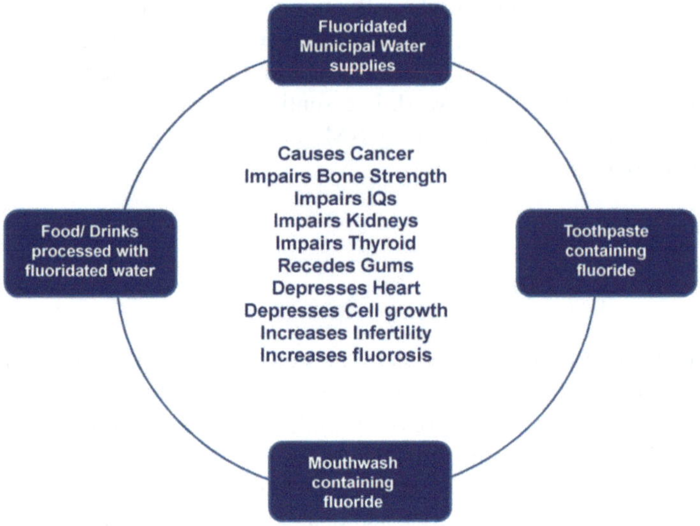


**Fig. 15.1** Different sources of fluoride (F) and its effects in the environment

reported to be due to lack of F-free water (Almas et al. 1999; Rashmi et al. 2013). In industries and agriculture, inorganic F compounds (e.g., HF, NaF, $CaF_2$, sodium hexafluorosilicate ($Na_2SiF_6$), and hexafluorosilicic acid ($H_2SiF_6$)) are widely used; as a result, anthropogenic contamination of the environment increases with F-containing compounds in biogeocenoses up to toxic levels (Table 15.1).

## 15.4 Nanotechnology for Fluoride Remediation

The term *nanomaterial* describes a particle that is on the nanoscale level or in the size range of 1–100 nm. The first scientific report explaining the nature of NPs was provided by Michael Faraday in his pioneering work "Experimental Relations of Gold to Light" in 1857. Nanobioremediation is a technology allowing removal of F from pollutant sites, using NMs synthesized with the aid of nanotechnology by a "green method" (Yadav et al. 2017). NPs include organic compounds (proteins, polysaccharides, viruses, etc.) and inorganic compounds (iron oxyhydroxides, aluminosilicates, metals, etc.) that occur naturally (Hough et al. 2011). These generally include fullerenes, metal clusters, and large molecules. NPs can be synthesized by a "green route," which is a nontoxic, eco-friendly, cost-effective, and simple operation (Prasad et al. 2018). The mechanism of metal NP synthesis is plant-assisted reduction due to phytochemicals. The various phytochemicals responsible for reduction are terpenoids, amides, flavones, aldehydes, ketones,

**Table 15.1** US Public Health Service (USPHS) recommendations for maximal allowable fluoride (F) intake in drinking water (USPHS 1962; Arif et al. 2014)

| Annual average maximal daily air temperature (°C) | Recommended F concentration (mg/L) | | | Maximal acceptable F concentration (mg/L) |
|---|---|---|---|---|
| | Lower | Optimal | Upper | |
| 10.0–12.0 | 0.9 | 1.2 | 1.7 | 2.4 |
| 12.1–14.6 | 0.8 | 1.1 | 1.5 | 2.2 |
| 14.7–17.7 | 0.8 | 1.0 | 1.3 | 2.0 |
| 17.8–21.4 | 0.7 | 0.9 | 1.2 | 1.8 |
| 21.5–26.2 | 0.7 | 0.8 | 1.0 | 1.6 |
| 26.3–32.5 | 0.6 | 0.7 | 0.8 | 1.4 |

and carboxylic acids (Jha et al. 2009; Joshi et al. 2018). A huge variety of biological resources such as microorganisms (bacteria, yeast, fungi, algae, and viruses) (Prasad et al. 2016) and plants can be utilized for NP synthesis (Mohanpuria et al. 2008; Prasad 2014).

### *15.4.1 Nanotubes*

Nanomaterials have great flexibility due to their unique properties and can produce several possible outcomes for F remediation from water sources. Carbon nanotubes (CNTs) are NMs that circle into a tube, and they include two categories: single-walled CNTs (SWCNTs) and multiwalled CNTs (MWNTs). Carbon NMs have been shown to be effective over a wide range of pH with favorable Langmuir or Freundlich isotherms (Mauter and Elimelech 2008). CNTs have several exclusive properties—such as their surface morphology, great water flux capacity, fine pore size, high conductivity, distinctive adsorption capability, and ability to be anchored to a functional group to amplify affinity and reactivity toward the target molecule—that make them a promising material for water remediation (Zhu et al. 2005). Unfunctionalized groups CNTs with Unfunctional groups tend to be impenetrable in water and become toxic. To enable CNTs to be easily separated from dispersion, several functional groups (e.g., hydroxyl, carboxyl, and amines) are attached and increase water solubility and biocompatibility. CNTs have been activated using $H_2O_2$, NaOH, $HNO_3$, and $KMnO_4$ in an oxidizing environment (Ruparelia et al. 2008). The adsorption capacity of MWCNTs can be increased, as oxidation occurs with nitric acid and thus a reactive site is constructed on the tip of the nanotube, resulting in an increased level of adsorption due to increased flexibility and active site reactivity. In most cases the strong interaction includes several driving forces such as a hydrophobic effect, π–π interaction, π–π electron–donor–acceptor interaction, electrostatic interaction, and hydrogen bonding acting simultaneously in a hexagonal array of carbon atoms in a graphite sheet, making CNTs a promising adsorbent material (Fig. 15.2).

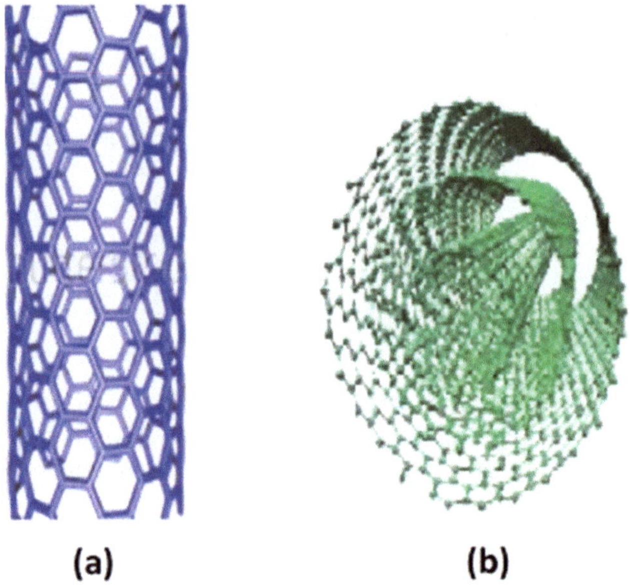


**Fig. 15.2** Structures of (**a**) single-walled carbon nanotubes (SWCNTs) and (**b**) multiwalled carbon nanotubes (MWCNTs) (Zhu et al. 2005)

## *15.4.2 Nanoscale Iron Nanoparticles*

Nanoparticles show vast diversity for both in situ and ex situ remediation, as they can be easily set up for treatment of polluted soils, sediments, and solid by-products in ex situ slurry reactors. Secondly, they can be used for treatment of water, wastewater, or gaseous process streams by attachment to solid surfaces such as carbon, zeolite, or membranes. Nowadays, iron NPs seem to be an effective component in nanoremediation. Borohydride is utilized as a reductant for producing iron NPs from Fe (II) and Fe (III) (Wang et al. 2006). NP zero-valent iron (nZVI) has a diameter range of 10–100 nm and shows a typical structure, with a surface area of 20–40 $m^2/g$; this gives 10–1000 times greater reactivity than granular Fe, which generally has a diameter of <1 $m^2/g$ (Wang and Zhang 1997). It has been shown to be efficient for treatment of diverse and widespread environmental pollutants such as chlorinated organic solvents, organochlorine pesticides, polychlorinated biphenyls, heavy metals, and radionuclides, in both ex situ and in situ applications (e.g., via direct injection into subsurface environments) (Li and Zhang 2006, 2007; Kanel et al. 2005; Uzum et al. 2009; Sun et al. 2006; Elebi et al. 2007; Mayo et al. 2007; Liu et al. 2005; Jahin 2014; Cundy et al. 2008) (Fig. 15.3).

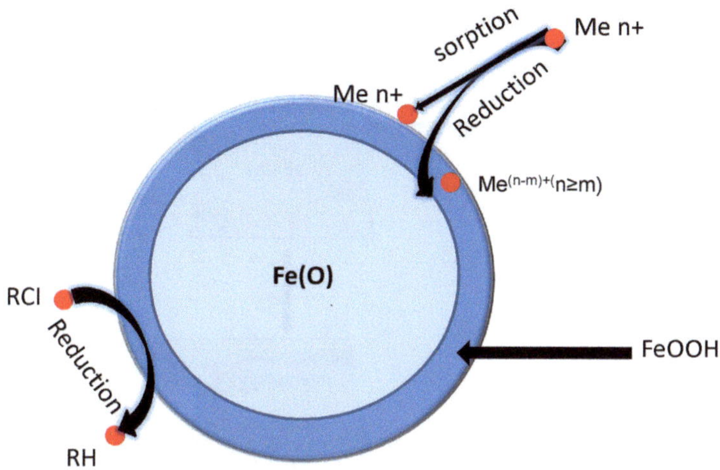


**Fig. 15.3** Schematic diagram of zero-valent iron (Li et al. 2006)

### *15.4.3 Graphene-Based Nanomaterials*

Graphene is a carbon sheet of $sp^2$-bonded carbon atoms densely packed in a honeycomb lattice (Geng et al. 2015). Moreover, the properties of GN can be changed according to the functional groups attached to it—such as carboxyl, hydroxyl, epoxy, or other organic groups—to facilitate its use in composite materials. The existence of surface functional groups and the large surface area of GN make it more attractive for defluoridation of water. In general, GO is synthesized from GN for utilization in F remediation. Recently, GN has been impregnated with numerous inorganic NPs such as Ag, Au, Pt, $TiO_2$, and ZnO (Qian et al. 2011). GN nanocomposites impregnated with Ag NPs has been synthesized through chemical method involving hydrazine, microwave-assisted reduction using sodium citrate as a reducing agent and the process is environmentally friendly and cost-efficient (Yuan et al. 2012). Two-dimensional GO materials have currently gained attention owing to several distinctive properties—such as chemical stability, high mechanical strength, thermal mobility, and flexibility—and are formed by the Hummers method using an oxidation process for graphite (Taherian et al. 2013; Gopalakrishnan et al. 2015; Lingamdinne et al. 2016) (Fig. 15.4).

## 15.5 Advantages and Disadvantages

Nanomaterials represent a major milestone in F remediation of the environment. The approach of synthesis of NMs by a green method makes this technology an effective method, as it is eco-friendly, nontoxic, selective, and cost effective for F remediation. The low toxicity and high adsorption ability of nanotubes makes

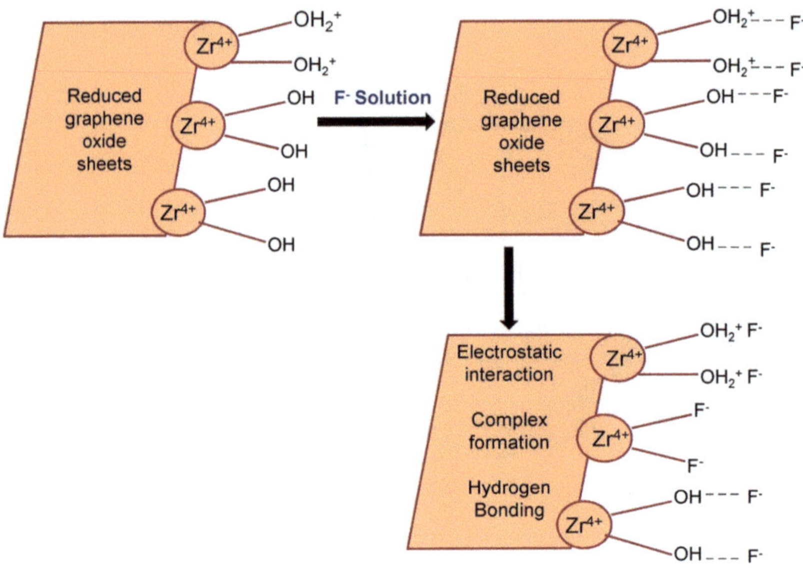


**Fig. 15.4** Schematic diagram illustrating the mechanism of remediation of fluoride (F) through reduced graphene oxide sheets. nanomaterial during remediation of fluoride (F) (Mohan et al. 2014)

them suitable for the process of F remediation from soil and water resources. Carbon-based NMs are extensively used for F removal because of their high conductivity stabilities, along with high reactivity. Another advantage is that NMs show fewer harmful effects than other forms of remediation, and such harmful effects should be evaluated and eradicated through further research. Leaching during the adsorption process is the major drawback of NMs and needs to be addressed in further research. Because of leaching, NMs are able to penetrate the food chain and get stored in plants, animals, and human beings through the food chain/food web. This major issue of leaching should be a concern, and accurate, dependable, eco-friendly, low cost, and clean resources for F remediation should be developed.

## 15.6 Conclusion

In the developing world, the hazard of waterborne infectious diseases is growing tremendously. The main intent of this chapter is to describe the utility of nanoparticles to elucidate potential issues such as fluoride remediation from pollutant water and more efficient recycling of it than through traditional methods. Nanobioremediation shows efficient potential to eradicate F from huge polluted sites in situ and ex situ, reducing the clean-up time and the contaminant concentration. Nanotechnology, in association with other technology, can change the face of research to deal with key challenges and also seems to be a promising approach for providing advances and innovative methods to clean up the ecosystem. The most approachable technology for the production of nanomaterials has been shown to be "green method" but some modifications are needed in this technology for further use. Furthermore, modifications in this technology are required to reduce or eliminate the disadvantages of the current technology.

**Acknowledgements** We acknowledge the Ministry of Human Resource and Development (MHRD), India, for providing financial assistance for the research project "Center for Excellence (CoE)—Water and Energy." The authors are deeply grateful to Prof. Aditya Shastri for providing research amenities and the Bioinformatics Centre, Banasthali Vidyapith, Rajasthan (India), for use of computational facilities.

## References

Almas K, Shakir ZF, Afzal M (1999) Prevalence and severity of dental fluorosis in Al-Qaseem province Kingdom of Saudi Arabia. Trop Dental J 22(85):44–47

Ando M, Tadano M, Asanuma S, Tamura K, Matsushima S, Watanabe T, Cao S (1998) Health effects of indoor fluoride pollution from coal burning in China. Environ Health Perspect 106(5):239

Ando M, Tadano M, Yamamoto S, Tamura K, Asanuma S, Watanabe T, Chen X (2001) Health effects of fluoride pollution caused by coal burning. Sci Total Environ 271(1–3):107–116

Arif M, Hussain J, Hussain I, Kumar S (2014) Fluoride toxicity and its distribution in groundwater of south east part of Nagaur district, Rajasthan, India. Int J Sci Res Agric Sci 1(6):110–117

Ayoob S, Gupta AK (2006) Fluoride in drinking water: a review on the status and stress effects. Crit Rev Environ Sci Technol 36:433–487

Bhatnagar A, Kumar E, Sillanpaa M (2011) Fluoride removal from water by adsorption—a review. Chem Eng J 171:811–840

Chaudhary V, Sharma M, Yadav BS (2009) Elevated fluoride in canal catchment soils of northwest Rajasthan, India. Fluoride 42(1):46–49

Cundy AB, Hopkinson L, Whitby RL (2008) Use of iron-based technologies in contaminated land and groundwater remediation: a review. Sci Total Environ 400:42–51

Elebi OC, Uzum C, Shahwan T, Erten HNA (2007) A radiotracer study of the adsorption behavior of aqueous $Ba^{2+}$ ions on nanoparticles of zero-valent iron. J Hazard Mater 148:761–767

Geng D, Wang H, Yu G (2015) Graphene single crystals: size and morphology engineering. Adv Mater 27(18):2821–2837

Gopalakrishnan A, Krishnan R, Thangavel S, Venugopal G, Kim SJ (2015) Removal of heavy metal ions from pharma-effluents using graphene-oxide nanosorbents and study of their adsorption kinetics. J Ind Eng Chem 30:14–19

Gupta VK, Tyagi I, Sadegh H, Shahryari-Ghoshekandi R, Makhlouf ASH, Maazinejad B (2015) Nanoparticles as adsorbent; a positive approach for removal of noxious metal ions: a review. Sci Technol Dev 34:195–214

Hough RM, Noble RRP, Reich M (2011) Natural gold nanoparticles. Ore Geol Rev 42(1):55–61

Islam M, Patel RK (2011) Thermal activation of basic oxygen furnace slag and valuation of its fluoride removal efficiency. Chem Eng J169:68–77

Jahin HS (2014) Fluoride removal from water using nanoscale zero-valent iron (NZVI). Int Water Technol J 4:173–182

Johnson DM, Deocampo DM, El-Mayas H, Greipsson S (2015) Induced phytoextraction of lead through chemical manipulation of switchgrass and corn; role of iron supplement. Int J Phytoremediation 17(12):1192–1203

Joshi N, Jain N, Pathak A, Singh J, Prasad R, Upadhyaya CP (2018) Biosynthesis of silver nanoparticles using *Carissa carandas* berries and its potential antibacterial activities. J Sol-Gel Sci Technol 86:682–689. https://doi.org/10.1007/s10971-018-4666-2

Jha AK, Prasad K, Prasad K, Kulkarni AR (2009) Plant system: nature's nanofactory. Colloids Surf B Biointerfaces 73(2):219–223

Kabiri K, Omidian H, Zohuriaan-Mehr MJ, Doroudiani S (2011) Superabsorbent hydrogel composites and nanocomposites: a review. Polym Compos 32(2):277–289

Kanel SR, Manning B, Charlet L, Choi H (2005) Removal of arsenic (III) from groundwater by nanoscale zero-valent iron. Sci Total Environ 39:1291–1298

Kratochvil D, Volesky B (1998) Advances in the biosorption of heavy metals. Trends Biotechnol 16(7):291–300

Kumari S, Khan S (2017) Defluoridation technology for drinking water and tea by green synthesized $Fe_3O_4/Al_2O_3$ nanoparticles coated polyurethane foams for rural communities. Sci Rep 7:8070–8081

Latha N, Gowri M (2014) Biosynthesis and characterisation of $Fe_3O_4$ nanoparticles using *Caricaya* papaya leaves extract. Int J Sci Res 3(11):1551–1556

Li XQ, Zhang WX (2006) Iron nanoparticles: the core–shell structure and unique properties for Ni (II) sequestration. Langmuir 22:4638–4642

Li XQ, Zhang WX (2007) Sequestration of metal cations with zero valent iron nanoparticles, a study with high resolution X-ray photoelectron spectroscopy (HR-XPS). J Phys Chem 111:6939–6946

Li XQ, Elliott DW, Zhang WX (2006) Zero-valent iron nanoparticles for abatement of environmental pollutants: materials and engineering aspects. Crit Rev Solid State Mater Sci 31(4):111–122

Lingamdinne LP, Koduru JR, Roh H, Choi YL, Chang YY, Yang JK (2016) Adsorption removal of Co (II) from waste-water using graphene oxide. Hydrometallurgy 165:90–96

Liu Y, Majetich SA, Tilton RD, Sholl DS, Lowry GV (2005) TCE dechlorination rates, pathways, and efficiency of nanoscale iron particles with different properties. Sci Total Environ 39:1338–1345

Matlochová A, Plachá D, Rapantová N (2013) The application of nanoscale materials in groundwater remediation. Pol J Environ Stud 22(5):1401–1410

Mauter MS, Elimelech M (2008) Environmental applications of carbon-based nanomaterials. Sci Total Environ 42(16):5843–5859

Mayo JT, Yavuz C, Yean S, Cong L, Shipley H, Yu W, Falkner J, Kan A, Tomson M, Colvin VL (2007) The effect of nanocrystalline magnetite size on arsenic removal. Sci Technol Adv Mater 8:71–75

Meenakshi, Maheshwari RC (2006) Fluoride in drinking water and its removal. J Hazard Mater 137:456–463

Mie R, Samsudin MW, Din LB, Ahmad A, Ibrahim N, Adnan SNA (2014) Synthesis of silver nanoparticles with antibacterial activity using the lichen *Parmotrema praesorediosum*. Int J Nanomed 9:121

Mohan S, Kumar V, Singh DK, Hasan SH (2016) Synthesis and characterization of rGO/ZrO 2 nanocomposite for enhanced removal of fluoride from water: kinetics, isotherm, and thermodynamic modeling and its adsorption mechanism. RSC Adv 6(90):87523–87538

Mohanpuria P, Rana NK, Yadav SK (2008) Biosynthesis of nanoparticles: technological concepts and future applications. J Nanopart Res 10(3):507–517

Ozsvath DL (2006) Fluoride concentrations in a crystalline bedrock aquifer Marathon County, Wisconsin. Environ Geol 50:132–138

Panigrahi S, Kundu S, Ghosh S, Nath S, Pal T (2004) General method of synthesis for metal nanoparticles. J Nanopart Res 6(4):411–414

Pizzo G, Piscopo MR, Pizzo I, Guiliana G (2007) Community water fluoridation and caries prevention: a critical review. Clin Oral Investig 11:189–193

Prasad R (2014) Synthesis of silver nanoparticles in photosynthetic plants. J Nanopart:963961. https://doi.org/10.1155/2014/963961
Prasad R (2017) Mycoremediation and environmental sustainability, vol 1. Springer, Cham
Prasad R (2018) Mycoremediation and environmental sustainability, vol 2. Springer, Cham
Prasad R, Jha A, Prasad K (2018) Exploring the Realms of Nature for Nanosynthesis. Springer International Publishing (ISBN 978-3-319-99570-0) https://www.springer.com/978-3-319-99570-0
Prasad R, Kumar V, Prasad KS (2014) Nanotechnology in sustainable agriculture: present concerns and future aspects. Afr J Biotechnol 13(6):705–713
Prasad R, Pandey R, Barman I (2016) Engineering tailored nanoparticles with microbes: quo vadis. WIREs Nanomed Nanobiotechnol 8:316–330. https://doi.org/10.1002/wnan.1363
Prasad R, Bhattacharyya A, Nguyen QD (2017) Nanotechnology in sustainable agriculture: recent developments, challenges, and perspectives. Front Microbiol 8:1014. https://doi.org/10.3389/fmicb.2017.01014
Qian Y, Wang C, Le ZG (2011) Decorating graphene sheets with Pt nanoparticles using sodium citrate as reductant. Appl Surf Sci 257(24):10758–10762
Rashmi G, Akhil S, Kusum G, Afifa Z, Monohar RK (2013) Dental fluorosis status in school children of Jaipur (Raj) India. IOSRJ Dental Med Sci 8:51
Ruparelia JP, Duttagupta SP, Chatterjee AK, Mukherji S (2008) Potential of carbon nanomaterials for removal of heavy metals from water. Desalination 232(1):145–156
Shameli K, Ahmad MB, Zamanian A, Sangpour P, Shabanzadeh P, Abdollahi Y, Zargar M (2012) Green biosynthesis of silver nanoparticles using *Curcuma longa* tuber powder. Int J Nanomed 7:5603–5610
Sun YP, Li XQ, Cao J, Zhang WX, Wang HP (2006) Characterization of zero-valent iron nanoparticles. Adv Colloid Interface Sci 120:47–56
Swarup D, Dwivedi SK (2002) Environmental pollution and effects of lead and fluoride on animal health. Indian Council of Agricultural Research Krishi Anusandhan Bhavan Pusa, New Delhi
Symonds R, Rose W, Reed M (1988) Contribution of $Cl^-$ and F bearing gases to the atmosphere by volcanoes. Nature 334:415–418
Taherian F, Marcon V, van der Vegt NF, Leroy F (2013) What is the contact angle of water on graphene? Langmuir 29(5):1457–1465
US Public Health Service (1962) Drinking water standards. Department of Health Education and Welfare, Washington, DC
Uzum C, Shahwan T, Eroglu AE, Hallam TB (2009) Synthesis and characterization of kaolinite-supported zero-valent iron nanoparticles and their application for the removal of aqueous $Cu^{2+}$ and $Co^{2+}$ ions. Appl Clay Sci 43:172–181
Vestergaard P, Jorgensen NR, Schwarz P, Mosekilde L (2008) Effects of treatment with fluoride on bone mineral density and fracture risk—a meta-analysis. Osteoporos Int 19:257–268
Wang CB, Zhang WX (1997) Synthesizing nanoscale iron particles for rapid and complete dechlorination of TCE and PCBs. Environ Sci Technol 31:2154–2156
Wang J, Ma T, Zhang Z, Zhang X, Jiang Y, Dong D, Zhang P, Li Y (2006) Fluoride in drinking water and its removal. J Hazard Mater 137:972
Yadav KK, Singh JK, Gupta N, Kumar V (2017) A review of nanobioremediation technologies for environmental cleanup: a novel biological approach. J Mater Environ Sci 8:740–757
Yuan W, Gu Y, Li L (2012) Green synthesis of graphene/Ag nanocomposites. Appl Surf Sci 261:753–758
Zare K, Najafi F, Sadegh H (2013) Studies of ab initio and Monte Carlo simulation on interaction of fluorouracil anticancer drug with carbon nanotube. J Nanostruct Chem 3(1):71
Zhu HY, Li JY, Zhao JC, Churchman GJ (2005) Photocatalysts prepared from layered clays and titanium hydrate for degradation of organic pollutants in water. Appl Clay Sci 28(1–4):79–88
Zhu S, Zhang J, Dong T (2009) Removal of fluorine from contaminated field soil by anolyte enhanced electrokinetic remediation. Environ Earth Sci 59(2):379–384

# Chapter 16
# Nanotechnology: A New Scientific Outlook for Bioremediation of Dye Effluents

**Monika Yadav and Suphiya Khan**

**Contents**



## 16.1 Introduction

Industrial wastewater generation and its treatment are a worldwide problem. Effluent is treated or untreated wastewater that flows out of a treatment plant, sewers, or industrial outfall. Although industries manufacture various useful products for mankind, they also generate different waste by-products in various forms that are responsible for the exaction of hazards and pollution. Most of the waste products are discharged in the soil and water bodies that ultimately pose a serious warning to mankind and routine functioning of the ecosystem (Chaves et al. 2016). Currently, the existence of hazardous pollutants in industrial waste effluents is the most serious environmental problem. The untreated wastewater discharge is greatly affecting marine ecosystem, livelihoods, and food chain. According to 2017 world water development report, it is estimated that worldwide 70–80% of industrial effluents

M. Yadav · S. Khan (✉)
Department of Bioscience and Biotechnology, Banasthali Vidyapith Tonk, Rajasthan, India



R. Prasad, E. Aranda (eds.), *Approaches in Bioremediation*, Nanotechnology in the Life Sciences, https://doi.org/10.1007/978-3-030-02369-0_16

are disposed into the environmental surroundings without any prior treatment. Main contributors of toxic industrial effluents are the by-products coming from the various industries such as textile, paper, dying, chemicals, fertilizers, pesticides, food processing, mining, etc. Dyes are organic colorants that contain unsaturated compound (chromophores) and functional group (auxochromes). The chromophores are responsible for color formation, and auxochromes intensify the color of dye. There are various chromophoric groups such as carbon monoxide (–CO), reactive oxygen (O), nitrogen (N-), nitrogen oxide (-$NO_2$), and quinonoid ($C_9H_{13}N_5O_3$), while auxochromes include aldehydes (–CHO), carboxylic group (–COOH), hydroxyl group (-OH), sulfonic acid (-$SO_3H$), and –$NHX_2$ (Solis et al. 2012). Dyes are divided into different categories based on chromophoric groups, viz., "azo (-N=N-), acidic, basic, disperse, reactive, and anthraquinone dyes." Natural dyes have constricted spectrum of colors and tendency to fade quickly when brought to sun exposure, and cleansing has restricted their application (Firmino et al. 2010). Conversely synthetic dyes provide an immense spectrum of colors that are bright and color fast. Among the synthetic dyes, azo dye is the most commonly used group. The disposal of effluents originating from various industries into water pools and circumambient is of major concern (Elango et al. 2017). Among these, the textile industry is considered as the largest contributor to dye wastewater, because it is censurable for two thirds of overall production of dye wastewater. During dyeing process, a large proportion approximately 50% of the initial dye load remains unfixed or unconstrained and released in environment as effluents (Singh and Arora 2011). It is estimated that 5000 tons of dyeing effluents are disposed directly or indirectly to the environment without any adequate treatment every year. Even at a minute concentration, colorants in wastewater are unpleasant and very threatening to aquatic flora, fauna, and human lives (Mahmoodi and Arami 2009). In India, dye production is estimated to be around 60,000 tones/year or 6.6% of the world production (CPCB, India). Generally, textile wastewater contains dye concentration in the vicinity of 1 g $L^{-1}$ (Tan et al. 2000). The color-free effluents' discharge to water bodies has made treatment of textile industrial wastewater a top priority. Therefore, it is of topmost importance to treat dye-containing effluents.

Conventionally, there are various chemical, physical, and biological methods for treatment of colorants from wastewater effluents including surface adsorption, membrane filtration membranes, coagulation, ozonation, photocatalyst-based oxidation, and Fenton process (Singh et al. 2012). Among these methods, adsorption process is most efficient and globally used as it is low cost, flexible, easy and produce less harmful by-products (Qu et al. 2013). Recent advances in nanotechnology offer incredible potential for the remediation of wastewater. Nano-bioremediation process involves the reduction of pollutants from contaminants by using nanoparticles, which is synthesized by green waste with the aid of nanotechnology. The green method has evolved remarkably to synthesize the novel nanomaterials that are stable, cheap, and eco-friendly. There are various conventional methods that are available to manufacture the nanomaterials, but the green route for synthesizing is more effective and advantageous. The ease of synthesis, less toxic nature, and environment-friendly approach of biological method make it more adoptable process

(Mie et al. 2014). Nowadays, nanomaterials have been proved as more efficient and more exploit for wastewater treatment because of their large surface area and economical synthesis. The oxide-based nanomaterials have been extensively explored for remediation of dye-containing effluents. The iron oxide nanoparticles are more favorable than other nanoparticles due to having super paramagnetism property. Super paramagnetic nanoparticles have many advantageous features such as biocompatibility, high surface area to volume ratio, less toxicity, chemically inert, and small diffusion resistance. The surface behavior of nanoparticles can be modified with some functional groups and organic or inorganic ions, which impart surfaces having good potential for removing or decolorizing the dye effluents (Huang et al. 2012). Thus, nanotechnology offers highly efficient and multifunctional processes that render high-geared and economical wastewater treatment techniques that are less dependent on large infrastructure (Qu et al. 2013).

## 16.2 Dyes and Their Classification

A dye is an organic colored substance that has strong or specific affinity with substrate on which it is being poured and may alter any crystal formation of the colored organic materials temporarily. The colorants adhere to their compatible substrate by forming complexes via covalent bond. Dyes are divided into various types according to their solubility, chemical structure, and application (Fig. 16.1). It is measured that over 10,000 different colorants are used in textile industries and around $7 \times 10^5$ tons of synthetic colorants are produced yearly.

## 16.3 Toxicology Effects of Dye

Textile dyes are important and globally being encountered directly or indirectly in many aspects of daily life. The first synthetic dye mauveine was invented by William Henry Perkin in 1856. The natural sources of colorants are plants, insects and molluscs, but narrow range of colors and tendency to fade quickly when exposed to the sun and cleansing have restricted their application (Shah et al. 2013). Conversely, synthetic colorants provide a broad spectrum of colors that are bright and color fast. After many years, various synthetic dyes were manufactured by workers such as fuchsine, auramine, benzidine, and naphthylamine and found that these are carcinogens for humans. It has been estimated that annually a million tons of synthetic dyes are produced at the international level. Textile wastewaters containing mixture of dyes generally have high biological oxygen demand (BOD), chemical oxygen demand (COD), total soluble solids (TSS), total dissolved solids (TDS), and alarming toxicities (Casieri et al. 2008). The textile industrial revolution aroused a great threat of pollution. The toxic effect of textile dyes can be classified into two types of effects, i.e., short-term (acute) effect and long-term (chronic) effects.

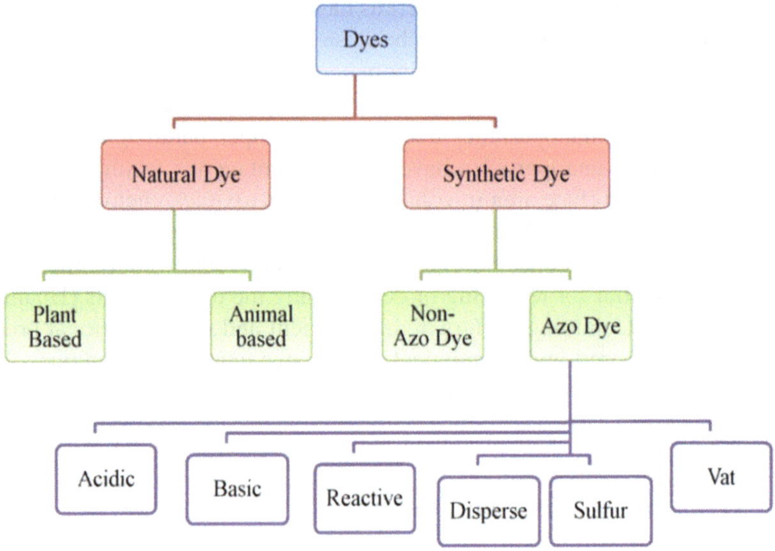


**Fig. 16.1** Classification of dyes

### *16.3.1 Acute Toxicity of Textile Dye*

Acute toxic effect involves skin irritation, skin sensitization, eyes irritation, oral ingestion, and inhalation. Acute toxic effects are mainly caused by reactive dyes and disperse dye (Tronnier 2002). The hydrophilic reactive dyes contain a group that forms a complex or covalent bond with the hydroxyl groups (-OH) present in the fibers during the dyeing activities. Reactive dyes are problematic to the workers who manufacture or handle the dyes. There are many evidences of toxicity caused by reactive dye such as contact dermatitis, allergic conjunctivitis, asthma, and other allergic reactions. The reactive dyes have capability to react with HSA (human serum albumin) and form a dye-HSA conjugate that acts as an antigen. The specific immunoglobulin E (IgE) produced by that antigen which forms histamine compounds that leads to human allergic reactions (Hunger and Sewekow 2003).

### *16.3.2 Chronic Toxicity of Textile Dye*

Genotoxic textile chemicals include dye that are mutagens, carcinogens, and teratogens and pose long-term health hazard (Table 16.1). Mutagen chemicals create mutations, but they may or may not be carcinogens to humans and animals. Teratogens are rarely used in textile industries because it causes birth defects in the offspring although teratogenicity is very uncommon in textile dyes. Some dyes such as 4-aminoazo dyes were found to be carcinogen to animals, and 2-naphthylamine (water soluble) cationic dye is a potent human bladder cancer-causing agent

(Myslak and Bolt 1988). Some dyes have capability to leach out an aromatic amine, a rodent carcinogen. Around 500 azo dyes form the aromatic amines after reduction of azo functional groups. Anthraquinone dye is the most important class of dyes that contains amino or methyl-amino groups that are mutagenic and carcinogenic, for example, 4-aminobiphenyl, benzidine, and toluidine, which are proven to be human carcinogens. Some disperse dyes such as disperse orange and disperse blue are

**Table 16.1** List of toxic dyestuffs

| S.No. | Compound name | CAS No. | Toxic properties | Usage |
|---|---|---|---|---|
| 1 | Disperse dye | | | |
| | Disperse Blue 1 | 2475-45-8 | Carcinogenic | Dyeing of polyester |
| | Disperse Orange 11 | 82-28-0 | | |
| | Disperse Yellow 3 | 2832-40-8 | Allergic | |
| | Disperse Yellow 23 | 6250-23-3 | | |
| | Disperse Orange 149 | 85136-74-9 | | |
| | Disperse Red 60 | 17418-58-5 | Mutagenic | |
| 2 | Acid Green 16 | 12768-78-4 | Mutagenic | Dyeing of cotton |
| 3 | Acid Red 26 | 3761-53-3 | Carcinogenic | Dyeing |
| 4 | Basic Green 4 | 10309-95-2 | Mutagenic | Dyeing |
| 5 | Basic Red 9 | 569-61-9 | Carcinogenic | Dyeing |
| 6 | Basic Violet 14 | 632-99-5 | Carcinogenic | Dyeing |
| 7 | Direct Black 38 | 1937-37-7 | Carcinogenic | Dyeing |
| 8 | Direct Blue 6 | 2602-46-2 | Carcinogenic | Dyeing |
| 9 | Direct Red 28 | 573-58-0 | Carcinogenic | Dyeing |
| 10 | Amines | | Carcinogenic | Dyeing of cotton, silk, wool |
| | Benzidine | | Bladder cancer | |
| | 2-Naphthylamine | | | |
| | 4-Aminodiphenyl | | | |
| | Aminoazotoluene | | | |
| 11 | Formaldehyde | | Suspected carcinogen, skin and respiratory sensitizer | Dye fixing |
| 12 | Alkylphenol ethoxylates | | Aquatic toxicity, endocrine disruption | Wetting, washing |
| 13 | Chlorophenol | | Cancer, skin sensitization | Dry cleaning |
| 14 | Heavy metals (Cd, Pb, Ni) | | Nephrotoxic | Dyes |
| 15 | Volatile organic compound | | Eye damage and respiratory sensitization | Printing |

carcinogen, while disperse violet is mutagenic to some extent (Hunger and Sewekow 2003). Aromatic amines are the precursors for many textile colorants and responsible for the carcinogenicity. Almost all nitrosamines are carcinogens. Hydrazines are applied in textile industries to produce heterocyclic coupling components having carcinogenic properties.

## 16.4 Nanotechnology for Textile Dye Effluent Remediation

Nanotechnology is the branch of science that addressed the dimensions and tolerance of less than 100 nm, especially the manipulation of matter on the basis of atoms, molecules, and supramolecules scale. Thus, nanotechnology is characterized by the use of very ultrafine manufactured particles (<100 nm) called ultrafine particles or nanomaterials. Nanomaterials are atoms or molecule aggregates with size range between 1 and 100 nm that can drastically modify their physiochemical properties compared with the bulk material. The more reactive forms of nanomaterials are membranes, nanowires, tubes, films, particles, quantum dots, and colloids. The nanomaterials are broadly classified into two groups of organic and inorganic nanoparticles. Organic nanomaterials include carbon nanoparticles, and inorganic nanomaterials include magnetic nanoparticles, noble metal nanoparticles (e.g., silver and gold), and semiconductor nanoparticles (e.g., titanium oxide and zinc oxide). The removal of pollutants from environment with the help of nanomaterials synthesized by using biological waste is called nano-bioremediation (Yadav et al. 2017).

In general, nano-bioremediation involves the use of nanomaterials either in in situ (within) or ex situ (off place) treatment of contaminated material (Latha and Gowri 2014). Due to the some negative sides of other traditional methods, green method is widely used for synthesis of nanoparticles. Biological route for synthesis of nanoparticles is eco-friendly, cost-effective, and stable and has grown markedly to create novel materials (Prasad et al. 2018). Microbial nanotechnology is also a newly emerged route to produce a nanoparticulate catalyst that involves the precipitation of transition metals such as palladium, gold, and iron on bacterial cell wall, resulting in the formation of bionano-material (Johnson et al. 2013). The efficiency of degradation of contaminants can be enhanced by combining different treatment technologies. Recently nano-bioremediation has been suggested as efficient, eco-friendly, and economically cheap alternative for both resource conservation and environmental remediation. The nanotechnology for wastewater treatment can be classified into three main categories on the basis of nature of nanomaterials: nano-adsorbents, catalysts, and membranes.

### *16.4.1 Nano-adsorbents*

In recent years, nanoparticles are being used as potent adsorbents. The nanoparticles due to having small size have high surface area to volume ratio which increases the adsorption capacity and chemical activity of nanomaterials for adsorption of metals on their surface (Kalfa et al. 2009). The two major factors that affect the adsorption phenomenon are adsorption coefficient $K_d$ and recitation partitioning of pollutants (Hu et al. 2010). The nanoparticles such as zinc oxides, magnesium oxide, titanium oxide, ferric oxide, and carbon-based nanomaterials have been used for contaminants adsorption (Tyagi et al. 2017). Nano-adsorbents possess two main properties that are innate surface and external functionalization. The large surface area to volume ratio, adsorption capacity, pH, temperature, chemical reactivity, location of atoms, incubation time, and high binding energy on surface are important factors that affect the adsorption rate (Khajeh et al. 2013). The nanoparticles used in remediation process of wastewater must be nontoxic and high adsorption and desorption capacity.

#### 16.4.1.1 Oxide-Based Nano-adsorbents

Oxide-based nano-adsorbents are organic and inorganic ultrafine materials which are usually synthesized by utilizing metals and nonmetals. These include titanium oxides, iron oxides, zinc oxides, magnesium oxides, and manganese oxides (Tyagi et al. 2017). The natural occurrence of metal and nonmetal oxides and its simple synthesis process make them cheap materials for the removal of wastewater contaminants. The iron oxide nanoparticles such as $\alpha$-$Fe_2O_3$ are found to remove most of the orange II (a common azo dye) at room temperature. The decolorization of dye is due to the electrostatic force of attraction between the iron oxide surface and orange II (Fig. 16.2). Furthermore, the "catalytic combustion at 300 °C in air for 3 h regenerate the iron oxide containing Orange II dye and the regenerated a-$Fe_2O_3$ material have almost the same adsorption rate" (Zhong et al. 2006). The $Fe_3O_4$ nanoparticles incorporated with polylysine ($Fe_3O_4$@GPTMS@P-Lys) absorb methylene blue, and Acid Red 18 was studied by Zhang et al. The copper oxide nanoparticles in association with activated carbon (Cu-NP-AC) under pH 2, contact time 25–30 min, adsorbent dosage (0.01–0.06 g), and temperature 333 K were found as potential adsorbent for removal of Acid Blue 129 (Nekouei et al. 2015). The ability of $MnO_5CuO_5Fe_2O_4$ nanospinels to adsorb brilliant green from textile effluents was analyzed by Hashemian et al. (2015). Amorphous metal oxide nanoparticle ($Fe_2O_3$, CoO, NiO) fabrication via a green process (laser irradiation in liquid) and their ability to remove methylene blue was tested by Li et al. (2015).

#### 16.4.1.2 Carbon-Based Adsorbents

Activated Carbon

Carbon-based nanomaterials have been proved as more advantageous than available materials due to having manageable pore size and large surface area. Activated carbon is the most frequently used method for dye decolorization by adsorption. Carbon materials are very effective for adsorbing cationic, reactive, disperse, and

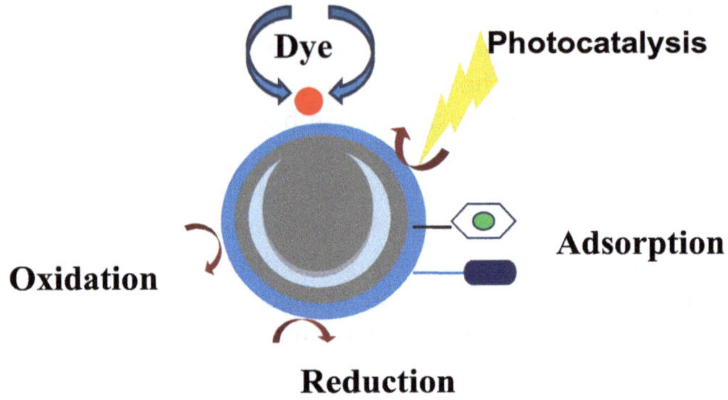


**Fig. 16.2** Different surface activities of oxide nanoparticles

acid dyes (Rao et al. 1994). Nano-MgO-encapsulated activated carbon was used for the removal of carcinogenic (CR) dye (Daniel and Syed 2015). Adsorption efficiency depends upon the type of carbon used and the characteristics of the effluents. Dye removal efficiency can be improved by using doses of carbon. This adsorption phenomenon has some drawbacks such as carbon that is expensive and has to be reactivated after treatment. Reactivation of activated carbon results in 10–15% loss of the sorbent.

Carbon Nanotube

Carbon nanotubes are emerging potent adsorptive materials that have a great potential to remove contaminants from wastewater through adsorption because of its high surface area, cylindrical hollow structure, hydrophobic walls, and flexible surfaces (Lu et al. 2006). From the researches, it has been analyzed that single-walled carbon nanotubes (SWCNTs) show higher adsorption capacity than multi-walled carbon nanotubes (MWCNTs). The hexagonal arrangements of carbon atoms allow them to interact with other molecules or atoms through hydrophobic and π-π electronic interaction. The batch experiment studies to analyze adsorption rate of cationic methylene blue (MB) and anionic orange II (OII), from aqueous solution by using multi-walled nanotubes (MWNTs) and carbon nanofibers (CNF) as adsorbents, were analyzed by Rodríguez and coworkers (Rodriguez et al. 2010). Yao et al. reported 41.63 mg/g adsorption capacity of carbon nanotubes (CNTs) for removal of methylene blue at 333 K. Some researchers performed a comparative study of multi-walled carbon nanotubes (MWCNTs) and single-walled carbon nanotubes (SWCNTs) as nanoadsorbents for the removal of reactive blue 4 (RB4) textile dye from aqueous solutions as a function of effects of pH, shaking time, and temperature (Machado et al. 2012). Magnetic-modified multi-walled carbon nanotubes (MWCNTs) were found effective for removal of cationic dyes crystal violet (CV), thionine (Th), Janus green B (JG), and methylene blue (MB). Impregnation of chitosan beads with carbon nanotubes was investigated as more efficient adsorbent for dye removal (Chatterjee et al. 2010).

### *16.4.2 Nano-catalysts*

The inorganic nano-catalysts such as semiconductor and metal oxides are gaining a noticeable attention for water and wastewater treatment. Different kinds of nano-catalysts are used for remediation, for example, photocatalysts, electrocatalysts, and Fenton-based nano-catalysts for improving chemical oxidation of pollutants (Table 16.2). Photocatalytic oxidation is the latest remediation technique. It utilizes ultraviolet (UV) illumination and titanium oxide ($TiO_2$) and generates low waste. The UV light helps to enhance the photocatalytic activity of $TiO_2$ during dye degradation. Among various nano-catalysts, $TiO_2$ is highly applied in photocatalytic reaction due to its high reactivity and chemical balancing (Akhavan 2009). Zinc oxide has been found to be well suited for photocatalysis of dye effluents, as it contains broad bandgap like $TiO_2$ (Lin et al. 2014). The use of nano-catalysts in chemical oxidation process has some advantages such as short period of time, target recalcitrant substances, and efficient transformation ability. The nano-catalysts are beneficial, but also have some negative aspects such as high cost of nanoparticles and inadequate reusability.

**Table 16.2** Photocatalytic efficiency of nano-catalysts for removal of dye from textile effluents

| Catalysts | Light region (nm) | Pollutant | Results | References |
|---|---|---|---|---|
| $Ti_2CO$ coating | 420 | Methylene blue (MB) | Removal of dye from 10 to 8.5 μ mol/L | Guan et al. (2016) |
| $TiO_2$/trititanate | UV-vis region | Rhodamine B (RB) | The degradation efficiency >91% was achieved | Chen et al. (2015) |
| AgBr/Zno | 410 | Methylene blue (MB) | Up to 87% MB was decomposed after 240 min | Dai et al. (2014) |
| 3D SnO | 365 | Methyl orange (MO) | Photocatalytic degradation of MO was 83% after 150 min | Cui et al. (2015) |
| ZnO nanorods | 365 | Rhodamine B (RB) | The quenched catalyzed improved the photocatalytic degradation of RB | Fang et al. (2015) |
| Zero-valent nanocopper | 420 | Methyl orange | 35% degradation of dye was achieved within 80 min | Liu et al. (2016) |
| Biomorphic $TiO_2$ photonic crystal | >420 | Methyl orange | Up to 30% removal of dye was observed within 4 h | Wang et al. (2016) |
| $CuFe_2O_4@C_3N_4$ | >420 | Orange II | Around 98% of orange II is removed within 210 min | Yao et al. (2015) |
| Polyaniline ZnO | Visible region | Methylene blue and malachite green | 99% dye degradation was achieved under sunlight | Eskizeybcket al. (2012) |
| $Y_2O_3Eu^{3+}$ | 350–400 | Methylene blue | Up to 90% degradation efficiency of MB was observed | Ramgopal et al. (2015) |
| ZnO/Zn | 365 | Methylene blue | 98% dye degradation was achieved | Lin et al. (2014) |

### *16.4.3 Nanofiltration Membranes*

Membrane filtration technology is the current advanced wastewater treatment technology. Nano-membranes are fabricated by nanoparticles to enhance the remediation efficiency in the aspect of permeability, catalytic reactivity, and fouling resistance (Zhang et al. 2015). Membrane filtration technology is simple, economical, effective remediation, and low space requirement. Nano-membrane filtration technology is highly used for removal of dye and other contaminants from wastewater effluents (Jie et al. 2015). The literature review clearly indicates the great potential of nanofiltration membrane for textile effluents treatment. Some researchers reported the NF membranes exhibit good performance for dye removal. Lopes et al. observed the maximum dye rejection up to 95–99% by using nanofiltration membrane NF45 and DK1073 (Lopes et al. 2005). The performance of nanofiltration membrane MPS 31 was investigated for dye retention that varied from 90% to 97%. Sungpet and coworkers investigated the higher efficiency of secondary layer formed by MPF 36 due to having larger MWCO in the presence of reactive dye and chloride (Sungpet et al. 2004). Tang and Chen hypothesized the dye rejection is independent on dye concentration. They found that at a pressure of 5 bar if the concentration of dye gradually increases from 92 to 1583 ppm, the dye rejection rate remains constant (Tang and Cheng 2002). The efficiency of membrane for decolorization of effluents is also dependent on cross-flow velocity. Although the nanofiltration is advantageous, membrane fouling is a weakness of filtration technique. Dyes introduce the undesirable flux decline by forming a colloidal fouling layer. The membrane fouling can be reversed by using feed pre-treatment modification of membrane surface and by controlling membrane cleaning.

## 16.5 Mechanism for Textile Dye Effluents Adsorption

Adsorption phenomenon of dye depends upon three main parameters, i.e., solution osmolarity, nature of adsorbents and adsorbates. Thus, the factors which affect the adsorption phenomenon are size of the adsorbate and adsorbent, charge on the surface of adsorbate and adsorbent, and pH and temperature of the solution. Generally, the redox chemical reactions take place on adsorbent surface, and the force responsible for adsorption includes electrostatic force of attraction, π-π bonding, hydrophobic interaction, and covalent bonding. There are various functional groups such as –OH, -COOH, and C-O which are present on adsorbent surface that provide active sites for binding of dye molecules on the surface (Shim et al. 2001). Figure 16.3 shows the ion exchange reaction as membrane coated with nanoparticles comes in contact with dye molecule. The proton ($H^+$) from water molecule after the dissociation of hydrogen bonding was exchanged with the dye molecule that finally leads the adsorption of dye on nanoparticles surface.

## 16.6 Advantages and Disadvantages

Nanoscience is an emerging technology that has great potential for effective treatment of wastewater. The high adsorption capacity, less waste by-product generation, and low toxicity of oxide-based nanoparticles and nanofiltration membrane make them more affirmative for removal of dye-containing textile effluents. The synthesis and utilization of nanoparticles have been widely used due to their small size (nano range), easy synthesis, modification ability, biocompatibility, and high surface area to volume ratio (Prasad et al. 2016, 2018). The green route for synthesis of nanoparticles makes them more accessible, eco-friendly, and low-cost adsorbents for dye removal Aziz et al. 2015. Along with the advantageous properties, the nanoscience has some drawbacks such as less stability, high reactivity with any functional group, antifouling of membranes, and leaching of by-products in environment that directly affect the flora and fauna. The advance researches are required to enhance the stability of nanoparticles by reducing their surface energy, high rejection rate, and antifouling resistant for nano-membranes. Thus, it is utmost important to make nanotechnology more clean, cheap, and environment-friendly.

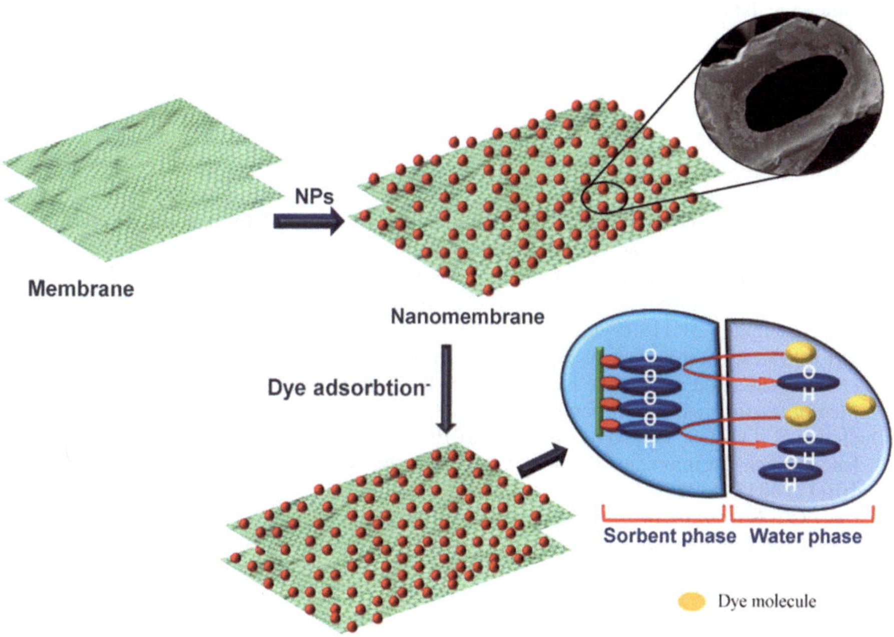


**Fig. 16.3** Mechanism for adsorption of textile dye effluents on nanoparticles

## 16.7 Conclusion

In present scenario, there is a great need of advanced technology for wastewater treatment to ensure better quality of water and eliminate chemical and biological contaminants. Thus, nanoscience is an emerging promising technology for advanced wastewater effluent treatment. This chapter illustrates the various nanomaterials that have been developed and investigated successfully for advanced treatment of wastewater effluents. The chapter focused the nanomaterials such as nano-adsorbents, nano-catalysts, and nanofiltration membrane for degradation of dye-containing wastewater effluents. The nano-adsorbents and nano-catalysts were found to have more specific pollutant removal efficiency. Indeed, nanomaterials are more efficient than any conventional method for advanced treatment of wastewater, but the technology has some negative side too that has to be resolved. Thus, more researches are required to re-evaluate the toxicity efficiency of advanced nanomaterials in order to make it more clean and eco-friendly.

**Acknowledgments** We acknowledge the ministry of human resource and development (MHRD), India, for providing financial assistance under the research project "Center for Excellence (CoE) – Water and Energy." Authors are deeply grateful to Prof. Aditya Shastri for giving the research amenities and the Bioinformatics Centre, Banasthali Vidyapeeth, Rajasthan (India) for use of computational facilities.

## References

Akhavan O (2009) Lasting antibacterial activities of Ag–TiO2/Ag/a-TiO2 nanocomposite thin film photocatalysts under solar light irradiation. J Colloid Interface Sci 336(1):117–124

Aziz N, Faraz M, Pandey R, Sakir M, Fatma T, Varma A, Barman I, Prasad R (2015) Facile algae-derived route to biogenic silver nanoparticles: synthesis, antibacterial and photocatalytic properties. Langmuir 31:11605–11612. https://doi.org/10.1021/acs.langmuir.5b03081

Casieri L, Varese G, Anastasi A, Prigione V, Svobodova K, Filippelo V (2008) Decolorization and detoxification of reactive industrial dyes by immobilized fungi *Trametes pubescens* and *Pleurotus ostreatus*. Folia Microbiol 53:44–52

Chatterjee S, Lee DS, Lee MW, Woo SH (2010) Enhanced molar sorption ratio for naphthalene through the impregnation of surfactant into chitosan hydrogel beads. Bioresour Technol 101(12):4315–4321

Chaves LCC, Navoni JA, de Morais Ferreira D, de Medeiros SB, da Costa TF, Petta RA, do Amaral VS (2016) Water mutagenic potential assessment on a semiarid aquatic ecosystem under influence of heavy metals and natural radioactivity using micronuclei test. Environ Sci Pollut Res 23(8):7572–7581

Chen CC, Lin CL, Chen LC (2015) Functionalized carbon nanomaterial supported palladium nano-catalysts for electrocatalytic glucose oxidation reaction. Electrochim Acta 152:408–416

Cui Y, Wang F, Iqbal MZ, Wang Z, Li Y, Tu J (2015) Synthesis of novel 3D SnO flower-like hierarchical architectures self-assembled by nano-leaves and its photocatalysis. Mater Res Bull 70:784–788

Dai K, Lv J, Lu L, Liu Q, Zhu G, Li D (2014) Synthesis of micro-nano heterostructure AgBr/ZnO composite for advanced visible light photocatalysis. Mater Lett 130:5–8

Daniel S, Syed PS (2015) Sequestration of carcinogenic dye in waste water by utilizing an encapsulated activated carbon with nano MgO. Int J Chem Technol Res 7:2235–2243

Elango G, Rathika G, Elango S (2017) Physico-chemical parameters of textile dyeing effluent and its impacts with case study. Int J Res Chem Environ 7(1):17–24

Eskizeybek V, Sarı F, Gülce H, Gülce A, Avcı A (2012) Preparation of the new polyaniline/ZnO nanocomposite and its photocatalytic activity for degradation of methylene blue and malachite green dyes under UV and natural sun lights irradiations. Appl Catal B Environ 119:197–206

Fang J, Fan H, Ma Y, Wang Z, Chang Q (2015) Surface defects control for ZnO nanorods synthesized by quenching and their anti-recombination in photocatalysis. Appl Surf Sci 332:47–54

Firmino PIM, da Silva MER, Cervantes FJ, dos Santos AB (2010) Colour removal of dyes from synthetic and real textile wastewaters in one-and two-stage anaerobic systems. Bioresour Technol 101(20):7773–7779

Guan S, Hao L, Lu Y, Yoshida H, Pan F, Asanuma H (2016) Fabrication of oxygen-deficient $TiO_2$ coatings with nano-fiber morphology for visible-light photocatalysis. Mater Sci Semicond Process 41:358–363

Hashemian S, Dehghanpor A, Moghahed M (2015) Cu0.5Mn0.5Fe2O4 nano spinels as potential sorbent for adsorption of Brilliant Green. J Ind Eng Chem 24:308–314

Hu H, Wang Z, Pan L (2010) Synthesis of monodisperse Fe3O4@ silica core–shell microspheres and their application for removal of heavy metal ions from water. J Alloys Compd 492(1–2):656–661

Huang J, Cao Y, Liu Z, Deng Z, Tang F, Wang W (2012) Efficient removal of heavy metal ions from water system by titanate nanoflowers. Chem Eng J 180:75–80

Hunger K, Sewekow U (2003) Health and safety aspects. In: Hunger K (ed) Industrial dyes: chemistry, properties, applications. Wiley, Weinheim, pp 625–641

Jie G, Kongyin Z, Xinxin Z, Zhijiang C, Min C, Tian C, Junfu W (2015) Preparation and characterization of carboxyl multi-walled carbon nanotubes/calcium alginate composite hydrogel nano-filtration membrane. Mater Lett 157:112–115

Johnson A, Merilis G, Hastings J, Palmer ME, Fitts JP, Chidambaram D (2013) Reductive degradation of organic compounds using microbial nanotechnology. J Electrochem Soc 160(1):4613–4651

Kalfa OM, Yalçınkaya Ö, Türker AR (2009) Synthesis of nano B2O3/TiO2 composite material as a new solid phase extractor and its application to preconcentration and separation of cadmium. J Hazard Mater 166(1):455–461

Khajeh M, Laurent S, Dastafkan K (2013) Nanoadsorbents: classification, preparation, and applications (with emphasis on aqueous media). Chem Rev 113(10):7728–7768

Latha N, Gowri M (2014) Bio synthesis and characterisation of $Fe_3O_4$ nanoparticles using *Caricaya papaya* leaves extract synthesis. Int J Sci Res 5:12–17

Li LH, Xiao J, Liu P, Yang GW (2015) Super adsorption capability from amorphousization of metal oxide nanoparticles for dye removal. Sci Rep 5:9028

Li W, Zhang Y, Tian G, Xie S, Xu Q, Wang L, Bu Y (2016) Fabrication of graphene-modified nano-sized red phosphorus for enhanced photocatalytic performance. J Mol Catal A Chem 423:356–364

Lin ST, Thirumavalavan M, Jiang TY, Lee JF (2014) Synthesis of ZnO/Zn nano photocatalyst using modified polysaccharides for photodegradation of dyes. Carbohydr Polym 105:1–9

Lopes CN, Petrus JCC, Riella HG (2005) Color and COD retention by nanofiltration membranes. Desalination 172:77–83

Lu C, Chiu H, Liu C (2006) Removal of zinc (II) from aqueous solution by purified carbon nanotubes: kinetics and equilibrium studies. Ind Eng Chem Res 45(8):2850–2855

Machado FM, Bergmann CP, Lima EC, Royer B, De Souza FE, Jauris IM, Fagan SB (2012) Adsorption of Reactive Blue 4 dye from water solutions by carbon nanotubes: experiment and theory. Phys Chem Chem Phys 14(31):11139–11153

Mahmoodi NM, Arami M (2009) Numerical finite volume modeling of dye decolorization using immobilized titania nanophotocatalysis. Chem Eng J 146(2):189–193

Mie R, Samsudin MW, Din LB, Ahmad A, Ibrahim N, Adnan SNA (2014) Synthesis of silver nanoparticles with antibacterial activity using the lichen *Parmotrema praesorediosum*. Int J Nanomed 9:121

Myslak ZW, Bolt HM (1988) Occupational exposure to azo dyes and risk of bladder cancer. Zbl Arbeitsmed 38:310–321

Nekouei F, Nekouei S, Tyagi I, Gupta VK (2015) Kinetic, thermodynamic and isotherm studies for acid blue 129 removal from liquids using copper oxide nanoparticle modified activated carbon as a novel adsorbent. J Mol Liq 201:124–133

Prasad R, Pandey R, Barman I (2016) Engineering tailored nanoparticles with microbes: quo vadis. WIREs Nanomed Nanobiotechnol 8:316–330. https://doi.org/10.1002/wnan.1363

Prasad R, Jha A, Prasad K (2018) Exploring the Realms of Nature for Nanosynthesis. Springer International Publishing (ISBN 978-3-319-99570-0) https://www.springer.com/978-3-319-99570-0

Qu X, Alvarez PJ, Li Q (2013) Applications of nanotechnology in water and wastewater treatment. Water Res 47(12):3931–3946

Ramgopal G, Vidya YS, Anantharaju KS, Prasad BD, Sharma SC, Prashantha SC, Nagabhushana H (2015) Bio-inspired synthesis of $Y_2O_3$: $Eu_3$+ red nanophosphor for eco-friendly photocatalysis. Spectrochim Acta A Mol Biomol Spectrosc 141:149–160

Rao KCLN, Krishnaiah K, Ashutush (1994) Color removal from a dye stuff industry effluent using activated carbon. Indian J Chem Technol 1:13–19

Rodríguez A, Ovejero G, Sotelo JL, Mestanza M, García J (2010) Adsorption of dyes on carbon nanomaterials from aqueous solutions. J Environ Sci Health A 45(12):1642–1653

Shah MP, Patel KA, Nair SS, Darji AM (2013) Isolation, identification and screening of dye decolorizing bacteria. Am J Microbiol Res 1(4):62–70

Shim JW, Park SJ, Ryu SK (2001) Effect of modification with $HNO_3$ and NaOH on metal adsorption by pitch-based activated carbon fibers. Carbon 39(11):1635–1642

Singh K, Arora S (2011) Removal of synthetic textile dyes from wastewaters: a critical review on present treatment technologies. Crit Rev Environ Sci Technol 41(9):807–878

Singh P, Iyengar L, Pandey A (2012) Bacterial decolorization and degradation of azo dyes in microbial degradation of xenobiotics. Springer, Berlin, pp 101–133

Solís M, Solís A, Pérez HI, Manjarrez N, Flores M (2012) Microbial decolouration of azo dyes: a review. Process Biochem 47(12):1723–1748

Sungpet A, Jiraratananon R, Luangsowan P (2004) Treatment of effluents from textile-rinsing operations by thermally stable nanofiltration membranes. Desalination 160:75–81

Tan BH, Teng TT, Omar AM (2000) Removal of dyes and industrial dye wastes by magnesium chloride. Water Res 34(2):597–601

Tang C, Chen V (2002) Nanofiltration of textile wastewater for water reuse. Desalination 143:11–20

Tronnier H (2002) Effect of textiles on human skın. SOFW J 128(4):8

Tyagi I, Gupta VK, Sadegh H, Ghoshekandi RS, Makhlouf ASH (2017) Nanoparticles as adsorbent; a positive approach for removal of noxious metal ions: a review. Sci Technol Dev 34(3):195–214

Wang Y, Xiong DB, Zhang W, Su H, Liu Q, Gu J, Zhang D (2016) Surface plasmon resonance of gold nanocrystals coupled with slow-photon-effect of biomorphic TiO2 photonic crystals for enhanced photocatalysis under visible-light. Catal Today 274:15–21

Yadav KK, Singh JK, Gupta N, Kumar V (2017) A review of nanobioremediation technologies for environmental cleanup: a novel biological approach. J Mater Environ Sci 8:740–757

Yao Y, Lu F, Zhu Y, Wei F, Liu X, Lian C, Wang S (2015) Magnetic core–shell CuFe2O4@ C3N4 hybrids for visible light photocatalysis of Orange II. J Hazard Mater 297:224–233

Zhang X, Lin B, Zhao K, Wei J, Guo J, Cui W, Li J (2015) A free-standing calcium alginate/polyacrylamide hydrogel nanofiltration membrane with high anti-fouling performance: preparation and characterization. Desalination 365:234–241

Zhong LS, Hu JS, Liang HP, Cao AM, Song WG, Wan LJ (2006) Self assembled 3D flowerlike iron oxide nanostructures and their application in water treatment. Adv Mater 18(18):2426–2431

# Chapter 17
# Carbon-Based Nanostructured Materials for Energy and Environmental Remediation Applications

**Shagufta Afreen, Rishabh Anand Omar, Neetu Talreja, Divya Chauhan, and Mohammad Ashfaq**

## Contents



S. Afreen
Dayanand Girls Post Graduate College, Kanpur, India

R. A. Omar
Centre for Environmental Science and Engineering, Indian Institute of Technology Kanpur, Kanpur, India

N. Talreja
Department of Bio-nanotechnology, Gachon University, Seongnam, South Korea

D. Chauhan (✉)
Department of Chemistry, Punjab University, Chandigarh, India

M. Ashfaq (✉)
School of Life Science, BS Abdur Rahaman Institute of Science and Technology, Chennai, India



R. Prasad, E. Aranda (eds.), *Approaches in Bioremediation*, Nanotechnology in the Life Sciences, https://doi.org/10.1007/978-3-030-02369-0_17

## 17.1 Introduction

Presently, global demands for technology are increasing continuously, despite the need to ensure the safety of the environment. These constant demands are exacerbating environmental issues such as global warming, climate change, natural resource depletion, waste disposal, loss of biodiversity, deforestation, ocean acidification, ozone layer depletion, acid rain, water pollution, and urban sprawl, which are responsible for environmental damage (Sadhu et al. 2018). These environmental problems have brought about serious issues associated with human and animal health. Another issue is related to energy, as the economic growth of a nation is usually associated with energy, which is an important input. Nonrenewable fossil fuels—such as coal, oil, and natural gas—are the major energy resources currently in use (Owusu and Sarkodie 2016). In this regard, protection of the environment and utilization of energy are two major important challenges that need to be addressed to realize a sustainable future.

Recently, nanomaterials—an open division of materials measuring less than 100 nm—have been widely used in various advanced applications such as electronics, biomedicine, energy storage, and environmental remediation (Prasad et al. 2016). Depending on their overall shape, these materials can be zero dimensional (0D), one dimensional (1D), two dimensional (2D), or three dimensional (3D). These materials show different properties from their bulk counterparts, due to their size (Manawi et al. 2018; Iijima 1991). The differences involve their pore size distribution, reactivity, surface functionality, tensile strength, conductivity, electrical properties, mechanical properties, optical functions, etc. Many benefits of nanotechnology depend on the fact that the properties of these materials can be easily modified at extremely small scales to perform specific functions, thus greatly extending the materials science toolkit (Khan et al. 2017). Different metals (e.g., copper, nickel, titanium, iron, and cobalt) or nonmetallic materials (especially carbon based materials, e.g., carbon, oxygen, nitrogen, and chlorine) have already been applied as adsorbents for environmental remediation and energy sources, and have greatly enhanced alternative options to meet the world's increasing demands to save the environment, as well as the need for energy.

Metal-based nanomaterials include nanowires, nanorods, and quantum dots. Among all of them, quantum dots are particularly promising for optical applications because of their higher extinction coefficients (Leatherdale et al. 2002). Similarly, nanorods also show potential applications in various technologies due to the reflectivity of the rods, which may be altered by changing the orientation with an applied electrical field. Nanowires may be used as additives in advanced composites, nanoscale metallic interconnects, and quantum devices, and as field-emitters, thereby being important in various electronic, optoelectronic, and nanoelectromechanical devices. Metal oxide–based nanoparticles such as titanium oxides, zinc oxides, magnesium oxides, manganese oxides, and ferric oxides are potential candidates to serve as adsorbents for environmental remediation (Gupta et al. 2015). Energy storage systems include electrochemical capacitors (ECs), batteries, and fuel cells (Winter and Brodd 2004). Metal-based electrode materials in these systems are very effective as they have high durability, high reactivity, and high electrical conductivity in comparison with other materials, but the low cycling stability of metal-based systems limits their application in these devices (Lukatskaya et al. 2016) In spite of their potential benefits, there are still some shortcomings—such as toxicity, corrosion, and stability—in the use of these materials as adsorbents for environmental remediation and energy applications.

Among the nanomaterials, carbon-based nanomaterials have attracted wide attention due to their unique structural features such as a large specific surface area with wide pore size distribution; a large aspect ratio; good thermal stability (Xin and Song 2015) in an inert atmosphere; intrinsic hydrophobic nature; presence of vast functional groups; fast mass transfer performance; high mechanical strength; and excellent electrical, mechanical, and optical properties, with ease of chemical modification due to the exposed surface and edges. These materials can be used in various applications such as sensors (Blasdel et al. 2015; Yan et al. 2014; Lipomi et al. 2011; Wujcik et al. 2013, 2014; Monty et al. 2013; Faraz et al. 2018), photovoltaics (Guo et al. 2010; Wei et al. 2007; Ross et al. 2009), field emission transistors (Javey et al. 2003; Wang et al. 2008; Xia et al. 2015), fuel cells (Li et al. 2003; Seger and Kamat 2009; Qie et al. 2012), supercapacitors (Stoller et al. 2008; Zhu et al. 2012a, 2014; Liu et al. 2010), composites (Ajayan et al. 1994; Stankovich et al. 2006; Zhao et al. 2010; Wang et al. 2014; Liu et al. 2015), biomaterials (Tang et al. 2010; Harrison & Atala 2007), and environmental remediation (Zhu et al. 2011, 2012b). Based on their size, shape, and dimensionality, carbon-based materials are classified into different categories: 0D carbon nanostructures (Buckminster fullerenes and carbon dots), 1D carbon nanostructures (carbon nanotubes (CNTs) and carbon nanofibers (CNFs)), 2D nanostructures (graphene), and 3D carbon nanostructures (carbon sponges) (Visakh 2016). Usually, the physical and chemical behaviors of these carbon nanomaterials are strongly determined by their structure and interfacial interactions with surrounding bulk materials. Each type of carbon material has its own pros and cons.

This chapter discusses the general contribution of carbon-based nanostructured materials in environmental remediation and energy storage applications. Different types of carbon nanostructured materials and their classification based on dimensionality and size are discussed. The contributions of the unique features of carbon nanostructures in two major applications—environmental remediation and energy storage—and future possibilities and challenges are discussed.

## 17.2 Dimensionality-Based Classification of Carbon Nanostructures

Carbon nanomaterials can be classified into four different dimensional categories: (1) 0D materials are defined as those that have three dimensions in the nanometer size range, such as fullerene and atomic clusters; (2) 1D materials consist of two dimensions in the nanometer size range, such as multilayers, nanotubes, nanorods, nanowires, and nanofibers; (3) 2D materials have one dimension in the nanometer size range, such as graphene; and (4) 3D nanomaterials contain equiaxed nanometer-sized grains, e.g., carbon sponges. Figure 17.1 shows the classification of carbon nanomaterials on the basis of dimensionality.

## 17.2.1 *Zero-Dimensional Materials*

### 17.2.1.1 Fullerenes

Buckminsterfullerene ($C_{60}$)—the first 0D carbon material to be discovered—was found in 1985 by Curl, Kroto, and Smalley while they were vaporizing graphite; they were awarded a Nobel Prize for this discovery in 1996. As Fuller had designed a spherical roof structure, this molecule was named fullerene. The diameter of the $C_{60}$ molecule is around 1 nm in range, and it is the smallest member of the fullerene family, representing an excellent 0D nanomaterial for nanoscience. After graphite and diamond, fullerene is the third form of carbon (Sarkar 2011). Fullerenes have unique characteristics, and thereby their properties are easily modified. Modification of fullerenes affects their electronic structure, solubility, and physical properties, thereby enabling their use in various applications, including photoenergy storage, environmental remediation, sensing, photovoltaics, and photoresists. Fullerenes can

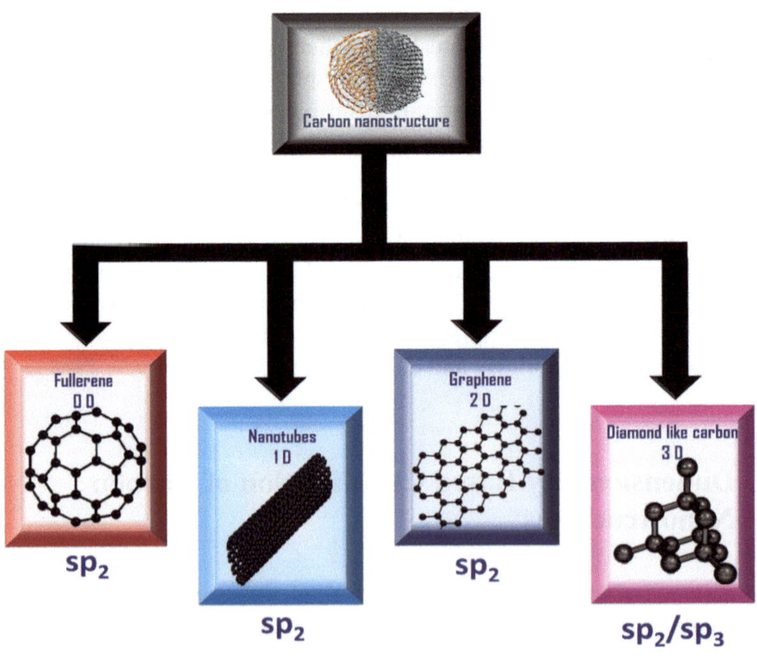


**Fig. 17.1** Classification of carbon nanomaterials on the basis of dimensionality

display a photochromic effect, which is a change in light transmission based on intensity (Bakry et al. 2007).

## 17.2.2 One-Dimensional Materials

### 17.2.2.1 Carbon Nanotubes and Nanofibers

CNTs and CNFs are the most important 1D nanostructures. Significant progress in the development of these materials has attracted attention due to their applicability in large areas. CNTs and CNFs have high mechanical strength and high chemical stability with a combination of unique electrical properties, making them potentially useful in a wide variety of applications (Sankararamakrishnan and Chauhan 2014).

A CNT has a tube-like structure and is the strongest 1D material, with a diameter of a few nanometers and a length of up to 18 cm (Javey and Kong 2009). CNTs were first synthesized in 1991 by Iijima, using an arc evaporation method. CNTs are very promising candidates in various fields such as electronics, agriculture, environmental remediation, and photovoltaics. Due to their tunable structure they can be easily modified according to the application required. Metallic CNTs have aroused a lot of research interest for their applicability as very-large-scale integration (VLSI) interconnects due to their high thermal stability, high thermal conductivity, and large current-carrying capability (Kaushik and Majumder 2015). CNT can carry a current density in excess of 103 $MA/cm^2$, which can enhance electrical performance as well as eliminating the electromigration reliability concerns that plague current nanoscale Cu interconnects (Wei et al. 2001).

CNFs have a great advantage over CNTs and its toxicity is insignificant (Ashfaq et al. 2013) compared with that of a CNT. Like CNTs, CNFs are potential candidates for various applications such as environmental remediation (Khare et al. 2013; Talreja et al. 2014), energy storage, drug delivery (Ashfaq et al. 2014), antibiotic materials (Ashfaq et al. 2016a), wound dressing (Ashfaq et al. 2016b), and agriculture (Ashfaq et al. 2017; Kumar et al. 2018).

## 17.2.3 Two-Dimensional Materials

### 17.2.3.1 Graphene

Graphene was first isolated in 2004 by Andre Geim and Kostya Novoselov, for which they were awarded a Nobel Prize. In contrast to the well-known instability of 2D materials, graphene is a stable 2D nanosheet due to its high thermal stability; its decomposition temperature is also higher than those of other 2D materials. Graphene contains $sp^2$-bonded carbon atoms with unique electronic properties due to a zero–band gap structure, band-tuning ability, extremely high carrier mobility, large surface area, thermal transport properties, and high chemical stability (Lu et al. 2013b); therefore, it can be applied in various applications such as energy storage devices, actuators, solar cells, environmental remediation, supercapacitors, and batteries. Graphene can be synthesized using various methods such as mechanical exfoliation, liquid phase exfoliation, and chemical vapor deposition (CVD). (Bhuyan et al. 2016). Among these methods, CVD of carbon atoms by use of a metal catalyst has attracted interest, as this method is reliable, viable, and suitable for large-scale production. There are various advantages of graphene grown by the CVD method, as CVD-grown graphene layers are very easy to transfer (Zheng et al. 2017).

## *17.2.4 Three-Dimensional Materials*

### 17.2.4.1 Carbon Sponges

A 3D nanostructure can be synthesized by modifying a 2D-sheet material into a 3D microporous structure by incorporation of individual chemically modified structures through various gelation technologies, using supramolecular interaction—mainly van der Waals forces, hydrogen bonding, dipole interactions, electrostatic interactions, and $\pi$–$\pi$ stacking. These structures are commonly known as hydrogels, aerogels, sponges, or forms. Common examples of 3D carbon-based architectures are carbon aerogels, sponges, and carbon aerosols. 3D carbon-based materials have various advantages over other-dimensional materials, such as high efficiency, strong mechanical strength, high electrical conductivity, and superior thermal, chemical, and electrochemical stability (Nasir et al. 2018; Kroto et al. 1985), which can be helpful for making economical devices especially for energy applications. Carbon aerogels and sponges demonstrate not only good selectivity in purification technology but also excellent adsorption and oil recovery ability (Wu et al. 2014; Moumita et al. 2016); thus, they are suitable for treatment of oil spills.

3D networks of carbon-based gels are also attractive for energy storage and conversion devices, sensors, and catalytic systems (Dutta et al. 2017). Moreover, several inherent merits—including large specific surface areas, stimuli-responsive properties, oxygen-containing functional groups, and conjugated domains—make them suitable candidates for removing pollutants from contaminated water. Various approaches have been adopted for preparation of 3D carbon-based architectures, which can be divided into four categories: (1) self-assembly using a chemical modification method; (2) template-guided approaches (Li et al. 2011); (3) solvothermal/organic sol–gel reactions (Feng and Li 2017); and (4) LightScribe patterning technology (Strong et al. 2012).

## 17.3 Environmental Remediation Applications

The persistent increase in pollution is a worldwide threat because of industrialization, urbanization, and also changing lifestyles, which produce smoke, poisonous gases, and fumes, released into the environment. Living organisms can be adversely affected by various types of contaminants such as biological, chemical, physical, and radiological contaminants. Therefore, the need for environmental cleanup is a major concern in the current scenario. Industrial activities are the main causes of pollutants, such as use of pesticides and fertilizers, or the release of other pollutants that destroy ecosystems.

Water contamination has become a leading concern over the last ten decades due to disposal of contaminated waste in the water system. These contaminants show adverse effects on humans, animals, and plants. The situation is becoming frightening due to increased industrialization and urbanization. Rapid industrialization activities include mining, fertilizer manufacturing and use, tanneries, metal plating, pesticide manufacturing and use, the paper industry, and battery manufacturing and use, which are leading sources of contamination of environmental systems (Ayangbenro and Babalola 2017; Mukherjee 2011; Ruomeng et al. 2017). Wastewater containing heavy metal ions is directly or indirectly released into bodies of water without any prior treatment (Ferdous et al. 2016). The most direct and most brutal impact is lack of proper sanitation, resulting in a lack of safe drinking water, which affects populations globally. Other threats include chemical toxicants (heavy metal ions, mainly chromium (Cr), zinc (Zn), mercury (Hg), copper (Cu), cadmium (Cd), arsenic (As), thallium (Tl), nickel (Ni), and lead (Pb)), various dyes, pharmaceutical waste (Khulbe & Matsuura 2018; Tchounwou et al. 2012), and exposure to pathogens, mainly bacteria. Effluent containing dyes contaminates water bodies, including groundwater, thereby making it unfit for agricultural and drinking purposes. Effluent containing dyes and pharmaceutical compounds contains various chemicals that may be toxic, carcinogenic, and mutagenic to living organisms, including humans and animals (Hai et al. 2007; Joshi et al. 2008; Saraswat et al. 2012; Talreja et al. 2016). Preventative measures that need to be implemented include reductions in use, recycling processes, or minimization of toxic chemicals, including chemical and biological contaminants. In this context, provision of safe drinking water and clean air is one of the biggest challenges nowadays.

### *17.3.1 Chemical Contaminants*

Contaminants may occur naturally or may be derived from human-related activities, such as industry, agriculture, pesticide use, and mining (Wuana and Okieimen 2011; Jaiswal et al. 2012). The presence of chemical contaminants in water systems has increased with time. However, the presence of contamination does not necessarily translate into adverse effects on human health. Epidemiological and toxicological studies of chemical contaminants are required for assessment of their effects on human health (Villanueva et al. 2014). The toxicity of chemical contaminants depends on the types of contaminants and its concentration. Exposure to larger amounts of contaminants may be dangerous for human health. Usually, such exposure occurs through ingestion (eating/drinking) or inhalation. Working in industries or living around industrial sites or dumping areas also pose risks of exposure. Additionally, certain lifestyles may increase exposure and adverse health effects due to hunting and gathering activities (Martin & Griswold 2009).

Certain chemical contaminants—mainly xenobiotic compounds (Awasthi et al. 2009; Singh et al. 2011) and heavy metal ions—are not biodegradable (Lakherwal 2014); therefore, they accumulate in living systems. Most heavy metal ions are recognized to be cytotoxic or carcinogenic in nature. Among all heavy metal ions, Zn is a trace element that is required for human health; it is mainly required for physiological functions of tissue and also regulates various biochemical processes (Prashanth et al. 2015). Nonetheless, excessive amounts of Zn may cause various health issues, including stomach cramps, skin irritations, vomiting, anemia, and nausea (Mustafa et al. 2011). Cu is also an essential element, which is involved in metabolism. However, excessive intake of Cu causes serious complications, including vomiting, convulsions, or even death (Ashish et al. 2013). Excessive amounts of Ni also cause severe complications, including lung and kidney-related diseases, pulmonary fibrosis, gastrointestinal distress, and dermatitis.

Arsenic (As) is considered one of the most hazardous metal contaminants and is found in most parts of the world. Various industries—mainly those involving wood preservation or manufacturing of semiconductors, chemicals, paints, etc.—are the main sources of As contamination (Jaishankar et al. 2014). Other than those sources, As can be released in larger quantities through volcanic activity, erosion of rocks, and forest fires. Ninety percent of industrial As is used in the wood-preserving industry. Animal feeding operations and use of fertilizers and pesticides may also release large amounts of As into the environment (Aktar et al. 2009). According to the US Environmental Protection Agency (EPA), the permissible limit of As in drinking water is 10 parts per billion (ppb). As is odorless and tasteless, is a known carcinogenic agent, and can cause various cancers—mainly skin, lung, liver, and bladder cancer (Sankararamakrishnan et al. 2016). Low-level exposure to As may cause vomiting, nausea, blood vessel damage, darkening of the skin, and a sensation of pins and needles, and it decreases the production of erythrocytes and white blood cells.

Cadmium (Cd) is a very toxic metal and is present in soils and rocks, including coal and mineral fertilizers. It is widely used in pigments, batteries, metal coatings, electroplating, and plastics. It is also a known carcinogenic agent (Jaishankar et al. 2014) and has been characterized by the EPA as a carcinogenic agent for humans. According to the EPA, the permissible limit of Cd in drinking water is 5 ppb. Exposure to Cd may cause vomiting, diarrhea, and severe stomach irritation, leading to kidney damage, lung damage, and bone fragility. Chronic exposure to Cd causes kidney dysfunction, and an even higher level of exposure leads to death.

Chromium (Cr) is found in rocks, animals, plants, and soil, and may occur in the form of a liquid, solid, or gas. It is extensively used in metal alloys such as stainless steel, protective coatings on metal (electroplating), magnetic tapes, and pigments for paints, cement, paper, rubber, composite floor coverings, and other materials (Martin & Griswold 2009). Soluble forms of Cr are used in wood preservatives. According to the EPA, the permissible limit of Cr in drinking water is 100 ppb. Cr exits in two states: trivalent Cr(III) and hexavalent Cr(VI). Cr(VI) is highly toxic compared with Cr(III). Cr(VI) affects human physiology and accumulates in the food chain, thereby leading to various health issues such as damage to the lining of the nose, nose ulcers, a runny nose, breathing problems (such as asthma, cough, shortness of breath, or wheezing), skin ulcers, and skin irritation. It can also cause liver, kidney, circulatory, and nerve tissue damage (Saha et al. 2011; Chauhan et al. 2012).

Lead (Pb) is widely used in batteries, ammunition, metal products such as solder and pipes, paints, gasoline, and X-ray shielding devices. It may damage the central nervous system, kidneys, liver, reproductive system, cellular processes, and brain functions. The symptoms of Pb exposure are muscle weakness, renal failure, hallucinations, headache, dizziness, and irritability. A higher level of exposure to Pb may cause miscarriage in pregnant women, while in men it may damage the organ responsible for the production of sperm (Bhartiya and Singh 2012).

Mercury (Hg) combines with other elements to produce metallic Hg, which is widely used to produce chlorine gas, caustic soda, batteries, switches, bulbs, and dental fillings, and is also used in thermometers (Devito and Brooks 2005). Hg is a known neurotoxin, which may damage the central nervous system (Kim and Kim 2012). Exposure to large amounts of Hg can cause chest pain, kidney dysfunction, pulmonary destruction, and dyspnea. According to the EPA, mercuric chloride and methyl mercury are possibly human carcinogenic agents. The permissible limit of Hg in drinking water is 2 ppb. Higher exposure to Hg may cause permanent damage to the brain, kidneys, and developing fetuses. The consequences for brain function may result in irritability, shyness, tremors, changes in vision or hearing, and memory problems. Low-level exposure may cause nausea, vomiting, diarrhea, high blood pressure, skin rashes, and eye irritation (Do et al. 2017).

Selenium (Se) is a trace mineral extensively distributed in rocks and soils. It is extensively used in electronics, nutritional supplements, the glass industry, plastics, paints, enamels, inks, rubber, pharmaceuticals, poultry, livestock, pesticides, rubber, antidandruff agents, fungicidal agents, and medical diagnostics (Martin & Griswold 2009). Trace amounts of Se are required for cellular function, but larger amounts are toxic (Jaishankar et al. 2014). Se plays a role in the functioning of the thyroid gland (Köhrle 1999). The tolerable upper intake level is 400 μg of Se/day. Consumption of larger amounts may lead to selenosis. The symptoms are neurological defects, loss of hair, and nail brittleness (Morris and Crane, 2013). Longer-term exposure can cause respiratory irritation, bronchial spasms, and coughing.

Silver (Ag) often combines with other elements, mainly sulfide, chloride, and nitrate. It is extensively used in antibacterial agents, disinfectants, dental filling, and electronics (Beyth et al. 2015). According to the EPA, Ag is not classifiable as a human carcinogenic agent. High levels of exposure may result in a condition called argyria—a blue-gray discoloration of the skin and other body tissues. Argyria appears to be a cosmetic problem, which may not be otherwise harmful to health. A high level of exposure to Ag may cause breathing problems, lung and throat irritation, and stomach pains (Martin & Griswold 2009). Skin contact with Ag may cause mild allergic reactions such as a rash or inflammation in some people. According to the EPA, the permissible limit of Ag in drinking water is 0.1 ppb (Kumar et al. 2011).

Fluorine (F) contamination is a major cause of morbidity globally. The F ion has been extensively used in dental treatment because of its anticariogenic and antimicrobial properties (Unosson et al. 2012). The antibacterial property of F is due to acidification of the bacterial cytoplasm. Formation of calcium fluoride, reduced hypersensitivity, osteoblast proliferation, and firmer bone anchorage are other roles of F in biological systems (Ullah and Zafar 2015). Therefore, F has been added to various biomaterials such as bioceramics, glass, composite materials, and surface coatings of dental implants. Incorporation of F may enhance the stability of biomaterials. On the other hand, high intake of F can give rise to dental fluorosis (an unattractive brown mottling of teeth) and skeletal fluorosis (which increases bone density and can eventually lead to fractures and crippling skeletal distortion).

### *17.3.2 Gaseous Contaminants*

Energy has strong impacts on the environment, especially when energy is produced from fossil fuels such as coal, oil, gas, and wood. The burning of fossil fuels results in fog, smog, and global warming, which cause declines in forests, vegetation, and human health. Humans may be affected by asthma and cancerous diseases (Cengel and Boles 2007; Slezakova et al. 2011). Heart and lung-related disease has also been observed due to air pollution and may lead to death (Foster and Kumar 2011; Nasrallah and Balling 1995; Pope et al. 2002). Nowadays, approximately 85% of the world's energy resources are obtained from combustion of fossil fuels. Several harmful elements and gases are emitted from vehicles, furnaces, stoves, and electrical power plants (Fan et al. 2008; Frey et al. 2009; Guo et al. 2006; Shen et al. 2007; South 1993). Generally, carbon monoxide (CO), unburned hydrocarbons (UHC) and nitrogen oxides (NOx) (Huo et al. 2010; Kwok et al. 2010; Liu et al. 2010; Romm 2006; Zhang et al. 2008; Zheng et al. 2009), sulfur oxides (SOx), particulate matter (e.g., $PM_{10}$ and $PM_{2.5}$), ammonia, and soot, in addition to volatile organic compounds (VOCs) (Duflo et al. 2008; Frankenberg et al. 2005; Greenstone and Gayer 2009; Reynolds et al. 2011), are the main combustion products of gas turbines, steam power, and vehicle engines. Whereas some of these gases occur naturally, such as carbon dioxide expelled in air from the lungs, they are serious pollutants when they come from burning of fossil fuels such as coal, oil, and natural gas. Figure 17.2 shows a schematic illustration of different types of gaseous contaminants.

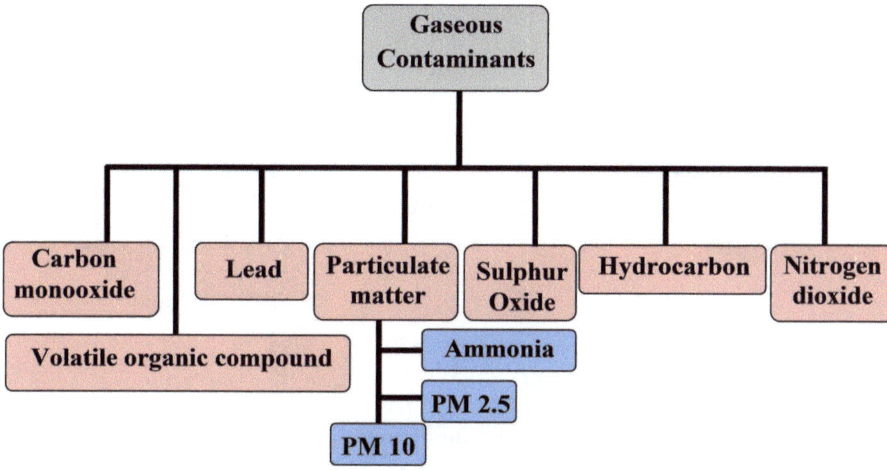


**Fig. 17.2** Different types of gaseous contaminants

### *17.3.3 Biological Contaminants*

Several organisms such as viruses, fungi, protozoa, dust mites, fleas, and bacteria contaminate environmental systems. These biological contaminants cause various diseases, thereby adversely affecting human health. Most such microorganisms—mainly bacteria—occur in nature and are released into the environment naturally, as well as synthetically via industrial farming and human sewage (Sivasubramaniam and Franks 2016). There are various sources of contamination, and bacteria are also found in soil and in the intestinal tracts of animals (Cabral 2010) (mainly coliform bacteria). Hence, coliform bacteria is found in the human gut but rarely causes adverse effects. However, dangerous toxin-producing coliform bacteria, such as *Escherichia coli* (*E. coli*), may cause bloody diarrhea and even kidney failure in humans (Chauhan et al. 2016).

Biological contaminants mainly cause airway diseases, which can induce immediate hypersensitivity (IgE) reactions, immunological reactions, and infection (Galli et al. 2008). Biological contaminants are potential toxins and irritants that may damage the function of organs, such as through irritant dermatitis, mycotoxin-induced flu-like symptoms, diarrhea, and cancer (usually from ingestion) (James and Seltzer 1994). *E. coli* and *Staphylococcus aureus* (*S. aureus*) bacteria are some of the most common biological contaminants in water systems, leading James & to hazardous effects on human health and causing several diseases associated with bacterial infection. These microorganisms also contaminate the air when they are released by living and dead organisms. Bacteria can also affect the skin, eyes, ears, and most body organs—mainly the nervous, respiratory, digestive, reproductive, urinary, and cardiac systems (Crosby 2017). Figure 17.3 shows the effects of various contaminants on human health.

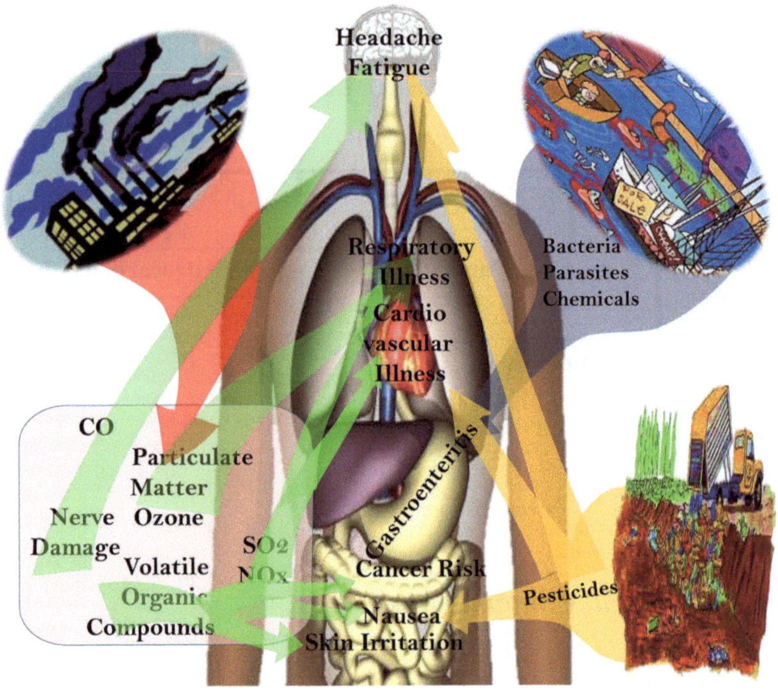


**Fig. 17.3** Effects of various contaminants on human health

In general, chemicals including heavy metal ions, gaseous contaminants, and biological (bacterial and viral) contaminants in the environment threaten the safety of drinking water and also cause various complications or even death.

### *17.3.4 Use of Carbon-Based Nanomaterials in Environmental Remediation Applications*

The introduction of nanotechnology has created enormous scope and opportunities for synthesis of various nanomaterials with unique characteristics to treat pollutants. Nanosized materials play crucial roles in environmental remediation applications such as the treatment of air, water, industrial and domestic wastewater, soil, and sediments. Using nanotechnology, various efficient, eco-friendly, and economically viable nanosized materials have been synthesized (Anjum et al. 2016), which have unique characteristics and potential ability to decontaminate industrial effluent and water bodies (surface, ground, and drinking water). Nanosized materials are categorized by their nature as nanoadsorbents, nanocatalysts, and nanomembranes.

Nanoadsorbent materials have been shown to efficiently remove pollutants from wastewater and air (Anjum et al. 2016). Various materials such as activated carbon,

CNTs, CNFs, graphene, metals and their oxides, silica, and clay have been widely used as adsorbent materials. Nanomaterials such as metal oxides and semiconductors have gained significant attention from researchers in the development of newer water remediation technologies. Several types of nanocatalyst are used for degradation of pollutants from wastewater, and some catalysts also have antimicrobial properties (Amin et al. 2014); therefore, they are also useful for removal of biological contaminants. Nanomembranes are widely used in treatment processes. Pressure-driven treatment of wastewater has been shown to be an ideal technology to remove contaminants from water. Nanomembranes are widely used in treatment of wastewater because of their small pore size, low cost, effectiveness, and eco-friendliness (Anjum et al. 2016). Nanomembranes are mainly developed by using metal nanoparticles, nonmetal particles, and carbon-based nanomaterials (Bhattacharya et al. 2016).

Carbon-based nanomaterials—mainly CNTs, CNFs, graphene, and fullerenes—are extensively used in removal of contaminants from air and water (Scida et al. 2011). Nonetheless, their small particle size, poor dispersion, and difficulty of separation are still concerns. To combat such issues, researchers have modified CNTs into multiwalled CNTs (MWCNTs). Magnetic MWCNTs have greater dispersion ability and can be easily separated from wastewater. Various studies (Khulbe & Matsuura 2018) have suggested that MWCNTs efficiently remove heavy metal ions such as Pb(II) and Mn(II). Some studies have suggested that surface-modified CNTs have better sorption ability in comparison with those that have no such modifications. CNFs are relatively newer materials produced using CVD techniques. CNFs are directly used in various applications such as environmental applications, sensors, and biological applications (Sharma et al. 2010). CNFs can be grown on solid substrates (activated carbon fibers or phenolic beads). CNFs grown on anisotropic conductive film (ACF) and phenolic resin beads also contain metal nanoparticles. The metal nanoparticles dispersed in such CNFs have potential ability to remove various contaminants such as As, Cr, Pb, F, vitamin $B_{12}$, and pharmaceutical effluent (Sharma et al. 2010).

Surface functionalization increases the overall adsorption process. Several processes have been used by different researchers, such as acidic treatment, functional group grafting, and metal impregnation. Surface functionalization alters different characteristics of the materials, such as the surface area, surface charge, hydrophobicity, and dispersion (Ghosh 2016). Acidic treatment of nanomaterials has been carried out using different types of acid such as $H_2O_2$, $H_2SO_4$, $HNO_3$, $KMnO_4$, and HCl (Dong et al. 2013; Mazov et al. 2012). Acidic treatment of nanomaterials removes their surface impurities and also allows adhesion of new functional groups to the surface, thereby increasing the sorption ability of the materials. Grafting of functional groups on the surface of nanomaterials is another process used to improve their surface characteristics, and is done using a plasma technique (Chen et al. 2001), chemical modification (Liu et al. 2007), or a microwave process (Zhang et al. 2014). Among all of these, the plasma technique is preferable because it requires less energy and is an eco-friendly process. Moreover, carbon-based nanomaterials modified by using metal or metal oxides show potential ability to remove heavy metal ions.

Graphene-based nanomaterials have unique characteristics such as a large surface area, mechanical strength, light weight, flexibility, and chemical stability, which makes them suitable candidates for various environmental applications (Ke & Wang 2016). Graphene and its oxide are used as adsorbents for removal of heavy metal ions (Yuxi & Gaoquan 2011; Peng et al. 2017). The presence of hydroxyl and carboxyl functional groups on the surface of graphene oxide increases its sorption ability (Zheng & Kim 2015; Minitha et al. 2018). Various researchers have used graphene-based hybrid nanomaterials, such as graphene with metal or metal oxides, in environmental remediation applications to enhance the efficiency of the adsorbent.

In general, carbon-based nanomaterials have potential ability to remove various contaminants from air and water. The sorption ability is enhanced by surface functionalization and also incorporation of metal nanoparticles. Therefore, dispersion of metal nanoparticles in carbon-based nanomaterials has potential ability to remove various contaminants from water and air.

## 17.4 Energy Applications

Carbon-based materials play wide roles in the development of advanced energy storage devices. Due to their large surface area, pore size distribution, mechanical properties, and electrical properties, these materials are considered some of the most promising ones and are involved in a wide range of research interests. Carbon nanomaterials such as CNTs, fullerenes, carbon dots, and graphene are finding relevant applications in the energy storage field, which are described in the following sections.

### *17.4.1 Dye-Sensitized Solar Cells*

Solar cells, also known as photovoltaic cells, are a low-cost solution for renewable energy and were discovered in 1991 by Regan and Graetzel. Solar cells can easily convert photons of a specific wavelength into electricity. Solar cells are mainly classified into two groups: (1) crystalline silicon; and (2) thin films. Common semiconductors used in first- and second-generation solar cells are mainly crystalline silicon, III–V compounds, cadmium telluride, and copper indium selenide/sulfide (Zulkifili et al. 2015; Yun et al. 2014; Zhenqing et al. 2005). The thin-film group includes economically viable energy conversion devices with an easy and simple synthesis/fabrication process. Some wide–band gap semiconductor electrodes are created by incorporation of dye molecules. Pt is supposed to be a widely used material in dye-sensitized solar cells (DSSCs). However the high cost and scarcity of platinum has encouraged the scientific community to search for other appropriate materials for the counter electrodes of DSSCs (Hwang et al. 2015; Wujcik & Monty 2013; O'Regan & Graetzel 1991).

Carbon-based materials have extraordinary properties, including high conductivity, good electrochemical activity, and low cost; therefore, they are considered favorable alternatives to expensive Pt in DSSCs. Among all carbon-based materials, graphene-based materials have been used as a medium to increase electron transportation and enhance the efficiency of solar energy conversion (Zheng and Zhang 2016). This supporting function (electron transportation) of graphene-based materials is also used in other applications such as batteries and fuel cells.

## *17.4.2 Supercapacitors*

Electrochemical energy storage devices, also referred to as supercapacitors, are potential next-generation energy storage devices for use in wide applications such as electric vehicles, hybrid electric vehicles, portable electronic devices, and backup power cells, due to various properties including their high power density, high cycling stability, and very short charging time (Lu et al. 2013a, b). Supercapacitors store energy mainly by two common mechanisms: (1) pseudocapacitance; and (2) electrochemical double-layer capacitance (EDLC). A pseudocapacitor mainly involves faradic reactions in charge transfer processes (Lee et al. 2018). Some metals and metal oxides or conducting polymers are good examples of the pseudocapacitance type of process. On the other hand, EDLC processes work through the charge accumulated at the electrode–electrolyte interface, using a fast adsorption/desorption process of electrolyte ions on large–surface area electrode materials, mainly carbon-based materials (Wu et al. 2017; Novoselov et al. 2004).

Carbon-based supercapacitors have various advantages over metal-based supercapacitor materials, such as high power density and high cycling stability, but their low energy density often limits their applications for use in batteries (Lukatskaya et al. 2016).

Compared with 0D and 1D nanostructures, 2D carbon-based materials have advantages for use as supercapacitors (Ke and Wang 2016). Among 2D materials, graphene is one of the most prominent structures. Graphene is a potential 2D material that has attracted more attention than other carbon materials for use in supercapacitors due to its high electrical conductivity, large surface area, and large surface to volume ratio (Jana et al. 2017). However, graphene sheets suffer some disadvantages due to their tendency to restack during utilization as electrode materials (Zhang et al. 2018). Researchers are still finding methods to solve this problem; one of the most promising methods is to introduce some functionality or metal doping over the graphene sheets to prevent restacking of the sheets. CNTs, which have a 1D structure, have been broadly used for EDLC. Small-diameter single-walled CNTs (SWCNTs) have a specific surface area of around 1000 $m^2/g$ and demonstrate capacitance of up to 100 F/g in nonaqueous/aqueous electrolytes (Lu and Dai 2010). CNTs offer an exposed surface to functionalize and high mechanical strength with excellent electrical properties that make them better candidates for energy storage, but have the disadvantage of moderate volumetric capacitance due to low density. Also, CNTs can be used as substrates for fabricating electrodes by decorating the CNT surface with a thin film of redox materials. Owing to their superior electrochemical performance, eco-friendliness, and lower production costs, carbon materials are still considered promising materials for EDLC and pseudocapacitance (Zuo et al. 2017).

### *17.4.3 Batteries*

The Li-ion battery is another form of energy storage material, which stores energy in the form of chemical energy. Li-ion batteries have advantages over capacitors due to their high energy and power density, and can significantly reduce greenhouse gas emissions (Zheng et al. 2017). Due to their high energy efficiency, they are used in various electronics applications. Recently, Li-ion batteries have received wide interest for use in both research and industries. Carbon materials are considered the best materials for Li-ion batteries. They can be easily lithiated to some extent. The structure of the carbon material usually defines some common factors such as the amount of lithium that is reversibly incorporated into the carbon lattice, the faradic losses during the first charge–discharge cycle, and the voltage profile during charging and discharging (Flandrois and Simon 1999). Carbon-based Li-ion batteries are supposed to be the best performers because they have higher energy density (which can reach 200 Wh/kg) than other metal-based materials (Julien et al. 2015). Carbon-based materials—including activated carbons, 1D CNTs, and 2D graphene nanosheets—are good candidates for technological applications for a variety of sustainable energy storage devices, as carbon materials possess various properties such as low cost, light weight, adaptable porosity, easy processability, and simplicity of chemical modification (Wang and Kaskel 2012). In general, the increasing specific surface area and pore size distribution of carbon structures allow better electrochemical capacitance performance in terms of both the power delivery rate and the energy storage capacity.

## 17.5 Conclusion and Future Prospects

This chapter has discussed the significance of carbon-based materials for energy conversion and environmental remediation applications. Various unique properties such as a large surface area, excellent pore size distribution, ease of porous texture modification, and chemical tunability of carbon-based materials make them suitable candidates for the aforementioned application. Functional groups on carbon nanostructures provide opportunities to tailor chemical functionalities and tune electronic properties. The unique heterogeneous electronic properties of some two-dimensional carbon nanostructures, due to the presence of $sp^2$ and $sp^3$ atoms, provide great opportunities to favor these materials for use in various electronic devices, as $sp^2$ domains are favorable sites for charge transfer in electronic applications. Their excellent usability for environmental applications has been demonstrated by tuning of their surface chemistry through tailoring of oxidation levels, surface modification, and implementation of substitutional doping. Hence, it can be concluded that carbon nanomaterials still have significant opportunities to be applied in future research on energy and environmental applications.

**Acknowledgments** The authors acknowledge support from NPDF, SERB, Department of Science and Technology, New Delhi, India in the form of a research grant (PDF/2016/003602).

## References

Ajayan PM, Stephan O, Colliex C, Trauth D (1994) Aligned carbon nanotube arrays formed by cutting a polymer resin-nanotube composite. Science 265:1212–1214

Aktar MW, Sengupta D, Chowdhury A (2009) Impact of pesticides use in agriculture: their benefits and hazards. Interdiscip Toxicol 2(1):1–12

Amin MT, Alazba AA, Manzoor U (2014) A review of removal of pollutants from water/wastewater using different types of nanomaterials. Adv Mater Sci Eng 2014:1–24

Anjum M, Miandad R, Waqas M, Gehany F, Barakat MA (2016) Remediation of wastewater using various nano-materials. Arab J Chem. https://doi.org/10.1016/j.arabjc.2016.10.004

Ashfaq M, Singh S, Sharma A, Verma N (2013) Cytotoxic evaluation of the hierarchal web of carbon micro-nanofibers. Ind Eng Chem Res 52:4672–4682

Ashfaq M, Khan S, Verma N (2014) Synthesis of PVA-CAP-based biomaterial in situ dispersed with Cu nanoparticles and carbon micro-nanofibers for antibiotic drug delivery applications. Biochem Eng J 90:79–89

Ashfaq M, Verma N, Khan S (2016a) Copper/Zinc bimetal nanoparticles-dispersed carbon nanofibers: a novel potential antibiotics material. Mater Sci Eng C 59:938–947

Ashfaq M, Verma N, Khan S (2016b) Highly effective Cu/Zn-carbon micro/nanofiber-polymer nanocomposite-based wound dressing biomaterial against the *P. aeruginosa* multi-and extensively drug-resistant strains. Mater Sci Eng C 77:630–641

Ashfaq M, Verma N, Khan S (2017) Carbon nanofibers as a micronutrient carrier in plants: efficient translocation and controlled release of Cu nanoparticles. Environ Sci Nano 4:138–148

Ashish B, Neeti K, Himanshu K (2013) Copper toxicity: a comprehensive study. Res J Recent Sci 2:58–67

Awasthi SK, Ashfaq M, Singh S (2009) Effect of glucose and chloramphenicol on ABS biodegradation by a bacterial consortium. Biol Med 1(2):15–19

Ayangbenro AS, Babalola OO (2017) A new strategy for heavy metal polluted environments: a review of microbial biosorbents. Int J Environ Res Public Health 14(1):1–94

Bakry R, Vallant RM, Najam-ul-Haq M, Rainer M, Szabo Z, Huck CW, Bonn GK (2007) Medicinal applications of fullerenes. Int J Nanomedicine 2(4):639–649

Beyth N, Houri-Haddad Y, Domb A, Khan W, Hazan R (2015) Alternative antimicrobial approach: nano-antimicrobial materials. Evid Based Complement Alternat Med 2015:246012. https://doi.org/10.1155/2015/246012

Bhartiya DK, Singh K (2012) Heavy metals accumulation from municipal solid wastes with different animal dung through vermicomposting by earthworm *Eisenia fetida*. World Appl Sci J 17(1):133–139

Bhattacharya K, Mukherjee SP, Gallud A, Burkert SC, Bistarelli S, Bellucci S, Bottini M, Star A, Fadeel B (2016) Biological interactions of carbon-based nanomaterials: from coronation to degradation. Nanomedicine 12(2):333–351

Bhuyan MSA, NizamUddin M, Bipasha MMIFA, Hossain SS (2016) Synthesis of graphene. Int Nano Lett 6(2):65–83

Blasdel NJ, Wujcik EK, Carletta JE, Lee KS, Monty CN (2015) Fabric nanocomposite resistance temperature detector. IEEE Sensors J 15:300–306

Cabral JPS (2010) Water microbiology: bacterial pathogens and water. Int J Environ Res Public Health 7(10):3657–3703

Cengel YA, Boles MA (2007) Thermodynamics: an engineering approach, 6th edn. McGraw-Hill, Boston

Chauhan D, Jaiswal M, Sankararamakrishnan N (2012) Removal of cadmium and hexavalent chromium from electroplating waste water using thiocarbamoyl chitosan. Carbohydr Polym 88:670–675

Chauhan D, Afreen S, Mishra S, Sankararamakrishnan N (2016) Synthesis, characterization and application of Zinc augmented aminated PAN nanofibers towards decontamination of chemical and biological contaminants. J Ind Eng Chem 55:50–64

Chen Q, Dai L, Gao M, Huang S, Mau A (2001) Plasma activation of carbon nanotubes for chemical modification. J Phys Chem B 105:618–622

Crosby C (2017) Human body organ systems. Hill & Ponton, PA. https://www.hillandponton.com/human-body-organ-systems/. Accessed

DeVito SC, Brooks WE (2005) Mercury. In: Kirk-Othmer encyclopedia of chemical technology. Wiley, Hoboken. https://doi.org/10.1002/0471238961.1305180304052209.a01.pub3

Do SY, Lee CG, Kim JY (2017) Cases of acute mercury poisoning by mercury vapor exposure during the demolition of a fluorescent lamp factory. Ann Occup Environ Med 29:19. https://doi.org/10.13140/RG.2.1.2733.6169

Dong C, Campell AS, Eldawud R, Perhinschi G, Rojanasakul Y, Dinu CZ (2013) Effects of acid treatment on structure, properties and biocompatibility of carbon nanotubes. Appl Surf Sci 264:261–268

Duflo S, Greenstone M, Hanna R (2008) Indoor air pollution, health and economic well-being. Surv Perspect Integr Envir Soc 1:1–9

Dutta S, Kim J, Ide Y, Kim JH, Hossain MSA, Bando Y, Yamauchi Y, Wu KCW (2017) 3D network of cellulose-based energy storage devices and related emerging applications. Mater Horiz 4:1–52

Fan S, Wang B, Tesche M, Engelmann R, Althausen A, Liu J (2008) Meteorological conditions and structures of atmospheric boundary layer in October 2004 over Pearl River Delta area. Atmos Environ 42:6174–6186

Faraz M, Abbasi A, Naqvi FK, Khare N, Prasad R, Barman I, Pandey R (2018) Polyindole/CdS nanocomposite based turn-on, multi-ion fluorescence sensor for detection of $Cr^{3+}$, $Fe^{3+}$ and $Sn^{2+}$ ions. Sensors Actuators B Chem 269:195–202. https://doi.org/10.1016/j.snb.2018.04.110

Feng SH, Li GH (2017) Hydrothermal and solvothermal syntheses. In: Xu R, Xu Y (eds) Modern inorganic synthetic chemistry, 2nd edn. Elsevier, Amsterdam, pp 73–104

Ferdous A, Maisha N, Sultana N, Ahmed S (2016) Removal of heavy metal from industrial effluents using Baker's yeast. AIP Conf. Proc. 1754(1). https://doi.org/10.1063/1.4958452

Flandrois S, Simon B (1999) Carbon materials for lithium-ion rechargeable batteries. Carbon 37:165–180

Foster A, Kumar N (2011) Health effects of air quality regulations in Delhi, India. Atmos Environ 45:1675–1683

Frankenberg C, Meirink JF, van Weele M, Platt U, Wagner T (2005) Assessing methane emissions from global space borne observations. Science 308:1010–1014

Frey HC, Zhai H, Rouphail NM (2009) Regional on road vehicle running emissions modeling and evaluation for conventional and alternative vehicle technologies. Environ Sci Technol 43:8449–8445

Galli SJ, Tsai M, Piliponsky AM (2008) The development of allergic inflammation. Nature 454(7203):445–454

Ghosh S (2016) Surface functionalized hybrid nanomaterials: implications in bio sensing and therapeutics surface chemistry of nano biomaterials. Appl Nano Biomater 3:1–32
Greenstone M, Gayer T (2009) Quasi-experimental and experimental approaches to environmental economics. J Environ Econ Manag 57:21–44
Guo H, Wang T, Blake DR, Simpson IJ, Kwok YH, Li YS (2006) Regional and local contributions to ambient non methane volatile organic compounds at a polluted rural/coastal site in Pearl River Delta, China. Atmos Environ 40:2345–2359
Guo CX, Yang HB, Sheng ZM, Lu ZS, Song QL, Li CM (2010) Layered graphene/quantum dots for photovoltaic devices. Angew Chem Int Ed 49:3014–3017
Gupta VK, Tyagi I, Sadegh H, Ghoshekandi RS, Makhlouf ASH, Maazinejad B (2015) Nanoparticles as adsorbent; a positive approach for removal of noxious metal ions: a review. Sci Technol Dev 34(3):195–214
Hai FI, Yamamoto K, Fukushi K (2007) Hybrid treatment systems for dye wastewater. Crit Rev Environ Sci Technol 37:315–377
Harrison BS, Atala A (2007) Carbon nanotube applications for tissue engineering. Biomaterials 28:344–353
Huo H, Zhang Q, Wang MQ, Streets DG, He K (2010) Environmental implication of electric vehicles in China. Environ Sci Technol 44:4856–4861
Hwang S, Batmunkh M, Nine MJ, Chung H, Jeong H (2015) Dye-sensitized solar cell counter electrodes based on carbon nanotubes. Chem Phys Chem 16(1:53–65
Iijima S (1991) Helical microtubules of graphitic carbon. Nature 354:56–58
Jaishankar M, Tseten T, Anbalagan N, Mathew BB, Beeregowda KN (2014) Toxicity, mechanism and health effects of some heavy metals. Interdiscip Toxicol 7(2):60–72
Jaiswal M, Chauhan D, Sankararamakrishnan N (2012) Copper chitosan nanocomposite: synthesis, characterization, and application in removal of organophosphorous pesticide from agricultural runoff. Environ Sci Pollut Res 19:2055–2062
James M, Seltzer JM (1994) Biological contaminants. J Allergy Clin Immunol 94:318–326
Jana A, Scheer E, Polarz S (2017) Synthesis of graphene–transition metal oxide hybrid nanoparticles and their application in various fields. Beilstein J Nanotechnol 8:688–714
Javey A, Kong J (2009) Carbon nanotube electronics. Springer, ISBN 978-0-387-36833-7, United States of America
Javey A, Guo J, Wang Q, Lundstrom M, Dai H (2003) Ballistic carbon nanotube field-effect transistors. Nature 424:654–657
Joshi T, Iyengar L, Garg SK (2008) Isolation identification and application of novel bacterial consortium TJ-1 for decolorization of structurally different azo dyes. Bioresour Technol 99:7115–7121
Julien C, Mauger A, Vijh A, Zaghib K (2015) Lithium batteries. Springer, Cham. https://doi.org/10.1007/978-3-319-19108-9_2
Kaushik BK, Majumder MK (2015) Carbon nanotube based VLSI interconnects: analysis and design. Springer, New Delhi
Ke S, Wang J (2016) Graphene-based materials for supercapacitor electrodes—a review. J Mater 2(1):37–54
Khan I, Saeed K, Khan I (2017) Nanoparticles: properties, applications and toxicities. Arab J Chem. https://doi.org/10.1016/j.arabjc.2017.05.011
Khare P, Talreja N, Deva D, Sharma A, Verma N (2013) Carbon nanofibers containing metal-doped porous carbon beads for environmental remediation applications. Chem Eng J 229:72–81
Khulbe K, Matsuura T (2018) Removal of heavy metals and pollutants by membrane adsorption techniques. Appl Water Sci 8:19. https://doi.org/10.1007/s13201-018-0661-6
Kim Y, Kim JW (2012) Toxic encephalopathy. Saf Health Work 3(4):243–256
Köhrle J (1999) The trace element selenium and the thyroid gland. Biochimie 81:527–533
Kroto HW, Heath JR, O'Brien SC, Curl RF, Smalley RE (1985) $C_{60}$: buckminsterfullerene. Nature 318:162

Kumar V, Talreja N, Deva D, Sankararamakrishnan N, Sharma A, Verma N (2011) Development of bi-metal doped micro and nano multi-functional polymeric adsorbent for the removal of fluoride and arsenic in waste-water. Desalination 282:27–38

Kumar R, Ashfaq M, Verma N (2018) Novel PVA/starch-encapsulated Cu/Zn bimetal nanoparticle carrying carbon nanofibers as a biodegradable and anti-reactive oxidative nanofertilizer. J Mater Sci 53(10):7150–7164

Kwok RHF, Fung JCH, Lau AKH, Fu JS (2010) Numerical study on seasonal variations of gaseous pollutants and particulate matters in Hong Kong and Pearl River Delta Region. J Geophys Res Atmos 115:D16308

Lakherwal D (2014) Adsorption of heavy metals: a review. IJERD 4(1):41–48

Leatherdale CA, Woo WK, Mikulec FV, Bawendi MG (2002) On the absorption cross section of CdSe nanocrystal quantum dots. J Phys Chem 106:7619–7622

Lee JSM, Briggs EM, Chang Hu C, Cooper AI (2018) Controlling electric double-layer capacitance and pseudocapacitance in heteroatom-doped carbons derived from hyper crosslinked microporous polymers. Nano Energy 46:277–289

Li W, Liang C, Zhou W, Qiu J, Zhou Z, Sun G, Xin Q (2003) Preparation and characterization of multiwalled carbon nanotube–supported platinum for cathode catalysts of direct methanol fuel cells. J Phys Chem 107:6292–6299

Li B, Jung HY, Wang H, Kim YL, Kim T, Hahm MG, Busnaina A, Upmanyu M, Jung YJ (2011) Ultra-thin SWNTs films with tunable, anisotropic transport properties. Adv Funct Mater 21:1810–1815

Lipomi DJ, Vosgueritchian M, Tee BCK, Hellstrom SL, Lee JA, Fox CH, Bao Z (2011) Skin like pressure and strain sensors based on transparent elastic films of carbon nanotubes. Nat Nanotechnol 6:788–792

Liu M, Yang Y, Zhu T, Liu Z (2007) A general approach to chemical modification of single-walled carbon nanotubes with peroxy organic acids and its application in polymer grafting. J Phys Chem C 111(6):2379–2385

Liu C, Yu Z, Neff D, Zhamu A, Jang BZ (2010) Graphene-based supercapacitor with an ultrahigh energy density. Nano Lett 10:4863–4868

Liu J, Huang J, Wujcik EK, Qiu B, Rutman D, Zhang X, Salazard E, Wei S, Guo Z (2015) Hydrophobic electrospun polyimide nanofibers for self-cleaning materials. Macromol Mater Eng 300:358–368

Lu W, Dai L (2010) Carbon nanotube supercapacitors. In: Marulanda JM (ed) Carbon nanotubes. IntechOpen. https://www.intechopen.com/books/carbon-nanotubes/carbon-nanotube-supercapacitors. Accessed

Lu G, Yu K, Wen Z, Chen J (2013a) Semiconducting graphene: converting graphene from semimetal to semiconductor. Nanoscale 5:1353–1368

Lu P, Xue D, Yang H, Liu Y (2013b) Supercapacitor and nanoscale research towards electrochemical energy storage. Int J Smart Nano Mater 4(1):2–26

Lukatskaya MR, Dunn B, Gogotsi Y (2016) Multidimensional materials and device architectures for future hybrid energy storage. Nat Commun 7:1–12647

Manawi YM, Ihsanullah ID, Samara A, Al-Ansari T, Atieh MA (2018) A review of carbon nanomaterials' synthesis via the chemical vapor deposition (CVD) method. Materials (Basel) 11(5):1–822

Martin S, Griswold W (2009) Human health effects of heavy metals. Cent Hazard Subst Res 785:532–6519

Mazov I, Kuznetsov VL, Simonova IA, Stadnichenko AI, Ishchenko AV, Romanenko AI, Anikeeva OB (2012) Oxidation behavior of multiwall carbon nanotubes with different diameters and morphology. Appl Surf Sci 258(17):6272–6280

Minitha CR, Anithaa VS, Subramaniam V, Kumar RTR (2018) Impact of oxygen functional groups on reduced graphene oxide-based sensors for ammonia and toluene detection at room temperature. ACS Omega 3(4):4105–4112

Monty CN, Wujcik EK, Blasdel NJ (2013) Flexible electrode for detecting changes in temperature, humidity, and sodium ion concentration in sweat. US Patent 20130197319A1

Morris JS, Crane SB (2013) Selenium toxicity from a misformulated dietary supplement, adverse health effects, and the temporal response in the nail biologic monitor. Nutrients 5(4):1024–1057

Moumita K, Jaehwan K, Junghwan O, Il-Kwon O (2016) Recent progress in multifunctional graphene aerogels. Front Mater 3:1–29

Mukherjee S (2011) Applied mineralogy: applications in industry and environment. Science Springer, Dordrecht

Mustafa S, Khan HM, Shukla I, Shujatullah F, Shahid M, Ashfaq M, Azam A (2011) Effect of ZnO nanoparticles on ESBL producing *Escherichia coli* & *Klebsiella* sp. East J Med 16:253–257

Nasir S, Hussein MZ, Zainal Z, Yusof NA (2018) Carbon-based nanomaterials/allotropes: a glimpse of their synthesis, properties and some applications. Materials 11:295

Nasrallah HA, Balling RC (1995) The heated debate: greenhouse predictions versus climate reality. Pacific Research Institute for Public Policy, San Francisco

Novoselov KS, Geim AK, Morozov SV, Jiang D, Zhang Y, Dubonos SV, Grigorieva IV, Firsov AA (2004) Electric field effect in atomically thin carbon films. Science 306:666–669

O'Regan B, Graetzel M (1991) A low-cost, high efficiency solar cell based on dye-sensitized colloidal $TiO_2$ films. Nature 353:737–740

Owusu PA, Sarkodie SA (2016) A review of renewable energy sources, sustainability issues and climate change mitigation. Cogent Eng 3(1):1–14

Peng W, Li H, Liu Y, Song S (2017) A review on heavy metal ions adsorption from water by graphene oxide and its composite. J Mol Liq 230:496–504

Pope CA III, Burnett RT, Thun MJ, Calle EE, Krewski D, Ito K (2002) Lung cancer, cardiopulmonary mortality, and long-term exposure to fine particulate air pollution. JAMA 287:1132–1141

Prasad R, Pandey R, Barman I (2016) Engineering tailored nanoparticles with microbes: quo vadis. WIREs Nanomed Nanobiotechnol 8:316–330. https://doi.org/10.1002/wnan.1363

Prashanth L, Kattapagari KK, Chitturi RT, Baddam VRR, Prasad LK (2015) A review on role of essential trace elements in health and disease. J NTR Univ Health Sci 4(2):75–85

Qie L, Chen WM, Wang ZH, Shao QG, Li X, Yuan LX, Hu XL, Zhang WX, Huang YH (2012) Nitrogen-doped porous carbon nanofiber webs as anodes for lithium ion batteries with a super high capacity and rate capability. Adv Mater 24:2047–2050

Reynolds CO, Kandlikar M, Badami GM (2011) Determinants of PM and GHG emissions from natural gas-fueled auto rickshaws in Delhi. Transp Res D Transp Environ 16:160–165

Romm J (2006) The car and fuel of the future. Energy Policy 34:2609–2614

Ross RB, Cardona CM, Guldi DM, Sankaranarayanan SG, Reese MO, Kopidakis N, Peet J, Walker B, Bazan GC, Drees M (2009) Endohedral fullerenes for organic photovoltaic devices. Nat Mater 8:208–212

Ruomeng Yu R, Yongzheng Shi Y, Dongzhi Yang D, Yaxin Liu Y, Jin Qu J, Zhong-Zhen Yu ZZ (2017) Graphene oxide/chitosan aerogel microspheres with honeycomb-cobweb and radially oriented microchannel structures for broad-spectrum and rapid adsorption of water contaminants. ACS Appl Mater Interfaces 9(26):21809–21819

Sadhu SD, Garg M, Kumar A (2018) Major environmental issues and new materials. In: Hussain CM, Mishra AK (eds) New polymer nanocomposites for environmental remediation. Elsevier, Amsterdam, pp 77–97

Saha R, Nandi R, Saha B (2011) Sources and toxicity of hexavalent chromium. J Coord Chem 64:1782–1806

Sankararamakrishnan N, Chauhan D (2014) Studies on the use of novel nano composite (CNT/chitosan/Fe(0)) towards arsenate removal. J Environ Res Develop 8:594–599

Sankararamakrishnan N, Chauhan D, Dwivedi J (2016) Synthesis of functionalized carbon nanotubes by floating catalytic chemical vapor deposition method and their sorption behavior toward arsenic. Chem Eng J 284:599–608

Saraswat R, Talreja N, Deva D, Sankararamakrisnan N, Sharma A, Verma N (2012) Development of novel in-situ nickel-doped, phenolic resin-based micro-nanoactivated carbon adsorbents for the removal of vitamin B-12. Chem Eng J 197:250–260

Sarkar J (2011) Characterization of benzoxaborole-based antifungal resistance mutations demonstrates that editing depends on electrostatic stabilization of the leucyl-tRNA synthetase editing cap. FEBS Lett 585(19):2986–2991
Scida K, Stege PW, Haby G, Messina GA, García CD (2011) Recent applications of carbon-based nanomaterials in analytical chemistry: critical review. Anal Chim Acta 691(1–2):6–17
Seger B, Kamat PV (2009) Electrocatalytically active graphene-platinum nanocomposites. Role of 2-D carbon support in PEM fuel cells. J Phys Chem 113:7990–7995
Sharma A, Verma N, Sharma A, Deva D, Sankararamakrishnan N (2010) Iron doped phenolic resin based activated carbon micro and nanoparticles by milling: synthesis, characterization and application in arsenic removal. Chem Eng Sci 65:3591–3601
Shen DS, Cheng SY, Liu L, Chen T, Guo XR (2007) An integrated MM5-CMAQ modeling approach for assessing transboundary PM10 contribution to the host city of 2008 Olympic Summer Games—Beijing, China. Atmos Environ 22:1237–1250
Singh S, Awasthi SK, Iyengar L, Ashfaq M, Singh P (2011) Mineralization of aminobenzenesulfonates by a newly isolated bacterial co-culture (AS1 & AS2). Biol Med 3(2):53–59
Sivasubramaniam D, Franks AE (2016) Bioengineering microbial communities: their potential to help, hinder and disgust. Bioengineered 7(3):137–144
Slezakova K, Pires JCM, Martins FG, Pereira MC, Alvim-Ferraz MC (2011) Identification of tobacco smoke components in indoor breathable particles by SEM-EDS. Atmos Environ 45:863–872
South DW (1993) Coal and $CO_2$: what are the options? Private Power Executive Sep–Oct:25–31
Stankovich S, Dikin DA, Dommett GHB, Kohlhaas KM, Zimney EJ, Stach EA, Piner RD, Nguyen ST, Ruoff RS (2006) Graphene-based composite materials. Nature 442:282–286
Stoller MD, Park S, Zhu Y, An J, Ruoff RS (2008) Graphene-based ultracapacitors. Nano Lett 8:3498–3502
Strong V, Dubin S, Kady MFE, Lech A, Wang Y, Weiller BH, Kaner RB (2012) Patterning and electronic tuning of laser scribed graphene for flexible all-carbon devices. ACS Nano 6(2):1395–1403
Talreja N, Verma N, Kumar D (2014) Removal of hexavalent chromium fromwater using Fe-grown carbon nanofibers containing porous carbon microbeads. J Water Process Eng 3:34–45
Talreja N, Verma N, Kumar D (2016) Carbon bead-supported ethylene diamine functionalized carbon nanofibers: an excellent adsorbent for salicyclic acid. CLEAN-Soil Air Water 44(11):1461–1470
Tang Z, Wu H, Cort JR, Buchko GW, Zhang Y, Shao Y, Aksay IA, Liu J, Lin Y (2010) Constraint of DNA on functionalized graphene improves its biostability and specificity. Small 6:1205–1209
Tchounwou PB, Yedjou CG, Patlolla AK, Sutton DJ (2012) Heavy metals toxicity and the environment. EXS 101:133–164
Ullah R, Zafar MS (2015) Oral and dental delivery of fluoride: a review. Res Rev Fluoride 48(3):195–204
Unosson E, Cai Y, Jiang X, Lööf J, Welch K, Engqvist H (2012) Antibacterial properties of dental luting agents: potential to hinder the development of secondary caries. Int J Dent 2012:1–7
Villanueva CM, Kogevinas M, Cordier S (2014) Assessing exposure and health consequences of chemicals in drinking water: current state of knowledge and research needs. Environ Health Perspect 122(3):213–221
Visakh PM (2016) Introduction for nanomaterials and nanocomposites: state of art, new challenges, and opportunities. In: Visakh PM, Morlanes MJM. Nanomaterials and nanocomposites: zero- to three-dimensional materials and their composites. Wiley-VCH, Weinheim, pp 1–20
Wang J, Kaskel S (2012) KOH activation of carbon-based materials for energy storage. J Mater Chem 22(45):23710–23725
Wang X, Ouyang Y, Li X, Wang H, Guo J, Dai H (2008) Room-temperature all-semiconducting sub-10-nm graphene nanoribbon field-effect transistors. Phys Rev Lett 100:206803

Wang Y, He Q, Qu H, Zhang X, Guo J, Zhu J, Zhao G, Colorado HA, Yu J, Sun L (2014) Magnetic graphene oxide nanocomposites: nanoparticles growth mechanism and property analysis. J Mater Chem 2:9478–9488

Wei JP, Srinivasan C, Han H, Valentine JS, Gralla EB (2001) Evidence for a novel role of copper–zinc superoxide dismutase in zinc metabolism. J Biol Chem 276(48):44798–44803

Wei J, Jia Y, Shu Q, Gu Z, Wang K, Zhuang D, Zhang G, Wang Z, Luo J, Cao A, Wu D (2007) Double-walled carbon nanotube solar cells. Nano Lett 7:2317–2321

Winter M, Brodd R (2004) What are batteries, fuel cells, and super capacitors? Chem Rev 104(10):4245–4269

Wu ZY, Li C, Liang HW, Zhang YN, Wang X, Chen JF, Yu SH (2014) Carbon nanofiber aerogels for emergent cleanup of oil spillage and chemical leakage under harsh conditions. Sci Rep 4:4079

Wu Z, Li L, Yan JM, Zhang XB (2017) Materials design and system construction for conventional and new concept super capacitors. Adv Sci 4:1600382

Wuana RA, Okieimen FE (2011) Heavy metals in contaminated soils: a review of sources, chemistry, risks and best available strategies for remediation. ISRN Ecology 402647:20

Wujcik EK (2013) Discovery of nanostructured material properties for advanced sensing platforms. Dissertation, University of Akron

Wujcik EK, Monty CN (2013) Nanotechnology for implantable sensors: carbon nanotubes and graphene in medicine. Wiley Interdiscip Rev Nanomed Nanobiotechnol 5:233–249

Wujcik EK, Blasdel NJ, Trowbridge D, Monty CN (2013) Ion sensor for the quantification of sodium in sweat samples. IEEE Sensors J 13:3430–3436

Wujcik EK, Wei H, Zhang X, Guo J, Yan X, Sutrave N, Wei S, Guo Z (2014) Antibody nanosensors: a detailed review. RSC Adv 4:43725–43745

Xia F, Farmer DB, Lin Y, Avouris P (2015) Graphene field-effect transistors with high on/off current ratio and large transport band gap at room temperature. Nano Lett 10:715–718

Xin W, Song Y (2015) Mesoporous carbons: recent advances in synthesis and typical applications. RSC Adv 5:83239–83285

Yan C, Wang J, Kang W, Cui M, Wang X, Foo CY, Chee KJ, Lee PS (2014) Highly stretchable piezo resistive graphene-nanocellulose nanopaper for strain sensors. Adv Mater 26:2022–2027

Yun S, Hagfeldt A, Ma T (2014) Pt-free counter electrode for dye-sensitized solar cells with high efficiency. Adv Mater 26(36):6210–6237

Yuxi X, Gaoquan S (2011) Assembly of chemically modified graphene: methods and applications. J Mater Chem 21:3311–3323

Zhang J, Wang T, Chameides WL, Cardelino C, Blake DR, Streets DG (2008) Source characteristics of volatile organic compounds during high ozone episodes in Hong Kong, Southern China. Atmos Chem Phys 8:4983–4996

Zhang Q, Zhang S, Zhang L (2014) Microwave-assisted modification of carbon nanotubes with biocompatible polylactic acid. J Mater Sci Chem Eng 2:7–12

Zhang Y, Gao Z, Song N, He J, Li X (2018) Graphene and its derivatives in lithium–sulfur batteries. Mater Today Energy 9:319–335

Zhao X, Zhang Q, Chen D, Lu P (2010) Enhanced mechanical properties of graphene-based poly (vinyl alcohol) composites. Macromolecules 43:2357–2363

Zheng Q, Kim JK (2015) Synthesis, structure, and properties of graphene and graphene oxide. In: Zheng Q, Kim JK. Graphene for transparent conductors: synthesis, properties and applications. Springer, New York, pp 29–94

Zheng X, Zhang L (2016) Photonic nanostructures for solar energy conversion. Energy Environ Sci 9:2511–2532

Zheng J, Shao M, Che W, Zhang L, Zhong L, Zhang Y (2009) Speciated VOC emission inventory and spatial patterns of ozone formation potential in the Pearl River Delta, China. Environ Sci Technol 43:8580–8586

Zheng JS, Zhang L, Shellikeri A, Cao W, Wu Q, Zheng JP (2017) A hybrid electrochemical device based on a synergetic inner combination of Li ion battery and Li ion capacitor for energy storage. Sci Rep 7:41910

Zhenqing X, Kumar A, Kumar A (2005) Amperometric detection of glucose using a modified nitrogen-doped nanocrystalline diamond electrode. J Biomed Nanotechnol 1(4):1–5

Zhu J, Wei S, Gu H, Rapole SB, Wang Q, Luo Z, Haldolaarachchige N, Young DP, Guo Z (2011) One-pot synthesis of magnetic graphene nanocomposites decorated with core@double-shell nanoparticles for fast chromium removal. Environ Sci Technol 46:977–985

Zhu J, Chen M, Qu H, Luo Z, Wu S, Colorado HA, Wei S, Guo Z (2012a) Magnetic field induced capacitance enhancement in graphene and magnetic graphene nanocomposites. Energy Environ Sci 6:194–204

Zhu J, Sadu R, Wei S, Chen DH, Haldolaarachchige N, Luo Z, Gomes JA, Young DP, Guo Z (2012b) Magnetic graphene nanoplatelet composites toward arsenic removal. ECS J Solid State Sci Technol 1:M1–M5

Zhu J, Chen M, Wei H, Yerra N, Haldolaarachchige N, Luo Z, Young DP, Ho TC, Wei S, Guo Z (2014) Magneto capacitance in magnetic micro tubular carbon nanocomposites under external magnetic field. Nano Energy 6:180–192

Zulkifili ANB, Kento T, Daiki M, Fujiki A (2015) The basic research on the dye-sensitized solar cells (DSSC). J Clean Energy Technol 3:382–387

Zuo W, Li R, Zhou C, Li Y, Xia J, Liu J (2017) Battery-super capacitor hybrid devices: recent progress and future prospects. Wiley-VCH, Weinheim

# Index





R. Prasad, E. Aranda (eds.), *Approaches in Bioremediation*, Nanotechnology in the Life Sciences, https://doi.org/10.1007/978-3-030-02369-0

**G**